GENERAL RELATIVITY

GENERAL RELATIVITY

Proceedings of the Forty Sixth Scottish Universities
Summer School in Physics,
Aberdeen, July 1995.

A NATO Advanced Study Institute.

Edited by

G S Hall – University of Aberdeen

J R Pulham – University of Aberdeen

Series Editor

P Osborne – University of Edinburgh

Copublished by
Scottish Universities Summer School in Physics &
Institute of Physics Publishing, Bristol and Philadelphia

Copyright © 1996

The Scottish Universities Summer School in Physics

No part of this book may be reproduced in any form
by photostat, microfilm or any other means without
written permission from the publishers.

British Library Cataloguing-in-Publication Data:

*A catalogue record for this book is available
from the British Library.*

*ISBN 0-7503-0-395-6 (hbk)
0-7503-0-419-7 (pbk)*

Library of Congress Cataloging-in-Publication Data are available.

Copublished by

SUSSP Publications
The Department of Physics, Edinburgh University,
The King's Buildings, Mayfield Road, Edinburgh EH9 3JZ, Scotland.

and

Institute of Physics Publishing, wholly owned by
The Institute of Physics, London.

Institute of Physics Publishing, Techno House, Redcliffe Way, Bristol BS1 6NX, UK.
US Editorial Office: Institute of Physics Publishing, The Public Ledger Building,
Suite 1035, 150 South Independence Mall West, Philadelphia, PA 19106, USA

Printed in Great Britain by J W Arrowsmith Ltd, Bristol

SUSSP Proceedings

1	1960	Dispersion Relations
2	1961	Fluctuation, Relaxation and Resonance in Magnetic Systems
3	1962	Polarons and Excitons
4	1963	Strong Interactions and High Energy Physics
5	1964	Nuclear Structure and Electromagnetic Interactions
6	1965	Phonons in Perfect and Imperfect Lattices
7	1966	Particle Interactions at High Energy
8	1967	Methods in Solid State and Superfluid Theory
9	1968	Physics of Hot Plasmas
10	1969	Quantum Optics
11	1970	Hadronic Interactions of Photons and Electrons
12	1971	Atoms and Molecules in Astrophysics
13	1972	Properties of Amorphous Semiconductors
14	1973	Phenomenology of Particles at High Energy
15	1974	The Helium Liquids
16	1975	Non-linear Optics
17	1976	Fundamentals of Quark Models
18	1977	Nuclear Structure Physics
19	1978	Metal Non-metal Transitions in Disordered Solids
20	1979	Laser-Plasma Interactions: 1
21	1980	Gauge Theories and Experiments at High Energy
22	1981	Magnetism in Solids
23	1982	Laser-Plasma Interactions: 2
24	1982	Lasers: Physics, Systems and Techniques
25	1983	Quantitative Electron Microscopy
26	1983	Statistical and Particle Physics
27	1984	Fundamental Forces
28	1985	Superstrings and Supergravity
29	1985	Laser-Plasma Interactions: 3
30	1985	Synchrotron Radiation
31	1986	Localisation and Interaction
32	1987	Computational Physics
33	1987	Astrophysical Plasma Spectroscopy
34	1988	Optical Computing

/continued

SUSSP Proceedings (continued)

35	1988	Laser-Plasma Interactions: 4
36	1989	Physics of the Early Universe
37	1990	Pattern Recognition and Image Processing
38	1991	Physics of Nanostructures
39	1991	High Temperature Superconductivity
40	1992	Quantitative Microbeam Analysis
41	1992	Spatial Complexity in Optical Systems
42	1993	High Energy Phenomenology
43	1994	Determination of Geophysical Parameters from Space
44	1994	Quantum Dynamics of Simple Systems
45	1994	Laser-Plasma Interactions 5: Inertial Confinement Fusion
46	1995	General Relativity
47	1995	Laser Sources and Applications
48	1996	High Power Electromagnetic Radiation
49	1997	Monitoring the Physics of Coastal Processes
50	1998	Semiconductor Opto-electronics
51	1998	Muon Science

Lecturers

Bernard Carr	QMW, University of London
Robert Geroch	University of Chicago
Ray d'Inverno	University of Southampton
Alberto Lobo	Universidad de Barcelona, Barcelona
Gernot Neugebauer	Friedrich Schiller Universitaet, Jena
Zoltán Perjés	Hungarian Academy of Sciences, Budapest
Malcolm Perry	University of Cambridge
Norna Robertson	University of Glasgow
Bernd Schmidt	Albert Einstein Institut, Potsdam
John Wainwright	University of Waterloo
Clifford Will	Washington University

Executive Committee

Prof G S Hall	University of Aberdeen	*Director and Editor*
Dr J E F Skea	University of Aberdeen	*Secretary*
Dr R C Clark	University of Aberdeen	*Treasurer*
Dr J R Pulham	University of Aberdeen	*Editor*
Dr B Haddow	Trinity College, Dublin	*Steward*

International Organising Committee

Prof G S Hall	University of Aberdeen
Prof M A H MacCallum	QMW, London
Prof G Neugebauer	Freidrich Schiller Universität, Jena
Prof Z Perjés	Hungarian Academy of Sciences, Budapest
Prof J Wainwright	University of Waterloo, Canada
Prof G F R Ellis	University of Cape Town, South Africa
Prof R P Geroch	University of Chicago
Prof B G Schmidt	Albert Einstein Institut, Potsdam

Preface

The forty-sixth Scottish Universities Summer School in Physics was held in the University of Aberdeen from July 16th until July 29th, 1995, the quincentenary year of the University. It was attended by 65 postgraduate and postdoctoral students from 22 countries and the courses, on a variety of current topics in general relativity, were delivered by 11 internationally renowned experts. The School was supported by NATO, the Scottish Universities, the EC, the NSF and Trinity College Cambridge. The interest and generosity of these institutions is very gratefully acknowledged.

Relativity Theory requires the student or researcher to have detailed knowledge of both mathematics and physics and, more recently, of computing. The main aim of the School was to ensure that a sufficient breadth as well as depth of courses in these topics was covered by experts in the field. The students mounted a poster session of high quality describing their work and prizes were awarded for the best three posters. These Proceedings contain the lecture notes of the courses given together with abstracts of some of the posters. The School also benefited greatly from a lecture on the life and work of James Clerk Maxwell delivered by Professor Irvine, the Principal of Aberdeen University. Maxwell was Professor of Natural Philosophy in Aberdeen from 1856 until 1860.

On the lighter side of things the members of the School took part in football matches, a barbecue, pub quizzes and visits to castles and a whisky distillery. A wonderful spirit of fellowship prevailed throughout.

As Director of the School, I would like to conclude by adding my own personal thanks to all those people whose efforts were indispensable in the smooth running of the administration of the School. The catering services in the University served us in a friendly and efficient manner—as did Mrs Margaret Riddoch in the School's Central Administration office, Dr Barry Haddow, Mrs Pat Brown, Mrs Françoise Clark and many other people whose names I never knew. But I wish here to record my special gratitude to the other members of the committee, Dr Roger Clark, Dr John Pulham and Dr Jim Skea without whose tireless efforts the School would have been no more than a good idea.

Graham Hall (Director)
Aberdeen, May 1996

Contents

Mathematics of General Relativity 1
Bernd Schmidt

Partial Differential Equations of Physics 19
Robert Geroch

Gravitostatics and Rotating Bodies 61
Gernot Neugebauer

A Guide to Basic Exact Solutions 83
Zoltán Perjés

Relativistic Cosmology 107
John Wainwright

Black Holes in Cosmology and Astrophysics 143
Bernard Carr

Source of Gravitational Waves 203
José Alberto Lobo

Detection of Gravitational Waves 223
Norna A Robertson

The Confrontation between General Relativity and Experiment.
A 1995 Update 239
Clifford M Will

Algebraic Computing in General Relativity 283
Ray d'Inverno

Numerical Computing in General Relativity 331
Ray d'Inverno

Quantum Gravity 377
Malcolm Perry

Poster Abstracts 407
Participants' Addresses 413
Index 419

Mathematics of General Relativity

Bernd Schmidt

Max-Planck-Institut für Gravitationsphysik
Potsdam

1 Introductory remarks

This article treats two topics. The first is the local existence theory of the vacuum field equations. The idea here is to close the gap between an introductory course on Relativity and papers on the Cauchy problem: Choquet–Bruhat and York (1980) or Fischer and Marsden (1979).

The second topic is the Newtonian limit. In the last few years various authors have obtained some very nice and interesting results which I would like to bring to a wider audience.

2 Analytic solutions of the vacuum field equations

In this section I shall describe how to demonstrate the existence of analytic solutions of the field equations
$$R_{ab} = 0. \tag{1}$$

Let us take any local coordinate system and write the Riemann tensor explicitly in terms of the metric:
$$R^a_{bcd} = 2\partial_{[c}\Gamma^a_{d]b} + \text{ terms in } g_{ab}, \partial_c g_{ab}. \tag{2}$$

The field equations form a quasilinear second order system for the unknown components of the metric tensor g_{ab}. Suppose you have any system of partial differential equations. How do you set about finding whether it has solutions at all? In the first place it is very helpful to try to determine analytic solutions. Consider, as a simple example, the wave equation
$$-\partial_0^2 f + \partial_1^2 f + \partial_2^2 f + \partial_3^2 f = -\partial_0^2 f + \Delta f = 0. \tag{3}$$

In analogy with the case of an ordinary differential equation, one can check that given f and $\partial_0 f$ as analytic functions in a neighbourhood of a point on the hyperplane $x^0 = 0$ one can determine a unique formal Taylor series satisfying the wave equation. First we determine all x^0-derivatives. The second derivative

$$\partial_0^2 f = \Delta f \qquad (4)$$

is known from the data via the wave equation. Differentiation of the equation with respect to x^0 determines all higher x^0-derivatives:

$$\partial_0^3 f = \partial_0 \Delta f = \Delta \partial_0 f \qquad (5)$$

$$\partial_0^4 f = \partial_0^2 \Delta f = \Delta \Delta f \qquad (6)$$

and so on. All the mixed x^0–x^α derivatives can be calculated from the above.

The remaining problem is the convergence of the formal power series. This is implied by a famous theorem by Cauchy and Sonja Kowalevskaya which guarantees convergence in the above situation.

We formulate the theorem for the following system

$$\partial_0^2 f^A = F^A(x^0, x^\alpha, f^B, \partial_a f^B, \partial_\alpha \partial_a f^B) \qquad (7)$$

where the indices take the ranges $A = 1, \ldots, N$, $a = 0, \ldots, n$ and $\alpha = 1, \ldots, n$.

Systems for which the Cauchy–Kowalevskaya theorem shows existence of analytic solutions have the property that one can solve for the highest derivative with respect to one coordinate, and no higher derivative with respect to the other coordinates appears. For the heat equation $\dot{f} = \Delta f$ the theorem does not apply.

Theorem 2.1 *In system (7) let F^A be an analytic function of all its arguments and give $f^A(0, x^\alpha)$ and $\partial_0 f^A(0, x^\alpha)$ as analytic 'initial data'. Then the system has a unique analytic solution.*

Proof A proof can be found in John (1978). □

This theorem tells us immediately that Maxwell's equations, the Yang–Mills equations and the equations of Galilei- or Lorentz-invariant hydrodynamics have unique analytic solutions determined by analytic data.

Now we can look at Einstein's field equations (1) from the point of view of the Cauchy–Kowalevskaya theorem. We must try to solve for the second derivatives with respect to one coordinate. In any coordinate system x^a, ($a, b = 0,1,2,3$ and $\alpha, \beta = 1,2,3$) we find that

$$\begin{aligned}
0 &= 2R_{\alpha\beta} = g^{00}\partial_{00}g_{\alpha\beta} + \text{ terms in } g_{ab}, \partial_c g_{ab}, \partial_a \partial_b g_{cd} \\
0 &= 2R_{\alpha 0} = g^{0\beta}\partial_{00}g_{\alpha\beta} + \ldots \\
0 &= 2R_{00} = g^{\alpha\beta}\partial_{00}g_{\alpha\beta} + \ldots
\end{aligned} \qquad (8)$$

The terms not listed contain no second derivatives with respect to x^0.

There are no equations for $\partial_{00}g_{0b}$. Hence the Cauchy–Kowalevskaya theorem cannot be used. A hypersurface for which not all the highest transversal derivatives can be calculated from the interior and lower transversal derivatives is called 'characteristic'. Hence we find that for (1) all hypersurfaces are characteristic (there was nothing special about x^0). This is not really surprising! Because of the tensorial nature of the equations we do not expect a uniqueness theorem to hold (we can make coordinate transformations which are the identity up to a certain order on the initial slice).

Suppose that g^{00} does not vanish (x^0=constant is spacelike or timelike). Then we can solve for $\partial_{00}g_{\alpha\beta}$. If we choose any analytic initial data $g_{\alpha\beta}$, $\partial_o g_{\alpha\beta}$ at $x^0 = 0$ and any analytic functions g_{0b} (with $g^{00} \neq 0$), the equations $R_{\alpha\beta} = 0$ have unique analytic solutions.

The equations $R_{0b} = 0$ are conditions for the initial data, 'the constraints', if we eliminate $\partial_0^2 g_{\alpha\beta}$. Suppose we choose data satisfying the constraints and evolve these data. Then we have determined a metric, and the question is whether the constraints are also satisfied outside the initial slice. This turns out to be so!

To see how this miracle happens, let us first find the geometrical meaning of the constraints. In any coordinate system the metric components satisfy the contracted Bianchi identities and so

$$\partial_0 G_a^0 + \partial_\beta G_a^\beta + (G\Gamma)_a = 0. \qquad (9)$$

The terms G_a^0 contain no second derivatives with respect to x^0 because in $\partial_\beta G_\alpha^\beta$ and $(G\Gamma)_a$ at most second x^0-derivatives occur. For $R_{\alpha\beta} = 0$ the equations $G_a^0 = 0$ are equivalent to $R_{0a} = 0$.

If $x^0 = 0$ is spacelike or timelike, the equations $G_a^0 = 0$ are called 'constraints' and they express conditions between g_{ab} and $\partial_0 g_{ab}$ on the hypersurface $x^0 =$ constant. (If $x^0 = 0$ is a null hypersurface, the equations $G_a^0 = 0$ are four of Bondi's 'main equations', which play a particular role in the characteristic initial value problem.) In this case these constraints imply coordinate independent relations between $g_{\alpha\beta}$, the inner metric of the hypersurface, and its second fundamental form $k_{\alpha\beta}$.

We can see this most easily if we introduce Gauss coordinates relative to $x^0 = 0$. Then we have $g_{0\beta} = 0$, $g_{00} = \pm 1$ and

$$ds^2 = \pm(dx^0)^2 + g_{\alpha\beta}(x^0, x^\sigma)dx^\alpha dx^\beta. \qquad (10)$$

The imbedding curvature, or second fundamental form, is

$$2k_{\alpha\beta} = \partial_0 g_{\alpha\beta}. \qquad (11)$$

The constraints are as follows, with all ∇-operations and traces with respect to $g_{\alpha\beta}$

$$0 = G_{\alpha 0} = \nabla^\beta(k_{\alpha\beta} - g_{\alpha\beta}k_\sigma^\sigma) \qquad \text{'Momentum constraints'} \qquad (12)$$
$$0 = 2G_{00} = k_{\alpha\beta}k^{\alpha\beta} - (k_\sigma^\sigma)^2 - {}^{(3)}R \qquad \text{'Hamiltonian constraint'}. \qquad (13)$$

The constraints are tensorial equations for a non-degenerate metric $g_{\alpha\beta}$ and a symmetric tensor field $k_{\alpha\beta}$ on the 3-manifold $x^0 = 0$.

The evolution equations for the second x^0-derivatives of $g_{\alpha\beta}$ in Gauss coordinates are

$$0 = R_{\alpha\beta} = -\partial_0 k_{\alpha\beta} + k_{\alpha\beta}k_\sigma^\sigma + {}^{(3)}R_{\alpha\beta}. \qquad (14)$$

Let us now show that the constraints are compatible with the evolution equations.

Theorem 2.2 *Let g_{ab} be non-degenerate and analytic in the coordinates (x^0, x^α). Suppose that the hypersurface $x^0 = 0$ is spacelike or timelike, that $R_{\alpha\beta} = 0$ in a neighbourhood of $x^0 = 0$ and that $G_a^0 = 0$ on $x^0 = 0$. Then $G_a^0 = 0$ in a neighbourhood of $x^0 = 0$.*

Proof The condition $R_{\alpha\beta} = 0$ implies that G_b^a can be expressed linearly and homogeneously in G_a^0

$$\begin{aligned}
G_0^0 &= \frac{1}{2} g^{00} R_{00} \\
G_\beta^0 &= g^{00} R_{0\beta} \\
G_\beta^\alpha &= g^{0\alpha} R_{0\beta} - \frac{1}{2} \delta_\beta^\alpha (g^{00} R_{00} + 2 g^{0\lambda} R_{0\lambda}) \\
G_0^\alpha &= g^{\alpha 0} R_{00} + g^{\alpha\lambda} R_{0\lambda}.
\end{aligned} \qquad (15)$$

The Bianchi-identities

$$0 = \nabla_a G_b^a = \partial_0 G_b^0 + \partial_\alpha G_b^\alpha - \Gamma_{ba}^s G_s^a + \Gamma_{as}^a G_b^s \qquad (16)$$

are therefore a linear, homogeneous system for G_b^0. For this system we can use the theorem of Cauchy–Kowalevskaya. As $G_b^0 = 0$ on $x^0 = 0$, we have the unique solution $G_b^0 \equiv 0$. □

Note that without analyticity there is no uniqueness theorem for the above equations.

We will now show the existence of local analytic solutions of $R_{ab} = 0$.

Theorem 2.3 *Suppose (x^0, x^β) is a local coordinate system. Suppose that we are given analytic data $g_{\alpha\beta}(0, x^\lambda)$ (non-degenerate), $\partial_0 g_{\alpha\beta}(0, x^\lambda)$ and arbitrary analytic functions $g_{0b}(x^0, x^\lambda)$ satisfying the constraints (12) and (13) on $x^0 = 0$. Suppose also that, on $x^0 = 0$, $g^{00} \neq 0$ and g_{ab} is non-degenerate.*

Then there exist unique analytic functions $g_{\alpha\beta}(x^c)$ which agree with the data on $x^0 = 0$, and the metric defined by $g_{\alpha\beta}$ and g_{0b} satisfies $R_{ab} = 0$.

Proof The theorem of Cauchy–Kowalevskaya will be used for the evolution equations $R_{\alpha\beta} = 0$ with data satisfying the constraints. By Theorem 2.2 above all the field equations hold. □

We still have to find analytic data which satisfy the constraints. Let me just give the simplest example.

Take $k_{\alpha\beta} = 0$ and let $h_{\alpha\beta}$ be any non-degenerate 3-metric. Then $^{(3)}R = 0$, the Hamiltonian constraint, holds for $g_{\alpha\beta} = \Phi^4 h_{\alpha\beta}$, provided Φ satisfies

$$h^{\alpha\beta} \nabla_\alpha \nabla_\beta \Phi - \frac{1}{8} R(h_{\alpha\beta}) \Phi = 0.$$

This is an elliptic or wave equation for Φ. Again we can show the existence of solutions with the help of the Cauchy–Kowalevskaya theorem.

So far we have constructed solutions in certain coordinates. It is easy to obtain statements which are coordinate independent.

Theorem 2.4 *Let S' be an analytic 3-manifold. Let h' be a non-degenerate metric and k' a symmetric tensor field, both defined on S', which are analytic and satisfy the constraints.*

Then there exists an analytic vacuum spacetime in which there is a hypersurface S with first and second fundamental forms h and k such that S, h, k are isometric to S', h', k'. All such spacetimes are isometric in a neighbourhood of S.

Proof If we choose $g_{00} = -1$, $g_{0\beta} = 0$ in Theorem 2.3, we obtain a solution in Gauss coordinates by introducing on S' local coordinates (x^β). As Gauss coordinates relative to a hypersurface are unique, we locally obtain a unique spacetime in which $x^0 = 0$ is isometric to a piece of S'.

If we cover S' with coordinates $V_{(j)}$ in which h' and k' have convergent power series, we obtain spacetimes with metrics $g_{(j)}$. With the obvious identification in Gauss coordinates we construct a unique spacetime. $\square$

Now we can also understand what different choices of 'lapse' and 'shift' g_{0b} mean. Using lapse and shift one can calculate the second fundamental form at the initial slice. At the metric determined using some lapse and shift perform the coordinate transformation to Gauss coordinate from the coordinates determined by the particular lapse and shift. Because of the uniqueness of the analytic solutions we obtain the same metric as we would had we evolved the invariant data in Gauss coordinates.

3 C^∞ solutions of the vacuum field equations

There are various reasons why, in physics, it is not enough to establish existence for analytic solutions only. For example, in an analytic solution the smallest part of the universe determines the whole universe. Therefore I want to outline the existence theory for solutions which are C^k or C^∞.

In Section 2, spacelike and timelike initial surfaces were treated in the same way. The Cauchy–Kowalevskaya theorem makes no distinction between the wave equation and the Laplace equation. This changes drastically for non-analytic solutions.

To construct such solutions one can try to consider sequences of analytic solutions with Cauchy data converging to smooth data. Under which conditions will the sequence of analytic solutions determined by the data converge to a smooth solution?

Here is an instructive example due to Hadamard: the functions

$$U_n(t,x) = \frac{1}{n^2}\sin(nt)\sin(nx) \qquad V_n(t,x) = \frac{1}{n^2}\sinh(nt)\sin(nx) \qquad (17)$$

satisfy the wave and Laplace equations respectively:

$$\ddot{U}_n = U_n'' \qquad \ddot{V}_n = -V_n''. \tag{18}$$

At the initial line $t = 0$ we have

$$U_n(0, x) = V_n(0, x) = 0 \qquad \partial_t U_n(0, x) = \partial_t V_n(0, x) = \frac{1}{n}\sin(nx). \tag{19}$$

The Cauchy data converge to zero as $n \to \infty$. For the wave equation, the solutions U_n converge to zero; for the Laplace equation, however, the solutions V_n blow up for any $t > 0$. The Cauchy problem is unstable for the elliptic equation.

We can only hope to approximate smooth solutions by analytic ones if we have a stable Cauchy problem. This is the guiding idea of 'hyperbolicity': partial differential equations are 'hyperbolic' if they have stable solutions for the initial value problem.

Unfortunately it is not so easy to decide whether certain equations are hyperbolic or not. Only for systems of linear equations with constant coefficients are necessary and sufficient conditions for hyperbolicity known (Gårding hyperbolicity). For general quasilinear systems there is no criterion that can always be applied. Consider as an example the equations (14). Because of the way in which the second spatial derivatives appear in the Ricci tensor, these equations are not hyperbolic according to any of the known definitions of hyperbolicity. Nevertheless the Cauchy problem has smooth, stable solutions.

Let me describe one class of systems for which one has a nice existence theory. The system

$$g^{rs}(F^B)\partial_{rs}F^A + H^A(F^B, \partial F^C) = 0 \tag{20}$$

consists of quasilinear coupled wave equations, if $g^{rs}(F^B)$ has Lorentz signature for the chosen initial data. It can be rewritten as a first order symmetric hyperbolic system (John 1978), about which you can read more in Geroch's article in these Proceedings. It can be shown that smooth initial data determine unique smooth solutions in a small neighbourhood of the initial surface.

The existence theory for such systems is too complicated to be considered here. Let me just give some hints. As mentioned before, we try to approximate by solutions that are already known. These are either analytic solutions, or solutions of systems with constant coefficients. Hence we have to approximate the coefficients in the equations and the initial data and try to control convergence. The basic tools are 'energy estimates' which can be derived for hyperbolic systems.

Let us return to the vacuum field equations, which can be written as

$$R_{ab} = {}^{(h)}R_{ab} + \frac{1}{2}(g_{ai}\partial_b\Gamma^i + g_{bi}\partial_a\Gamma^i) \tag{21}$$

where $\Gamma^i := \Gamma^i_{rs}g^{rs}$ and ${}^{(h)}R_{ab} := -(\frac{1}{2})g^{rs}\partial_{rs}g_{ab} + H_{ab}(g, \partial g)$.

The equations ${}^{(h)}R_{ab} = 0$ form a system of the form (20). Initial conditions $g_{ab}(0, x^\beta)$ and $\partial_0 g_{ab}(0, x^\beta)$ determine locally unique C^∞ or C^k solutions, provided that $x^0 = 0$ is spacelike for the data. Now it is crucial that the initial hypersurface is spacelike. The equations ${}^{(h)}R_{ab} = 0$ consist of 10 evolution equations for all 10 metric components. The domain of dependence is as for a wave equation with metric g_{ab}.

The equation $^{(h)}R_{ab} = 0$ implies $R_{ab} = 0$ if, for the metric determined via the evolution equations $^{(h)}R_{ab} = 0$, the equation $\Gamma^i = 0$ holds. The geometrical meaning of this condition is

$$g^{ab}\nabla_a \nabla_b (x^i) = \Gamma^i = 0. \tag{22}$$

This is a linear wave equation for the coordinates. Therefore there exist in any spacetime local coordinates (x^i) with $\Gamma^i = 0$, known as 'harmonic coordinates'.

How can we arrange that $\Gamma^i = 0$ holds for a solution of $^{(h)}R_{ab} = 0$? If this is possible at all it can be achieved through a particular choice of the data. The equations $0 = \nabla^b G_{ab}$ and $0 = {}^{(h)}R_{ab}$ imply wave equations for Γ^i,

$$g^{ab}\partial_{ab}\Gamma^i + B^{is}_r \partial_s \Gamma^r = 0 \tag{23}$$

where B^{is}_r is a function of g and ∂g.

The equations $\Gamma^i(0, x^\beta) = \partial_0 \Gamma^i(0, x^\beta) = 0$ imply that $\Gamma^i(x^0, x^\beta) \equiv 0$ because of the uniqueness of the solutions (again the initial slice has to be spacelike). Hence 'harmonicity propagates'.

For the initial conditions $g_{ab}(0, x^\beta)$, $\partial_0 g_{ab}(0, x^\beta)$, such that $x^0 = 0$ is spacelike and

$$\Gamma^i(0, x^\beta) = \partial_0 \Gamma^i(0, x^\beta) = 0$$

holds, we obtain a solution of $R_{ab} = 0$. (The derivatives $\partial_0 \Gamma^i(0, x^\beta) = 0$ can be calculated using the evolution equations.)

Next we shall show that such data exists:

Theorem 3.1 *Let h, k be any solution of the constraints. Then there exist $g_{ab}(0, x^\beta)$, $\partial_0 g_{ab}(0, x^\beta)$ such that $\Gamma^i(0, x^\beta)=0$ and $g_{\alpha\beta}(0, x^\beta)=h_{\alpha\beta}$, $\partial_0 g_{\alpha\beta}(0, x^\beta)=k_{\alpha\beta}$ hold. $^{(h)}R_{ab}=0$ on $x^0=0$ implies $\partial_0 \Gamma^i(0, x^\beta)=0$.*

Proof In some coordinate system the solution of the constraints is $h_{\alpha\beta}(x^\gamma)$, $k_{\alpha\beta}(x^\gamma)$. We choose first

$$g_{00}(0, x^\gamma) = -1, \quad g_{0\beta}(0, x^\gamma)=0, \quad g_{\alpha\beta}(0, x^\gamma)=h_{\alpha\beta}, \quad \partial_0 g_{\alpha\beta}(0, x^\gamma)=k_{\alpha\beta}. \tag{24}$$

The condition that $\Gamma^i=0$ holds on $x^0=0$ uniquely determines the ∂_0-derivatives of g_{0b}.

$$\begin{aligned}
0 &= \Gamma^i = (\sqrt{-g})^{-1}\partial_a[(\sqrt{-g})g^{ab}\partial_b x^i] = \partial_a g^{ai} + \tfrac{1}{2}g^{ai}g^{rs}\partial_a g_{rs} \\
0 &= g_{ij}\Gamma^i = -g^{ai}\partial_a g_{ij} + \tfrac{1}{2}g^{rs}\partial_j g_{rs} \\
0 &= g_{ij}\Gamma^i = -g^{0i}\partial_0 g_{ij} - g^{\alpha i}\partial_\alpha g_{ij} + \tfrac{1}{2}g^{rs}\partial_j g_{rs} \\
0 &= g_{ij}\Gamma^i = -g^{00}\partial_0 g_{0j} - g^{\alpha i}\partial_\alpha g_{ij} + \tfrac{1}{2}\{g^{00}\partial_j g_{00} + g^{\rho\sigma}\partial_j g_{\rho\sigma}\}
\end{aligned}$$

For $j=0$ and $j=\tau$ we obtain

$$\begin{aligned}
0 &= -g^{00}\partial_0 g_{00} - g^{\alpha i}\partial_\alpha g_{i0} + \tfrac{1}{2}\{g^{00}\partial_0 g_{00} + g^{\rho\sigma}\partial_0 g_{\rho\sigma}\} \\
0 &= -g^{00}\partial_0 g_{0\tau} - g^{\alpha i}\partial_\alpha g_{i\tau} + \tfrac{1}{2}\{g^{00}\partial_\tau g_{00} + g^{\rho\sigma}\partial_\tau g_{\rho\sigma}\}.
\end{aligned} \tag{25}$$

The last two equations show that $\partial_0 g_{0a}$ is uniquely determined.

Next we want $\partial_0 \Gamma^i = 0$. The equations

$$R_{ab} = {}^{(h)}R_{ab} + \tfrac{1}{2}(g_{ai}\partial_b \Gamma^i + g_{bi}\partial_a \Gamma^i) \tag{26}$$

imply, using ${}^{(h)}R_{ab} = 0$, $\Gamma^a = \partial_\beta \Gamma^a = 0$ for $x^0 = 0$ ($g_{00} = -1$, $g_{0a} = 0$), that

$$G_a^0 = -\tfrac{1}{2} g_{ab}\partial_0 \Gamma^b. \tag{27}$$

Hence, if the constraints are satisfied, we have finally $\partial_0 \Gamma^m = 0$. □

With this theorem and the general existence theorem for quasilinear wave equations we obtain a local solution in particular coordinates. The geometrical version is now obvious.

Theorem 3.2 *Let S' be a smooth 3-manifold, h' a Riemannian metric, and k' a symmetric tensor field on S'. Both fields are smooth and satisfy the constraints.*

Then there exists a smooth vacuum spacetime in which there is a hypersurface S, and a diffeomorphism from S to S', such that its first and second fundamental forms are isometric to h', k'. Any two such solutions are isometric in a neighbourhood of S.

Proof Take a covering V_i of S by local charts. From the above considerations we obtain local solutions $\overset{(i)}{g_{ab}}$.

Suppose we choose two coordinate systems (x^α) and $(x^{\alpha'})$ on some $V_{(i)}$. The coordinates (x^α) determine a unique metric g_{ab} by the construction in Theorem 3.1. Now we define harmonic functions relative to g_{ab}, $y^{a'}$, which agree with $(x^{\beta'})$ and $(\partial_0 x^{\beta'})$ on $x^0 = 0$. If we transform g_{ab} to these coordinates, we obtain metric components ${}^y g_{a'b'}$ which satisfy the same equations and initial values as $g_{a'b'}$. Hence, because of the uniqueness of the solution of the quasilinear system we have ${}^y g_{a'b'} = g_{a'b'}$.

Because we now know that the initial data determine local unique solutions, we can introduce local Gauss coordinates and patch the solutions together. □

The harmonic coordinate condition was the first coordinate condition which was used to make the equations hyperbolic. Later, other possibilities were found. The 'conformal field equations' (Friedrich 1985) are hyperbolic in any coordinate system.

A unique global spacetime determined by the data can be constructed with the help of Zorn's lemma by patching together local solutions. These spacetimes have the property that all the causal past-directed curves from a point in the future of the Cauchy surface intersect the Cauchy surface just once. A spacetime which contains a hypersurface with this property is called *globally hyperbolic*.

A globally hyperbolic solution of Einstein's field equation may still have an extension. This extension is, however, not determined by the Cauchy data. This brings us to the concept of the *domain of dependence*, a central notion in the theory of smooth solutions of hyperbolic equations.

Take a point p in the future of the Cauchy surface S of some globally hyperbolic solution. All the past directed causal curves will intersect the Cauchy surface in some set A. The solution at the point p depends only on the data in A. The points in which the solution is determined by the data in A is called $D(A)$, the domain of dependence of A.

So far we have considered only the vacuum field equations. For matter described by classical fields (Maxwell, Yang–Mills fields) we have essentially the same situation as in the vacuum case.

For phenomenological matter like perfect fluids, the situation is similar. One also obtains local solutions of the Cauchy problem. There is, however, the problem that we cannot treat the boundary of the fluid. This is not only a problem in relativity, we face the same difficulty in Galilei-invariant hydrodynamics even without gravity. For certain equations of state this problem can be handled. (Rendall 1992a)

4 Global questions

In this section I want to describe some global results and problems. Let me first discuss the Christodoulou–Klainerman theorem. In its roughest form it states:

Cauchy data near Minkowski data determine solutions near Minkowski space.

In particular, the solutions have complete timelike and null geodesics and a causal structure similar to Minkowski space in the sense that the solutions have a null infinity.

This theorem sounds obvious, but before it was proved (and it is a quite long and complicated proof) we did not really know whether it would be true. Take as a warning the example of 1-dimensional hydrodynamics (Galilei invariant). It is known that in this case all solutions have shocks (singularities); the smaller the data, the later the shock occurs.

A second example of a global existence theorem was proved by Friedrich (1986) in the case of a positive cosmological constant. The de Sitter solution is the solution with maximal symmetry. Data near de Sitter data define solutions which become asymptotically de Sitter in the past as well as in the future.

Friedrich's method of 'regular conformal field equations' has shown itself to be very useful in establishing the existence and asymptotic properties of solutions of Einstein's field equations

$$\tilde{R}_{ab} = \Lambda \tilde{g}_{ab}. \tag{28}$$

I shall try to outline this method and some results achieved by it. References can be found in (Friedrich 1994).

Motivated by Penrose's treatment of null infinity via a conformal rescaling of the spacetime metric $\tilde{g}_{ab}$ the aim is to use the 'unphysical metric'

$$g_{ab} = \Omega^2 \tilde{g}_{ab} \tag{29}$$

together with the conformal factor Ω as the unknown fields. Rewriting of (28) in terms of the unphysical metric and Ω leads to the equation

$$\tilde{R}_{ab} = R_{ab} + 2\Omega^{-1}\nabla_a\nabla_b\Omega + g_{ab}g^{cd}\left(\Omega^{-1}\nabla_c\nabla_d\Omega - 3\Omega^{-2}\nabla_c\Omega\nabla_d\Omega\right) = \Lambda\Omega^{-2}g_{ab}. \tag{30}$$

This equation is singular for $\Omega = 0$, *i.e.* precisely at those points (at infinity) where we want to understand its consequences.

The 'regular conformal field equations' are a system of equations (for the conformal factor, the rescaled metric, the non-physical Ricci tensor and the rescaled Weyl tensor defined by $d^a{}_{bcd} := \Omega^{-1} C^a{}_{bcd}$) which are equivalent to (30) and regular in the sense that no factor of Ω^{-1} occurs in the equations and that Ω does not appear in the principle part of the differential operator associated with the equations (the terms with the highest derivatives). The Bianchi identities are part of the equations.

These equations are not only regular, but can furthermore be split into a system of symmetric hyperbolic evolution equations and constraints which are compatible with the evolution. This allows us to study various Cauchy and characteristic initial value problems with an initial hypersurface on which Ω vanishes, and hence to prescribe data at points which are at infinity in the physical spacetime. It is worthwhile to note that the evolution equations are, in a certain sense, hyperbolic in any coordinate system.

I shall now consider the three cases $\Lambda > 0$, $\Lambda < 0$, and $\Lambda = 0$ and describe some of the results obtained.

The de Sitter solution has a positive cosmological constant and is geodesically complete. It can be conformally imbedded into the Einstein universe and its boundary consists of two spacelike hypersurfaces $\mathscr{I}^+$ and $\mathscr{I}^-$ on which Ω vanishes. Therefore this imbedding defines a solution of the conformal field equations. One can now analyse the Cauchy problem to find out which data can be given on a spacelike hypersurface on which $\Omega = 0$ and $d\Omega \neq 0$. The result is that a positive definite metric is freely specifiable together with the electric part of the rescaled conformal Weyl tensor. de Sitter data determine uniquely the de Sitter solution. If we take data sufficiently near these data, general theorems on the stability of solutions of symmetric hyperbolic systems on compact domains imply that these solutions exist on a sufficiently large domain and reach a second hypersurface on which Ω vanishes. Translating this result to physical spacetime, we have constructed a solution which is geodesically complete and asymptotically de Sitter in the past as well as in the future.

Suppose that we would like to prove the same theorem, working in physical spacetime. Changing the data from de Sitter data on some Cauchy surface would, by general theorems, only give a solution on some compact part of the de Sitter spacetime. To obtain a solution which is geodesically complete, special estimates would be needed and no general method is known to obtain such estimates. Thanks to the conformal equations no such estimates are needed.

Recently Friedrich (1994) considered the anti–de Sitter spacetime ($\Lambda < 0$). Again it can be conformally imbedded into the Einstein universe. Its boundary is timelike with the topology $S^2 \times \mathbf{R}$. To prove existence of solutions which are asymptotically anti–de Sitter, Friedrich solved a boundary initial value problem for the conformal field equations with a timelike boundary. This is the first general initial boundary value problem in the context of Einstein's equations which has found a complete treatment. Translating again to physical spacetime the solution behaves asymptotically like anti–de Sitter near $\mathscr{I}^+$, which is timelike.

For de Sitter and anti–de Sitter space the boundaries of the conformal imbeddings into the Einstein universe are smooth hypersurfaces, and it is possible to pose regular initial or boundary value problems. This is different for the conformal imbedding of Minkowski space into the Einstein universe. Besides the smooth null hypersurfaces at infinity, $\mathscr{I}^+$ and $\mathscr{I}^-$, there are the vertices of these null cones, I^0 and $I^\pm$, where Ω vanishes. Examples like the Schwarzschild spacetime show that in general the conformal structure will be singular at spacelike infinity I^0.

Postponing the problem of spacelike infinity, the hyperboloidal Cauchy problem was studied. Prescribing almost Minkowski data on a hypersurface intersecting $\mathscr{I}^+$, Friedrich showed the existence of solutions which are geodesically future-complete and have a regular point I^+ as future timelike infinity. So far no such result has been obtained by working with the equations in spacetime.

To treat the usual Cauchy problem by conformal techniques one has to analyse the singularity at spacelike infinity enforced by positive ADM-mass. Investigations are under way to decide which class of data will evolve into spacetimes with smooth $\mathscr{I}^+$, and will finally resolve the general structure of spatial infinity.

These are the only global theorems that I am aware of. They should really be considered just as a starting point. What happens if one increases the data? It is to be hoped that numerical solutions will provide some hints and inspiration.

About 20 years before the theorem by Christodoulou-Klainerman the first singularity theorems were proved by Hawking and Penrose (compare Clarke and Schmidt (1977)). The structure of these theorems is that the following 4 conditions are incompatible:

(a) $G_{ab}V^aV^b \geq 0$,

(b) a causality condition, for example 'globally hyperbolic',

(c) 'a strong gravitational field', for example a 'closed trapped surface',

(d) spacetime is geodesically complete.

The usual interpretation is that (a)–(c) imply that the spacetime contains a incomplete causal geodesic.

It is typical for all these theorems that the field equations enter only via the inequality (a). Hence the theorems are of a geometrical nature. Therefore it is not surprising that almost nothing can be said about the nature of the incompleteness. We can only expect more detailed insight into the structure of the singularity if we really use the field equations.

Again numerical relativity might help. There are indications that the singularity of vacuum solutions with compact Cauchy surfaces may have a rather simple structure (Berger and Moncrief 1993).

5 The frame theory

Let me first describe a common formulation of Newton's and Einstein's theory of gravity developed in (Ehlers 1981). Consider the following objects

$$(M^4, \ g_{ab}, \ h^{ab}, \ \Gamma^a_{bc}, \ T^{ab}, \ \lambda) \tag{31}$$

where g_{ab}, h^{ab}, T^{ab} are smooth tensor fields on M^4, Γ^a_{bc} is a connection on M^4 and λ is a real constant. We 'move indices' with h^{ab}, g_{ab} and use the notation

$$v^a_\bullet = h^{ab} v_b \qquad w^\bullet_a = g_{ab} w^b$$

We demand the following axioms:

Metric axioms

A1. At each point of M^4 there exists a vector V^a, called *timelike*, with $g_{ab} V^a V^b > 0$.

A2. At each point the quadratic form $\omega_a \to h^{ab} \omega_a \omega_b$ is positive definite on the kernel $\{\omega_a : \omega_a V^a = 0\}$ of any timelike vector V^a.

A3. $g_{ab} h^{ac} = -\lambda \delta^c_b$.

Connection axioms

A4. $g_{ab;c} = 0 \qquad h^{ab}{}_{;c} = 0$

A5. $R^a{}_{b \bullet d}{}^c = R^c{}_{d \bullet b}{}^a$

Dynamical axioms

A6. $T^{ab}{}_{;b} = 0$

A7. $R_{ab} = 8\pi (T^{\bullet \bullet}_{a\ b} - \frac{1}{2} g_{ab} T^{c \bullet}_{\ c})$

Matter content, 'perfect fluid'

A8. $T^{ab} = (\rho + \lambda p) U^a U^b + p h^{ab} \qquad g_{ab} U^a U^b = 1$

The constant λ is called the 'causality constant' and can be thought of as c^{-2}. In the case $\lambda > 0$ the relations A1–A3 imply that g_{ab} has Lorentzian signature $(+ - - -)$ and we have Einstein's theory.

We will now try to see that $\lambda = 0$ is the 4-dimensional formulation of Newton's theory which was given by Cartan (1923) and Friedrichs (1927).

Let us derive some consequences of the axioms:

(a) The first two metric axioms imply that there is always a basis in the tangent space of any point such that

$$h^{ab} = \text{diagonal}\,(0,1,1,1) \qquad g_{ab} = \text{diagonal}\,(1,0,0,0)\,. \tag{32}$$

Hence we can always find a covector field t_a with $g_{ab} = t_a t_b$ and $h^{ab} t_b = 0$. Now $g_{ab;c} = 0$ gives $t_{a;b} = 0$, so t_a is a gradient and there exists a function t satisfying $t_{,a} = t_a$.

(b) The hypersurfaces $t = $ constant are totally geodesic: Let v^a be a vector in such a hypersurface. Then $t_a v^a = 0$ and hence $(t_a v^a)_{;b} = t_a v^a_{;b}$. The derivative of v^a in the direction w^b which is $v^a_{;b} w^b$ is also in $t = $ constant because we have $t_a v^a_{;b} w^b = 0$.

(c) Axiom A2 shows that a Riemannian metric h_{ab} is defined in the cotangent space of $t =$ constant. Therefore we have its inverse metric, γ_{ab}, defined in the tangent space of $t =$ constant. We have $\gamma_{ab}\omega_\bullet^a\omega_\bullet^b = h^{ab}\omega_a\omega_b$.

(d) The connection induced in $t =$ constant is the Levi-Civita connection of the metric $\gamma_{\alpha\beta}$: Suppose v_a and w_a are parallel propagated along a curve $x^b(s)$ with tangent $\dot{x}^b$. Then we have $(h^{ab}v_a w_b)\dot{} = 0$.

(e) The curvature tensor $R^a{}_{bcd}$ of the spacetime connection can be restricted to the $t =$ constant hypersurface and this tensor is the curvature tensor of the connection induced on the hypersurface. (This is a general property of totally geodesic hypersurfaces.)

(f) The Ricci identity for t_a implies $t_a R^a{}_{bcd} = 0$. In coordinates adapted to the hypersurface $t =$ constant we can see that $R_{\alpha\beta}$ is the Ricci tensor of the inner connection. The field equations show that this tensor vanishes, hence the metric on the hypersurfaces is flat.

(g) As the submanifolds $t =$ constant are locally Euclidean, one can introduce 'orthonormal' local coordinates t, x^α such that

$$g_{ab} = \text{diag}\,(1,0,0,0) \qquad h^{ab} = \text{diag}\,(0,1,1,1). \tag{33}$$

In these coordinates we obtain $\Gamma^\alpha_{\beta\gamma} = 0$ because $t =$ constant is flat. The equation $\Gamma^0_{ab} = 0$ follows from $t_{,a;b} = 0$. The equation $\Gamma^\alpha{}_{0\beta} = -\Gamma^\beta{}_{0\alpha} =: \omega_{\alpha\beta}$ follows from $h^{ab}_{;c} = 0$.

(h) If we write in a 3-dimensional vector notation using $\mathbf{g} = (g^\alpha)$, $g^\alpha = -\Gamma^\alpha_{00}$ and $\boldsymbol{\omega} = (\omega_{23}, \omega_{31}, \omega_{12})$, the geodesic equation for freely falling test particles assumes the form

$$\ddot{\mathbf{x}} = \mathbf{g} + 2\dot{\mathbf{x}} \times \boldsymbol{\omega}. \tag{34}$$

(i) The axiom A5, together with the dynamical laws, reduces for a perfect fluid to the equations:

$$\nabla \cdot \boldsymbol{\omega} = 0 \qquad \nabla \times \mathbf{g} + 2\dot{\boldsymbol{\omega}} = 0 \tag{35}$$

$$\nabla \times \boldsymbol{\omega} = 0 \qquad \nabla \cdot \mathbf{g} - 2\boldsymbol{\omega} \cdot \boldsymbol{\omega} = -4\pi\rho \tag{36}$$

and the (generalised) Euler equations

$$\dot{\rho} + \nabla \cdot (\rho \mathbf{V}) = 0 \tag{37}$$

$$\rho(\dot{\mathbf{V}} + \mathbf{V} \cdot \nabla \mathbf{V} - \mathbf{g} - 2\mathbf{V} \times \boldsymbol{\omega}) = -\nabla p \tag{38}$$

where $\mathbf{V} = (U^\alpha)$ denotes the 3-velocity of the fluid.

One recognizes that the case $\lambda = 0$ almost gives Newton's theory. Almost, for in (34–38) both $\mathbf{g}$ and $\boldsymbol{\omega}$ depend on t and x^α.

(j) If $\boldsymbol{\omega} = 0$ we have Newtonian theory. This is implied by the condition $R^{\alpha\beta}{}_{\gamma\delta} = 0$. If it holds, and if non-rotational orthonormal coordinates are used, the connection components have the form

$$\Gamma^a_{bc} = t_{,a} t_{,c} h^{ad} U_{,d} \tag{39}$$

where $\mathbf{g} = -\nabla U$ and U is the Newtonian potential. This formulation exhibits the nature of U as a 'connection potential'.

6 The Newtonian Limit

The frame theory allows a very natural definition of Newtonian limits.

Definition 6.1 *A 1-parameter family of solutions of the frame theory, $0 \leq \lambda < \lambda_0$,*

$$(M^4, \ g_{ab}(\lambda), \ h^{ab}(\lambda), \ \Gamma^a_{bc}(\lambda), \ T^{ab}(\lambda)) \tag{40}$$

with fields defined on the same manifold M^4 has a Newtonian limit if all fields (together with a number of partial derivatives) have limits

$$(M^4, \ g_{ab}(0), \ h^{ab}(0), \ \Gamma^a_{bc}(0), \ T^{ab}(0)). \tag{41}$$

The families defined by the following metrics have a Newtonian limit:

(a) **Minkowski space**

$$g_{ab}dx^a dx^b = dt^2 - \lambda(dx^2 + dy^2 + dz^2). \tag{42}$$

(b) **Schwarzschild metric**

$$g_{ab}dx^a dx^b = (1 - \frac{2M}{r}\lambda)dt^2 - \lambda\{(1 - \frac{2M}{r}\lambda)^{-1}dr^2 + r^2(d\theta^2 + \sin^2\theta d\phi^2)\}. \tag{43}$$

In a similar way one can treat the case of a static fluid solution with its exterior Schwarzschild field.

(c) **Plane Waves**

More surprising is the case of a family of plane waves which have a Newtonian limit (Ehlers 1981):

$$g_{ab}dx^a dx^b = dt^2 - \lambda(dx^2 + dy^2 + dz^2) + P(dt - \sqrt{\lambda}\,z)^2 \tag{44}$$

where $P = \lambda\left[(x^2 - y^2)A''_+(t - \sqrt{\lambda}\,z) + xy A''_\times(t - \sqrt{\lambda}\,z)\right]$ and A_+ and $A_\times$ are arbitrary. The Newtonian limit has the gravitational potential

$$\Phi = \frac{1}{4}\left[(x^2 - y^2)A''_+(t) + xy A''_\times(t)\right] \tag{45}$$

which is a time dependent, not asymptotically flat, field with quadrupole structure. The Newtonian motion of test particles imitates the motion of particles in the plane gravitational wave in Einstein's theory.

(d) **Robinson-Walker cosmologies**

$$g_{ab}dx^a dx^b = dt^2 - \lambda R^2 \frac{d\boldsymbol{\xi}\cdot d\boldsymbol{\xi}}{1 - \frac{\lambda}{4}E\boldsymbol{\xi}\cdot\boldsymbol{\xi}} \tag{46}$$

The Newtonian limit in this case gives the isotropic cosmologies studied by Heckmann and Schücking (1959).

If one tries to construct such families using the field equations, one faces the following problem. Consider the system of two coupled wave equations

$$-\frac{1}{c^2}\ddot{\phi} + \Delta\phi = F(\phi,\psi)$$
$$-\ddot{\psi} + \Delta\psi = G(\phi,\psi). \qquad (47)$$

For $c \to \infty$ we obtain a mixed hyperbolic elliptic system. We expect a similar behaviour for Einstein's equations coupled to matter. For time independent systems the matter equation as well as the equation for the gravitational field should be elliptic, so that this case should be easier to study.

There is a very nice result by Heilig (1994). He studied the problem of existence for stationary, axisymmetric rotating fluid solutions. First he rewrote Lichtenstein's (1933) original proof of the Newtonian problem in terms of modern mathematics. Lichtenstein's paper contains the first existence proof and was completely forgotten. In a second paper Heilig (1995) shows the existence of a λ-family of rotating fluids in Einstein's theory which have a Newtonian limit.

Heilig's proof is based on an implicit function theorem argument. He starts with a Newtonian solution and studies the linearisation in the parameter to check the assumptions of the implicit function theorem in Banach spaces. The variables and the form of the field equations used were given by Ehlers (unpublished) and they are elliptic for $\lambda > 0$ as well as for $\lambda = 0$. This result shows that a proper treatment of the Newtonian limit has not only philosophical interest but can actually lead to results in Einstein's theory.

What can we expect in time dependent situations? Ehlers' formulation of the field equations mentioned above shows that the differential equations change from hyperbolic to elliptic. No general theorems are available which would give the existence of solutions with a Newtonian limit. Lottermoser (1992) has studied families of solutions of the constraints with a Newtonian limit. He could, however, say nothing about the limit at later times. At present there is no existence theory for time dependent isolated fluid bodies. Hence we can not even begin to investigate the question of the Newtonian limit.

The first theorem for a time dependent situation was proved by Rendall (1993). He takes as a matter model a collisionless gas described by a distribution function $f(x^a, p^b)$, which defines the energy momentum tensor

$$T^{ab} = \int f(x^d, p^s) p^a p^b |g|^{1/2} \frac{1}{p^0} dp^1 dp^2 dp^3. \qquad (48)$$

The Vlasov equation is

$$p^a \frac{\partial f}{\partial x^a} - \Gamma^a_{bc} p^b p^c \frac{\partial f}{\partial p^a} = 0. \qquad (49)$$

For the coupled Vlasov-Einstein system, he shows the existence of a 1-parameter family of asymptotically flat solutions with a Newtonian limit. Any Newtonian solution can appear in such a family.

7 Post-Newtonian approximations

Post-Newtonian approximations are important in deriving observational consequences from Einstein's theory. The 4-dimensional formulation of Newton's theory and the frame theory should lead to a better understanding of such approximations.

Winicour (1983) studied 'Newtonian gravity on the null cone' in this spirit. In a sequence of papers he comes up with a derivation of the 'quadrupole formula'. I have no time to treat this here.

Rendall (1992b) studies the solvability of post-Newtonian equations. He considers a family:

$$g_{ab}(\epsilon) = g^0_{ab}(\epsilon) + \epsilon g^1_{ab}(\epsilon) + \ldots \epsilon^r g^r_{ab}(\epsilon) \tag{50}$$

and assumes that:

(a) $g_{ab}(\epsilon)$ is a Lorentz metric for $\epsilon > 0$.

(b) $h^{ab} = \epsilon^2 g^{ab}(\epsilon)$ has an expansion in ϵ.

(c) $\Gamma^a_{bc}(\epsilon)$ has an expansion in ϵ.

(d) $g_{ab}(0)$ has rank 1 and is negative semidefinite at some point.

(e) $h^{ab}(0)$ has rank 3 and is negative semidefinite at some point.

(f) $G^{ab}(\epsilon)$, the Einstein tensor of $g_{ab}(\epsilon)$, has an expansion in ϵ.

Then using ϵ-dependent coordinate transformations one obtains

$$g_{ab}dx^a dx^b = (-1 + 2U\epsilon^2 + O(\epsilon^3))(dx^0)^2 + \delta_{\alpha\beta}(\epsilon^2 + O(\epsilon^3))dx^\alpha dx^\beta + O_{0\alpha}(\epsilon^4)dx^\alpha dx^0 \tag{51}$$

If we have $G^{ab} = 8\pi G T^{ab}$ for a perfect fluid, U satisfies the Poisson equation and $T^{ab}{}_{;b} = 0$ implies the hydrodynamical equations. Hence we have in leading order the Newtonian limit. Note that the flatness of the space section is a consequence of the assumptions.

For higher orders one obtains a hierarchy of field equations from $G^{ab} = 8\pi T^{ab}$ and equations of motions from $T^{ab}{}_{;b} = 0$. The field equations in order ϵ^1 are pure gauge. The order ϵ^2 defines the first post-Newtonian equation, the order ϵ^3 is gauge, the order ϵ^4 defines the second post-Newtonian equation. Rendall (1992) has shown that the third post-Newtonian equations, order ϵ^6, cannot be solved globally. At this order gravitational radiation becomes relevant and one expects post-Newtonian approximations to hold only in the near zone.

The metric (51) looks very similar to the usual starting point of post-Newtonian expansions. Using $x^0 = \epsilon t$ and $\tilde{g}_{ab} = \epsilon^{-2} g_{ab}$ we obtain

$$\tilde{g}_{ab}dx^a dx^b = (-1 + 2U\epsilon^2 + O(\epsilon^3))(dt)^2 + (\delta_{\alpha\beta} + O(\epsilon))dx^\alpha dx^\beta + O_{0\alpha}(\epsilon^3)dx^\alpha dt \tag{52}$$

This looks like a post-Minkowskian expansion. If we solve the linearised field equations, we obtain the Poisson equation if we make some 'slow motion' assumption, but $T^{ab}{}_{;b} = 0$ implies free motion of the fluid in leading order. This inconsistency is typical for all approaches to obtaining the Newtonian limit via linearised theory. Special tricks are necessary to get around this difficulty.

Traditionally, post-Newtonian expansions are developed for isolated systems. They are, however, equally important in Cosmology. As the isotropic fluid cosmologies lead in Einstein's and Newton's theory to the same ordinary differential equation for the scale factor, in astrophysics the Newtonian picture is mostly used for the 'standard cosmological model'. In the context of structure formation, perturbations are studied which can best be understood if one treats them as post-Newtonian. The 4-dimensional formulation of Newtonian theory allows us to do this in a natural way.

References

Berger B and Moncrief V, 1993, Numerical investigation of cosmological singularities, *Phys Rev* D, **48** 4676.
Cartan H, 1923, *Ann Ecole Norm* **40** 325.
Choquet–Bruhat Y and York J, 1980, The Cauchy Problem, In: *General Relativity and Gravitation*, Vol. 1, ed Held A, Plenum Press New York.
Christodoulou D and Klainerman S, 1993, The Global Nonlinear Stability of Minkowski Space, *Annals of Mathematical Studies*, Princeton Univ. Press.
Clarke C and Schmidt B, 1977, Singularities: the State of the Art, *GRG* **8** 129.
Ehlers J, 1981, The Newtonian limit of general relativity, In: *Classical mechanics and relativity: relationships and consistency*, ed Ferrarese G, Naples 1991.
Fischer A, Marsden J, 1979, The initial value problem and dynamical formulation of general relativity, in: *General Relativity*, eds Hawking S, Israel W, Cambridge University Press.
Friedrichs K O, 1927, *Math Ann* **98** 566.
Friedrich H, 1985, On the Hyperbolicity of Einstein's and Other Gauge Field Equations, *Commun Math Phys* **100** 525.
Friedrich H, 1986, Existence and structure of past asymptotically simple solutions of Einstein's field equations with positive cosmological constant, *JGP* **3** 101.
Heckmann O and Schücking E, 1959, *Encycl of Physics*, VIII, Springer Verlag Berlin.
Heilig U, 1994, On Lichtenstein's Analysis of Rotating Newtonian Stars, *Ann Henri Poincaré Physique Theoretique* **60** 457.
Heilig U, 1995, On the Existence of Rotating Stars in General Relativity, *Commun Math Phys* **166** 457.
John F, 1978, *Partial Differential Equations*, Springer Verlag New York.
Lichtenstein L, 1933, *Gleichgewichtsfiguren rotierender Flüssigkeiten*, Springer Berlin.
Lottermoser M, 1992, A Convergent Post–Newtonian Approximation for the Constraint Equations in General Relativity, *Ann de L'Institut H Poincaré* **57** 279.
Rendall A, 1992a, The initial value problem for a class of general relativistic fluid bodies. *J Math Phys*, **33** 1047.
Rendall A, 1992b, On the definition of post–Newtonian approximations, *Proc Roy Soc Lond* **438** 341.
Rendall A, 1993, The Newtonian limit of asymptotically flat solutions of the Vlasov–Einstein system, *Commun Math Phys* **163** 89.
Winicour J, (1983), Newtonian gravity on the null cone, *J Math Phys* **24** 1193.

Partial Differential Equations of Physics

Robert Geroch

Enrico Fermi Institute
Chicago

1 Introduction

The physical world is traditionally organized into various systems: electromagnetism, perfect fluids, Klein-Gordon fields, elastic media, gravitation, *etc.* Our descriptions of these individual systems have certain features in common: use of fields on a fixed space-time manifold M, a geometrical interpretation of the fields in terms of M, partial differential equations on these fields, an initial-value formulation for these equations. Yet beyond these common features there are numerous differences of detail: some systems of equations are linear, and some are not; some have constraints, and some do not; some arise from Lagrangians, and some do not; some are first-order, and some higher-order. Systems also differ in other respects, *e.g.* as to what fields they need as background, what interactions they permit (or require). It almost seems as though, in the end, every physical system has its own special character.

It might be useful to have a systematic treatment of the fields and equations that arise in the description of physical systems. Thus, there would be a general definition of a 'field', and a general form for a system of partial differential equations on such fields. The treatment would consist of a framework sufficiently broad to encompass the systems found in nature, but no broader. One would, for example, treat the initial-value formulation once and for all within this broad framework, with the formulations for individual physical systems emerging as special cases. In a similar way, one would treat—within a quite general context—constraints, the geometrical character of physical fields, how some systems require other fields as a background, how interactions operate, *etc.* The goal of such a treatment would be to get a better grip on the structural features of the partial differential equations of physics. Here are two examples of issues on which such a treatment might shed light. What, if any, is the physical basis on which the various fields on the manifold M are grouped into separate physical systems? Thus, for instance, the fields (n, F_{ab}, ρ, u^a) are grouped into (n, ρ, u^a) (a perfect fluid), and (F_{ab}) (electromagnetic field). By 'physical basis', we mean in terms of the dynamical equations on these fields. A second issue is this: how does it come about that the fields of general relativity are singled out as those for which diffeomorphisms on M are gauge? On the face of it, this singling-out seems surprising, for the diffeomorphisms act equally well on all the physical fields on M.

We shall here discuss, in a general, systematic way, the structure of partial differential equations describing physical systems. We take it as given that there is a fixed, four-dimensional manifold M of 'space-time events', on which all the action takes place. Thus, for instance, we are not considering discrete models. Further, physical systems are to be described by certain 'fields' on M. These may be more general than mere tensor fields: our framework will admit spinors, derivative operators, and perhaps other field-types not yet thought of. But we shall insist, largely for mathematical convenience, that the set of field-values at each space-time point be finite-dimensional. We shall further assume that these physical fields are subject to systems of partial differential equations. That is, we assume, among other things, that 'physics is local in M'. Finally, we shall demand that the partial differential equations be first-order (*i.e.* involving no higher than first space-time derivatives of the fields), and quasilinear (*i.e.* linear in those first derivatives). That the equations be first-order is no real assumption: higher-order equations can, and will, be cast into first-order form by introducing new auxiliary fields. (Thus, to treat the Klein-Gordon equation on scalar field ψ, we introduce an auxiliary vector field playing the role of $\nabla \psi$.) It is my feeling that this is more than a mere mathematical device: the auxiliary fields tend to have direct physical significance. It is not so clear what we are actually assuming when we demand quasilinearity. It is certainly possible to write down a first-order system of partial differential equations that is not even close to being quasilinear (*e.g.* $(\partial \psi/\partial x)^2 + (\partial \psi/\partial t)^2 = \psi^2$). But all known physical systems seem to be described by quasilinear equations, and it is anyway hard to proceed without this demand. In any case, 'first-order, quasilinear' allows us to cast all the partial differential equations into a convenient common form, and it is on this common form that the program is based. A case could be made that, at least on a fundamental level, all the 'partial differential equations of physics' are hyperbolic—that, for example, elliptic and parabolic systems arise in all cases as mere approximations of hyperbolic systems. Thus, Poisson's equation for the electric potential is just a facet of a hyperbolic system, Maxwell's equations.

In Section 2, we introduce our general framework for systems of first-order, quasilinear partial differential equations for the description of physical systems. The physical fields become cross-sections of an appropriate fibre bundle, and it is on these cross-sections that the differential equations are written. So, for instance, the coefficients in these equations become certain tensor fields on the bundle space. This framework, while broad in its reach, is not particularly useful for explicit calculations. The remaining sections describe various structural features of these system of partial differential equations. A 'hyperbolization' (Section 3) is a casting of the system of equations (or, commonly, a subsystem of that system) into what is called symmetric, hyperbolic form. To such a form there is applicable a general theorem on existence and uniqueness of solutions. This is the initial-value formulation. The constraints (Section 4) represent a certain subsystem of the full system, the equations of which play play a dual role: providing conditions that must be satisfied by initial data, and leading to differential identities on the equations themselves. The constraints are integrable if these 'differential identities' really are identities; and complete if the constraint subsystem, together with the subsystem involved in the hyperbolization, exhausts the full system of equations. The geometrical character of the physical fields has to do with how they 'transform', *i.e.* with lifting diffeomorphisms on M to diffeomorphisms on the bundle space—see section 5. Combining all the systems of physics into one master bundle B, then the full set of equations on this bundle will be M-diffeomorphism invariant. This diffeomorphism invariance requires an appropriate adjustment in the initial-value formulation for this combined system. Finally, we turn (Section 6) to the relationships between the various physical fields, as reflected in their differential equations. Physical fields on space-time can interact on two broad levels: dynamically (through their derivative-terms), and kinematically (through terms algebraic in the fields). Roughly speaking, two fields are part of the same physical system if their derivative-terms cannot be separated into individual equations; and one field is a background for another if the former appears algebraically in the derivative-terms of the latter. The kinematical (algebraic) interactions are the more familiar couplings between physical systems.

It is the examples that give life to this general theory. We have assembled, in Appendix A, a variety of standard examples of physical systems: the fields, the equations, the hyperbolizations, the constraints, the background fields, the interactions, *etc.* We shall refer to this material frequently as we go along. Thus, this is not the standard type of appendix (to be read later, if at all, by those interested in technical details), but rather is an integral part of the general theory. Indeed, it might be well to review this material first as a kind of introduction. Appendix B contains a statement and an outline of the proof of the theorem on existence and uniqueness of solutions of symmetric, hyperbolic systems of partial differential equations.

All in all, this subject forms a pleasant comingling of analysis, geometry, and physics.

2 Preliminaries

Fix, once and for all, a smooth, (connected) four-dimensional manifold M. The points of M will be interpreted as the events of space-time, and, thus, M itself will be interpreted as the space-time manifold. We do not, as yet, have a metric, or any other geometrical structure, on M.

We next wish to introduce various types of 'fields' on M. To this end, let $b \xrightarrow{\pi} M$ be a smooth fibre bundle[1] over M. That is, b is some smooth manifold (called the *bundle manifold*) and π is some smooth mapping (called the *projection mapping*); and these are such that locally (in M) b can be written as a product in such a way that π is the projection onto one factor[2]. An example is the tangent bundle of M: here, b is the eight-dimensional manifold of all tangent vectors at all points of M, and π is the

---

[1] See, *e.g.* Steenrod (1954). Note that, in contrast to what is done in this reference, we introduce no Lie group acting on b.

[2] This means, in more detail, that, given any point $x \in M$, there exists an open neighborhood U of x, a manifold F, and a diffeomorphism ζ from $U \times F$ to $\pi^{-1}[U]$ such that $\pi \circ \zeta$ is the projection of $U \times F$ to its first factor.

mapping that extracts, from a tangent vector at a point of M, the point of M. That the local-product condition holds, in this example, is seen by expressing tangent vectors in terms of their components with respect to a local basis in M. Returning to the general case, for any point x of M, the *fibre* over x is the set of points $\pi^{-1}(x)$, *i.e.* the set of points $\kappa \in b$ such that $\pi(\kappa) = x$. It follows from the local-product condition that each fibre is a smooth submanifold of b, and that all the fibres are diffeomorphic with each other. In the example of the tangent bundle, for instance, the fibre over point $x \in M$ is the set of all tangent vectors at x. Next, let A be any smooth submanifold of M. A *cross-section* over A is a smooth mapping $A \xrightarrow{\phi} b$ such that $\pi \circ \phi$ is the identity mapping on A. Thus, a cross-section assigns, to each point x of A, a point of the fibre over x. Typically, A will be of dimension four (*i.e.* an open subset of M), or three.

We interpret the fibre over x as the space of allowed physical states at the space-time point x, *i.e.* as the space of possible field-values at x. Then the bundle manifold b is interpreted as the space of all field-values at all points of M. A cross-section over the submanifold A becomes a field, defined at the points of A. In most, but not all, examples (Appendix A) b will be a tensor bundle. Thus, for electromagnetism b is the ten-dimensional manifold of all antisymmetric, second-rank tensors at all points of M. For general relativity, by contrast, b is the fifty-four-dimensional manifold a point of which is comprised of a point of M, a Lorentz-signature metric at that point, and a torsion-free derivative operator at that point. In both of these examples, the projection π merely extracts the point of M.

It is convenient to introduce the following notation. Denote tensors in M by lower-case Latin indices; and tensors in b by lower-case Greek indices. Then, at any point $\kappa \in b$, we may introduce mixed tensors, where Latin indices indicate tensor character in M at $\pi(\kappa)$, and Greek indices tensor character in b at κ. For example, the derivative of the projection map is written $(\nabla \pi)_\alpha{}^a$, *i.e.* it sends tangent vectors in b at point κ to tangent vectors in M at $\pi(\kappa)$. The derivative of a cross-section, ϕ, over a four-dimensional region of M is written $(\nabla \phi)_a{}^\alpha$; and we have, from the defining property of a cross-section,

$$(\nabla \phi)_a{}^\alpha (\nabla \pi)_\alpha{}^b = \delta_a{}^b. \qquad (1)$$

A vector λ^α at $\kappa \in b$ is called *vertical* if it is tangent to the fibre through κ, *i.e.* if it satisfies $\lambda^\alpha (\nabla \pi)_\alpha{}^b = 0$. Elements of the space of vertical vectors at κ will be denoted by primed Greek superscripts. Thus, $\lambda^{\alpha'}$ means 'λ is a tangent vector in b, a vector which, by the way, is vertical'. Elements of the space dual to that of the vertical vectors will be denoted by primed Greek subscripts. Thus, $\mu_{\alpha'}$ means 'μ is a linear mapping from vertical vectors in b to the reals'. More generally, these primed indices may appear in mixed tensors. Note that we may freely remove primes from superscripts (*i.e.* ignore the verticality of an index), and add primes to subscripts (*i.e.* restrict the mapping from all tangent vectors to just vertical ones), but not the reverses. As an example of this notation, we have $(\nabla \pi)_{\alpha'}{}^a = 0$.

To illustrate these ideas, consider electromagnetism. Then a typical point of the bundle manifold b is $\kappa = (x, F_{ab})$, where x is a point of M and F_{ab} is an antisymmetric tensor at x. A tangent vector λ^α in b at κ can be represented as an 'infinitesimal change[3] in both the point x of M and the antisymmetric tensor F_{ab}'. Given such

---

[3] By 'infinitesimal change in the point of...', we mean 'tangent vector to a curve in...'.

a λ^α, the combination $\lambda^\alpha(\nabla\pi)_\alpha{}^a$ is that tangent vector in M at x represented by just the 'change in x'-part of λ^α (ignoring the 'change in F_{ab}' part). Such a λ^α is vertical provided its 'change in x' vanishes; so, a vertical vector is represented simply as an infinitesimal change in the antisymmetric tensor F_{ab}, with x fixed. In this example, we might introduce the field $\mu_{\alpha'ab} = \mu_{\alpha'[ab]}$ on b, which takes any such vertical vector, $\lambda^{\alpha'}$, and returns, as $\lambda^{\alpha'}\mu_{\alpha'ab}$, the change in the tensor F_{ab} at x.

A *connection* on the fibre bundle $b \xrightarrow{\pi} M$ is a smooth field $\gamma_a{}^\alpha$ on b satisfying

$$\gamma_a{}^\alpha(\nabla\pi)_\alpha{}^b = \delta_a{}^b.$$

Given a connection $\gamma_a{}^\alpha$, those vectors at $\kappa \in b$ that can be written in the form $\xi^a \gamma_a{}^\alpha$ for some ξ^a are called *horizontal*. Of course, there exist many possible connections, and so many such notions of 'horizontality'. It follows directly from these definitions that, fixing a connection, every tangent vector in b at κ can be written, uniquely, as the sum of a horizontal and a vertical vector, *i.e.* that every vector can be split into its horizontal and vertical parts. We may incorporate this observation into the notation by allowing ourselves the operations with primes that were previously prohibited. In the presence of a fixed connection, $\gamma_a{}^\alpha$, we may affix a prime to a Greek superscript (by taking the vertical projection); and, in a similar way, we may remove a prime from a Greek subscript. For example, we have $\gamma_a{}^{\alpha'} = 0$. Note that in every case the removal and subsequent affixing of a prime leaves a tensor unchanged (but not so for affixing a prime and its subsequent removal.)

Again consider, as an example, the case of electromagnetism. (Any other (non-scalar) tensor bundle would be similar.) Fix any smooth derivative operator ∇_a on the manifold M. Then this ∇_a will give rise to a connection $\gamma_a{}^\alpha$ on b, in the following manner. For $\kappa = (x, F_{ab})$ any point of b, and ξ^a any tangent vector in M at x, let $\lambda^\alpha = \xi^a \gamma_a{}^\alpha$ be that tangent vector in b at κ represented as follows: 'The infinitesimal change in x is that dictated by ξ^a, while the infinitesimal change in F_{ab} is that resulting from parallel transport, via ∇_a, of F_{ab} from x along ξ^a.' We thus specify the combination $\xi^a \gamma_a{}^\alpha$ for every ξ^a, and so the tensor $\gamma_a{}^\alpha$ itself. Note that we have $\lambda^\alpha(\nabla\pi)_\alpha{}^a = \xi^a$, which shows that the $\gamma_a{}^\alpha$ so defined is indeed a connection. So, the horizontal vectors at κ in this example are those for which 'the infinitesimal change in F_{ab} is exactly that resulting from parallel transport'. Clearly, every tangent vector in b can be written, uniquely, as the sum of a vertical vector and such a horizontal vector. While every derivative operator on M gives rise, as above, to a connection on b, there are many other connections on b (corresponding roughly to 'non-linear parallel transport').

We shall not routinely make use of a connection in what follows, for two reasons. First, for some fields, such as the derivative operator of general relativity, we have no natural connection. Second, even when there is a natural connection (*e.g.* for electromagnetism), that connection will itself be a dynamical variable. It is awkward having one dynamical field playing a crucial role in the kinematics of another.

We now wish to write down a certain class of partial differential equations on cross-sections. To this end, let $k_A{}^m{}_\alpha$ and j_A be smooth fields on b. Here, the index 'A' lives in some, as yet unspecified, vector space. Normally, this vector space will be some tensor product involving tensors in M and in b, *i.e.* 'A' will merely stand for some combination of Latin and Greek indices. But, at least in principle, this could be some newly constructed vector space attached to each point of b, in which case we would have to introduce a new fibre bundle, with base space b, to house it. Consider now the partial differential equation

$$k_A{}^m{}_\alpha (\nabla \phi)_m{}^\alpha + j_A = 0 \qquad (2)$$

where $U \xrightarrow{\phi} b$ is a smooth cross-section over some open subset U of M. This equation is to hold at every point $x \in U$, where k and j are evaluated 'on the cross-section', i.e. at $\phi(x)$. Note that this is a first-order equation on the cross-section, linear in its first derivative. The 'number of unknowns' at each point is the dimension of the fibre; the 'number of equations' the dimension of whatever is the vector space in which the index 'A' lives. The coefficients in this equation, $k_A{}^m{}_\alpha$ and j_A, are functions on the bundle manifold b, i.e. these coefficients may 'depend on both the point of M and the field-value ϕ'.

Apparently, every system of partial differential equations describing a physical system in space-time can be cast into the form of (2). Various examples are given in Appendix A. Many, such as those for a perfect fluid, the electromagnetic field, or the charged Dirac particle, are already packaged in the appropriate form. Others must be brought into this form by introducing auxiliary fields. In the Klein-Gordon case, for example, we must augment the scalar field ψ by its space-time derivative, ψ_a, resulting in a bundle space with five-dimensional fibres. We then obtain, on (ψ, ψ_a), a first-order system of equations of the form (2). For general relativity, the fibre over $x \in M$ consists of pairs (g_{ab}, ∇_a), where g_{ab} is a Lorentz-signature metric and ∇_a a torsion-free derivative operator at x. The curvature tensor arises in (2) as the derivative of the derivative operator. Let us agree that *all* first-order equations on the fields (even those that follow from differentiating other equations) are to be included in (2). Thus, for example, (36) is included for the Klein-Gordon system. Note that the only structure we are imposing on the physical fields at this stage is a differentiable structure, as carried by the manifold b. If you wish to utilize any additional features on these fields—for example the ability to add fields, to multiply them by numbers, to multiply them by each other, *etc.*—then this must be introduced, separately and explicitly, as additional structure on the bundle space b. For example, for electromagnetism, but not for a perfect fluid, each fibre has the additional structure of a vector space

The present formulation of partial differential equations carries with it a certain gauge freedom. Let $\lambda_A{}^m{}_b$ be any smooth field on b. Then (2) remains invariant under adding to $k_A{}^m{}_\alpha$ the expression $\lambda_A{}^m{}_b (\nabla \pi)_\alpha{}^b$ and, at the same time, adding to j_A the expression $-\lambda_A{}^m{}_m$. That is, the solutions ϕ of (2) before these changes in k and j are precisely the same as the solutions after. Note that $k_A{}^m{}_{\alpha'}$, (i.e. what results from contracting $k_A{}^m{}_\alpha$ only with vertical v^α) is gauge-invariant. Furthermore, this tensor exhausts the gauge-invariant information, in the following sense: given any field $\hat{k}_A{}^m{}_\alpha$ satisfying $\hat{k}_A{}^m{}_{\alpha'} = k_A{}^m{}_{\alpha'}$, then there exists one and only one gauge transformation that sends $k_A{}^m{}_\alpha$ to $\hat{k}_A{}^m{}_\alpha$. This gauge freedom reflects the idea that 'the horizontal part' of the α-contraction in (2) does not really involve the derivative of the cross-section, by virtue of the identity (1). Thus, the components of $k_A{}^m{}_\alpha$ that participate in this part of the α-contraction are not contributing to the dynamics. It would be most convenient if we could somehow circumvent this gauge freedom, e.g. by rewriting (2) to involve only the gauge-invariant part, $k_A{}^m{}_{\alpha'}$, of $k_A{}^m{}_\alpha$. Unfortunately, this cannot be done in any natural way in general. But it can be done in the presence of some fixed connection, $\gamma_a{}^\alpha$, on the bundle b. In fact, given a connection, we may always achieve through gauge a

$k_A{}^m{}_\alpha$ that is 'vertical in α' in the sense that it annihilates every horizontal vector h^α. Furthermore, this requirement on $k_A{}^m{}_\alpha$ completely exhausts the gauge freedom. Indeed, given $k_A{}^m{}_\alpha$ and connection $\gamma_a{}^\alpha$, then the gauge transformation with $\lambda_A{}^m{}_b = -\gamma_b{}^\alpha k_A{}^m{}_\alpha$, uniquely, does the job. It will sometimes be convenient, when a connection is available, to exploit this gauge-choice.

3 Hyperbolizations

A key feature of the partial differential equations of physics is their initial-value formulation, *i.e.* their formulation in terms of initial data and 'time'-evolution. It turns out that this formulation can be carried out in a rather general setting. This is the subject of the present, and much of the following, section.

Fix a partial differential equation of the form (2), so we have in particular fixed smooth fields $k_A{}^m{}_\alpha$ and j_A on b. By a *hyperbolization* of (2), we mean a smooth field $h^A{}_{\alpha'}$ on b such that

(i) the field $h^A{}_{\alpha'} k_A{}^m{}_{\beta'}$ on b is symmetric in α', β', and

(ii) for each point $\kappa \in b$, there exists a covector n_m in M at $\pi(\kappa)$ such that the tensor $n_m h^A{}_{\alpha'} k_A{}^m{}_{\beta'}$ at κ is positive-definite.

Note that the definition involves only $k_A{}^m{}_{\alpha'}$, and neither j_A nor the rest of k. Thus, in particular, the definition is gauge-invariant. Note also that the hyperbolizations at a point $\kappa \in b$ form an open subset of a vector space. For $h^A{}_{\alpha'}$ any hyperbolization, and $v^{\alpha'}$ any nonzero (vertical) vector at a point, the combination $v^{\alpha'} h^A{}_{\alpha'}$ at that point must be nonzero. (This follows by contracting the positive-definite tensor in (ii) with $v^{\alpha'} v^{\beta'}$.) But this implies, in turn, that the dimension of the space equations in (2) (that of the index 'A') must be greater than or equal to the dimension of the space of unknowns (that of the index 'α'). So, if this dimensionality criterion fails, then there can be no hyperbolization. But suppose this criterion is satisfied. Can we then guarantee a hyperbolization? The answer is no. In fact, there is no known, practical procedure, given a general partial differential equation (2), for finding its hyperbolizations, or, indeed, for even determining whether or not one exists. (This is essentially a little algebra problem: given a tensor $k_A{}^m{}_{\alpha'}$ at a point, what are the tensors $h^A{}_{\alpha'}$ at that point with $h^A{}_{\alpha'} k_A{}^m{}_{\beta'}$ symmetric?) In practice, hyperbolizations are found, in sufficiently low dimensions, by solving explicitly the algebraic equations inherent in (i) and (ii) ; and, in higher dimensions, by guessing. Physical considerations frequently suggest candidates.

Consider again the example of electromagnetism (Appendix A). We have already remarked that, at point $\kappa = (x, F_{ab})$ of the bundle space b, a typical vertical vector, which we now write $\delta\phi^{\alpha'}$, is represented by an infinitesimal change, δF_{ab}, in the electromagnetic field at x. Since the left sides of Maxwell's equations, (30) and (31), consist of a vector and a third-rank, antisymmetric tensor, the index 'A' lies in the eight-dimensional vector space of such objects. That is, a typical vector in this space can be written $\sigma^A = (s^a, s^{abc})$, with $s^{abc} = s^{[abc]}$. (Note that, since dim 'A' = 8 is greater than dim 'α'=6, our dimensionality criterion above is satisfied.) The fields $k_A{}^m{}_\alpha$ and j_A

are to be read off by comparing Maxwell's equations, (30) and (31), with the general partial differential equation (2). We thus obtain

$$k_A{}^m{}_{\beta'}\sigma^A n_m \delta\hat{\phi}^{\beta'} = s^a(n^b \delta\hat{F}_{ab}) + s^{abc}(n_{[a}\delta\hat{F}_{bc]}). \tag{3}$$

Here, we have represented $k_A{}^m{}_{\beta'}$ by giving the scalar that results from contracting away its indices, on vectors σ^A, n_m, and $\delta\hat{\phi}^{\beta'}$. The field j_A of (2), on the other hand, depends on gauge. If we choose for the gauge that determined by the derivative operator ∇_a on M used in Maxwell's equations (30), (31), then we have $j_A = 0$. Now fix any vector t^a at x, and consider the tensor $h^A{}_{\alpha'}$ given, in (32), as the A-index vector that results from the contraction $h^A{}_{\alpha'}\delta\phi^{\alpha'}$. Substituting this vector for σ^A in (3), we obtain

$$\begin{aligned} h^A{}_{\alpha'}\delta\phi^{\alpha'} k_A{}^m{}_{\beta'} n_m \delta\hat{\phi}^{\beta'} &= \delta F^a{}_m t^m (n^b \delta\hat{F}_{ab}) - \tfrac{3}{2} t^{[a}\delta F^{ab]}(n_{[a}\delta\hat{F}_{bc]}) \\ &= 2 t^a n^b \left(\delta F_{(a}{}^m \delta\hat{F}_{b)m} - \tfrac{1}{4} g_{ab} \delta F_{mn} \delta\hat{F}^{mn}\right). \end{aligned} \tag{4}$$

It now follows, provided only that the vector t^a is chosen timelike, that the $h^A{}_{\alpha'}$ of (32) is a hyperbolization. Indeed, condition (i) follows from the fact that the last expression in (4) is symmetric under interchange of δF_{ab} and $\delta\hat{F}_{ab}$; and condition (ii) follows from the fact that, whenever n_m is timelike with $t^m n_m < 0$, the last expression in (4) defines a positive-definite quadratic form in δF_{ab}. Thus, every timelike vector field t^a on M gives rise to a hyperbolization of Maxwell's equations. In fact, this family exhausts the hyperbolizations in the Maxwell case.

The situation is similar for many other physical examples—see Appendix A. Thus, the hyperbolizations of the Klein-Gordon equation are characterized by two vector fields on M, and those for the perfect-fluid equation by two scalar fields. Even dissipative fluids can be described by equations admitting a hyperbolization (see, e.g. Liu et al. 1986, Geroch and Lindblom 1991, Geroch 1995). There are only two physical examples, as far as I am aware, for which there exist no hyperbolization. One is Einstein's equation, for which the lack of a hyperbolization is related to the diffeomorphism-invariance of the theory; and the other is dust. We shall return to each of these examples later.

Fix a hyperbolization, $h^A{}_{\alpha'}$, of (2). For each point $\kappa \in b$, denote by s_κ the collection of all covectors n_m in M at $\pi(\kappa)$ such that the tensor $n_m h^A{}_{\alpha'} k_A{}^m{}_{\beta'}$ is positive-definite. Then s_κ is a nonempty (by condition (ii)), open, convex cone. The physical interpretation of these cones will turn out to be the following. The tangent vectors p^a in M at $\pi(\kappa)$ such that $p^a n_a > 0$ for all $n_a \in s_\kappa$ represent the 'signal-propagation directions' of the physical field. Note that these p^a form a closed, nonempty, convex cone at each point, the 'dual cone' of s_κ. These cones depend not only on the point x of M, but also in general on the value of the field at x, i.e. on where we are in the fibre over x. In cases in which there is more than one hyperbolization, these cones could also depend on which hyperbolization has been selected. But it turns out that, for most physical examples, these cones are essentially independent of hyperbolization. Thus, in the case of electromagnetism, the signal propagation directions p^a consist of all timelike and null vectors lying in one of the two halves of the light cone. In the case of a perfect fluid, the p^a form the 'sound cone'. Is it possible to isolate, via a definition, the crucial algebraic feature of $k_A{}^m{}_{\alpha'}$ in such physical examples that is responsible for hyperbolization-independent cones?

Suppose that included among the various physical fields on M is a spacetime metric g_{ab}. In that case, we say that the system (2) is *causal* if all the signal-propagation directions are timelike or null. This is equivalent to the condition that each s_κ includes all timelike vectors lying within one of the two halves of the light cone. A perfect fluid, for example, is causal provided its sound speed, $\frac{\partial p}{\partial \rho} + \frac{n}{\rho+p}\frac{\partial p}{\partial n}$, does not exceed the speed of light.

Fix a hyperbolization $h^A{}_{\alpha'}$ of (2). This hyperbolization leads, as we now explain, to an initial-value formulation. By *initial data* we mean a smooth, three-dimensional submanifold S of M, together with a smooth cross-section, $S \xrightarrow{\phi_0} b$, over S, such that, for every point $x \in S$, a normal n_m to S at x lies in the cone $s_{\phi_0(x)}$. In other words, we must specify the physical state of the system at each point of the three-dimensional manifold S, in such a way that, at every point of S, all signal-propagation directions are transverse to S. Note that the role of the cross-section, ϕ_0, in this definition is to determine the cone within which the normal to S must lie. Thus, a change of cross-section, keeping S fixed, could destroy the initial-data character of (S, ϕ_0). (Changing the hyperbolization could, in principle, also change the initial-data character, but, as we remarked earlier, it generally does not.) As an example of these definitions we have: 'If we have on M a spacetime metric g_{ab} with respect to which (2) is causal, then every (S, ϕ_0), with S spacelike, is initial data.'

We may now summarize the fundamental existence-uniqueness theorem as follows. Given initial data (S, ϕ_0), there exists, in a suitable neighborhood U of S, one and only one cross-section, $U \xrightarrow{\phi} b$, such that

$$\phi|_S = \phi_0 \qquad (5)$$

and

$$h^A{}_{\beta'}[k_A{}^m{}_\alpha (\nabla \phi)_m{}^\alpha + j_A] = 0. \qquad (6)$$

Condition (5) ensures that the field ϕ, specified over the neighborhood U of S, agree, on S itself, with the given initial conditions, ϕ_0. Condition (6) ensures that the field ϕ satisfies a certain partial differential equation derived from (2) (specifically, by contracting it with $h^A{}_{\beta'}$). In short, the theorem states that we can 'solve' the partial differential equation (6), uniquely, subject to any given initial conditions. There is given in Appendix B a more detailed version of this theorem (including more information regarding the neighborhood U), and a sketch of the proof. This version, in particular, supports our interpretation of the cones s_κ in terms of signal-propagation.

Since every solution of (2) is automatically a solution of (6), the theorem above guarantees local uniqueness of the solutions of any system, (2), of partial differential equations admitting a hyperbolization. Thus, for most systems of interest in physics, initial data lead to a unique local solution. Furthermore, if the hyperbolization $h^A{}_{\alpha'}$ is invertible (which holds, by the way, if and only if dim 'A' = dim 'α'), then (6) is equivalent to (2). In this case, e.g. for a perfect fluid, the theorem also guarantees local existence of solutions of (2). But in many physical examples, electromagnetism included, $h^A{}_{\alpha'}$ is not invertible, so part of (2) is lost in the passage to (6). In these cases, we cannot guarantee, directly from the theorem, local existence of solutions of (2). The fate of these 'lost equations' is the subject of the following section.

Let us now return briefly to the example of dust (see Appendix A). With the traditional choice of fields—ρ (mass density) and u^a (unit, timelike four-velocity)—the dust equations, (68) and (69), admit no hyperbolization. This is perhaps surprising, for this system 'obviously' has an initial-value formulation. It turns out that, if we introduce the auxiliary field $w_a{}^b = \nabla_a u^b$, then the corresponding system of equations on this new set of fields, $(\rho, u^a, w_a{}^b)$, does admit a hyperbolization. It is not clear what, if any, is the physical meaning of this modification. Furthermore, the hyperbolization it produces is apparently lost on coupling the dust with gravitation via Einstein's equation. (This behavior is a consequence of the appearance of a Riemann tensor in the equations on $(\rho, u^a, w_a{}^b)$.) What is going on physically in this example?

4 Constraints

Fix a partial differential equation of the form (2), so we have in particular fixed smooth fields $k_A{}^m{}_\alpha$ and j_A on b. While much of the material of this section finds application to the initial-value formulation, we require at this stage no specific hyperbolization—nor, indeed, even the existence of one.

A *constraint* at point $\kappa \in b$ is a tensor c^{An} at κ such that

$$c^{A(n} k_A{}^{m)}{}_{\alpha'} = 0. \tag{7}$$

Note that the definition is gauge-invariant, and that the constraints at κ form a vector space. A number of examples, for various physical systems, is given in Appendix A. For instance, the equations for a perfect fluid admit only the zero constraint; those for Klein-Gordon, a ten-dimensional vector space of constraints; and those for general relativity an eighty-four dimensional vector space. Maxwell's equations, on the other hand, admit a two-dimensional vector space of constraints: the general c^{An} is given, in this case, by (33), where x and y are arbitrary numbers. To check that this c^{An} does indeed satisfy (7), combine it with the $k_A{}^m{}_{\alpha'}$ for Maxwell's equations given by (3), to obtain

$$c^{An} k_A{}^m{}_{\alpha'} \delta \hat{\phi}^{\alpha'} = x \delta \hat{F}^{nm} + y \epsilon^{nmbc} \delta \hat{F}_{bc}. \tag{8}$$

Now symmetrize both sides over n, m. Each constraint, as we shall see, plays two distinct roles: it signals a differential condition that must be imposed on initial data for (2), as well as a differential identity involving (2). In the case of Maxwell's equations, for example, the first role is reflected in the familiar 'spatial constraint equations', $\nabla \cdot \mathbf{E} = 0$, $\nabla \cdot \mathbf{B} = 0$. The second role is reflected in the fact that identities result from taking the divergence and curl, respectively, of Maxwell's equations, (30) and (31).

We begin with the first role. Fix constraint c^{An}. Let $U \xrightarrow{\phi} b$ be any solution of (2), defined in open $U \subset M$, and let S be any three-dimensional submanifold of U. Consider now the equation

$$n_a c^{Aa}[k_A{}^m{}_\alpha (\nabla \phi)_m{}^\alpha + j_A] = 0 \tag{9}$$

at points x of S, where n_a is a normal to S at x and the coefficients are evaluated at $\kappa = \phi(x)$. We first note that (9) holds on S, for it is a consequence of (2). We next claim that the left side of (9) involves only $\phi_0 = \phi|_S$, i.e. only ϕ restricted to S. To see this, first note that ϕ_0 alone determines $(\nabla \phi)_m{}^\alpha$ at points of S up to addition of a term of the form $n_m v^{\alpha'}$. But such a term annihilates $n_a c^{Aa} k_A{}^m{}_\alpha$, by the defining equation, (7), for a constraint, and so does not contribute to the left side of (9). What we have shown, then, is that (9) is a 'constraint equation': it is a differential equation on cross-sections over S that must be satisfied by every restriction to S of a solution of (2). In the Maxwell case, for example, the two independent constraints give rise, via (9), to the vanishing of the divergence of the electric and magnetic fields. Note that the (S, ϕ_0) above need not be initial data: we have as yet introduced no hyperbolization.

We next introduce a notion of 'sufficiently many' constraints. We say the constraints are *complete* if, for any point $\kappa \in b$ and any nonzero covector n_n at $\pi(\kappa) \in M$, we have

$$\dim(c^{An} n_n) + \dim(v^{\alpha'}) = \dim(\sigma^A). \tag{10}$$

The first term[4] is the dimension of the space of all vectors of the indicated form, as c^{An} runs over all constraints at κ. The second term is the dimension of the space of vertical vectors, *i.e.* the dimension of the fibres. The last term is the dimension of the space of equations represented by (2). Equation (10) means, roughly speaking, that there are at least as many equations as unknowns in (2), and that any excess is taken up entirely by constraint equations, (9). This interpretation will be made more precise shortly.

The constraints are complete for the vast majority of physical examples—see Appendix A. Thus (10) reads: for the perfect-fluid equations, '$0 + 5 = 5$'; for Maxwell's equations, '$2 + 6 = 8$'; for the Klein-Gordon equations, '$6 + 5 = 11$'. For Einstein's equation the constraints are not complete: (10) reads '$64 + 50 = 110$'. This, as we shall see later, is related to the diffeomorphism-invariance of the theory[5]. Is there some simple characterization of those tensors $k_A{}^m{}_{\alpha'}$ that yield complete constraints?

The second role of a constraint is in signalling a differential identity involving (2). The idea here is very simple. Equation (2) is to hold at every point x of some open subset U of M. Taking the x-derivative, ∇_n, of this equation, and contracting with any constraint c^{An}, we obtain an equation involving the first and second derivatives of the cross-section. But, as it turns out, the second-derivative term drops out, by virtue of (7), and so we are left with an algebraic—in fact, quadratic—equation in the first derivative, $(\nabla \phi)_m{}^\alpha$, of the cross-section. That is, we obtain an integrability condition for (2). If this integrability condition holds as an algebraic consequence of (2), then we say our constraint is integrable.

Unfortunately, all this becomes somewhat more complicated when written out explicitly. Fix any (torsion-free) derivative operator, ∇_α, on the manifold b, such that the derivative of every vertical vector field is vertical. (Such always exists, at least locally, by the local-product character of the fibre bundle.) Extend[6] this operator to mixed fields on b by demanding $\nabla_\beta (\nabla \pi)_\alpha{}^m = 0$. Then the operator 'derivative along the cross-section' is $(\nabla \phi)_n{}^\alpha \nabla_\alpha$. Applying this operator to (2), and contracting with any constraint c^{An}, we obtain

---

[4]Note that this term can be—and in examples (such as Klein-Gordon) frequently is—less than the dimension of the vector space of constraints.

[5]We remark that there exist examples (although apparently no physically interesting ones) of a tensor $k_A{}^m{}_\alpha$ admitting a hyperbolization, but whose constraints are not complete.

[6]This is done as follows. Any field ξ^m can be written in the form $(\nabla \pi)_\beta{}^m \xi^\beta$, uniquely but for the freedom to add to ξ^β any vertical vector field. Now define $\nabla_\alpha \xi^m = (\nabla \pi)_\beta{}^m \nabla_\alpha \xi^\beta$, noting that the right side is invariant under this freedom. Note that this extension of ∇_α to fields with Latin indices is unique.

$$c^{An}(\nabla_\beta k_A{}^m{}_\alpha)(\nabla\phi)_n{}^\beta(\nabla\phi)_n{}^\alpha + c^{An}(\nabla_\beta j_A)(\nabla\phi)_n{}^\beta = 0. \tag{11}$$

In the derivation of (11), there arises initially the term $[c^{An}k_A{}^m{}_\alpha]\,[(\nabla\phi)_n{}^\beta \nabla_\beta(\nabla\phi)_m{}^\alpha]$, involving the second derivative of the cross-section. To see that this term vanishes, first note that the index 'α' in the second factor is vertical (contracting with $(\nabla\pi)_a{}^s$), and so only the antisymmetrization of this factor over n,m contributes (by definition of a constraint), yielding zero (by the torsion-free character of ∇_α). Equation (11) is our integrability condition. We say that constraint c^{An} is *integrable* if (11) is an algebraic consequence of (2). What this means, in more detail, is that the left side of (11) is some multiple of the left side of (2) plus some multiple of the difference between the two sides of the identity (1), where each of these two multiplying factors is an expression linear in $(\nabla\phi)_a{}^\alpha$. Writing this out and equating powers of $(\nabla\phi)_a{}^\alpha$, we conclude that the constraint c^{An} is integrable if and only if there exist tensors $\sigma^{Am}{}_\alpha$ and $\sigma^a{}_b{}^m{}_\alpha$, with $\sigma^a{}_a{}^m{}_\alpha = 0$, such that

$$\nabla_\alpha(c^{Am}\tilde{k}_A{}^n{}_\beta) + \nabla_\beta(c^{An}\tilde{k}_A{}^m{}_\alpha)$$
$$= \sigma^{Am}{}_\alpha \tilde{k}_A{}^n{}_\beta + \sigma^{An}{}_\beta \tilde{k}_A{}^m{}_\alpha + \sigma^m{}_s{}^n{}_\beta(\nabla\pi)_\alpha{}^s + \sigma^n{}_s{}^m{}_\alpha(\nabla\pi)_\beta{}^s \tag{12}$$

where we have set $\tilde{k}_A{}^m{}_\alpha = k_A{}^m{}_\alpha + \tfrac{1}{4}j_A(\nabla\pi)_\alpha{}^m$. Applying a prime to both 'α' and 'β' in this equation, and using (7), we obtain

$$2\nabla_{[\alpha'}(c^{Am}k_{|A|}{}^n{}_{\beta']}) = \sigma^{Am}{}_{\alpha'}k_A{}^n{}_{\beta'} + \sigma^{An}{}_{\beta'}k_A{}^m{}_{\alpha'}. \tag{13}$$

This part of (12) is manifestly gauge-invariant (involving only $k_A{}^m{}_{\alpha'}$, and not j_A or the rest of k), and independent of the derivative operator ∇_α (involving only the 'vertical curl'). What remains of (12) is essentially one scalar relation, expressing the divergence of $c^{An}j_A$ in terms of other fields. Is there some simple way to write this remaining relation, e.g. a way that separates its physical content from the gauge freedom inherent in (k,j), ∇_α, and the σ's? In electromagnetism, to take one example, (13) is satisfied with $\sigma^{Am}{}_{\alpha'} = 0$. What remains of (12) in this example is just the vanishing of the divergence of the electric charge-current.

Failure of integrability would mean that we have somehow failed to include in (2) all the relevant conditions on the first derivative of the cross-section. The standard procedure, in such circumstances, is, first, to enlarge (2) to encompass the additional conditions on $(\nabla\phi)_m{}^\alpha$. Then look for any additional constraints arising from this enlargement, and if any of these fail to be integrable, enlarge (2) further, *etc*. Unfortunately, it is not clear, in the present general context, how to implement this procedure. How do you 'enlarge' a system, (2), linear in $(\nabla\phi)_m{}^\alpha$, to encompass a quadratic relation (11)?

Nowhere in this section so far have we introduced a hyperbolization. It is perhaps striking that so much of the subject of constraints can be carried out at this level, for it is largely in their interaction with hyperbolizations that constraints come to the fore. We turn now to this interaction. Fix, therefore, a hyperbolization, $h^A{}_{\alpha'}$, for (2).

Let c^{An} be a constraint. Then Equation (9) holds for the restriction, $\phi_0 = \phi|_S$, of any solution, ϕ_0, of (2) to any three-dimensional submanifold, S, of M. So, in particular, this equation holds when (S, ϕ_0) are initial data, *i.e.* when the normal n_a to S at each point κ lies in the cone $s_{\phi_0(x)}$. Thus, given initial data, (S, ϕ_0), we have no hope of finding a corresponding solution of (2) unless those data satisfy (9) for every constraint c^{An}. Equations (9) become constraint equations on initial data.

Next, fix $\kappa \in b$ and $n_a \in s_\kappa$. Then, we claim, for any constraint c^{An} and any (vertical) vector $v^{\alpha'}$, we can have $n_a c^{Aa} = v^{\alpha'} h^A{}_{\alpha'}$ only if each side is zero. Indeed, this equality implies $(n_a c^{Aa}) k_A{}^m{}_{\beta'} v^{\beta'} n_m = (v^{\alpha'} h^A{}_{\alpha'}) k_A{}^m{}_{\beta'} v^{\beta'} n_m$. But the left side vanishes (by definition of a constraint), while vanishing of the right side implies $v^{\alpha'} = 0$ (by $n_a \in s_\kappa$). What we have shown, in other words, is that the subspace of vectors of the form $n_a c^{Aa}$ with c^{Aa} a constraint, and that of vectors of the form $v^{\alpha'} h^A{}_{\alpha'}$ with $v^{\alpha'}$ vertical, have only the zero vector in common. But this implies that the left side of (10) is less than or equal to the right side. That is, in the presence of a hyperbolization, 'half' of (10) is automatic. Now suppose that the constraints are complete, *i.e.* that the full equality (10) holds. It then follows that our two independent subspaces, $\{n_a c^{Aa}\}$ and $\{v^{\alpha'} h^A{}_{\alpha'}\}$, in fact span the space of all vectors of the form σ^A. What this means, in geometrical terms, is that the 'constraint components' of (2) (the results of contracting it with vectors of the form $n_a c^{Aa}$), and the 'dynamical components' of (2) (the result of contracting it with $h^A{}_{\alpha'}$), together comprise the whole of (2). Completeness, in the presence of a hyperbolization, means the absence of any 'stray equations' in (2).

Finally, we return to the issue, raised in the previous section, of when there is an initial-value formulation for the full equation (2). Fix a hyperbolization $h^A{}_{\alpha'}$ for this equation, and suppose that its constraints are both complete and integrable. Let (S, ϕ_0) be initial data, and suppose that these data satisfy all the constraint equations of the type (9) (for if not, then there is certainly no evolution of these data). By the general existence-uniqueness theorem (Section 3 and Appendix B), there exists a solution, $U \xrightarrow{\phi} b$, of the evolution (6), with $\phi|_S = \phi_0$, where U is an appropriate neighborhood of the three-dimensional submanifold S of M. We now claim that, under certain conditions, this cross-section ϕ satisfies our full system, (2), of partial differential equations. It is convenient, for purposes of this paragraph, to introduce upper-case Greek indices to lie in the vector space of constraints; so, in this notation, we have a single constraint tensor, $c^{Aa}{}_\Gamma$. Denote the left side of (2), evaluated on the cross-section ϕ, by I_A. Thus, we have $h^A{}_{\alpha'} I_A = 0$ everywhere in U (by (6)), and $I_A = 0$ on S (by completeness); and we wish to show $I_A = 0$ everywhere in U. To this end, consider the expression

$$(c^{Am}{}_\Gamma + \sigma^{m\alpha'}{}_\Gamma h^A{}_{\alpha'}) \nabla_m I_A \tag{14}$$

where $\sigma^{m\alpha'}{}_\Gamma$ is any field on b. We claim that this expression is, everywhere in U, a multiple of I_A. Indeed, the first term in parentheses leads to such a multiple since, by integrability, $c^{Am}{}_\Gamma \nabla_m I_A$ is a multiple of I_A; and the second term also leads to such a multiple, using $h^A{}_{\alpha'} I_A = 0$ and differentiating by parts. To summarize, we have shown so far that I_A vanishes on S, and satisfies a certain first-order, quasilinear (in fact, linear) partial differential equation arising from the expression (14). Since this differential equation clearly has as one solution $I_A = 0$, we can conclude that $I_A = 0$ in a neighborhood of S if we can show local uniqueness of its solutions.

The most direct way to prove local uniqueness of solutions of a partial differential equation is to show that it admits a hyperbolization. In the present instance, tensor

$h^{A\Gamma}$ is a hyperbolization of the differential equation resulting from (14), for some choice of $\sigma^{m\alpha'}{}_A$, provided $h^{A\Gamma}$ has the following property: 'The expression $h^{A\Gamma}c^{Bm}{}_\Gamma v_A w_B$, for all v, w with $v_A h^A{}_{\alpha'} = w_A h^A{}_{\alpha'} = 0$, is symmetric under interchange of v and w, and, contracted with some n_m, is positive-definite.' When, in terms of the original $k_A{}^m{}_{\alpha'}$ and its hyperbolization $h^A{}_{\alpha'}$, does such a hyperbolization $h^{A\Gamma}$ exist? In physical examples (*e.g.* in electromagnetism) such an $h^{A\Gamma}$ does indeed exist, and so we have in these examples uniqueness of solutions of the equation resulting from (14), and so an initial-value formulation for the full system (2). Is there any simple, reasonably general, condition on $k_A{}^m{}_{\alpha'}, h^A{}_{\alpha'}$ that will guarantee existence of a hyperbolization $h^{A\Gamma}$? Are there interesting cases in which uniqueness of solutions of the differential equation arising from (14) must be shown by some other method?

5 The combined system: diffeomorphisms

In the preceding sections, we have been analyzing the structure of the partial differential equation describing a single physical system. This analysis was to be applied separately to the electromagnetic field, a perfect fluid, or whatever. However, in the real world, all these systems coexist on M, normally in interaction with each other. We now consider the system that results from combining all these subsystems.

Once again, we have a smooth fibre bundle, $B \xrightarrow{\Pi} M$, over the four-dimensional space-time manifold M. Now, however, the fibre in B over $x \in M$ represents the possible values at x of all possible physical fields in the universe. Thus, this fibre would include an antisymmetric tensor (for electromagnetism), a Lorentz metric and derivative operator (for gravity), two scalar and one vector field (for a perfect fluid), *etc.* Note that we are implicitly assuming that the space that results from combining these fields is finite-dimensional, and that we are somehow capable of 'finding' it. One or both of these assumptions may be incorrect. In any case, we imagine that we have constructed such a bundle. Again, a cross-section, $M \xrightarrow{\Phi} B$, of B represents an assignment of a complete physical state (of everything) to each point of space-time, *i.e.* a statement of the entire dynamics of the universe. And, again, we impose on such cross-sections the general first-order, quasilinear partial differential equation,

$$K_A{}^m{}_\alpha (\nabla \Phi)_m{}^\alpha + J_A = 0. \tag{15}$$

In the analogous equation for a single system, (2), we allowed the coefficients, $k_A{}^m{}_\alpha$ and j_A, to be arbitrary (smooth) fields on the bundle manifold b. That is, we allowed these coefficients to depend on both 'the point of the space-time manifold M and the value of the field assigned to that point'. But, in the context of this combined system, an explicit dependence of the coefficients, $K_A{}^m{}_\alpha$ and J_A, on the point of M is, we suggest, inappropriate. After all, we identify the points of the manifold M, not by somehow 'perceiving them directly', but rather more indirectly, by observing the various physical fields on M. So, for instance, the physical distinction between two points, x and y, of M rests on the difference between the values of some physical field at x and at y. (This issue did not arise in the context of a single system, for there x-dependence of k and j could arise through other physical fields, not included in the dynamics of (2).) In any case, we expect that the coefficients of K and J in (15) will depend explicitly only on the fibres of B, with any dependence on the point of M arising only implicitly through the cross-section. Unfortunately, this expectation, at least as it is stated above, does not make mathematical sense! The problem is that our fibre bundles are not naturally products, and so there is no such thing as 'a function only of the fibre-variables, independent of the base-space variables'. We must therefore proceed in a different way.

We now demand that, as part of the physical content of the bundle B, there be given on it the following additional structure. To each diffeomorphism $\mathcal{D}$ on the manifold M, there is to be assigned[7] a lifting of it to a diffeomorphism, $\hat{\mathcal{D}}$, on the manifold B. By 'lifting', we mean that $\hat{\mathcal{D}}$ must satisfy $\Pi \circ \hat{\mathcal{D}} = \mathcal{D} \circ \Pi$, i.e. that $\hat{\mathcal{D}}$ must take entire fibres to fibres, such that the induced diffeomorphism on M is precisely the original $\mathcal{D}$. We further require of these liftings that they respect the group structure of the M-diffeomorphisms, i.e. that $\widehat{(\mathrm{id}_M)} = (\mathrm{id}_B)$ and $\widehat{(\mathcal{D} \circ \mathcal{D}')} = \hat{\mathcal{D}} \circ \hat{\mathcal{D}}'$. In short, we must specify how the physical fields 'transform' under diffeomorphisms on M. In the examples of Appendix A, the fibres consist of tensors, spinors, derivative operators, etc, and on such geometrical objects there is a natural action of M-diffeomorphisms. Indeed, we claim that it is this 'transformation behavior' that endows such fields with a geometrical content in terms of M. For instance, given a point of the tangent bundle of M, the direction in M in which the vector 'points' can be read out from the M-diffeomorphisms whose lifts leave this point invariant. Consider now the bundle B that results from combining all the examples of Appendix A. Lift diffeomorphisms from M to this B by combining these liftings for all the individual examples.

We now have the machinery to express the idea that the coefficients in (15) be 'functions only of the physical fields'. We demand that, for every diffeomorphism $\mathcal{D}$ on M, its lifting $\hat{\mathcal{D}}$ leave $K_A{}^m{}_\alpha$ and J_A invariant (noting that this makes sense for fields having indices in both B and M), up to gauge. It then follows that, for Φ any cross-section satisfying (15) and $\mathcal{D}$ any diffeomorphism on M, the 'transformed cross-section', $\hat{\mathcal{D}} \circ \Phi \circ \mathcal{D}$, is also a solution.

The 'infinitesimal version' (with diffeomorphisms replaced by vector fields) of all this is the following. To each (smooth) vector field ξ^a on M, there is to be assigned a lifting of it to a vector field, $\hat{\xi}^\alpha$, on B. By 'lifting' we mean that $\hat{\xi}^\alpha (\nabla \Pi)_\alpha{}^a = \xi^a$. We require of these liftings that they be linear (i.e. that $\widehat{(c\xi + \eta)} = c\hat{\xi} + \hat{\eta}$, for c constant), and Lie-bracket preserving (i.e. that $\widehat{[\xi, \eta]} = [\hat{\xi}, \hat{\eta}]$). Invariance of the coefficients in (15) under these infinitesimal diffeomorphisms now becomes[8]

$$\mathcal{L}_{\hat{\xi}} K_A{}^m{}_\alpha = \Lambda_A{}^m{}_b (\nabla \Pi)_\alpha{}^b, \qquad \mathcal{L}_{\hat{\xi}} J_A = -\Lambda_A{}^m{}_m \qquad (16)$$

for some field $\Lambda_A{}^m{}_b$ on B, and for every vector field ξ^a on M. We shall further assume that $\hat{\xi}^\alpha$ results from ξ^a through the action of some differential operator[9]:

---

[7]This demand rules out most gauge theories other than electromagnetism (Appendix A). What happens in these theories is that to each diffeomorphism on M is assigned a number of liftings to B. Indeed, the collection of all liftings assigned to the identity diffeomorphism on M is called the gauge group. Presumably, much of what follows could be generalized to include such gauge theories.

[8]The infinitesimal version of the transformation of solutions of (15) under diffeomorphisms becomes that, for every ξ^a, its lifting be a linearized solution of (15).

[9]It is possible that this assumption follows already from the general properties above of the liftings of diffeomorphisms.

$$\hat{\xi}^\alpha = \delta^{\alpha' m_1 \cdots m_s}{}_r \nabla_{m_1} \cdots \nabla_{m_s} \xi^r + \cdots. \tag{17}$$

In (17), we have written out only the highest-order term. The coefficient of this term, $\delta^{\alpha' m_1 \cdots m_s}{}_r = \delta^{\alpha'(m_1 \cdots m_s)}{}_r$, is some smooth field on B, independent of the derivative operator employed in (17). Note that the index 'α' of δ is vertical, as follows from the definition of a lifting. As an example, consider the system resulting from combining all the examples of Appendix A. In this case the highest order appearing in the expression, (17), for $\hat{\xi}^\alpha$ is $s = 2$, and this order occurs only for the derivative operator of general relativity. A typical vertical vector in the bundle of derivative operators is given by $\delta \phi^{\alpha'} = \delta \Gamma^a{}_{bc}$—see Appendix A. Then the δ of (17) becomes

$$\delta^{\alpha' m_1 m_2}{}_r = \delta^a{}_r \delta^{m_1}{}_{(b} \delta^{m_2}{}_{c)} \tag{18}$$

reflecting the action of a Lie derivative on a derivative operator.

The equation, (15), for the combined system can never admit any hyperbolization. To see this, first note that, given any three-dimensional submanifold S of M, there always exists a diffeomorphism $\mathcal{D}$ on M that is the identity in an arbitrarily small neighborhood of S, but not outside that neighborhood. But now, were there a hyperbolization, then the transformation of solutions Φ of (15) by such a diffeomorphism would violate the uniqueness theorem for hyperbolic systems (Section 3 and Appendix B). Another way to see this is to note that (16) and (17) together imply

$$K_A{}^{(m}{}_\alpha \delta^{|\alpha'| m_1 \cdots m_s)}{}_r = 0 \tag{19}$$

which implies in turn that, for any n_m, the tensor $K_A{}^m{}_{\alpha'} n_m$ is annihilated on contraction with $\delta^{\alpha' m_1 \cdots m_s}{}_r n_{m_1} \cdots n_{m_s}$. But a nonzero vertical vector annihilating $K_A{}^m{}_{\alpha'} n_m$ precludes a hyperbolization.

How, in light of this observation, are we ever to recover an initial-value formulation in physics (*i.e.* in (15))? The answer is that we must formulate a modified version of the initial-value formulation, in which, given suitable initial data, solutions of (15) will always exist, but will be unique only up to the diffeomorphism freedom. We now describe one scheme (suggested by general relativity) to implement this program. There may well be others. The idea is to supplement (15) on the cross-section Φ with a second system of equations, of the form

$$\nu_{a\alpha} (\nabla \Phi)_b{}^\alpha + \nu_{ab} = 0 \tag{20}$$

where $\nu_{a\alpha}$ and ν_{ab} are smooth fields on B. We wish to so arrange matters that first (15) and (20), taken together on B, have an initial-value formulation, and second that (20) can always be achieved, in a suitable sense, via the diffeomorphism freedom.

We first consider the issue of an initial-value formulation for (15), (20). A hyperbolization for the system (15), (20) consists of fields $H^A{}_{\alpha'}$ and $I_{\alpha'}{}^{ma}$ on B such that the expression

$$H^A{}_{\alpha'} K_A{}^m{}_{\beta'} + I_{\alpha'}{}^{ma} \nu_{a\beta'} \tag{21}$$

is symmetric in indices α', β', and is positive-definite on contraction with some covector n_m at each point. What are the constraints? Let us demand that $\nu_{a\alpha}$ have rank four (*i.e.*

that $\xi^a \nu_{a\alpha} = 0$ implies $\xi^a = 0$). Then (20) represents a sixteen-dimensional vector space of additional equations on Φ. The general constraint for (20) is given by $c^{An} = x^{abn}$, with $x^{abn} = x^{a[bn]}$. Thus, the dimension of the vector space of constraints is twenty-four, while, for any fixed n_n, the dimension of the space of all $c^{An}n_n$ is just twelve. Now consider (10), the condition for completeness of the constraints. As we have just seen, the act of supplementing (15) by (20) increases the first term in (10) by twelve, the second term by zero, and the third term by sixteen. Thus, in order that the system (15), (20) be complete, the original system, (15), must have yielded a left side of (10) exceeding the right side by exactly four. Finally, integrability of the constraints arising from (20) yields a system of partial differential equations on the coefficients $\nu_{a\alpha}$ and ν_{ab}. The geometrical content of these equations is that (20) be precisely the requirement that the cross-section Φ lie within a certain submanifold V of B. This V has codimension four (i.e. dimension four less than that of B), and its tangent vectors are those ξ^α with $\xi^\alpha \nu_{a\alpha} = 0$.

We may summarize the discussion above as follows. Let there be given a field $\nu_{a\alpha'}$ on B satisfying the following three conditions:

(i) For some fields H and I, the expression (21) is symmetric, and, contracted with some n_m, is positive-definite.

(ii) At each point of B, $\nu_{a\alpha'}$ has rank four, so the space of vectors $v^{\alpha'}$ at each point with $v^{\alpha'}\nu_{a\alpha'} = 0$ has codimension four, and

(iii) these vector spaces can be integrated to give submanifolds, of codimension four, within each fibre.

Given such a field $\nu_{a\alpha'}$, choose any submanifold V of B, of codimension four, such that V intersects each fibre of B in one of the submanifolds in (iii) above. Now restrict the cross-section Φ to this submanifold V, yielding an equation of the form (20). By construction, the system (15), (20) admits an initial-value formulation.

We next turn to the issue of 'achieving' equation (20) via diffeomorphisms. Let Φ be any solution of (15), and V any submanifold of B, as described above, such that, over some three-submanifold S of M, the cross-section Φ lies within V. We wish to find a diffeomorphism on M that sends the entire cross-section Φ to lie within the submanifold V. We may choose our diffeomorphism, together with its first $(s-1)$ derivatives, to be the identity on S, but it must begin to differ from the identity on evolution off S. We shall be able continually to adjust Φ, via a diffeomorphism, to lie within V provided we can generate, via (17), any $\xi^{\alpha'}$ transverse to V. But this is precisely the statement that the operator

$$\nu_{a\alpha'}\delta^{\alpha' m_1 \cdots m_s}{}_r \nabla_{m_a} \cdots \nabla_{m_s} \xi^r \tag{22}$$

be hyperbolic. By 'hyperbolic' here, we mean that the system that results from introducing as auxiliary fields the first $(s-1)$ derivatives of ξ^a admits a hyperbolization in the sense of Section 3. See the discussion at the end of Appendix A.

To summarize, we may recover an initial-value formulation[10] for (15) provided we can find a field $\nu_{a\alpha'}$ on B satisfying conditions (i)-(iii) above, together with: (iv) the operator in (22) is hyperbolic. As an example (in fact, *the* example) of this scheme, consider Einstein's equation (for the derivative operator) in general relativity[11] In this example $\nu_{a\alpha'}$ is given by

$$\nu_{a\alpha'}\delta\phi^{\alpha'} = g_{ab}g^{cd}\delta\Gamma^b{}_{cd}. \tag{23}$$

This ν indeed satisfies the four conditions listed above. First note that, for Einstein's equation, the left side of (10) does indeed exceed the right side by four. The $H^{A'}{}_\alpha$ of (20) is given by

$$H^A{}_{\alpha'}\delta\phi^{\alpha'} = (s^{cdrs}\delta\Gamma^{[a}{}_{rs}t^{b]} + 2g^{(d|[a}s^{b]|c)rs}\delta\Gamma^p{}_{rs}g_{pq}t^q, \; -\frac{1}{2}s^{abcd}\delta\Gamma^r{}_{cd}g_{rs}t^s) \tag{24}$$

for any t^a timelike, and $s^{abcd} = s^{(cd)(ab)}$ positive-definite in its two index pairs. Using (18) and (23), the operator of (22) is just the wave operator. Note that the field $\nu_{a\alpha'}$ of (23) is diffeomorphism-invariant. Thus, the breaking of the diffeomorphism invariance takes place solely through the choice of $\nu_{a\alpha}$, *i.e.* the choice of submanifold V. Are there any other $\nu_{a\alpha'}$'s that work for general relativity?

The diffeomorphisms, of course, act simultaneously on all the fields in the combined bundle B. What, then, is the feature that singles out the field ∇_a of general relativity (as opposed, say, to the field F_{ab} of electromagnetism) to be the one for which the diffeomorphisms are taken to be the 'gauge'? The answer, we suggest, is the following. Let n_m be any covector in M which, contracted into the expression (21), yields a positive-definite quadratic form. Now apply this quadratic form to the vertical vector

$$v^{\alpha'} = \delta^{\alpha' m_1 \cdots m_s}{}_a n_{m_1} \cdots n_{m_s}\xi^a. \tag{25}$$

Then the first term arising from (21) vanishes, by (19), and so, by positive-definiteness, we must have $\nu_{a\alpha'}v^{\alpha'}$ nonzero. We may restate this observation as follows: that field whose diffeomorphism-behavior involves the highest number of derivatives of ξ^a in (17) (*i.e.* that is represented by vertical vectors of the form (25)) is also the field restricted by the gauge-fixing equation, (20) (corresponding to vertical vectors not annihilating $\nu_{a\alpha'}$). In the case of the combined bundle B resulting from the examples of Appendix A, the highest number of derivatives of ξ^a arising from diffeomorphism-behavior is $s = 2$, and the field having this behavior is, of course, the derivative operator ∇_a. In this manner, the derivative operator of general relativity acquires the diffeomorphisms as its gauge group. These remarks have the following curious consequence. Suppose that, at some time in the future, there were introduced a new physical field, having order $s = 3$ in (17). Then, apparently, that new field would take the diffeomorphisms as its gauge group; leaving no 'gauge freedom' in general relativity.

---

[10] There is a somewhat more elegant, if less accessible, way to formulate this. Consider $s = 2$ in (17). Modify the bundle B to include, in each fibre, a copy of the twenty-dimensional manifold of all tensors $\rho_a{}^b$ at all points of M. Think of a cross-section of this modified bundle as including a diffeomorphism on M and a 'candidate' $(\rho_a{}^b)$ for the derivative of this diffeomorphism. (15) and (20) can be combined as a hyperbolic system on this modified bundle. The 'unfolding of the diffeomorphism' is then already incorporated in the modified bundle.

[11] For recent work on the initial-value formulation of Einstein's equation, see, *e.g.* Fritelli *et al.* (1994), Choquet-Bruhat (1996).

6 Physical systems: interactions

We adopt the view that the combined bundle B, with its partial differential equation (15) and action of the diffeomorphism group, comprises all there is in the (nonquantum) physical world. But, by contrast, we do not view our world as such a single entity. Rather, it appears to be divided into various 'physical systems'. For example, one such system is comprised of the electromagnetic field F_{ab} alone; another, of the four-velocity u^a, mass density ρ, and particle-number density n for a perfect fluid. We do not, for example, organize these four fields into one system (F_{ab}, n), and another (u^a, ρ). We then think of these individual systems as 'interacting' with each other. Interactions take place on a number different of levels. Consider, for example, the electromagnetic field F_{ab}. In the absence of a space-time metric, there is no natural choice for the electromagnetic field tensor (F_{ab}, or F^{ab}, or some density?); and, even if some such choice were made, there is no way to write down Maxwell's equations. (Note that neither of these assertions is true with the roles of the metric and electromagnetic field reversed.) This is one level of interaction. On a different level is the interaction of a charged fluid on the electromagnetic field through the appearance of the the fluid charge-current in Maxwell's equations. These structural features of the world—the notion of physical systems and their various levels of interaction—must somehow be 'derived' from (15) on B, at least if our view of the primacy of (15) is to be maintained. How all this comes about is the subject of this section.

Let $b \xrightarrow{\pi} M$ be a fibre bundle. By a *quotient bundle* of b, we mean a smooth manifold $\hat{b}$, together with smooth mappings $b \xrightarrow{\zeta} \hat{b} \xrightarrow{\hat{\pi}} M$, such that

(i) $\hat{\pi} \circ \zeta = \pi$, and

(ii) $\hat{b} \xrightarrow{\hat{\pi}} M$ is a fibre bundle (over M), and $b \xrightarrow{\zeta} \hat{b}$ is a fibre bundle (over $\hat{b}$).

Thus, a quotient 'inserts a manifold $\hat{b}$ between b and M, in such a way that there is created a fibre bundle on each side of $\hat{b}$'. The following example will illustrate both the mathematical structure and the types of applications we have in mind. Let b be the bundle whose fibre over $x \in M$ consists of pairs (g_{ab}, F_{ab}), where g_{ab} is a Lorentz-signature metric at x, and F_{ab} an antisymmetric tensor (the electromagnetic field). Now let $\hat{b} \xrightarrow{\hat{\pi}} M$ be the bundle whose fibres include only the Lorentz metrics, and let $b \xrightarrow{\zeta} \hat{b}$ be the mapping that 'ignores F_{ab}'. This $b \xrightarrow{\zeta} \hat{b} \xrightarrow{\hat{\pi}} M$, we claim, is a quotient bundle. Furthermore, it reflects the natural relationship between F and g, i.e. that it is meaningful to discard F_{ab} while retaining g_{ab}, but not to discard g_{ab} while retaining F_{ab}.

Let $b \xrightarrow{\zeta} \hat{b} \xrightarrow{\hat{\pi}} M$ be a quotient bundle. Fix any cross-section, $M \xrightarrow{\hat{\phi}} \hat{b}$, of $\hat{b}$. We now construct, using this $\hat{\phi}$, a new fibre bundle, $\check{b} \xrightarrow{\check{\pi}} M$, as follows. For the manifold $\check{b}$, we take the submanifold $\zeta^{-1}[\hat{\phi}[M]]$ of b (i.e. the set of points of b lying above the fixed cross-section $\hat{\phi}$ of $\hat{b}$); and for the mapping $\check{\pi}$ we take the restriction to the submanifold $\check{b}$ of the projection π. The bundles $\hat{b} \xrightarrow{\hat{\pi}} M$ and $\check{b} \xrightarrow{\check{\pi}} M$ represent a kind of 'splitting' of the original bundle $b \xrightarrow{\pi} M$. For instance, we have (dim fibre $\hat{b}$) + (dim fibre $\check{b}$) = (dim fibre b). Every cross-section, ϕ, of b yields both a cross-section of $\hat{b}$ (namely, $\hat{\phi} = \zeta \circ \phi$), and a cross-section of $\check{b}$ (namely, ϕ). And, conversely, cross-sections of $\hat{b}$ and $\check{b}$ combine to form a cross-section of b. But, and this is the key point of the construction, the bundle $\check{b}$ requires for its very existence a given cross-section of $\hat{b}$. We illustrate this construction with our earlier example (with b-fibres (g_{ab}, F_{ab}), and $\hat{b}$-fibres (g_{ab})). Fix a cross-section of $\hat{b}$, i.e. fix a metric field $\tilde{g}_{ab}$ on M. Then the submanifold $\check{b}$ of b consists of all $(x, \tilde{g}_{ab}, F_{ab})$. That is, the metric at each $x \in M$ is required to be the fixed $\tilde{g}_{ab}$ there, while the electromagnetic field F_{ab} at x remains arbitrary. So, $\check{b}$ in this example is the bundle over M of electromagnetic fields, in the presence of the fixed background metric $\tilde{g}_{ab}$. Clearly, a cross-section of the bundle $\hat{b}$ (metric field $\tilde{g}_{ab}$), together with a cross-section of $\check{b}$ (an electromagnetic field in the presence of background metric $\tilde{g}_{ab}$), yield a cross-section of the original bundle b; and conversely.

The notion of a quotient bundle captures the idea of one set of physical fields serving as the kinematical background for another. Thus, the metric serves as the kinematical background field for the electromagnetic field; and the metric and electromagnetic fields together serve as the kinematical background for a charged scalar field. We turn now from kinematics to dynamics.

Fix a fibre bundle $b \xrightarrow{\pi} M$, and a quotient bundle thereof, $b \xrightarrow{\zeta} \hat{b} \xrightarrow{\hat{\pi}} M$. Then, as we have just seen, the cross-sections of b 'split', in the sense that specifying a cross-section ϕ of b is equivalent to specifying a cross-section $\hat{\phi}$ of $\hat{b}$, together with a cross-section $\check{\phi}$ of the bundle $\check{b}$ derived from $\hat{\phi}$. Next, let there be specified a system of quasilinear, first-order partial differential equations, (2), on cross-sections ϕ of b. When does this equation also split into separate equations on the cross-sections, $\hat{\phi}$ and $\check{\phi}$, that comprise ϕ? We are here concerned only with the first term in (2)—the dynamical part of the differential equation. The remainder will be discussed shortly. So, fix a field $k_A{}^m{}_{\alpha'}$ on b; we wish to split it into corresponding fields on $\hat{b}$ and $\check{b}$. There is a natural choice for the field on $\check{b}$, namely the restriction to the submanifold $\check{b}$ of b of the given field $k_A{}^m{}_{\alpha'}$ on b. On taking this restriction, only some of the components of $k_A{}^m{}_{\alpha'}$ survive, namely, those represented by contraction of $k_A{}^m{}_{\alpha'}$ with vectors $v^{\alpha'}$ tangent to the submanifold $\check{b}$. The remaining components, those lost under this restriction, must now be recovered from a suitable field, $\hat{k}_{\hat{A}}{}^m{}_{\hat{\alpha}'}$ on $\hat{b}$. This recovery will occur provided that

(i) the pullback of $\hat{k}_{\hat{A}}{}^m{}_{\hat{\alpha}'}$ from $\hat{b}$ to b is a linear combination of $k_A{}^m{}_{\alpha'}$ on b, and

(ii) this pullback, together with the restriction of $k_A{}^m{}_{\alpha'}$ to $\check{b}$, exhausts $k_A{}^m{}_{\alpha'}$.

The first condition means, in more detail, that, for some field $\mu_{\hat{A}}{}^A$ on b, we have

$$(\nabla \zeta)_{\alpha'}{}^{\hat{\alpha}'} \hat{k}_{\hat{A}}{}^m{}_{\hat{\alpha}'} = \mu_{\hat{A}}{}^A k_A{}^m{}_{\alpha'} \tag{26}$$

where the right side is evaluated at $\kappa \in b$, the left side at $\zeta(\kappa) \in \hat{b}$. The left side of (26) is the pullback of $\hat{k}_{\hat{A}}{}^m{}_{\hat{\alpha}'}$ from $\hat{b}$ to b via the mapping ζ; the right side, some linear combination (with coefficients μ) of $k_A{}^m{}_{\alpha'}$. This condition guarantees that the dynamical part of the differential equation to be imposed on $\hat{b}$ will come from the dynamical part of the differential equation originally given on b. The second condition means, in more detail, that, for any vector σ^A such that the restriction of $\sigma^A k_A{}^m{}_{\alpha'}$ to $\check{b}$ vanishes, we have $\sigma^A = \tau^{\hat{A}} \mu_{\hat{A}}{}^A$ for some $\tau^{\hat{A}}$. In other words, what is lost on restriction (to $\check{b}$) must be regained via the pullback (from $\hat{b}$). Note that such a $\hat{k}$ on $\hat{b}$, if it exists, is unique.

Given fibre bundle $b \xrightarrow{\pi} M$, and field $k_A{}^m{}_{\alpha'}$ on b, by a *reduction* of this we mean a quotient bundle $\hat{b}$, with field $\hat{k}_{\hat{A}}{}^m{}_{\hat{\alpha}'}$ on $\hat{b}$, satisfying the two conditions above (in the case of (ii), for every cross-section $\hat{\phi}$). To illustrate this definition, we return again to our earlier example. On the metric-electromagnetic bundle b above, introduce the field $k_k{}^m{}_{\alpha'}$ arising from the metric-Maxwell equations:

$$\nabla_a g_{bc} = 0 \qquad \nabla^b F_{ab} = 0 \qquad \nabla_{[a} F_{bc]} = 0. \tag{27}$$

Let $\hat{b}$ be the quotient bundle above (in which only the metric g_{ab} is retained). On $\hat{b}$, introduce the field $\hat{k}_{\hat{A}}{}^m{}_{\hat{\alpha}'}$ arising from the first equation in (27). (Here, we are making essential use of the fact that no electromagnetic field appears in this equation.) This $\hat{b}$, $\hat{k}_{\hat{A}}{}^m{}_{\hat{\alpha}'}$, we claim, is a reduction. Indeed, condition (i) follows from the fact that, whenever cross-section (g_{ab}, F_{ab}) of the bundle b satisfies the full system (27), then the corresponding cross-section g_{ab} of $\hat{b}$ satisfies the first of these equations. Condition (ii) follows from the fact that, given a field g_{ab} (cross-section of $\hat{b}$) satisfying the first equation, and then an F_{ab} (cross-section of $\check{b}$) satisfying, in the presence of that g_{ab} as background, the last two equations, then the full set, (g_{ab}, F_{ab}), satisfies the full system (27.) Note, by contrast, that there is no reduction with the roles of the electromagnetic field and the metric reversed.

Let us now return to the combined bundle B, with its partial differential equation (15). This $(B, K_A{}^m{}_{\alpha'})$ will, of course, have numerous reductions; and the $(\hat{B}, \hat{K}_{\hat{A}}{}^m{}_{\hat{\alpha}'})$ that result may have further reductions; and so on. Any fields lost (*i.e.* incorporated in bundle $\check{B}$) in such a reduction will be called a *physical system*; while the fields remaining (in the bundle $\hat{B}$) will be called *background fields* for that physical system. These definitions, we suggest, capture the way in which fields are grouped together in physics, and the sense in which the equations for some fields require as prerequisites other fields. For example, the fields (u^a, ρ, n) for a perfect fluid form a physical system (but not, *e.g.* (ρ, n) alone); with background the space-time metric. The charged Dirac field is a physical system, with background the electromagnetic and metric fields together; and the electromagnetic field is a physical system, with background the metric field. This definition also produces a couple of minor surprises. For the Klein-Gordon equation, the scalar and vector fields, ψ and ψ_a, form separate physical systems, the former having no background fields; the latter, just the metric field. (Thus, neither of these has the other as background!) Similarly, the metric and derivative operator of general relativity form separate physical systems, the metric having no background, the derivative operator the metric as background. Note, from the examples of Appendix A, that the metric is a background for virtually every physical system (the sole exceptions being those, such as the Klein-Gordon ψ, having such a wide variety of hyperbolizations that every covector in M lies in the cone s_κ for at least one of them). This feature presumably reflects the fact that, in order that there be a hyperbolization for the combined system (15), it is not enough merely that there be a hyperbolization for each individual physical system making up B. These individual hyperbolizations must also be such that they have in their various s_κ a common n_m. In order to achieve this, the individual physical systems must somehow arrange to 'communicate' with each other what their dynamics is. That communication takes place by sharing the space-time metric g_{ab} as a background field.

So far, we have focused exclusively on the 'dynamical part' of (15)—the $K_A{}^m{}_{\alpha'}$. We turn now to the remainder of this equation—the J_A. Roughly speaking, whenever that portion of (15) that specifies the dynamics of one physical system has its J_A depending on the fields of another, then we say that the second system interacts on the first. However, we must exercise some care in formulating this idea precisely. For instance, J_A is not gauge-invariant (and, indeed, can, by a suitable gauge transformation, be made to vanish); and furthermore it is unclear what 'depends on' is to mean for fields on a bundle space.

Consider again (2). Fix a reduction of that system, so we have a quotient bundle $b \xrightarrow{\zeta} \hat{b} \xrightarrow{\hat{\pi}} M$, together with a field $k_A{}^m{}_{\alpha'}$ on b, satisfying conditions (i) and (ii) above. Any cross-section, $\hat{\phi}$, of $\hat{b}$ gives rise to a new bundle, $\check{b}$; and this $\hat{\phi}$, together with a cross-section $\check{\phi}$ of $\check{b}$, specifies a cross-section ϕ of the original bundle b. Furthermore, the dynamical term $k_A{}^m{}_{\alpha'}$ on b splits into corresponding dynamical terms on $\hat{b}$ and $\check{b}$. We say that the physical system represented by cross-sections $\check{\phi}$ of $\check{b}$ *interacts on* the physical system represented by cross-sections $\hat{\phi}$ of $\hat{b}$ provided there is no similar way to split the entire partial differential equation (2) on b. Thus, in more detail, the $\check{b}$-system interacts on the $\hat{b}$-system provided there exist no fields $\hat{k}_{\hat{A}}{}^m{}_{\hat{\alpha}}$, $\hat{j}_{\hat{A}}$ on $\hat{b}$ such that

$$(\nabla \zeta)_\alpha{}^{\hat{\alpha}} \hat{k}_{\hat{A}}{}^m{}_{\hat{\alpha}} = \mu_{\hat{A}}{}^A k_A{}^m{}_\alpha \qquad \hat{j}_{\hat{A}} = \mu_{\hat{A}}{}^A j_A \tag{28}$$

up to gauge. This is to be compared with (26). We are merely strengthening condition (ii) of the definition of a reduction to require an entire partial differential equation on $\hat{b}$ whose pullback is a linear combination of the given partial differential equation on b. In order words, if we can write the equations on $\hat{b}$ in such a way that 'no $\check{b}$-fields are involved', then $\check{b}$ does not interact on $\hat{b}$.

As an example, consider an electromagnetic field and uncharged perfect fluid, in the presence of a background metric and derivative operator. Then the electromagnetic field does not interact on the perfect fluid. We can find a reduction of this system in which the electromagnetic field is carried in $\check{b}$, the perfect fluid in $\hat{b}$; and a system of equations on $\hat{b}$ (the perfect-fluid equations, described by $\hat{k}_{\hat{A}}{}^m{}_{\hat{\alpha}}$ and $\hat{j}_{\hat{A}}$) not involving the electromagnetic field. Similarly, the perfect fluid does not interact on the electromagnetic field. However, if this is a charged perfect fluid, then each of these physical systems interacts on the other; the perfect fluid on the electromagnetic field through the charge-current term in Maxwell's equations; the electromagnetic field on the perfect fluid through the Lorentz-force term in the fluid equations. As a second example, note that the derivative operator interacts on virtually every physical system through the use of this derivative operator in writing out field equations; conversely virtually every physical system interacts on the derivative operator through Einstein's equation.

We remark that the definition is manifestly gauge-invariant. Note that, as the definition is formulated, 'interacts on' is not defined at all unless we have an appropriate reduction. Thus, for example, we are not permitted even to ask whether the metric interacts on the electromagnetic field. (Perhaps it would be more natural to extend the definition so that background fields for a given physical system automatically interact on that system.) Finally, we remark that 'interact on' need not be reciprocal; It is possible for system A to interact on B, but not B on A. It is not difficult to construct an example, *e.g.* with A a Klein-Gordon system and B a perfect fluid. Introduce an additional term, involving the Klein-Gordon fields, on the right side of the equation, (43), of fluid particle-number conservation. Conservation of fluid stress-energy is not thereby disturbed. However, I am not aware of any such examples arising naturally in physics. Should this observation be elevated to a general principle?

We are, in a real sense, finished at this point. The world is described, once and for all, by the combined bundle B, with its cross-sections subject to (15). In particular, whatever interactions there are in the world have already been included in the term J_A of this equation. However, it is traditional to think of interactions in a somewhat different way—to think of them as capable of being 'turned on and off' by some external agency (presumably, us). Consider, for example, the electromagnetic-charged fluid system. The fields are F_{ab}, n, ρ, and u^a; and there appears, on the right side of the Maxwell equation (30), a term enu^a (charge-current), and, on the right side of the fluid stress-energy conservation equation (41) a term $enF^a{}_m u^m$ (Lorentz force). Here, e is some fixed number (charge per particle). In the traditional view for this particular system, we think of this system as arising, not full-blown in its final form, but rather in two distinct steps. First, introduce the system with 'no interaction' ($e = 0$), and then 'turn on the interaction' by adjusting e to its correct value.

This traditional view may be expressed, in the present general framework, as follows. On the combined bundle B, there is to be specified a 'basic version' of the dynamical equations, (15)—a version in which 'all interactions that can be turned off have been'. Thus, the 'electromagnetic interaction' has been turned off in the basic version: the only physical system that interacts on the electromagnetic field is the derivative operator. The 'gravitational interaction' has been turned off: no field interacts on the derivative operator. For more complicated interactions, e.g. those of contact forces between materials, it may not always be clear how this 'turning off' is to be carried out. The derivative operator generally survives into the basic version to interact with most physical systems. Physically, this phenomenon is a reflection of the equivalence principle. Mathematically, it arises because it is difficult to eliminate the derivative operator and still lift diffeomorphisms (i.e. 'maintain covariance') as in Section 5. This role of the derivative operator is related to the fact that the behavior of ∇_a under an infinitesimal diffeomorphism, given in (17), involves the second derivative of the vector field ξ^a, whereas other physical fields involve only the first derivative. To see this, consider a physical field, such as the F_{ab} for electromagnetism, having $s = 1$ in (17). Consider that portion of the combined equation, (15), referring to the electromagnetic field, and apply to it the infinitesimal diffeomorphism generated by ξ^a. Then the term '$(\nabla\Phi)_m{}^\alpha$' acquires a second derivative of ξ^a. So, invariance of (15) under infinitesimal diffeomorphisms can be maintained only if a second derivative of ξ^a appears elsewhere in the electromagnetic portion of the equation, i.e. in the term J_A. That is, some physical field must have, for its transformation behavior (17) under infinitesimal diffeomorphisms, $s = 2$. That field is the derivative operator.

In addition to this basic version of (15), there is to be given some 'interaction fields', i.e. some fields δJ_A on B. We are free to add, to the left side of (15), any linear combination, with constant coefficients (the coupling constants), of these δJ_A. This is the step of turning on the interactions. One such δJ_A, for instance, is that which inserts the fluid charge-current nu^a into (30), and the Lorentz force $nF^a{}_m u^m$ into (41). Another inserts the trace-reversed combined stress-energies of all physical systems into[12] Einstein's equation, (50). It is not entirely clear how much of this traditional view is psychological and how much physical. One way to argue that it has physical content might be to produce a general construction, given only the full combined (15), that yields the basic version of this equation, as well as the appropriate δJ_A.

These δJ_A do not span, at each point of the bundle B, the space of vectors of type 'σ_A'. That is, there are algebraic restrictions on the allowed interactions. These restrictions appear to be an essential part of the physical content of the systems under considerations. Disallowed, for example, are interactions on the electromagnetic field that represent a magnetic charge-current; on a perfect fluid that provide a source for particle-number conservation; and on the metric in (48). See Appendix A for further examples.

The basic version of (15) on B must certainly be a viable system of equations, and so in particular its constraints must be complete[13] and integrable. What happens to viability of this system on turning on some interaction δJ_A? Completeness of the constraints will not change, for this involves only the dynamical term, that with coefficient $K_A{}^m{}_{\alpha'}$ in (15). But integrability could be destroyed by turning on such an interaction. Indeed, the necessary and sufficient condition that integrability be retained for fixed constraint c^{An} of (15) under addition to J_A a term δJ_A is that, for some τ^A and $\tau_a{}^b$ with $\tau_a{}^a = 0$, we have

$$\nabla_\alpha(c^{Am}\delta J^A) = \sigma^{Am}{}_\alpha \delta J_A + \tau^A(K_A{}^m{}_\alpha + \frac{1}{4}(J_A + \delta J_A)(\nabla \pi)_\alpha{}^m) + \tau^m{}_s(\nabla \pi)_\alpha{}^s \qquad (29)$$

where $\sigma^{Am}{}_\alpha$ is the tensor that appears in (12) for the basic version of (15). The proof of this assertion is straightforward. Demand that the integrability condition, (12), be satisfied for both $K_A{}^m{}_\alpha$, J_A and $K_A{}^m{}_\alpha$, $J_A + \delta J_A$. Thus, any viable candidate δJ_A for an interaction that can be 'turned on' must satisfy the condition of (29).

Equation (29) is apparently a rather severe restriction on the interactions allowed in nature. The main reason for this is its nonlinearity: given δJ_A and $\delta J'_A$, for each of which there exist τ's satisfying (29), we have no guarantee that there exist τ's that work for their sum. Here is a physical example of this behavior. Let δJ_A insert a perfect-fluid stress-energy into Einstein's equation, and $\delta J'_A$ insert an electromagnetic interaction into the perfect-fluid equations. Then each of these modifications of (15) preserves integrability of the gravitational constraint, but their sum, $\delta J_A + \delta J'_A$, does not. To achieve integrability in this example, it is necessary to add to this sum a term that also inserts the electromagnetic stress-energy into Einstein's equation. Indeed, it is not even obvious, from (29), that whenever δJ_A preserves integrability in (15), then so does $2\delta J_A$!

So, finding a collection of interaction expressions, δJ_A, that can be turned on in any linear combination, preserving all the while integrability of the constraints of (15) does not appear to be easy. So, how does nature accomplish this? Is there, for example, some simple, general criterion that can be applied to the δJ_A's to guarantee integrability?

---

[12]Note that these individual stress-energies cannot be inserted one at a time, with separate δJ_A, for to do so would violate integrability of the constraints of the combined system.

[13]That is, complete in the sense appropriate to the diffeomorphism freedom inherent in (15), namely that the left side of (10) exceeds the right side by four.

Appendix A Examples

Electromagnetism

The field is an antisymmetric tensor field, $F_{ab} = F_{[ab]}$, on M, with background metric g_{ab}. The equations are

$$\nabla^b F_{ab} = 0 \tag{30}$$

$$\nabla_{[a} F_{bc]} = 0. \tag{31}$$

Thus, the fibres of the bundle b are six-dimensional, a typical vertical vector being given by $\delta\phi^{\alpha'} = \delta F_{ab} = \delta F_{[ab]}$. A typical vector in the space of these equations is given by $\sigma^A = (s^a, s^{abc})$, where s^a is a vector (the coefficient of (30)), and s^{abc} an antisymmetric third-rank tensor (the coefficient of (31)). Thus, the equation-space is eight-dimensional.

The general hyperbolization at a point is given by

$$h^A{}_{\alpha'} \delta\phi^{\alpha'} = (\delta F^a{}_m t^m,\ -\frac{3}{2} t^{[a} \delta F^{bc]}) \tag{32}$$

where t^a is an arbitrary timelike vector. The general constraint at a point is given by

$$c^{An} = (x g^{an},\ y \epsilon^{abcn}) \tag{33}$$

where x and y are numbers. Thus, the constraints form a two-dimensional vector space, while the space of vectors of the form $c^{An} n_n$, for fixed n_n, is also two-dimensional. The constraints are complete and integrable.

The allowed interactions are

$$\delta j_A = (j_a,\ 0) \tag{34}$$

i.e. no magnetic charge-current is allowed. These will preserve integrability provided $\nabla^a j_a = 0$.

Klein-Gordon

The fields consist of a scalar field, ψ, and a vector field, ψ_a, on M, with background metric g_{ab}. The equations are

$$\nabla_a \psi = \psi_a \tag{35}$$
$$\nabla_{[a} \psi_{b]} = 0 \tag{36}$$
$$\nabla^a \psi_a = 0. \tag{37}$$

Thus, the fibres of the bundle b are five-dimensional, a typical vertical vector being given by $\delta\phi^{\alpha'} = (\delta\psi, \delta\psi_a)$. A typical vector in the space of equations is given by $\sigma^A = (s^a, s^{ab}, s)$ (the respective coefficients of (35)–(37)), where s^{ab} is antisymmetric. Thus, the equation-space is eleven-dimensional.

The general hyperbolization at a point is given by

$$h^A{}_{\alpha'}\delta\phi^{\alpha'} = (-w^a\delta\psi,\ -t^{[a}\delta\psi^{b]},\ \frac{1}{2}t^a\delta\psi_a) \tag{38}$$

where t^a is any timelike vector (say, future-directed), and w^a any vector not past-directed timelike or null. In order that this hyperbolization be causal, we must require in addition that w^a be future-directed timelike or null. The general constraint at a point is given by

$$c^{An} = (x^{an},\ y^{abn},\ 0) \tag{39}$$

where x^{an} and y^{abn} are arbitrary antisymmetric tensors. Thus, the constraints form a ten-dimensional vector space, while the space of vectors of the form $c^{An}n_n$, for fixed n_n, is six-dimensional. The constraints are complete and integrable. It turns out that there are two separate physical systems (in the sense of Section 6) here. The first involves the field ψ alone: the equation is (35), the hyperbolizations those generated by w^a in (38), and the constraints those generated by x^{an} in (39). There are no background fields for this physical system. The other involves the field ψ_a: the equations are (36), (37), the hyperbolizations generated by t^a in (38), and the constraints generated by y^{abn} in (39). The background field is the metric.

The situation regarding allowed interactions is not entirely clear (reflecting, perhaps, a certain lack of physical context for this system). Certainly, one allowed interaction is

$$j_A = (0,\ 0,\ -m^2\psi) \tag{40}$$

where m is a number. This results in the massive Klein-Gordon system. (Note that the 'tachyon equation'—the result of letting m be imaginary in (40)—admits a hyperbolization). It seems likely that all allowed interactions have zeros in the first two entries of (40), for otherwise it is difficult to preserve integrability. (Passage to a charged Klein-Gordon system is not turning on an interaction in our sense, for it requires an entirely new bundle b: this example is discussed later in this Appendix.) Whether there are allowed other interactions of the form (40), but with different third entries on the right (for all of which, incidentally, all the constraints are integrable), is unclear.

Perfect Fluid

The fields consist of two scalar fields, n and ρ, and a unit timelike vector field, u^a, on M, with background metric g_{ab}. The equations are

$$(\rho + p)u^m\nabla_m u^a + (g^{ab} + u^a u^b)\nabla_b p = 0 \tag{41}$$
$$u^m\nabla_m \rho + (\rho + p)\nabla_m u^m = 0 \tag{42}$$
$$u^m\nabla_m n + n\nabla_m u^m = 0 \tag{43}$$

where $p(n,\rho)$ is some fixed function (the function of state). The first two equations are the components of conservation of $T^{ab} = (\rho + p)u^a u^b + pg^{ab}$, the third conservation of $N^a = nu^a$.

Thus, the fibres of the bundle b are five-dimensional, a typical vertical vector being given by $\delta\phi^{\alpha'} = (\delta n, \delta\rho, \delta u^a)$, with δu^a (because of unit-ness of u^a) orthogonal to u^a. A typical vector in the space of equations is given by $\sigma^A = (s_a, s, \hat{s})$ (respective coefficients of (41)–(43)), with s_a orthogonal to u^a (reflecting that the left side of (41) is). The equation-space is five-dimensional.

This physical system admits no hyperbolization unless

$$\rho + p > 0, \quad \left(\frac{\partial p}{\partial \rho} + \frac{n}{\rho + p}\frac{\partial p}{\partial n}\right) > 0. \tag{44}$$

Physically, this is the requirement that inertial mass and sound speed both be positive. So, the fibres of the bundle b must be suitably restricted to achieve (44) everywhere. The most general hyperbolization at a point is then given by

$$h^A{}_{\alpha'}\delta\phi^{\alpha'} = x\left\{\left((\rho + p)\frac{\partial p}{\partial \rho} + \frac{n}{\rho + p}\frac{\partial p}{\partial n}\right)\delta u_a, \frac{\partial p}{\partial \rho}\delta p, \frac{\partial p}{\partial n}\delta p\right\}$$
$$+ y\left\{\left(\delta n - \frac{n}{\rho + p}\delta\rho\right)\left(0, \frac{n}{\rho + p}, -1\right)\right\}, \tag{45}$$

where x and y are any numbers with $xy > 0$, and where we have set $\delta p = (\frac{\partial p}{\partial n})\delta n + (\frac{\partial p}{\partial \rho})\delta\rho$. These hyperbolizations are all causal provided

$$\frac{\partial p}{\partial \rho} + \frac{n}{\rho + p}\frac{\partial p}{\partial n} \leq 1 \tag{46}$$

(*i.e.* physically, provided the sound-speed does not exceed light-speed). If (46) fails, then none are causal. There are no constraints: this follows, for example, from existence of a hyperbolization and equality of the dimension of the space of fields and that of equations.

The allowed interactions (*e.g.* electromagnetic, contact-force, *etc.*) are given by

$$j_A = (j_a, j, 0). \tag{47}$$

That is, arbitrary sources are allowed in the equation of stress-energy conservation, but none is allowed in the equation of particle-number conservation.

Gravitation

The fields consist of a symmetric, Lorentz-signature metric, g_{ab}, together with a (torsion-free) derivative operator, ∇_a, on M. The equations are

$$\nabla_a g_{bc} = 0 \tag{48}$$
$$R_{ab(c}{}^m g_{d)m} = 0 \tag{49}$$
$$R_{m(ab)}{}^m = 0. \tag{50}$$

Here, $R_{abc}{}^d$, with symmetries $R_{abc}{}^d = R_{[ab]c}{}^d$ and $R_{[abc]}{}^d = 0$, is the curvature tensor (*i.e.* the 'derivative of the derivative operator'), defined by the condition that

$$\nabla_{[a}\nabla_{b]}\xi_c = \frac{1}{2}R_{abc}{}^d\xi_d \tag{51}$$

for every covector field ξ_c on M.

Thus, the fibres of the bundle b are fifty-dimensional, a typical vertical vector being given by $\delta\phi^{\alpha'} = (\delta g_{ab}, \delta\Gamma^a{}_{bc})$, where $\delta g_{ab} = \delta g_{(ab)}$ (ten dimensions), and $\delta\Gamma^a{}_{bc} = \delta\Gamma^a{}_{(bc)}$ (forty dimensions). The latter represents a first-order change in the derivative operator, whose effect is to replace $\nabla_a \xi_b$ by $\nabla_a \xi_b + \delta\Gamma^m{}_{ab}\xi_m$. A typical vector in the space of equations is $\sigma^A = (s^{abc}, s^{abcd}, s^{ab})$ (the respective coefficients of (48)-(50)), where $s^{abc} = s^{a(bc)}$, $s^{abcd} = s^{[ab](cd)}$, and $s^{ab} = s^{(ab)}$. Thus, the equation space has dimension 110 $(= 40 + 60 + 10)$.

This system consists of two separate physical systems (in the sense of Section 6). For the first, the field is the metric g_{ab}, and the equation (48), with no background field. Its most general hyperbolization at a point is given by

$$h^A{}_{\alpha'}\delta\phi^{\alpha'} = x^{abcmn}\delta g_{mn} \qquad (52)$$

where x^{abcmn} has symmetries $x^{abcmn} = x^{a(mn)(bc)}$, and is such that $n_a x^{abcmn}$ is positive-definite in the index pairs 'b,c' and 'm,n', for some n_a. The general constraint for this physical system at a point is given by

$$c^{An} = x^{nabc} \qquad (53)$$

where $x^{nabc} = x^{[na](bc)}$. Thus, the vector space of constraints has dimension sixty, while the space of vectors of the form $c^{An}n_n$, for fixed n_n, has dimension thirty. These constraints are complete and, by virtue of (49), integrable. For the other physical system, the field is the derivative operator, the equations are (49), (50), and the background field is the space-time metric g_{ab}. This physical system has no hyperbolization, because of the diffeomorphism freedom. (But it does have a 'hyperbolization up to diffeomorphisms'. See (24)). The most general constraint for this system is given by

$$c^{An} = \left(x^{nabcd} + 2x^{[a}g^{b][c}g^{d)n} - g^{cd}x^{[a}g^{b]n}, \quad x^{(a}g^{b)n} - \tfrac{1}{2}g^{ab}x^n\right) \qquad (54)$$

where $x^{nabcd} = x^{[nab](cd)}$. Thus, the vector space of constraints has dimension fifty-four, while the space of vectors of the form $c^{An}n_n$, for fixed n_n, has dimension thirty-four. These constraints are integrable, but not complete. (This feature, again, is related to diffeomorphism-invariance—see Section 5).

The most general allowed interaction, apparently, is

$$\delta j_A = (0, \ 0, \ T_{ab} - \tfrac{1}{2}T^m{}_m g_{ab}) \qquad (55)$$

where $T_{ab} = T_{(ab)}$ (the stress-energy of matter). In order that this interaction preserve integrability of the constraints, we must have conservation, $\nabla^b T_{ab} = 0$. This is a very severe restriction on the fields contributing to T_{ab}, and the interactions between those fields.

Spin-s Systems

The field is a totally symmetric, 2s-rank spinor, $\psi^{A\cdots D} = \psi^{(A\cdots D)}$, on M, with background metric g_{ab}. Here, $s = \tfrac{1}{2}, 1, \tfrac{3}{2}, 2, \cdots$. The equation is

$$\nabla^{A'}{}_A \psi^{AB\cdots D} = 0. \qquad (56)$$

Thus, the fibres of the bundle b are $(4s+2)$-dimensional. (This, and all subsequent, reference to dimension means *real* dimension.) A typical vertical vector is given by $\delta\phi^{\alpha'} = \delta\psi^{A\cdots D} = \delta\psi^{(A\cdots D)}$. A typical vector in the space of equations is given[14] by $\sigma^A = s_{A'B\cdots D} = s_{A'(B\cdots D)}$. Thus, the equation-space is $8s$-dimensional.

The most general hyperbolization at a point is given by

$$h^A{}_{\alpha'}\delta\phi^{\alpha'} = t_{B'\cdots D'B\cdots D}\delta\bar\psi_{A'}{}^{B'\cdots D'} \tag{57}$$

where $t_{B'\cdots D'B\cdots D} = \bar{t}_{(B'\cdots D')(B\cdots D)}$, a Hermitian quadratic form on symmetric, $(2s-1)$-rank spinors, is positive-definite. The general constraint is given by

$$c^{An} = \epsilon^{N'A'}\epsilon^{N(B}x^{C\cdots D)} \tag{58}$$

where $x^{C\cdots D}$ is an arbitrary symmetric, $(2s-2)$-rank spinor. Thus, the constraints form a $(4s-2)$-dimensional vector space, while the space of vectors of the form $c^{An}n_n$, for fixed n_n, also has dimension $(4s-2)$. Note that, for $s = 1/2$, there are no constraints. These constraints are always complete, but they are integrable if and only if either $s \leq 1$, or the metric g_{ab} is conformally flat. (This is the famous 'inconsistency of the higher-spin equations'.) Except for the cases $s = 1$ (electromagnetism), and $s = 2$ (linearized gravity), it isn't clear what are the allowed interactions. For charged spin-s fields, see later in this appendix.

Elastic Solid

The fields consist of a scalar field ρ, a unit timelike vector field u^a, and a symmetric tensor field h_{ab} of signature $(0, +, +, +)$ satisfying $h_{ab}u^b = 0$, with background metric g_{ab}. The equations are

$$u^m\nabla_m h_{ab} + 2(\nabla_{(a}u^m)h_{b)m} = 0 \tag{59}$$
$$\rho u^m\nabla_m u^a + q^a{}_b\nabla_m\tau^{bm} = 0 \tag{60}$$
$$u^m\nabla_m\rho + \rho\nabla_m u^m + \tau^{mn}\nabla_m u_n = 0 \tag{61}$$

where we have set $q_{ab} = g_{ab} + u_a u_b$, the 'spatial metric'. Here, τ^{ab} is some fixed algebraic function of h_{ab}, u^a, and g_{ab}, satisfying $\tau^{ab}u_b = 0$ and $\tau^{ab} = \tau^{(ab)}$. The physical meaning of these equations is the following. The field ρ is the mass density, and u^a the material four-velocity. The field h_{ab} represents a sort of 'natural spatial geometry' for the material. Thus, we interpret the combination $h_{ab} - q_{ab}$ (natural geometry minus actual geometry) as the strain on the material; and (59) (which is just $\mathcal{L}_u h_{ab} = 0$) as requiring that the material carry along with it its natural geometry. The field τ^{ab} represents the stress of the material. Thus, we interpret $\tau^{ab}(h_{ab})$ as the stress-strain relation[15], and the combination $\rho u^a u^b + \tau^{ab}$ as the material stress-energy. (60), (61) are precisely conservation of this stress-energy.

---

[14]The notation is a little awkward here. The index on the left is in the space of equations; those on the right, in spinor space.

[15]Note that the stress-strain relation need not be linear. It would perhaps be natural to require that τ^{ab} vanish when $h_{ab} = q_{ab}$, but that requirement is not needed for what follows.

This system admits a hyperbolization if and only if $\rho > 0$; and in addition the tensor $\tau^{abcd} = \partial \tau^{ab}/\partial h_{bc}$ is symmetric under interchange of the index pairs a,b and c,d and positive-definite in these pairs. In this case, the most general hyperbolization is given by

$$h^A{}_{\alpha'}\delta\phi^{\alpha'} = x\left[2\delta\rho - \tilde{h}^{cm}(\rho q^d{}_m + \tau^d{}_m)\delta h_{cd}\right]\left(-\tilde{h}^{an}(\rho q^b{}_n + \tau^b{}_n),\ 0,\ 2\right)$$
$$+ y\left(\tau^{abcd}\delta h_{cd},\ 2h^{am}\delta u_m,\ 0\right) \quad (62)$$

where x and y are numbers with $xy > 0$, and $\tilde{h}^{ab} = \tilde{h}^{(ab)}$ is defined by $u_b \tilde{h}^{ab} = 0$ and $\tilde{h}^{am}h_{bm} = q^a{}_b$. These hyperbolizations are all causal if and only if

$$h_{mn}\tau^{ambn} \leq \frac{1}{2}\rho q^{ab} \quad (63)$$

which means, physically, that no acoustic wave-speed exceed the speed of light. This system has no constraints. If we generalize this system to allow the stress τ^{ab} to depend, not only on the natural geometry h_{ab}, but also on the mass density ρ, then the equations (59)–(61) never admit a hyperbolization. It seems peculiar that such a physically benign generalization would preclude a hyperbolization. What is going on here?

Presumably, the most general allowed interaction is given by

$$j_A = (0,\ j_a,\ j). \quad (64)$$

That is, interactions are allowed that exchange energy-momentum with the environment, but not that modify Lie transport of the natural geometry.

Special Relativity

The fields consist of a symmetric, Lorentz-signature metric g_{ab}, together with a derivative operator ∇_a, on M. The equations are

$$\nabla_a g_{bc} = 0 \quad (65)$$

$$R_{abc}{}^d = 0 \quad (66)$$

where $R_{abc}{}^d$ is the curvature tensor, given by (51).

The bundle space b is this case is identical to that for general relativity, and so in particular the fibres have dimension fifty. But now the equations are different. A typical vector in the space of equations is $\sigma^A = (s^{abc}, s^{abc}{}_d)$ (respective coefficients of (65), (66)), where $s^{abc} = s^{a(bc)}$, $s^{abc}{}_d = s^{[ab]c}{}_d$, and $s^{[abc]}{}_d = 0$. Thus, the equation-space has dimension 120 ($= 40 + 80$).

This system consists of two separate physical systems (in the sense of Section 6). The first, with field the metric, is identical to the similar system for general relativity. Thus, the hyperbolizations are given by (52), the constraints by (53). For the second system, the field is the derivative operator, and the equation (66). This system has no hyperbolizations, because of the diffeomorphism freedom. The most general constraint for this system is given by

$$c^{An} = x^{nabc}{}_d \quad (67)$$

with $x^{nabc}{}_d = x^{[nab]c}{}_d$ and $x^{[nabc]}{}_d = 0$. Thus, the vector space of constraints has dimension sixty, while the space of vectors of the form $c^{An}n_n$, for fixed n_n, has dimension forty-eight[16] The constraints are integrable, but not complete.

Apparently, no interactions whatever are permitted in (65), (66). Note that passage to general relativity is not 'turning on an interaction', because this is a change in the dynamical part of (66).

Dust

The fields consist of a scalar field, ρ, and a unit timelike vector field, u^a, on M, with background metric g_{ab}. The equations are

$$u^m \nabla_m u^a = 0 \qquad (68)$$
$$u^m \nabla_m \rho + \rho \nabla_m u^m = 0. \qquad (69)$$

Thus, the fibres of the bundle b are four-dimensional, a typical vertical vector being given by $\delta\phi^{\alpha'} = (\delta u^a, \delta\rho)$, with $u_a \delta u^a = 0$. A typical vector in the space of equations is $\sigma^A = (s_a, s)$ (respective coefficients of (68), (69)), with $s_a u^a = 0$. The equation-space is four-dimensional.

Remarkably enough, this system admits no hyperbolization. To see this, note that the most general candidate for a hyperbolization at a point is

$$h^A{}_{\alpha'} \delta\phi^{\alpha'} = (x_{ab}\delta u^b + x_a \delta\rho, \; y_a \delta u^a + y\delta\rho) \qquad (70)$$

for some tensors x_{ab}, x_a, y_a, and y. Combining this with the $k_A{}^m{}_{\alpha'}$ from (68), (69), we obtain

$$\begin{aligned} h^A{}_{\alpha'} \delta\phi^{\alpha'} k_A{}^m{}_{\beta'} \delta\hat\phi^{\beta'} &= u^m[x_{ab}\delta\hat u^a \delta u^b + x_a \delta\hat u^a \delta\rho + y_a \delta\hat\rho \delta u^a + y\delta\hat\rho\delta\rho] \\ &\quad + \rho\delta\hat u^m[y_a \delta u^a + y\delta\rho]. \end{aligned} \qquad (71)$$

We see that this is symmetric in $\delta\phi^{\alpha'}$ and $\delta\hat\phi^{\beta'}$ if and only if y, y_a, and x_a all vanish, and x_{ab} is symmetric. But these conditions preclude positive-definiteness of (71). This lack of a hyperbolization is discussed briefly at the end of Section 3. There are no constraints.

Any δj_A is, apparently, allowed as an interaction.

Charged Fields

Fix on the manifold M a smooth, antisymmetric tensor field F_{ab} having vanishing curl ($\nabla_{[a} F_{bc]} = 0$). We assume that M is simply connected, and that F has been so chosen

---

[16]Here is a mystery. For this system, the left side of (10) exceeds the right side by eight. But we might have expected, from the fact that (65), (66) have an initial-value formulation up to diffeomorphisms, an excess of four. What is the explanation for this discrepency?

that its integral over every compact 2-surface in M vanishes[17].

We wish to introduce the notion of tensors on M of charge e (some fixed number) and index type $t^{a\cdots}{}_{b\cdots}$ (some fixed arrangement of contravariant and covariant M-indices). To this end, fix a reference point x_o of M, and consider any point x of M. Consider the collection of all pairs, $(A_a, t^{a\cdots}{}_{b\cdots})$, where A_a is a vector field on M with $\nabla_{[a} A_{b]} = F_{ab}$ (existence of which is guaranteed by the assumptions above), and $t^{a\cdots}{}_{b\cdots}$ is a complex tensor at x. Two such pairs, $(A_a, t^{a\cdots}{}_{b\cdots})$, and $(A'_a, t'^{a\cdots}{}_{b\cdots})$, are taken as equivalent if

$$t'^{a\cdots}{}_{b\cdots} = \exp\left(ie \int (A'_m - A_m) ds^m\right) \; t^{a\cdots}{}_{b\cdots} \qquad (72)$$

where the integral on the right is over any curve from x_o to x. (Independence of the choice of curve is guaranteed by the conditions, above, on A_a and A'_a.) This is an equivalence relation. The equivalence classes, called *charge-e tensors* at x, form a vector space of (complex) dimension 4^s, where s is the total number of indices of $t^{a\cdots}{}_{b\cdots}$. A charge-e tensor field is an assignment, to each point $x \in M$, of a charge-e tensor at x. Thus, a charge-e tensor field can be represented by a pair $(A_a, t^{a\cdots}{}_{b\cdots})$ of fields on M, where pairs $(A_a, t^{a\cdots}{}_{b\cdots})$ and $(A'_a, t'^{a\cdots}{}_{b\cdots})$ are identified provided (72) holds for all x.

The discussion above was predicated on the choice of a fixed reference point, $x_o \in M$. Let us now change to a new reference point, $\tilde{x}_o \in M$. Fix a smooth curve γ from x_o to $\tilde{x}_o$. Then we identify a charge-e tensor at x, defined via reference point x_o, with charge-e tensor at x, defined via reference point $\tilde{x}_o$, provided these have respective representatives, $(A_a, t^{a\cdots}{}_{b\cdots})$ and $(A_a, \tilde{t}^{a\cdots}{}_{b\cdots})$, with $\tilde{t}^{a\cdots}{}_{b\cdots} = \exp(ie \int_\gamma A_m ds^m)\, t^{a\cdots}{}_{b\cdots}$. Note that this is independent of representative. But it does depend on choice of the curve γ: a change in γ changes this identification by an overall phase (the same for each point x). Thus, charged tensor fields (independent of reference point) make sense only up to an overall constant phase.

Now let there be given on M a space-time metric g_{ab}, with corresponding derivative operator ∇_a. We extend the action of this derivative operator to charged fields as follows. Given a charge-e tensor field, choose any representative $(A_a, t^{a\cdots}{}_{b\cdots})$, of it, and let its derivative be the charge-e tensor field with representative $(A_a, (\nabla_c - ieA_c)t^{a\cdots}{}_{b\cdots})$, noting that this is independent of the original choice of representative.

The charged Klein-Gordon system consists of charge-e fields ψ and ψ_a, with background fields the metric g_{ab} and electromagnetic field F_{ab}. The equations are the same as (35)–(37), where, of course, ∇_a now is the derivative operator on charged fields, except that there appears on the right of (36) the term $ie\psi F_{ab}$. (Failure to include this term would destroy integrability of the constraints.) Similarly, the charged spin-s system consists of a charge-e spinor field, $\psi^{A\cdots D}$, satisfying (56).

In order to fit these systems into the present framework, we must fix a reference point x_o, for otherwise the freedom to multiply fields by an overall constant phase makes it impossible to find a bundle to house the charged fields. The discussion of hyperbolizations and constraints then goes through in a manner identical to that of

---

[17] Vanishing of the 2-surface integrals of F means physically that all wormholes manifest zero net magnetic charge. In fact, we could allow here integral multiples of a certain fundamental magnetic charge, but, for simplicity, do not. The assumption of simple-connectedness of M avoids our having to consider Aharonov-Bohm effects. Again, we could relax this assumption, at the cost of somewhat complicating the discussion below.

the uncharged case, with one exception. The constraints for the equation for charged spin-s fields, for $s \geq 1$ and $F_{ab} \neq 0$, fail to be integrable. One further complication arises. There is no natural way to lift diffeomorphisms from M to the bundle spaces b associated with any of these charged fields (although there does, of course, exist a lift 'up to overall constant phase'). The reason for this is our having fixed a reference point, x_o, in order to introduce b.

Kinetic Theory

Let M, g_{ab} be a time-oriented space-time. Fix a nonnegative number m, and denote by Γ the seven-dimensional manifold of all pairs (x, p_a), where $x \in M$, and p_a is a future-directed vector at x with $p_a p^a = -m^2$. Thus, Γ is a fibre bundle over M. Denote by Γ_x the fibre over $x \in M$, so each Γ_x is a three-manifold. The field for kinetic theory is a nonnegative function f on the manifold Γ; and the equation is

$$p^a \nabla_a f = \mathcal{C}(f). \tag{73}$$

Here, $\mathcal{C}$ is, for each $x \in M$, a mapping from nonnegative functions h on Γ_x to functions on Γ_x, satisfying

$$\int \mathcal{C}(h) p^a = 0 \qquad \int \mathcal{C}(h) p^a p^b = 0 \tag{74}$$

for every h, where the integrals are over Γ_x using the natural volume element on this mass shell. Equation (73) is to hold for each $(x, p_a) \in \Gamma$, with $\mathcal{C}$ evaluated on the restriction of f to Γ_x. The physical interpretation of these equations is the following. The nonnegative function f is the distribution function (of particle position-momentum) for mass-m particles, and (73) is Boltzmann's equation. The $\mathcal{C}$ in (73) is the collision function; and (74) is local conservation, in collisions, of particle-number and energy-momentum.

This system does not, strictly speaking, fall within the framework of Section 2, for the space of allowed 'field values' at each point of M (nonnegative functions on Γ_x) is infinite-dimensional. But, if we agree to ignore this one defect, there results a nice example of our framework. A typical vertical vector is given by $\delta\phi^{\alpha'} = \delta f$, a function on Γ_x. A typical vector in the space of equations is $\sigma^A = s$, a function on Γ_x. The most general hyperbolization of (73) is given by

$$h^A{}_{\alpha'} \delta\phi^{\alpha'} = \mu \, \delta f \tag{75}$$

where μ is any positive function on Γ_x. There are no constraints. Presumably, there is allowed as an interaction in (73) any function δj on Γ such that, for every $x \in M$, $\int (\delta j) p^a = 0$, where this integral is over Γ_x. That is, interactions may not violate local particle-number conservation, but are otherwise arbitrary.

Lagrangian Systems

Fix a fibre bundle $\hat{b} \xrightarrow{\hat{\pi}} M$. By a (first-order) *Lagrangian* on the bundle $\hat{b}$, we mean a smooth function L as follows: L is a function on pairs, $(\kappa, \zeta_a{}^\alpha)$, where κ is a point of the bundle space $\hat{b}$, and $\zeta_a{}^\alpha$ is a tensor at κ satisfying $\zeta_a{}^\alpha (\nabla \pi)_\alpha{}^b = \delta_a{}^b$; and, for each such pair, $L(\kappa, \zeta_a{}^\alpha)$ is a density[18] in M at the point $\pi(\kappa)$. Such a Lagrangian gives rise to a system of partial differential equations on cross-sections $\hat\phi$ of the bundle $\hat b$. In order to write these equations explicitly, it is convenient to introduce a derivative operator ∇_α on mixed fields on $\hat b$, such that the derivative of every vertical vector field is vertical, and $\nabla_\alpha(\nabla\pi)_\beta{}^a = 0$. Then, e.g. the operator 'derivative along the cross-section $\hat\phi$' is given by $(\nabla\hat\phi)_a{}^\alpha \nabla_\alpha$. (See the discussion just preceding (11).) Written in terms of this operator, Lagrange's equation becomes[19]

$$\frac{\partial^2 L}{\partial \zeta_m{}^{\alpha'} \partial \zeta_n{}^{\beta'}}(\nabla\hat\phi)_m{}^\mu \nabla_\mu((\nabla\hat\phi)_n{}^\beta) + (\nabla\hat\phi)_m{}^\beta \nabla_\beta(\frac{\partial L}{\partial \zeta_m{}^{\alpha'}}) - \nabla_{\alpha'} L = 0. \qquad (76)$$

The coefficients in (76) are to be evaluated at $\zeta_a{}^\alpha = (\nabla\hat\phi)_a{}^\alpha$. This equation is of course independent of the choice of derivative operator. It is also unchanged under adding to L any function of the form $(\nabla_\alpha v^a)\zeta_a{}^\alpha$ (a 'total divergence'), where v^a is any M-density field on $\hat b$. Note that the spaces of equations and unknowns for (76) have the same dimension, namely, that of the fibres of the bundle $\hat b$.

In order to cast (76) into first-order form, we introduce auxiliary field $\zeta_a{}^\alpha$, subject to $\zeta_a{}^\alpha(\nabla\pi)_\alpha{}^b = \delta_a{}^b$; and we supplement (76) with the additional equations

$$(\nabla\hat\phi)_a{}^\alpha = \zeta_a{}^\alpha \qquad (77)$$

$$\zeta_{[b}{}^\beta \nabla_{|\beta|} \zeta_{a]}{}^\alpha = 0. \qquad (78)$$

The system (77), (78), (76) is closely analogous to the Klein-Gordon system, (35)–(37).

The fibres of the bundle b appropriate to this system have dimension $5n$, where n is the dimension of the fibres of $\hat b$. A typical vertical vector in the bundle b is given by $\delta\phi^{\alpha'} = (\delta\hat\phi^{\alpha'}, \delta\zeta_a{}^{\alpha'})$. Here, $\delta\hat\phi^{\alpha'}$, a vertical vector in $\hat b$, represents an infinitesimal change in the value of the cross-section $\hat\phi$ of $\hat b$, while $\delta\zeta_a{}^{\alpha'}$ represents an infinitesimal change in the tensor $\zeta_a{}^\alpha$. A typical vector in the space of equations is $\sigma^A = (s^a{}_{\alpha'}, s^{ab}{}_{\alpha'}, s^{\alpha'})$, with[20] $s^{ab}{}_{\alpha'} = s^{[ab]}{}_{\alpha'}$ (respective coefficients of (77), (78), (76)) Thus, the equation-space has dimension $11n$ ($= 4n + 6n + n$).

Set

$$S^{ab}{}_{\alpha'\beta'} = \frac{\partial^2 L}{\partial \zeta_{(a}{}^{\alpha'} \partial \zeta_{b)}{}^{\beta'}} \qquad (79)$$

the coefficient of the first term in (76). The system (76)–(78) admits a hyperbolization at a point if and only if the tensor $S^{ab}{}_{\alpha'\beta'} v^{\alpha'} v^{\beta'}$ is Lorentz-signature for every nonzero $v^{\alpha'}$, and, in addition, there exists a tangent vector t^a and a covector n_a that are both timelike for every one of these Lorentz metrics. Note that this is a rather severe condition on $S^{ab}{}_{\alpha'\beta'}$. When it is satisfied, the most general hyperbolization is given by

$$h^A{}_{\alpha'} \delta\phi^{\alpha'} = (-w^a{}_{\alpha'\beta'}, -t^{[a} S^{b]m}{}_{\alpha'\beta'} \delta\zeta_m{}^{\beta'}, \tfrac{1}{2} t^m \delta\zeta_m{}^{\alpha'}) \qquad (80)$$

where t^a is such a common timelike vector, and $w^a{}_{\alpha'\beta'} = w^a{}_{(\alpha'\beta')}$ has the property that $(t^b n_b) n_a w^a{}_{\alpha'\beta'}$ is positive-definite for some such common timelike covector n_a. (Such a

---

[18]That is, an antisymmetric, fourth-rank M-tensor, whose indices we shall suppress.

[19]Note that the Greek index of '$\partial/\partial \zeta_m{}^{\alpha'}$' acquires a prime, as a consequence of the fact that $\zeta_a{}^\alpha(\nabla\pi)_\alpha{}^b = \delta_a{}^b$.

[20]Note that the difference between the two sides of (77), as well as the left side of (78), is automatically vertical. Thus, a prime must be attached to the Greek subscripts of $s^a{}_{\alpha'}$ and $s^{ab}{}_{\alpha'}$.

$w^a{}_{\alpha'\beta'}$ always exists, e.g. that given by $t^a G_{\alpha'\beta'}$ with $G_{\alpha'\beta'}$ positive-definite.) Given a space-time metric g_{ab}, this system is causal if and only if every n_a lying in one half of the light cone of g_{ab} will serve as a 'common timelike covector' above. Equation (80) should be compared with (38), giving the Klein-Gordon hyperbolizations. The most general constraint for the system (76)–(78) is given by

$$c^{An} = (x^{na}{}_{\alpha'},\ y^{nab}{}_{\alpha'},\ 0) \qquad (81)$$

where $x^{na}{}_{\alpha'} = x^{[na]}{}_{\alpha'}$ and $y^{nab}{}_{\alpha'} = y^{[nab]}{}_{\alpha'}$. Thus, the constraints form a vector space of dimension $10n$, while the space of vectors of the form $c^{An}n_n$, for fixed n_n, has dimension $6n$. Equation (81) should be compared with (39), giving the Klein-Gordon constraints. These constraints are complete and integrable.

This system only allows those interactions that preserve its Lagrangian character, i.e. that result from some change in the Lagrangian. This change must be so chosen to leave invariant the tensor $S^{ab}{}_{\alpha'\beta'}$ of (79) (this being the coefficient of the dynamical term in (76)). The most general such change is given by $\delta L = W^m{}_\alpha \zeta_m{}^\alpha + W$, where $W^m{}_\alpha$ and W are arbitrary smooth fields on the bundle space $\hat{b}$. That is, these W's are allowed to depend on the κ-variables, but not on the $\zeta_a{}^\alpha$. This change in the Lagrangian L results in a term

$$\delta j_A = (0,\ 0,\ 2(\nabla_{[\alpha'} W_{\beta]}{}^m)\zeta_m{}^\beta - \nabla_{\alpha'} W) \qquad (82)$$

in (2). Thus, the allowed interactions on a Lagrangian system are very special indeed. In particular, the δj_A can be at most linear in the field $\zeta_a{}^\alpha$.

One example to which the discussion above can be applied is the Klein-Gordon system, with Lagrangian $L = \frac{1}{2}(\nabla_a \psi)(\nabla^a \psi)$. But there exists an alternative Lagrangian formulation for this system, starting from $L = -\frac{1}{2}\psi_a \psi^a + \psi^a \nabla_a \psi$. This alternative Lagrangian involves the full set of Klein-Gordon fields, ψ, ψ_a (as opposed to just ψ), and yields equations that are automatically first-order (as opposed to second-order). It turns out that such an alternative Lagrangian formulation is available quite generally. We sketch below how this comes about.

Recall that the points of the bundle $\hat{b}$ are denoted κ, and that a Lagrangian on $\hat{b}$ consists of a certain function $L(\kappa,\ \zeta_a{}^\alpha)$. Fix such a Lagrangian. Consider now the bundle b, whose points are pairs $(\kappa,\ \zeta_a{}^\alpha)$. Let us now introduce, on this new bundle b, the following Lagrangian:

$$\tilde{L} = \left(\frac{\partial L}{\partial \zeta_a{}^\alpha}\right)(\nabla \hat{\phi})_a{}^\alpha - L. \qquad (83)$$

The Lagrange equations arising from (83) are precisely (76) and (77), except that in (76) we must replace '$(\nabla \hat{\phi})_a{}^\alpha$' everywhere by '$\zeta_a{}^\alpha$'. Thus, we obtain from the Lagrangian (83) a first-order system from the start. But this system, despite the fact that its spaces of unknowns and equations have the same dimension, admits no hyperbolization. In fact, this system has constraints—essentially those given by the $x^{na}{}_{\alpha'}$ in (81). These constraints are not integrable, but their integrability conditions are (78), which are linear in first derivatives of the fields (as opposed to quadratic, which is what happens in the general case, (11)). So, we may include these integrability conditions with the other equations of our system. The result is (76)–(78), a system that admits a hyperbolization and has complete, integrable constraints. That is, the result is a system with an initial-value formulation.

Higher-Order Systems

It is conceivable that some physical phenomena may be described by higher-order systems of partial differential equations (*e.g.* arising from a Lagrangian of higher order). We describe briefly the conversion of such systems to first-order form, and their resulting initial-value formulation.

Consider a quasilinear, s^{th}-order system of partial differential equations, which we may write in the form

$$k_A{}^{m_1\cdots m_s}{}_{\alpha'}\nabla_{m_1}\cdots\nabla_{m_s}\phi^{\alpha'} + j_A = 0 \tag{84}$$

where $k_A{}^{m_1\cdots m_s}{}_{\alpha'} = k_A{}^{(m_1\cdots m_s)}{}_{\alpha'}$ and j_A are functions of point of M, the field ϕ, and, at most, its first $(s-1)$ derivatives. To achieve this form, we have introduced a connection in the bundle of which ϕ is a cross-section. (The coefficient k in (84), but not j, is independent of this choice.) As an example, Lagrange's equation for a higher-order Lagrangian takes the form (84), with the index 'A' replaced by 'β'', with $k_{\beta'}{}^{m_1\cdots m_s}{}_{\alpha'}$ symmetric in β', α', and with s even. Let us now, in order to achieve a first-order system, introduce auxiliary fields, $\phi_a{}^{\alpha'}$, $\phi_{ab}{}^{\alpha'}$, ..., $\phi_{a_1\cdots a_{s-1}}{}^{\alpha'}$, each symmetric in its Latin indices, and their corresponding equations,

$$\nabla_{(a_1}\phi_{a_2\cdots a_i)}{}^{\alpha'} = \phi_{a_1\cdots a_i}{}^{\alpha'} \tag{85}$$

$$\nabla_{[a}\phi_{a_1]a_2\cdots a_i}{}^{\alpha'} = \mu_{aa_1\cdots a_i}{}^{\alpha'}. \tag{86}$$

Here, $i = 1, \ldots, (s-1)$, and the μ on the right in (86) is a certain function of the lower-order ϕ's. Our first-order system consists of (84) with the derivative-term replaced by '$\nabla_{m_1}\phi_{m_2\cdots m_s}{}^{\alpha'}$', and (85), (86). The constraints for this system are always complete, and, by virtue of the choice of μ in (86), integrable.

When does this system admit a hyperbolization? No simple, general criterion is known, but the following remarks will at least suggest a possible line of attack. First note that the equations on $\phi_{a_1\cdots a_i}{}^{\alpha'}$ for $i < (s-1)$ (namely, (86), for the case $i < (s-1)$, and (87) for $i < (s-1)$) always admit a hyperbolization (in a manner similar to that of the Klein-Gordon system, (35) and (36)). What remains is the field $\phi_{a_1\cdots a_{s-1}}{}^{\alpha'}$, and its equations, (86) (in the case $i = (s-1)$) and (84)). Let $t^A{}_{\alpha'}$ be any tensor such that $t^A{}_{\alpha'}k_A{}^{m_1\cdots m_s}{}_{\beta'}$ is symmetric in α', β', and let $t^{a_1\cdots a_{s-1}}$ be any totally symmetric tensor. Consider the tensor given by

$$\begin{aligned}P_{\alpha'}{}^{m_1\cdots m_{s-1}an_a\cdots n_{s-1}}{}_{\beta'} &= \frac{1}{2}t^A{}_{\alpha'}t^{\{m_1\cdots m_{s-1}}k_A{}^{n_1\cdots n_{s-1}\}a}{}_{\beta'}\\ &\quad -\frac{1}{2}t^A{}_{\alpha'}t^{a\{m_1\cdots m_{s-2}}k_A{}^{m_{s-1}n_1\cdots n_{s-1}\}}{}_{\beta'}\end{aligned} \tag{87}$$

where '$\{m_1\cdots m_{s-1}n_1\cdots n_{s-1}\}$' means 'add all $2(s-1)$ terms that result from cyclic permutations of these indices, and then symmetrize the result over $m_1\cdots m_{s-1}$ and over $n_1\cdots n_{s-1}$'. Then, as is easily checked directly, this P has the properties

$$P_{\alpha'}{}^{m_1\cdots m_{s-1}an_1\cdots n_{s-1}}{}_{\beta'} = P_{(\alpha'}{}^{(n_1\cdots n_{s-1})a(m_1\cdots m_{s-1})}{}_{\beta')} \tag{88}$$

$$P_{\alpha'}{}^{m_1\cdots m_{s-1}(an_1\cdots n_{s-1})}{}_{\beta'} = t^{m_1\cdots m_{s-1}}t^A{}_{\alpha'}k_A{}^{n_1\cdots n_{s-1}a}{}_{\beta'}. \tag{89}$$

It follows from (88) that the differential operator $P_{\alpha'}{}^{m_1\cdots m_{s-1}an_1\cdots n_{s-1}}{}_{\beta'}\nabla_a\phi_{n_1\cdots n_{s-1}}{}^{\beta'}$ is automatically symmetric; and, from (89), that this differential operator is a linear combination of the left sides of (86) and (84). Thus, we obtain in this way a symmetrization of our first-order system. This symmetrization is actually a hyperbolization provided we can choose $t^A{}_{\alpha'}$ and $t^{a_1\cdots a_{s-1}}$ such that there exists at each point a covector n_a for which the quadratic form $n_a P_{\alpha'}{}^{m_1\cdots m_{s-1}an_1\cdots n_{s-1}}{}_{\beta'}$ is positive-definite (*i.e.* is positive when applied to any nonzero $\delta\phi_{m_1\cdots m_{s-1}}{}^{\alpha'}$).

When can this positive-definiteness condition be achieved? For $s = 1$ (*i.e.* when (84) is already first-order), this condition just repeats the definition of a hyperbolization of a first-order system. The complete solution for $s = 2$ is given in the discussion of Lagrangian systems in this Appendix. The cases $s > 2$ are considerably more difficult. Those for odd and even s appear to be rather different in character. Are there any simple, general conditions on $k_A{}^{m_1\cdots m_s}{}_{\alpha'}$, for $s > 2$, that imply existence of $t^A{}_{\alpha'}$ and $t^{a_1\cdots a_{s-1}}$ yielding, by the construction above, a hyperbolization? Are there any other symmetrizations of the first-order system (84)–(86)?

It is very likely (depending in part on how the derivative in (85) is structured) that these higher-order fields would, under infinitesimal diffeomorphisms, pick up in (17) higher derivatives of the generating vector field ξ^a. Thus, as discussed in Section 5, such fields may usurp the diffeomorphism gauge group from general relativity.

Appendix B Existence and uniqueness of solutions of symmetric, hyperbolic systems

Fix a smooth, four-dimensional manifold M, and a finite-dimensional vector space F (tensors over which will be denoted, respectively, by Latin and Greek indices). We are interested in F-valued functions on M, $M \xrightarrow{\phi} F$. Consider, on such functions, the partial differential equation

$$k_\alpha{}^m{}_\beta(x,\phi^\gamma(x))\nabla_m\phi^\beta + j_\alpha(x,\phi^\gamma(x)) = 0 \qquad (90)$$

where $x \in M$, and $k_\alpha{}^m{}_\beta = k_{(\alpha}{}^m{}_{\beta)}$ and j_α are smooth functions on $M \times F$. This is just (2), modified as follows: (i) we have written the bundle space as a product, and imposed a vector-space structure on the fibres (which can always be done locally); (ii) we have chosen the gauge so that the last index of k is vertical, and then, since now all Greek indices are vertical, have suppressed all primes; and (iii) we have multiplied (2) through by a suitable hyperbolization, as in (6). By *initial data* for (90), we mean a three-dimensional submanifold S of M, together with a smooth mapping $S \xrightarrow{\phi_0} F$, such that, for every point x of S, the tensor $n_m\, k_\alpha{}^m{}_\beta(x,\phi_0(x))$ is positive-definite, where n_m is a normal to S at x. (This agrees with the definition of Section 3.)

The fundamental theorem on existence and uniqueness of solutions of symmetric, hyperbolic partial differential equations may now be stated as follows (John 1982, Lax 1955, Friedrichs 1954).

Theorem *Let $S \xrightarrow{\phi_0} F$ be initial data for (90). Then:*

(a) *For some open neighborhood U of S, there exists a smooth solution, $U \xrightarrow{\phi} F$, of (90), with $\phi|_S = \phi_0$.*

(b) *For $U \xrightarrow{\phi} F$ and $U' \xrightarrow{\phi'} F$ two such solutions (so $\phi|_S = \phi'|_S = \phi_0$), there exists a neighborhood $\hat{U} \subset U \cap U'$ of S in which $\phi = \phi'$.*

(c) *Any $\hat{U} \subset U \cap U'$ will serve for part (b) provided it can be covered by a smooth, one-parameter family, S_t, of three-submanifolds of U such that (i) one of the S_t is S itself, (ii) each S_t coincides with S outside of a compact subset of S, and (iii) for every t, both ϕ and ϕ', restricted to S_t, are initial data.*

Part (a) of this theorem is existence: it guarantees a solution of (90), satisfying the given initial conditions, in some neighborhood of the initial surface S. Part (b) is uniqueness: it guarantees that two solutions, each defined in some neighborhood of S, must coincide in some common subneighborhood, $\hat{U}$. Part (c) strengthens part (b) by guaranteeing a certain minimum 'size' for $\hat{U}$: part (b) holds for any $\hat{U}$ that can be covered by a one-parameter family, S_t, of surfaces that result from deforming S within compact regions, provided each S_t can serve as an initial-data surface[21]. Thus, given $U \xrightarrow{\phi} F$ and $U' \xrightarrow{\phi'} F$, we can generate a $\hat{U}$ that works for part (b) by taking compact deformations of S within the intersection $U \cap U'$, stopping as soon as positive-definiteness of $n_m \, k_\alpha{}^m{}_\beta(x, \phi(x))$ or $n_m \, k_\alpha{}^m{}_\beta(x, \phi'(x))$ fails. It follows in particular from part (c) that the solution at a point of this $\hat{U}$ depends only on the data in a certain compact region of S. This observation is the basis of our discussion of 'signal-propagation directions' in Section 3. We remark that there are also results that strengthen part (a), by guaranteeing a certain minimum 'size' for its neighborhood U, but these are more complicated and less useful than the strengthening of part (b), as given by part (c) above. Of course, solutions of (90) can, and often do, evolve to become singular.

A sketch of the proof of the theorem follows.

Fix an alternating tensor, $\epsilon_{abcd} = \epsilon_{[abcd]}$, on M (to facilitate integration and to allow us to take divergences of vector fields), and a positive-definite metric $G_{\alpha\beta}$ on F (to facilitate taking norms). We now derive an inequality, (92), that we shall use three times in what follows. We first note that, for any fields ϕ^α, ϕ'^α, $\overline{k}_\alpha{}^m{}_\beta = \overline{k}_{(\alpha}{}^m{}_{\beta)}$, $\overline{k}'_\alpha{}^m{}_\beta = \overline{k}'_{(\alpha}{}^m{}_{\beta)}$, $\overline{j}_\alpha$, and $\overline{j}'_\alpha$ on M, we have

$$\nabla_m[(\overline{k}_\alpha{}^m{}_\beta + \overline{k}'_\alpha{}^m{}_\beta)(\phi^\alpha - \phi'^\alpha)(\phi^\beta - \phi'^\beta)]$$
$$= [\nabla_m(\overline{k}_\alpha{}^m{}_\beta + \overline{k}'_\alpha{}^m{}_\beta)](\phi^\alpha - \phi'^\alpha)(\phi^\beta - \phi'^\beta)$$
$$- 2(\phi^\alpha - \phi'^\alpha)[\overline{k}_\alpha{}^m{}_\beta - \overline{k}'_\alpha{}^m{}_\beta]\nabla_m(\phi^\beta + \phi'^\beta)$$
$$- 4(\phi^\alpha - \phi'^\alpha)[\overline{j}_\alpha - \overline{j}'_\alpha]$$
$$+ 4(\phi^\alpha - \phi'^\alpha)[\overline{k}_\alpha{}^m{}_\beta \nabla_m \phi^\beta + \overline{j}_\alpha - \overline{k}'_\alpha{}^m{}_\beta \nabla_m \phi'^\beta - \overline{j}'_\alpha] \quad (91)$$

---

[21]This $\hat{U}$ lies within the domain of dependence of S, suitably defined. It is merely for convenience that we characterize $\hat{U}$ using deformations of S, rather than curves whose tangents are propagation directions, as defined in Section 3.

Partial Differential Equations of Physics

where we have everywhere suppressed the variable x. To prove (91), expand the left side, and note that all terms cancel. Next, let S_t ($t \in [0, t_o]$) be a smooth family of three-submanifolds of M, each of which coincides with $S = S_0$ outside a compact subset of S, and on each of which $n_m(\overline{k}_\alpha{}^m{}_\beta + \overline{k}'_\alpha{}^m{}_\beta)$ is positive-definite, where n_n is the normal to S_t in the direction of increasing t. Denote by V the union of the S_t, so V has a boundary consisting of S and S_{t_o}. Now multiply (91) by $\exp(-2t/\tau)$, where τ is a positive number, and integrate over V. Integrating the left side by parts, the resulting volume integral involves the expression $(1/\tau)(\nabla_m t)(\overline{k}_\alpha{}^m{}_\beta + \overline{k}'_\alpha{}^m{}_\beta)$. But, by construction, this tensor is positive-definite, and so this volume integral will, provided τ is chosen sufficiently small, dominate the integral of the first term on the right. For the remaining three terms on the right in (91), use the Schwarz inequality. There results

$$\|(\phi - \phi')e^{-t/\tau}\| \leq \sqrt{\sigma\tau}\left(\int_S (\overline{k}_\alpha{}^m{}_\beta + \overline{k}'_\alpha{}^m{}_\beta)(\phi^\alpha - \phi'^\alpha)(\phi^\beta - \phi'^\beta)\epsilon_{mabc}dS^{abc}\right)^{1/2}$$
$$+2\sigma\tau\|(\overline{k}-\overline{k}')\nabla(\phi+\phi')e^{-t/\tau}\| + 4\sigma\tau\|(\overline{j}-\overline{j}')e^{-t/\tau}\|$$
$$+4\sigma\tau\|(\overline{k}\nabla\phi+\overline{j})e^{-t/\tau}\| + 4\sigma\tau\|(\overline{k}'\nabla\phi'+\overline{j}')e^{-t/\tau}\| \qquad (92)$$

where we have suppressed indices. The first term on the right in (92) is a surface term arising from the integration by parts (the surface term at S_{t_o} having been absorbed into the inequality). In (92), $\|\ \|$ means 'the square root of the integral of the square of the indicated field over V' (i.e. the L²-norm), and τ and σ are positive constants such that

$$\tau\nabla_m(\overline{k}_\alpha{}^m{}_\beta + \overline{k}'_\alpha{}^m{}_\beta) \leq (\nabla_m t)(\overline{k}_\alpha{}^m{}_\beta + \overline{k}'_\alpha{}^m{}_\beta) \geq \frac{1}{\sigma}G_{\alpha\beta} \qquad (93)$$

everywhere in V. (Note that such constants exist, by compactness of V and positive-definiteness of the middle expression in (93).) The inequality (92) is our final result. It asserts that two fields, ϕ and ϕ', are close to each other in V (left side of (92)) provided that they are close on the initial surface S (first term on the right), that each approximately satisfies an equation of the form (90) (fourth and fifth terms on the right), and that their respective coefficients, $(\overline{k}, \overline{j})$ and $(\overline{k}', \overline{j}')$, in this equation are close (second and third terms on the right). It is in this derivation of the inequality (92) that we make crucial use of the symmetric, positive-definite character of the coefficients $k_\alpha{}^m{}_\beta$, and the geometrical conditions of part (c) of the theorem.

We first prove uniqueness (parts (b) and (c) of the theorem). Let ϕ and ϕ' be two solutions, as in part (b), and let $\hat{U} \subset U \cap U'$ and the family S_t be as in part (c). Apply inequality (92), with $\overline{k}_\alpha{}^m{}_\beta = k_\alpha{}^m{}_\beta(x, \phi(x))$, $\overline{k}'_\alpha{}^m{}_\beta = k_\alpha{}^m{}_\beta(x, \phi'(x))$, $\overline{j}_\alpha = j_\alpha(x, \phi(x))$, and $\overline{j}'_\alpha = j_\alpha(x, \phi'(x))$. Then the first term on the right vanishes (by initial conditions), and the last two terms on the right vanish (by (90)). But, for the remaining two terms, each of $|k(\phi) - k(\phi')|$ and $|j(\phi) - j(\phi')|$ is bounded by a multiple (namely, the least upper bound of $|\partial k/\partial \phi|$ and $|\partial j/\partial \phi|$, respectively) of $|\phi - \phi'|$. So, choosing τ sufficiently small, the sum of these two remaining terms on the right of (92) is less than the left side of this inequality. We thus conclude that $\|(\phi - \phi')e^{-t/\tau}\| = 0$, and so that $\phi = \phi'$.

We next turn to existence. This is carried out in two steps.

Consider first the equation

$$\overline{k}_\alpha{}^m{}_\beta \nabla_m \phi^\beta + \overline{j}_\alpha = 0 \qquad (94)$$

with fixed fields $\bar{k}_\alpha{}^m{}_\beta(x)$, $\bar{j}_\alpha(x)$. This is just (90), but with the coefficients evaluated on a fixed background field. Fix a family S_t of surfaces, and a region V, as in the derivation of (92). Fix also a field $S \xrightarrow{\phi_0} F$ on S, and a positive number ϵ. We wish to show existence of a smooth ϕ in V, with $\phi|_S = \phi_0$, that is an 'ϵ-approximate' solution of (94), i.e. that is such that the square root of the integral over V of the square of the left side is less than or equal to ϵ. We may set $\phi_0 = 0$ (by replacing ϕ by $\phi + \hat{\phi}$ in (94), where $\hat{\phi}|_S = \phi_0$, and then absorbing into $\bar{j}_\alpha$ the extra term, $\bar{k}_\alpha{}^m{}_\beta \nabla_m \hat{\phi}^\beta$, thus created). Then, to show existence of an ϵ-approximate solution of (94), it suffices to show that fields of the form $\bar{k}_\alpha{}^m{}_\beta \nabla_m \phi^\beta$, with ϕ smooth and vanishing on S, are L^2-dense in V. But for this, in turn, it suffices to show that given any square-integrable field ψ^α on V, such that

$$\int_V \psi^\alpha \bar{k}_\alpha{}^m{}_\beta \nabla_m \phi^\beta = 0 \tag{95}$$

for every smooth ϕ vanishing on S, we must have $\psi^\alpha = 0$. So, fix such a ψ^α. Were this ψ smooth, then we could proceed as follows. Integrate the left side of (95) by parts. First choosing ϕ to have support in the interior of V, we obtain (from the volume integral) that $\nabla_m(\bar{k}_\alpha{}^m{}_\beta \psi^\beta) = 0$; and then choosing ϕ to be nonzero on S_{t_o}, we obtain (from the surface term) that $\psi^\alpha|_{S_{t_o}} = 0$. But these two together imply that $\psi^\alpha = 0$. Indeed, setting $\bar{k}'_\alpha{}^m{}_\beta = \bar{k}_\alpha{}^m{}_\beta$, $\bar{j}_\alpha = (\nabla_m \bar{k}_\alpha{}^m{}_\beta)\psi^\beta$, $\bar{j}'_\alpha = 0$, $\phi = \psi$, $\phi' = 0$ in (92), we obtain

$$\|\psi e^{-t/\tau}\| \leq \sqrt{\sigma \tau} \int_{S_{t_o}} \bar{k}_\alpha{}^m{}_\beta \psi^\alpha \psi^\beta \epsilon_{mabac} \, dS^{abc}$$
$$+ 4\sigma\tau \|(\nabla \bar{k})\psi e^{-t/\tau}\| + 4\sigma\tau \|\nabla(\bar{k}\psi) e^{-t/\tau}\| \tag{96}$$

while, for τ sufficiently small, the second term on the right is less than or equal to the left side. But, unfortunately, ψ^α need not be smooth, but only square-integrable. We therefore proceed as follows. Let h be a nonnegative, smooth, symmetric, two-point function on M, such that $h(x,y)$ vanishes for x and y sufficiently separated, and, for each fixed y, the integral of $h(x,y)$ over M has value one. Set $\hat{\psi}(x) = \int_{M_y} h(x,y)\psi(y)$, where the integration variable is $y \in M$. This is a smooth approximation to ψ. Now take a sequence of such h's such that 'h, together with its derivative, approaches a delta-function and its derivative', in the sense that

$$\int_{M_y} h(x,y) f(y) \to f(x), \qquad \nabla \int_{M_y} h(x,y) f(y) \to \nabla f(x) \tag{97}$$

for every smooth function f on M. Then the left side of (95), with ψ replaced by $\hat{\psi}$, approaches zero, and so, it can be shown, each of the first and third terms on the right in (96), with ψ replaced by $\hat{\psi}$, approaches zero. It now follows from (96) (again, choosing τ sufficiently small) that $\|\hat{\psi} e^{-t/\tau}\|$ approaches zero, and so that $\psi^\alpha = 0$. Thus, we have shown existence of an ϵ-approximate solution of (94), in a certain region V, with given initial data.

We now adopt an iterative procedure. Begin with any smooth field ϕ_1 on V, satisfying the initial condition, $\phi_1|_S = \phi_0$. Choose $\epsilon_2 > 0$, set $\bar{k}_\alpha{}^m{}_\beta = k_\alpha{}^m{}_\beta(x, \phi_1(x))$, $\bar{j}_\alpha = j_\alpha(x, \phi_1(x))$ in (94), and find an ϵ_2-approximate solution, ϕ_2, of that equation satisfying the initial condition. Then choose $\epsilon_3 > 0$, set $\bar{k} = k(\phi_2)$, $\bar{j} = j(\phi_2)$ in (94), and find an ϵ_3-approximate solution, ϕ_3, of that equation satisfying the initial condition. Continuing in this way, with $\epsilon_i \to 0$, we obtain a sequence of fields, $\phi_1, \phi_2, \ldots$. Each of these satisfies the initial condition, and each approximately satisfies (94) (better, as i increases) with its predecessor as background. We wish to show that the ϕ_i converge to the desired solution. To this end, set $\phi = \phi_i$, $\phi' = \phi_{i-1}$, $\overline{k} = k(\phi_{i-1})$, $\overline{k}' = k(\phi_{i-2})$, $\overline{j} = j(\phi_{i-1})$, $\overline{j}' = j(\phi_{i-2})$ in (92), to obtain

$$\begin{aligned}\|\phi_i - \phi_{i-1}\| &\leq \sigma \tau e^{t_0/\tau}(2\|(k(\phi_{i-1}) - k(\phi_{i-2}))\nabla(\phi_i + \phi_{i-1})\| \\ &\quad + 4\|j(\phi_{i-1}) - j(\phi_{i-2})\| + 4(\epsilon_i + \epsilon_{i-1})) \\ &\leq \sigma \tau e^{t_0/\tau}(\alpha\|\phi_{i-1} - \phi_{i-2}\| + 4(\epsilon_i + \epsilon_{i-1}))\end{aligned} \quad (98)$$

where we used $e^{-t_0/\tau} \leq e^{-t/\tau} \leq 1$ in the first step; and set

$$\alpha = 2\,\mathrm{lub}\left|\frac{\partial k}{\partial \phi}\right|\,\mathrm{lub}\,|\nabla(\phi_i + \phi_{i-1})| + 4\,\mathrm{lub}\left|\frac{\partial j}{\partial \phi}\right|$$

in the second. Suppose for a moment that ϕ_i and $\nabla \phi_i$ were uniformly bounded (*i.e.* there is a single constant that bounds all the ϕ_i throughout V, and similarly for $\nabla \phi_i$). Then σ (via (93)) and α (above) would remain bounded as i increases, and so, by choosing τ, t_o, and the ϵ_i sufficiently small[22] in (98), we would guarantee convergence of $\sum \|\phi_i - \phi_{i-1}\|$, and so convergence of the ϕ_i to some field ϕ on V. But, unfortunately, we cannot, at this stage, guarantee uniform boundedness of even the ϕ_i, much less of their derivatives. Indeed, all we control is a certain average of the ϕ_i, represented by the left side of (98). We therefore proceed as follows. Taking the x-derivative of (90), and introducing a new field $\phi_a{}^\alpha(x)$ to represent '$\nabla\phi$', we obtain

$$k_\alpha{}^m{}_\beta \nabla_m(\phi_a{}^\beta) + \phi_m{}^\beta(\nabla_a k_\alpha{}^m{}_\beta + \phi_a{}^\gamma \frac{\partial}{\partial \phi^\gamma} k_\alpha{}^m{}_\beta) + \nabla_a j_\alpha + \phi_a{}^\gamma \frac{\partial}{\partial \phi^\gamma} j_\alpha = 0 \quad (99)$$

The combination of (90) and (99) is a quasilinear, first-order system[23] of partial differential equations for the fields ϕ^α, $\phi_a{}^\alpha$. Furthermore, it inherits from (90) its hyperbolicity (since the coefficient of the derivative-term in (99) is the same k as in (90)). Now continue in this way, taking successive derivatives of (90), introducing successive fields, $\phi_{a\cdots c}{}^\alpha$, to represent the higher derivatives of ϕ^α, and obtaining successively larger hyperbolic systems. Consider the system that results after taking the fourth derivative—so the fields are now $\phi^\alpha, \phi_a{}^\alpha, \phi_{ab}{}^\alpha, \phi_{abc}{}^\alpha, \phi_{abcd}{}^\alpha$ and the equations (90) and its first four derivatives; now apply to this entire system the iterative procedure above. Then the left side of (98) will include a term $\|\phi_{i\,abcd}{}^\alpha - \phi_{i-1\,abcd}{}^\alpha\|$. We now apply a Sobolev inequality (Adams 1975), which asserts that $\|\nabla\nabla\nabla\phi\|$ provides a uniform bound on ϕ and its first two[24] derivatives. From this, combined with (98), it can be shown that $\phi_i{}^\alpha, \phi_{i\,a}{}^\alpha, \phi_{i\,ab}{}^\alpha, \phi_{i\,abc}{}^\alpha, \phi_{i\,abcd}{}^\alpha$ all converge, to some fields $\phi^\alpha, \phi_a{}^\alpha, \phi_{ab}{}^\alpha, \phi_{abc}{}^\alpha, \phi_{abcd}{}^\alpha$. The ϕ^α that results from this procedure is the desired solution of (90). To show that

---

[22] It is here that, by having to choose t_o sufficiently small, we restrict the size of the neighborhood U for part (a) of the theorem. (In the large-t region of V, the ϕ_i could fail to converge, indicating that there would be the final solution would become singular.) This is the starting point for deriving a strengthening of part (a) guaranteeing a minimum size to its neighborhood U.

[23] For initial data on S, we take, for ϕ^α, the given ϕ_0 and, for $\phi_a{}^\alpha$, '$\nabla_a \phi^\alpha$', as computed from ϕ_0 and (90) evaluated on S.

[24] We need, at this point, a uniform bound on the second derivative of ϕ because there appears in our equations, by this point, a term '$(\nabla\nabla\phi)(\nabla\nabla\phi)$', and we have on $\nabla\nabla\nabla\phi$ only an L^2 bound.

this ϕ^α is smooth, take still higher derivatives of (90), and proceed as above, introducing as new fields still higher derivatives of ϕ, and applying the iterative procedure above to the resulting hyperbolic system. One must check that, in the demonstration of existence for this succession of hyperbolic systems, the number t_o (which governs the size of the region V) can remain bounded away from zero. There results convergence of $\phi_{i\,a\cdots c}{}^\alpha = \nabla_a \cdots \nabla_c \phi_i{}^\alpha$, and so smoothness of $\phi^\alpha = \lim \phi_i{}^\alpha$.

Acknowledgement

I wish to thank Vivek Iyer for numerous discussions on these topics.

References

Adams R A, 1975, *Sobolev Spaces*, Academic Press.
Choquet-Bruhat Y and York J W, 1996(*to appear*), *CR Acad Sci Paris*.
Friedrichs K O, 1954, *Comm Pure and Applied Maths* **7** 345.
Fritelli S and Ruela O, 1994, *Comm Math Phys* **266** 221.
Geroch R, 1995, *J Math Phys* **36** 4226.
Geroch R and Lindblom L, 1991, *Ann Phys (NY)* **207** 394.
John F, 1982, *Partial Differential Equations*, Springer-Verlag.
Lax P D, 1955, *Comm Pure and Appl Maths* **8** 615.
Liu I S, Muller I and Ruggeri T, 1986, *Ann Phys (NY)* **169** 191.
Steenrod, 1954, *The Topology of Fibre Bundles*, Princeton University Press, Princeton.

Gravitostatics and Rotating Bodies

Gernot Neugebauer

Research Unit, Theory of Gravitation
Max Planck Society, Jena

1 Introduction

Good textbooks on Electrodynamics such as Jackson's 'Classical Electrodynamics' start with four or five chapters on electrostatics and magnetostatics devoted to the time-independent fields of stationary sources. The main topic is the mathematical and physical discussion of the linear Poisson equation and, especially, the solution of the classical boundary value problems for that equation. No doubt, a corresponding procedure would be just as desirable for systematic representations of gravity. However, gravitostatics including stationary phenomena such as rotating bodies is a nonlinear theory from the very beginning. For a long time, no mathematical algorithm for a systematic treatment of its basic equations was known. The situation changed in 1978, when several authors (Maison 1978, Belinski and Zakharov 1978, Harrison 1978, Herlt 1978, Hoenselaers et al. 1979, Neugebauer 1979, 1980, Hauser and Ernst 1979, 1980, Alekseev 1980) discovered that Einstein's vacuum equations for stationary and axially symmetric gravitational fields can be 'linearized' in terms of a 'Linear Problem' and solved by means of the so-called Inverse Scattering Method. This method even turned out to be the suitable instrument for tackling boundary value problems. Thus, the future book writers are equipped with some mathematical tools to formulate the chapters on gravitostatics. The following three lectures are meant to discuss some aspects of this project. In particular, we intend

(i) to reformulate Einstein's gravitostatics in terms of 'minimal' surfaces (Section 2),

(ii) to discuss the generation of exact vacuum solutions by means of the Inverse (Scattering) Method (Section 3),

(iii) to solve the boundary value problem for the simplest rotating body (Section 4).

Unfortunately, there is little hope of finishing the desired text book in the near future. As a rule, the material following the chapters on electrostatics deals with time-dependent phenomena, electromagnetic waves and radiation. For some of the corresponding chapters in the theory of gravitation one hardly knows what the headings would be.

2 Gravitostatics

2.1 Introduction and definitions

The most fascinating subject in Gravitostatics is the description of rotating bodies. In particular, the discovery of millisecond pulsars (Backer *et al.* 1982) and the prediction of Black Holes have stimulated the investigation of such objects with strong gravitational fields, for which a general relativistic treatment seems to be desirable. However, rigorous results about rotating bodies are relatively rare in general relativity. Amongst other things this is due to the difficulties with 'free surface problems' already known from Newton's theory of gravitation (the shape of the surface of a rotating self-gravitating fluid ball is a 'compromise' between gravitational and centrifugal forces and not known a priori) and the specific complexity of the differential equations of Einstein's theory. Though there are powerful (soliton) techniques to generate (formal) stationary axisymmetric solutions of the *vacuum* equations, no algorithm for integrating the field equations for the interior of the body is available.

This lecture is meant to review a 'minimal' surface approach to the rotating body problem which leads to a purely geometrical and coordinate-free description for both the interior and exterior region.

The basic structure is a (pseudo-) Riemannian space V_p (p for 'Potential Space', not to be confused with spacetime!) whose 'minimal' surfaces correspond to the axisymmetric stationary solutions of Einstein's field equations for perfect fluids.

In this section we will compile the fundamental equations and symmetries. After that we will develop the 'minimal' surface concept and discuss the fundamental equations for rotating bodies in General Relativity.

Stars and galaxies are complex thermodynamical systems, *i.e.* the energy momentum tensor is rather complicated. Nevertheless, in many cases it is sufficient to model the star matter by a one-component perfect fluid. In this case the set of state variables solely consists of the energy density $\varepsilon(x)$, the pressure $p(x)$, the four-velocity $u_i(x)$ and the gravitational field $g_{ik}(x)$ $(i,k = 1,2,3,4)$. These fields are solutions to the Einstein equations

$$R_{ik} - \frac{1}{2} R\, g_{ik} = -\kappa_0 \left((\varepsilon + p)\, u_i u_k + p g_{ik} \right) \qquad (1)$$

where κ_0 is Einstein's gravitational constant. Here the four-velocity u_i has a norm equal to -1 (we take $c=1$):

$$u_i u^i = -1 \qquad (2)$$

We may assume that the matter distribution and the gravitational field of rotating stars are stationary as well as axially symmetric. This is a crucial point of our analysis.

Though physical intuition tells us that a resting fluid ball should be a sphere (what about resting solids?) and a steadily rotating ball should take the shape of a body of revolution (otherwise it should emit gravitational radiation and cannot be stationary), a rigorous proof for the axisymmetry of compact stationary fluid matter distributions is not yet available. An interesting contribution comes from Lindblom (1976) who was able to show that the equilibrium configuration of a viscous fluid must be axisymmetric. An axisymmetric and stationary metric admits a 2-dimensional Abelian group of motions G_2,

$$\xi_{i;k} + \xi_{k;i} = 0, \qquad \eta_{i;k} + \eta_{k;i} = 0, \qquad \xi^i{}_{,k}\eta^k - \eta^i{}_{,k}\xi^k = 0 \qquad (3)$$

$$\xi_i \xi^i < 0 \qquad \eta^i \eta_i > 0$$

with Killing vector fields ξ^i and η^i. (A comma denotes the partial derivative and a semicolon the covariant derivative.)

From these equations we conclude that we can choose a coordinate system such that the Killing vectors take the form $\xi^i = \delta^i_4$ and $\eta^i = \delta^i_3$ and the metric tensor $g_{ik}(x)$ is independent of the time coordinate $x^4 = t$ and the space (azimuthal) coordinate $x^3 = \phi$.

The space-like Killing vector η^i generating axial symmetry has closed (compact) trajectories and vanishes on the rotation axis.

To describe a rotational motion, the four-velocity must be a linear combination of the group generators ξ^i and η^i,

$$u^i = e^{-V}(\xi^i + \Omega \eta^i) \qquad (4)$$

where, in general, the coefficients V and Ω depend on the coordinates. Obviously,

$$e^{2V} = -(\xi_i + \Omega \eta_i)(\xi^i + \Omega \eta^i).$$

The conditions (3) and (4) together with the field equations (1) imply (Kundt and Trümper 1966)

$$\varepsilon_{iklm}\eta^i \xi^k \xi^{l;m} = 0 = \varepsilon_{iklm}\xi^i \eta^k \eta^{l;m} \qquad (5)$$

i.e. the space-time of rotating fluid bodies admits 2-spaces orthogonal to the 2-dimensional group orbits formed by the Killing trajectories. (ε_{iklm} is the Levi-Civita tensor. Condition (5) excludes 'coils', where the source has a toroidal topology.) Then, in the adapted coordinate system, the space-time line element can be written in the form (Lewis 1932, Papapetrou 1966)

$$ds^2 = g_{AB}\, dx^A dx^B + g_{\mu\nu}\, dx^\mu dx^\nu$$
$$\xi^i = \delta^i_4 \qquad \eta^i = \delta^i_3 \qquad (6)$$

where the non-vanishing components of the metric tensor g_{AB} ($A, B=1, 2$) and $g_{\mu\nu}$ ($\mu, \nu=3, 4$) only depend on the coordinates (x^1, x^2) which label the points on the 2-spaces orthogonal to the Killing trajectories.

Stationarity and axisymmetry of the star matter distribution means invariance of the interior variables $\varepsilon(x)$, $p(x)$ and $u^i(x)$ under the symmetry group,

$$\varepsilon_{,i}\xi^i = 0 = \varepsilon_{,i}\eta^i, \qquad p_{,i}\xi^i = 0 = p_{,i}\eta^i \qquad (7)$$

$$u^i{}_{,k}\xi^k - \xi^i{}_{,k}u^k = 0 = u^i{}_{,k}\eta^k - \eta^i{}_{,k}u^k. \tag{8}$$

From (3), (4) and (8) one obtains

$$V_{,i}\xi^i = 0 = V_{,i}\eta^i, \qquad \Omega_{,i}\xi^i = 0 = \Omega_{,i}\eta^i \tag{9}$$

which means that V and Ω in the adapted coordinates (6) depend on x^1, x^2 alone. The integrability conditions of the Einstein equations (1),

$$T^{ik}{}_{;k} = 0,$$

i.e. the Euler equations, reduce to

$$-p_{,i} = (\varepsilon + p)\left(V_{,i} + e^{-V}\eta_k u^k \Omega_{,i}\right) \tag{10}$$

or

$$dp = -(\varepsilon + p)(dV + e^{-V}\eta_k u^k\, d\Omega). \tag{11}$$

Hence, ε and p must be functions of V and Ω. For *rigidly* rotating bodies, the angular velocity Ω is a constant. In this case ε must be a function of p, $\varepsilon = \varepsilon(p)$, and this equation of state determines the function $p = p(V)$ (and $\varepsilon = \varepsilon(V)$) via the differential equation

$$\varepsilon(p) + p = -\frac{dp}{dV}. \tag{12}$$

The equation of state must be chosen such that the resulting pressure function has a zero V_0, $p(V_0) = 0$. Then

$$V = V_0 \tag{13}$$

may describe the star surface at which the pressure coincides with the vanishing pressure of the vacuum region. V_0 determines the relative red shift $z = \exp(-V_0) - 1$ of photons emitted from the poles of the rotating star and received by an observer at infinity.

As an illustration consider an incompressible material, $\varepsilon = \varepsilon_0 = $ constant. Here we get from (12)

$$p(V) = \begin{cases} \varepsilon_0(e^{V_0 - V} - 1) & \text{inside} \\ 0 & \text{outside} \end{cases} \tag{14}$$

It should be emphasized that p is always continuous at $V=V_0$ but need not be differentiable there. In our mathematical analysis we consider p as a well-defined function of V from which $\varepsilon(V)$ can be derived via (12).

2.2 'Minimal' Surfaces

We define four invariant gravitational potentials, the Newtonian potential U, the gravitomagnetic potential A, the axis potential W and the superpotential α:

$$e^{2U} = -\xi_i\xi^i, \qquad\qquad A = -e^{-2U}\eta_i\xi^i$$
$$W^2 = (\eta_i\xi^i)^2 - \eta_k\eta^k\xi_i\xi^i, \qquad e^{-2\alpha} = e^{-2U}W_{,i}W^{,i} \tag{15}$$

In terms of these potentials, V takes the form

$$e^{2V} = e^{2U}\left([1 + \Omega A]^2 - \Omega^2 W^2 e^{-4U}\right) \tag{16}$$

and the line element (6) reads (Weyl 1917, Lewis 1932, Papapetrou 1966)

$$ds^2 = e^{-2U}\left(e^{2\alpha}W_{,C}W_{,D}\,h^{CD}h_{AB}\,dx^A dx^B + W^2 d\phi^2\right) - e^{2U}(dt + A\,d\phi)^2 \qquad (17)$$

$$\xi^i = \delta^i_4 \qquad \eta^i = \delta^i_3$$

where U, A, W, α only depend on $x^A (A = 1, 2)$. The function U is the generalization of the Newtonian gravitational potential for weak fields. The function A indicates the rotation of the source and plays the same role for rotating masses as the azimuthal component of the electromagnetic vector potential does for rotating charges. For rotating stars, its order of magnitude compared with the centrifugal potential, $2\Omega A/\Omega^2 R^2$ (where R is the radius of the star), varies between 0.2 (rapidly rotating neutron stars) and 10^{-7} (Sun). In a Minkowski space, $W=\rho$ (with ρ, ζ, ϕ as cylindrical coordinates) measures a distance from the symmetry axis (ζ-axis) of a body of revolution. The term 'super potential' comes from chiral field theory. Obviously, V is the Newtonian potential of a co-rotating observer.

Substituting the metric (17) into the field equations (1) with the four-velocity as in (4), one obtains after a straightforward calculation a set of second order nonlinear (semi-linear) partial differential equations.

Surprisingly, these differential equations turn out to be the Euler-Lagrange equations of a variational principle describing the 'minimal' 2-surfaces Σ,

$$\Sigma: \quad U = U(x^1, x^2), \quad A = A(x^1, x^2), \quad W = W(x^1, x^2), \quad \alpha = \alpha(x^1, x^2), \qquad (18)$$

of a four-dimensional pseudo-Riemannian space V_p with the line element

$$V_p: \quad dS^2 = -2d\alpha\,dW + 2W\,dU^2 - \frac{e^{4U}}{2W}dA^2 - 2\kappa_0 W e^{2\alpha - 2U} p(V)\,dW^2 \qquad (19)$$

V_p is a mathematical construction and should not be confused with space-time.

In addition, it can be shown that the interior metric γ_{AB} ($A, B = 1, 2$) of the 'minimal' surfaces

$$dS_2^{\,2} = \gamma_{AB}\,dx^A dx^B, \qquad dS_2^{\,2} > 0$$

$$\gamma_{AB} = -(\alpha_{,A}W_{,B} + \alpha_{,B}W_{,A}) + 2WU_{,A}U_{,B} - \frac{e^{4U}}{2W}A_{,A}A_{,B}$$

$$\qquad - 2\kappa_0 W e^{2\alpha - 2U} p(V) W_{,A} W_{,B} \qquad (20)$$

is conformally equivalent to the space-time submetric h_{AB}

$$\gamma_{AB} = \lambda\,h_{AB} \qquad (21)$$

so that h_{AB} in (17) can be replaced by γ_{AB},

$$dS^2 = e^{-2U}\left(e^{2\alpha}W_{,C}W_{,D}\gamma^{CD}\gamma_{AB}\,dx^A dx^B + W^2 d\phi^2\right) - e^{2U}(dt + A d\phi)^2 \qquad (22)$$

Thus we have found the following formulation of the gravitostatics of rotating bodies (Neugebauer and Herlt 1984).

The (exterior and interior) axisymmetric and stationary gravitational fields of rigidly rotating bodies are the 'minimal' 2-surfaces in the pseudo-Riemannian Potential Space V_p (19). Their interior 2-metric γ_{AB} (20) is conformally equivalent to the 2-metric of the space-time slices $t = $ constant, $\phi = $ constant in (22).

A generalization of this theorem for nonrigid rotation was given by Kramer (1988). Thus we are led to the following algorithm for the treatment of gravitostatic problems

(i) Choose the equation of state $\varepsilon = \varepsilon(p)$ (the matter model) and calculate $p = p(V)$ from (12), cf. (14).

(ii) Choose a suitable parametrization of the 2-surface Σ (18) and substitute it into the line element (19) to read off the components γ_{AB} (20) of the 2-surface. Note that minimal parametrizations, as for instance $U = U(W,\alpha)$, $A = A(W,\alpha)$, $x^1 = W$, $x^2 = \alpha$, are possible.

(iii) Write down the area $\mathcal{A}$ of Σ,

$$\mathcal{A} = \int_\Sigma d^2x \sqrt{\det \gamma} \qquad (23)$$

where $\det \gamma$ is the determinant of the 2-tensor γ_{AB}, and postulate that

$$\delta \mathcal{A} = 0 \qquad (24)$$

Equation (24) is the condition that Σ be 'minimal'. The quotation marks indicate that we cannot expect surfaces with a really minimal area. Numerical experiments show a 'saddle-point' behaviour of the solutions of the 'stationarity' condition (24).

(iv) Formulate the Euler-Lagrange equations of the variational principle (23,24), for instance

$$\frac{\partial \sqrt{\det \gamma}}{\partial U} - \left(\frac{\partial \sqrt{\det \gamma}}{\partial U_{,A}}\right)_{,A} = 0 \qquad etc. \qquad (25)$$

(v) Find solutions of the Euler-Lagrange equations $U = U(x^1, x^2)$, $A = A(x^1, x^2)$, $W = W(x^1, x^2)$, $\alpha = \alpha(x^1, x^2)$ and insert them and their gradients $U_{,A}$ etc. in the expression (20) to get the interior metric (first fundamental form) γ_{AB} of the 'minimal' surfaces.

(vi) Write down the space-time line element (the gravitational fields) (22).

There is no doubt that step (v) is the true problem. To tackle it one needs well-posed boundary conditions or additional symmetries.

2.3 Two examples: TOV Equations and Ernst equation

Spherical symmetry of a resting body ($\Omega = 0$) leads to considerable simplifications. In this case, the metric (17) contains spherical surfaces

$$d\Omega^2 = d\theta^2 + \sin^2\theta \, d\phi^2 \qquad (0 \leq \theta \leq \frac{\pi}{2},\ 0 \leq \phi \leq 2\pi)$$

and must therefore have the form

$$ds^2 = a(r)\,dr^2 + r^2(d\theta^2 + \sin^2\theta\,d\phi^2) - b(r)\,dt^2. \tag{26}$$

We may choose a parametrization $\alpha = \alpha(x^1, x^2)$, $A = 0$, $U = x^1$, $W = x^2$. Comparing (26) and (22) one reads off

$$e^{-2\alpha} = 1 + W^2 f(U) \tag{27}$$

or, equivalently,

$$\begin{aligned} e^{-2\alpha} &= 1 + B(U)\sin^2\theta \\ W &= A(U)\sin\theta \end{aligned} \tag{28}$$

for the parametrization of the corresponding minimal surfaces. The Euler-Lagrange equations can be reformulated to take the form of a first order system of differential equations ($V \equiv U!$),

$$\begin{aligned} A'(U) &= A\sqrt{\frac{1+B}{B - \kappa_0 A^2 p(U) e^{-2U}}} \\ B'(U) &= \left(4\kappa_0 A^2 p(U)\, e^{-2U} - 2B\right)\sqrt{\frac{1+B}{B - \kappa_0 A^2 p(U)\, e^{-2U}}} \end{aligned} \tag{29}$$

The boundary (regularity) condition on the axis of symmetry ($\theta = 0, \pi$) is satisfied from the beginning. To ensure $W=0$ in the centre ($U = U_c$), which is part of the axis of symmetry $W=0$, and $\alpha=0$ at infinity ($U=0$), where (22) becomes a Minkowski line element in cylindrical coordinates, one has to consider the conditions

$$A(U_c) = 0 \qquad B(0) = 0. \tag{30}$$

The equations (29,30) are the reformulated Tolman-Oppenheimer-Volkov equations for (resting) spherically symmetric star models. For given equations of state ($p(U)$ given!) numerical solutions are easy to find. An analytic solution is known for incompressible fluids ($\varepsilon = $ constant), cf. (14),

$$e^{-2\alpha} = 1 + \frac{1}{12}\kappa_0\varepsilon e^{-4U} W^2 \begin{cases} 1 - \left(3e^{V_0} - 4e^U\right)^2 & \text{inside} \\ \dfrac{\left(1 - e^{2U}\right)^4}{\left(1 - e^{2V_0}\right)^3} & \text{outside} \end{cases} \tag{31}$$

$$A^2 = \begin{cases} \dfrac{3e^{2U}\left[1 - \left(3e^{V_0} - 2e^U\right)^2\right]}{\kappa_0\varepsilon} & \text{inside} \\ \dfrac{3e^{2U}\left(1 - e^{2V_0}\right)^3}{\kappa_0\varepsilon\left(1 - e^{2U}\right)^2} & \text{outside} \end{cases} \tag{32}$$

According to (13), the surface of the star is given by $U = V_0$. Equation (31) is a 'minimal' surface form of the famous (interior and exterior) Schwarzschild solution, whose standard form

$$ds^2 = \begin{cases} \dfrac{dr^2}{1-\frac{1}{3}\kappa_0\varepsilon r^2} + r^2 d\Omega^2 - \frac{1}{4}\left(3e^{V_0} - \sqrt{1-\frac{1}{3}\kappa_0\varepsilon r^2}\right)^2 dt^2 & \text{inside} \\[2ex] \dfrac{dr^2}{1-\frac{2m}{r}} + r^2 d\Omega^2 - \left(1-\frac{2m}{r}\right) dt^2 & \text{outside} \end{cases} \quad (33)$$

with

$$m = \frac{\kappa_0 M}{8\pi} = \frac{\sqrt{3}}{2\sqrt{\kappa_0\varepsilon}}\left(1 - e^{2V_0}\right)^{3/2}, \tag{34}$$

follows from the coordinate transformation

$$r = e^{-U} A(U). \tag{35}$$

Then the coordinate radius r_0 of the star is

$$r_0 = e^{-V_0} A(V_0). \tag{36}$$

The condition (30) and the limitation $e^{V_0} < 1$ ($V_0 < 0$) yield the restriction

$$1 > e^{V_0} \geq \frac{1}{3} \tag{37}$$

and, correspondingly,

$$r_0 > \frac{9}{8} 2m \tag{38}$$

i.e. to exist as a static object the star must conceal its Schwarzschild radius $r = 2m$.

To obtain the field equations for the exterior region of a rotating star ($p = 0$), we choose the Lewis-Weyl parametrization

$$\gamma_{AB} = \lambda \delta_{AB} \qquad (\lambda \text{ a conformal factor}) \tag{39}$$

and denote the coordinates by $x^1 = \rho$, $x^2 = \zeta$. (Note that any two-dimensional line element $dS_2^2 = \gamma_{AB} dx^A dx^B$ with $\det \gamma > 0$ can be transformed into the line element $dS_2^2 = \lambda\left(dx^{1\,2} + dx^{2\,2}\right)$.) Comparing (39) and (20) we have

$$\sqrt{\det \gamma} = \lambda = -\alpha_{,A} W_{,A} + W U_{,A} U_{,A} - \frac{e^{4U}}{4W} A_{,A} A_{,A}$$

and can write down the Euler-Lagrange equations to the Lagrangian $\sqrt{\det \gamma}$, see equations (23) and (25),

$$(WU_{,A})_{,A} = -\frac{e^{4U}}{2W} A_{,A} A_{,A} \tag{40}$$

$$\left(\frac{e^{4U}}{W} A_{,A}\right)_{,A} = 0 \tag{41}$$

$$W_{,A,A} = 0 \tag{42}$$

$$\alpha_{,A,A} = -U_{,A} U_{,A} - \frac{e^{4U}}{4W^2} A_{,A} A_{,A} \tag{43}$$

The third equation has the general solution

$$W = \frac{1}{2}(f(z) + \overline{f(z)}), \qquad z = \rho + i\zeta \tag{44}$$

where $f(z)$ is holomorphic in z (the bar denotes complex conjugation). Preserving the parametrization (39) one can perform a transformation $f(z) \to z$. Therefore

$$W = \rho \tag{45}$$

without loss of generality. We must keep in mind that condition (39), $\gamma_{12}=\gamma_{21}=0$, means

$$\alpha_{,z} = 2\rho(U_{,z})^2 - \frac{e^{4U}}{2\rho}(A_{,z})^2 \tag{46}$$

which implies (43) if equations (40) and (41) hold. Moreover, the integrability condition

$$\alpha_{,z,\bar{z}} = \alpha_{,\bar{z},z} \tag{47}$$

is automatically satisfied if the equations (40) and (41),

$$(\rho U_{,A})_{,A} = -\frac{e^{4U}}{2\rho} A_{,A} A_{,A} \tag{48}$$

$$\left(\frac{e^{4U}}{\rho} A_{,A}\right)_{,A} = 0 \qquad (x^A = \rho, \zeta) \tag{49}$$

are valid. Thus we come to the following conclusion. To calculate the vacuum fields of rotating bodies one has simply to solve the system (48, 49) of two semi-linear partial differential equations. Afterwards one may calculate α from (46) via the path integral

$$\alpha(\rho, \zeta) = \int_{P_0}^{P(\rho,\zeta)} (\alpha_{,\rho'} d\rho' + \alpha_{,\zeta'} d\zeta') \tag{50}$$

where the result $\alpha(\rho, \zeta)$ is independent of the path of integration connecting P_0 and the arbitrary point $P(\rho, \zeta)$. Finally, the space-time line element (22) takes the Weyl-Lewis-Papapetrou form

$$ds^2 = e^{-2U}\left(e^{2\alpha}(d\rho^2 + d\zeta^2) + \rho^2 d\phi^2\right) - e^{2U}(dt + A d\phi)^2 \tag{51}$$

The discussion of the remaining equations (48, 49) becomes more comfortable after the duality rotation

$$\frac{e^{4U}}{\rho} A_{,\rho} = b_{,\zeta}, \qquad \frac{e^{4U}}{\rho} A_{,\zeta} = -b_{,\rho} \tag{52}$$

which automatically satisfies (49) provided that

$$(\rho e^{-4U} b_{,\rho})_{,\rho} + (\rho e^{-4U} b_{,\zeta})_{,\zeta} = 0. \tag{53}$$

Then (48) takes the form

$$(\rho U_{,A})_{,A} = -\frac{\rho e^{-4U}}{2} b_{,A} b_{,A}. \tag{54}$$

Introducing the Ernst function $f = e^{2U} + ib$, we may write the system (53, 54) in the form
$$(\text{Re} f)(\rho f_{,A})_{,A} = \rho f_{,A} f_{,A} \tag{55}$$
or in the more compact form
$$(\text{Re} f)\, \Delta f = (\nabla f)^2, \qquad \frac{\partial f}{\partial \phi} = 0 \tag{56}$$
where Δ is the 3-dimensional Laplacian operator.

The Ernst equation (56) turned out to be important in many areas of physics and mathematics (nonlinear σ-models, theory of instantons, geometry). In the subsequent lectures we will deal exclusively with that equation.

The 'minimal' surface formalism can be used for numerical calculations of rapidly rotating neutron stars (Neugebauer and Herold, Herold and Neugebauer 1992).

3 Inverse Scattering Method

3.1 The Linear Problem

The Ernst equation (56) is the integrability condition of the Linear Problem
$$\begin{aligned}
\Phi_{,z} &= \left\{ \begin{pmatrix} C & 0 \\ 0 & D \end{pmatrix} + \lambda \begin{pmatrix} 0 & C \\ D & 0 \end{pmatrix} \right\} \Phi \\
\Phi_{,\bar{z}} &= \left\{ \begin{pmatrix} \bar{D} & 0 \\ 0 & \bar{C} \end{pmatrix} + \frac{1}{\lambda} \begin{pmatrix} 0 & \bar{D} \\ \bar{C} & 0 \end{pmatrix} \right\} \Phi
\end{aligned} \tag{57}$$

where $\Phi = \Phi(\lambda, z, \bar{z})$ is a 2×2 matrix depending on the spectral parameter
$$\lambda = \sqrt{\frac{K - i\bar{z}}{K + iz}} \qquad (K \text{ a complex constant}) \tag{58}$$

as well as on the complex coordinates $z = \rho + i\zeta$, $\bar{z} = \rho - i\zeta$, whereas C and D are functions of $z, \bar{z}$ (ρ, ζ) alone. Indeed, from $\Phi_{,z\bar{z}} = \Phi_{,\bar{z}z}$ and the formulae
$$\lambda_{,z} = \frac{\lambda}{4\rho}(\lambda^2 - 1), \qquad \lambda_{,\bar{z}} = \frac{1}{4\rho\lambda}(\lambda^2 - 1) \tag{59}$$

it follows that a certain matrix polynomial in λ has to vanish. This yields a set of first order differential equations for C and D. It turns out that a part of those equations has the solution
$$D = \frac{f_{,z}}{f + \bar{f}}, \qquad C = \frac{\bar{f}_{,z}}{f + \bar{f}} \tag{60}$$

Having eliminated C and D in the remaining equations one obtains the (axisymmetric) Ernst equation (55). Vice versa, if f is a solution to the Ernst equation, the matrix Φ calculated from (57) does not depend on the path of integration. The idea of the Inverse Method is to discuss Φ, for fixed values of z and $\bar{z}$, as a holomorphic function of λ and to get C and D afterwards. This is an 'inverse' procedure compared with the 'normal' way which consists of the solution of differential equations with given coefficients. The term 'scattering' comes from the solution theory of the Korteweg-de Vries equation whose Linear Problem has partially the form of the (time-independent) Schrödinger equation. In this case, the calculation of the coefficients is equivalent to the construction of the Schrödinger potential from the scattering data.

We try to find a matrix $\Phi(\lambda, z, \bar{z})$ with the following properties

(a) $$\Phi_{,z}\Phi^{-1} = P + \lambda Q, \qquad \Phi_{,\bar{z}}\Phi^{-1} = R + \frac{1}{\lambda}S$$

(b) $$\Phi = \begin{pmatrix} \psi(\lambda) & \psi(-\lambda) \\ \chi(\lambda) & -\chi(-\lambda) \end{pmatrix}$$

(c) $$\overline{\psi\left(1/\bar{\lambda}\right)} = \chi(\lambda)$$

(d) $$\psi(\lambda = -1, z, \bar{z}) = \chi(\lambda = -1, z, \bar{z}) = 1 \quad \text{in the limit } K \to \infty.$$

Here P, Q, R, S are arbitrary 2×2 matrices which do not depend on λ (or K). Obviously, the conditions (b)–(d) ensure the coincidence of these matrices with the structure prescribed in (57) (diagonal and off-diagonal form, $\bar{R}_{11} = P_{22}$ etc.). Hence, any Φ satisfying (a)–(d) is a solution of (57), as well. Moreover, at $\lambda = 1$, Φ is closely related to the desired Ernst potential f. Indeed, from (60),

$$f = \chi(\lambda = 1, z, \bar{z}). \tag{61}$$

The deciding question is now: How do we find Φ's satisfying the first condition? We will present two examples.

3.2 Bäcklund transformations

Starting with a known 'seed' solution of the Ernst equation f_0 we may calculate from (57) and (60) the corresponding 'seed' matrix Φ_0. Then

$$\Phi = T\Phi_0, \tag{62}$$

where

$$T = K^{-1}(K + iz)^{n/2} \sum_{s=0}^{n} Y_s \lambda^s \tag{63}$$

is again a solution of the Linear Problem (57). The Darboux matrix T consists essentially of a 2×2 matrix polynomial in λ. The transformation (62,63) provides a so-called Bäcklund transformation of any Ernst potential f_0 into a new Ernst potential f ($f_0 \to \Phi_0 \to \Phi \to f = \chi(\lambda = 1, z, \bar{z})$).

The 'critical' points of the Bäcklund transformation (62,63) are the zeros $\lambda = \lambda_i$ ($K = K_i$) of the determinant of T, which is also a polynomial in λ. The condition $\det T(\lambda_i) = 0$ implies that $\det \Phi(\lambda_i) = 0$,

$$\det T(\lambda_i) = 0 \quad \Rightarrow \quad \det \Phi(\lambda_i) = 0 \qquad (i = 1 \ldots 2n) \tag{64}$$

$$\lambda_i^2 = \frac{K_i - i\bar{z}}{K_i + iz}$$

Surprisingly, $\Phi_{,z}\Phi^{-1}$ and $\Phi_{,\bar{z}}\Phi^{-1}$ have no singularities at the zeros λ_i. A detailed discussion shows that $\Phi_{,z}$ has a single pole of first order at $\lambda = \infty$ and this property holds for $\Phi_{,z}\Phi^{-1}$, as well. Hence, from Liouville's theorem, we may conclude that $\Phi_{,z}\Phi^{-1}$ has indeed the structure postulated in (a). Analogously, $\Phi_{,\bar{z}}\Phi^{-1}$ has a pole of first order at $\lambda = 0$. We will now construct Φ in the particular case $n = 2$. The second condition in (a)–(d) prescribes a special structure of the coefficients Y_s, which leads to

$$\Phi = K^{-1}(K + iz)\left[\begin{pmatrix} a & 0 \\ 0 & b \end{pmatrix} + \begin{pmatrix} 0 & c \\ d & 0 \end{pmatrix}\lambda + \begin{pmatrix} k & 0 \\ 0 & l \end{pmatrix}\lambda^2\right]\Phi_0. \tag{65}$$

Because of (64), $\Phi(\lambda_i)$ must have a null eigenvector b_i in the zeros λ_i,

$$\Phi(\lambda_i)b_i = 0, \qquad b_i = \begin{pmatrix} c_i \\ d_i \end{pmatrix}. \tag{66}$$

The Linear Problem (57) tells us that c_i and d_i have to be constants. Introducing the abbreviation

$$B_i = \begin{pmatrix} C_i \\ D_i \end{pmatrix} = \Phi_0 \begin{pmatrix} c_i \\ d_i \end{pmatrix} \tag{67}$$

and using (62), (65), (66), (67) and the second condition in (a)–(d) we obtain the following linear algebraic system:

$$\left[\begin{pmatrix} a & 0 \\ 0 & b \end{pmatrix} + \begin{pmatrix} 0 & c \\ d & 0 \end{pmatrix}\lambda_k + \begin{pmatrix} k & 0 \\ 0 & l \end{pmatrix}\lambda_k^2\right]\begin{pmatrix} C_k \\ D_k \end{pmatrix} = 0 \quad (k = 1, 2)$$

$$\left[\begin{pmatrix} a & 0 \\ 0 & b \end{pmatrix} + \begin{pmatrix} 0 & c \\ d & 0 \end{pmatrix} + \begin{pmatrix} k & 0 \\ 0 & l \end{pmatrix}\right]\begin{pmatrix} 1 \\ -1 \end{pmatrix} = \begin{pmatrix} 1 \\ -1 \end{pmatrix} \tag{68}$$

$$\left[\begin{pmatrix} a & 0 \\ 0 & b \end{pmatrix} + \begin{pmatrix} 0 & c \\ d & 0 \end{pmatrix}\lambda + \begin{pmatrix} k & 0 \\ 0 & l \end{pmatrix}\lambda^2\right]\begin{pmatrix} \psi_0 \\ \chi_0 \end{pmatrix} - \frac{K}{K + iz}\begin{pmatrix} \psi \\ \chi \end{pmatrix} = 0.$$

To get χ we merely need the second rows (there are four linear equations for the four quantities b, d, l and χ). Via Cramer's rule we obtain

$$\chi(\lambda, z, \bar{z}) = K^{-1}(K + iz)\,\chi_0(\lambda, z, \bar{z})\,\frac{\begin{vmatrix} 1 & 1 & 1 \\ \alpha_0 \lambda & \alpha_1 \lambda_1 & \alpha_2 \lambda_2 \\ \lambda^2 & \lambda_1^2 & \lambda_2^2 \end{vmatrix}}{\begin{vmatrix} 1 & 1 & 1 \\ -1 & \alpha_1 \lambda_1 & \alpha_2 \lambda_2 \\ 1 & \lambda_1^2 & \lambda_2^2 \end{vmatrix}} \tag{69}$$

where

$$\alpha_0 = \frac{\psi_0}{\chi_0}, \qquad \alpha_i = \frac{C_i}{D_i} \qquad (i = 1, 2). \tag{70}$$

The third condition in (a)–(d) restricts the K_i, α_i,

$$K_i = \bar{K}_i, \quad \alpha_i \bar{\alpha}_i = 1 \quad \text{or} \quad K_i = \bar{K}_k, \quad \alpha_i \bar{\alpha}_k = 1 \quad (i \neq k). \tag{71}$$

According to (61), we get from (69)

$$f = \frac{\begin{vmatrix} f_0 & 1 & 1 \\ \bar{f}_0 & \alpha_1\lambda_1 & \alpha_2\lambda_2 \\ f_0 & \lambda_1{}^2 & \lambda_2{}^2 \end{vmatrix}}{\begin{vmatrix} 1 & 1 & 1 \\ -1 & \alpha_1\lambda_1 & \alpha_2\lambda_2 \\ 1 & \lambda_1{}^2 & \lambda_2{}^2 \end{vmatrix}}. \tag{72}$$

For $f_0 = 1$ (Minkowski space) the α_i become constants ($\alpha_0 = 1$). This solution of the Ernst equation is known as the Kerr-NUT solution. Thus the Bäcklund transformation (72) can be interpreted as a superposition of a Kerr-NUT solution to any seed solution f_0. To get the 'pure' Kerr solution we have to make the choice

$$\begin{aligned} K_1 &= \bar{K}_1 = -m\cos\phi, & \alpha_1 &= \bar{\alpha}_1^{-1} = -e^{-i\phi} \\ K_2 &= \bar{K}_2 = m\cos\phi, & \alpha_2 &= \bar{\alpha}_2^{-1} = e^{i\phi} \end{aligned} \tag{73}$$

where m and ϕ are real constants. Then (72) can be cast in the form

$$f = \frac{r_1 e^{-i\phi} + r_2 e^{i\phi} - 2m\cos\phi}{r_1 e^{-i\phi} + r_2 e^{i\phi} + 2m\cos\phi} \tag{74}$$

where

$$r_A{}^2 = (K_A - i\bar{z})(K_A + iz) \qquad (A = 1, 2).$$

Equation (74) is the Kerr solution in Weyl-Lewis-Papapetrou coordinates. From this form it can be transformed to Boyer-Lindquist coordinates (r, θ) which are related to $z = \rho + i\zeta$ and $\bar{z}$ by

$$\rho = \sqrt{r^2 - 2mr + a^2}\sin\theta, \qquad \zeta = (r - m)\cos\theta, \qquad a = m\sin\phi$$

Calculating a and α via (46) and (52) one arrives at the Boyer-Lindquist form of the Kerr solution

$$\begin{aligned} ds^2 &= \left(1 - \frac{2mr}{r^2 + a^2\cos^2\theta}\right)^{-1}\left[\left(r^2 - 2mr + a^2\cos^2\theta\right)\right. \\ &\quad \times \left.\left(d\theta^2 + \frac{dr^2}{r^2 - 2mr + a^2}\right) + \left(r^2 - 2mr + a^2\right)\sin^2\theta\, d\phi^2\right] \\ &\quad - \left(1 - \frac{2mr}{r^2 + a^2\cos^2\theta}\right)\left(dt + \frac{2mra\sin^2\theta}{r^2 - 2mr + a^2\cos^2\theta}d\phi\right)^2 \end{aligned} \tag{75}$$

which describes a rotating Black Hole with the mass (parameter) m (the total mass M is $8\pi m/\kappa_0$) and the angular momentum per mass $a = m\sin\phi$. The remarkable effect of the nonlinear Bäcklund transformation is to act as a nonlinear creating operator generating the rotating Black Hole from the Gravitational Vacuum (Minkowski space, $f_0 = 1$).

The general Bäcklund Transformation (n arbitrary but even) $f_0 \to f$ (Neugebauer 1980) is a slight generalization of (72),

$$f = \frac{\begin{vmatrix} f_0 & 1 & 1 & \cdots & 1 \\ \bar{f}_0 & \alpha_1\lambda_1 & \alpha_2\lambda_2 & \cdots & \alpha_n\lambda_n \\ f_0 & \lambda_1{}^2 & \lambda_2{}^2 & \cdots & \lambda_n{}^2 \\ \bar{f}_0 & \alpha_1\lambda_1{}^3 & \alpha_2\lambda_2{}^3 & \cdots & \alpha_n\lambda_n{}^3 \\ \vdots & \vdots & \vdots & \ddots & \vdots \\ f_0 & \lambda_1{}^n & \alpha_2\lambda_2{}^n & \cdots & \alpha_n\lambda_n{}^n \end{vmatrix}}{\begin{vmatrix} 1 & 1 & 1 & \cdots & 1 \\ -1 & \alpha_1\lambda_1 & \alpha_2\lambda_2 & \cdots & \alpha_n\lambda_n \\ 1 & \lambda_1{}^2 & \lambda_2{}^2 & \cdots & \lambda_n{}^2 \\ -1 & \alpha_1\lambda_1{}^3 & \alpha_2\lambda_2{}^3 & \cdots & \alpha_n\lambda_n{}^3 \\ \vdots & \vdots & \vdots & \ddots & \vdots \\ 1 & \lambda_1{}^n & \alpha_2\lambda_2{}^n & \cdots & \alpha_n\lambda_n{}^n \end{vmatrix}}. \tag{76}$$

It describes the nonlinear superposition of the seed solution f_0 and $n/2$ Kerr-NUT solutions. For $f_0 = 1$ and coinciding parameters

$$\begin{aligned} K_1 = K_3 = K_5 = \cdots = K_{n-1}, & \quad \alpha_1 = \alpha_3 = \alpha_5 = \cdots = \alpha_{n-1} \\ K_2 = K_4 = K_6 = \cdots = K_n, & \quad \alpha_2 = \alpha_4 = \alpha_6 = \cdots = \alpha_n \end{aligned} \tag{77}$$

it contains the Tomimatsu-Sato solutions which can therefore be interpreted as the 'superposition' of $n/2$ identical Kerr solutions.

Bäcklund transformations were first used to generate soliton solutions of nonlinear wave equations (as, for instance, in sine-Gordon theory). In analogy, we may call the superposition of $n/2$ Kerr solutions the 'multi-soliton' solution of the Ernst equation.

3.3 Riemann-Hilbert problems

The author wants to emphasize that the work of the following sections has been done in close collaboration with Reinhard Meinel (Jena). In particular, the rotating disk solution was published in Neugebauer and Meinel (1993, 1994, 1995).

A further possibility for satisfying the first condition in (a)–(d) of Section 3.1 is to consider a closed contour Γ in the complex λ-plane (*cf.* Figure 1) and to postulate that $\Phi(\lambda)$ be regular inside and outside the contour and have a given jump along the contour, *i.e.*

$$\lambda \in \Gamma: \quad \Phi_i(\lambda) = \Phi_e(\lambda) C(K) \tag{78}$$

The subscripts 'i' and 'e' label the interior and exterior values of Φ on Γ and the 2×2 matrix $C(K)$, which depends on $K(K \in \Gamma)$ alone, characterizes the jump. The task of constructing $\Phi(\lambda)$ from the jump matrix is called a matrix Riemann-Hilbert Problem.

A *regular* Riemann problem is characterized by $\det \Phi(\lambda) \neq 0$. It can be shown that the condition (a) in Section 3.1 is automatically satisfied for functions $\Phi(\lambda, z, \bar{z})$

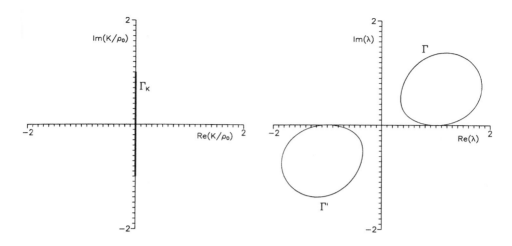

Figure 1. *The contour in the K-plane and in the λ-plane. ρ_0 is the coordinate radius of the disk.*

satisfying a Riemann-Hilbert problem. In the following section we will demonstrate that $C(K)$ can be calculated from the boundary values of a (well-posed) boundary value problem. For a regular Riemann-Hilbert problem the jump matrix C will take the form

$$\lambda \in \Gamma: \quad C = \begin{pmatrix} a(K) & 0 \\ b(K) & 1 \end{pmatrix} \tag{79}$$

with the consequence

$$\lambda \in \Gamma: \quad \chi_i(\lambda) = a(K)\chi_e(\lambda) - b(K)\chi(-\lambda). \tag{80}$$

A similar relation holds for ψ. Defining an auxiliary function $\xi(\lambda)$ satisfying a regular Riemann-Hilbert problem with the same contour ($\xi(-1) = 1$, $\xi \neq 0$),

$$\lambda \in \Gamma: \quad \xi_i(\lambda) = a(K)\,\xi_e(\lambda) \tag{81}$$

we get from (80)

$$\frac{\chi_i(\lambda)}{\xi_i(\lambda)} = \frac{\chi_e(\lambda)}{\xi_e(\lambda)} - \frac{b(K)}{\xi_i(\lambda)}\chi(-\lambda). \tag{82}$$

It is well known that scalar Riemann-Hilbert problems with an additive jump can be solved by Cauchy integrals. Here we have

$$\log \xi(\lambda) = \frac{1}{2\pi i} \int_\Gamma \left(\frac{1}{\lambda' - \lambda} - \frac{1}{\lambda' + 1} \right) \log a(K')\, d\lambda' \tag{83}$$

$$\frac{\chi(\lambda)}{\xi(\lambda)} = \left\{ 1 - \frac{1}{2\pi i} \int_\Gamma \left(\frac{1}{\lambda' - \lambda} - \frac{1}{\lambda' + 1} \right) \frac{b(K')}{\xi_i(\lambda')} \chi(-\lambda')\, d\lambda' \right\}. \tag{84}$$

For $\lambda \to -\lambda$, equation (84) becomes a linear integral equation for $\chi(-\lambda)$. Note that $a(K')$, $b(K')$ are considered to be given, so that $\xi(\lambda)$ can be calculated from (83) and inserted in the kernel of the linear integral equation. Once calculated, $\chi(-\lambda)$ may be substituted into (84) to yield $\chi(\lambda, z, \bar{z})$ and, finally, the Ernst function

$$f = \chi(\lambda = 1, z, \bar{z}).$$

Comparing the mathematical consequences of Riemann-Hilbert problems with Bäcklund transformations, we arrive at linear integral equations instead of linear algebraic equations. General boundary value problems lead to non-regular Riemann-Hilbert problems with $\det \Phi(\lambda_i) = 0$, so that linear integral equations must be combined with Bäcklund transformations. Boundary value problems in terms of linear integral equations have also been discussed by Alekseev (1987,1993).

4 The rigidly rotating disk of dust

4.1 Introduction

Dust matter consists of volume elements ('particles') which interact only through gravitational forces. Dust is a particular case of a perfect fluid with $p = 0$, cf. Equation (1). The energy-momentum balance $T^{ik}{}_{;k} = 0$ implies that the dust 'particles' move on the geodesics of their own field,

$$\dot{u}^i = 0. \tag{85}$$

As in Newtonian theory we expect the only finite mass configuration rotating rigidly around a symmetry axis to be an infinitesimally thin disk where the centrifugal forces balance the gravitational attraction. There are two arguments in favour of a general-relativistic treatment of self-gravitating rotating disks of dust:

(i) So far, no exact global solution of Einstein's equations describing a rotating star or any other rotating isolated matter distribution is known.

(ii) Disk models play an important role in astrophysics (galaxies, accretion disks). In view of exotic objects (such as quasars), general-relativistic disk models could become important.

What we try to find is the general relativistic generalization of the classical Maclaurin disk (the flattened limit of the Maclaurin spheroids). It should be mentioned that Bardeen and Wagoner (1969, 1971) developed a powerful series expansion technique for an approximate solution of the problem. We use the Ernst equation outside the disk. According to (4), the only constants entering our analysis are the constant angular velocity Ω and the central red-shift parameter V_0. The radius ρ_0 of the disk and the surface density σ_p corresponding to the baryonic mass cannot be prescribed beforehand but have to be calculated from the solution.

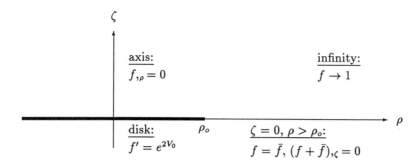

Figure 2. *Boundary conditions. f' is the Ernst potential in the co-rotating frame of reference defined by $\rho' = \rho$, $\zeta' = \zeta$, $\phi' = \phi - \Omega t$, $t' = t$ ($u^{i'} = e^{-V_0}\delta_4^{i'}$).*

4.2 The boundary conditions

The boundary consists of the upper symmetry axis, the surface of the disk, the lower symmetry axis and a circle at infinity connecting the negative and the positive part of the axis. The boundary conditions along this closed curve in a (ζ,ρ) slice may be taken from Figure 2. The axis condition expresses the axisymmetry ($|\eta^i| \to 0$: $A = 0$, $A_{,\zeta} = 0$, $b_{,\rho} = 0$ cf. (18,55)) and at infinity ($\rho^2 + \zeta^2 \to \infty$) the space-time is Minkowskian ($b \to 0$, $U \to 0$). The boundary condition in the disk is a consequence of the geodesic motion (85) and of the field equations (1) themselves. Note that the Ernst potential in the co-rotating system f' also satisfies the Ernst equation.

4.3 The solution

It turns out that a regular Riemann-Hilbert problem applies. To find the contour Γ we have to take into consideration that the (vacuum) Ernst equation fails in the dust region $\zeta = 0$, $0 \leq \rho \leq \rho_0$. On the other hand, structure (a) of Section 3.1 breaks down if $\lambda = 0$, $\lambda = \infty \in \Gamma$ $\left(\lambda^2 = \frac{K-\zeta-i\rho}{K-\zeta+i\rho}\right)$. Hence, the excluded domain in the K-plane is

$$\rho = |\text{Im}K|, \quad \zeta = \text{Re}K \quad \forall K \in \Gamma_K \tag{86}$$

where Γ_K (corresponding to Γ in the K-plane —see Figure 1) is given by

$$\Gamma_K: \quad -\rho_0 \leq \text{Im}K \leq \rho_0, \quad \text{Re}K = 0. \tag{87}$$

For an arbitrary pair ρ, ζ (ρ, ζ outside the disk) the contour Γ in the λ-plane consists of two disconnected curves Γ, Γ' (λ is a square root). The special structure (79) of the jump results from the assumption that ψ, χ should jump on Γ only (and not on Γ'). This assumption turns out to be sufficient to solve the boundary value problem. In order to get the jump coefficients $a(K)$ and $b(K)$ we have to integrate the Linear Problem (57) along the boundary (axis-disk-axis-infinity). Here λ becomes very simple

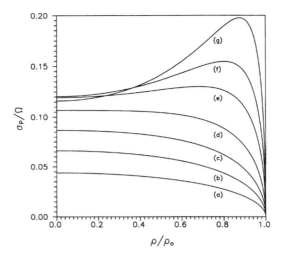

Figure 3. *Radial distribution of the surface mass density for the μ parameter values (a) 0.1, (b) 0.25, (c) 0.5, (d) 1.0, (e) 2.0, (f) 3.0, (g) $\mu = \mu_o = 4.62966\ldots$. For small values of μ one recognizes the Newtonian profile $\propto (1-\rho^2/\rho_0{}^2)^{1/2}$ of the Maclaurin disk. Note that $\rho_0 \to 0$ as $\mu \to \mu_0$.*

and one can take advantage of the boundary values. Surprisingly, the linear integral equation (84) can be solved and the Ernst potential can be represented explicitly in terms of hyperelliptic functions (Neugebauer and Meinel 1995).

We will discuss some properties of the solution. The underlying analytic expressions may be found in Neugebauer and Meinel (1994). First note that the invariant surface mass density σ_p in the disk is given by

$$\sigma_p = \frac{1}{2\pi} V_{,i} n^i \bigg|_{\zeta=0^+} \tag{88}$$

where n^i is the unit normal vector perpendicular to the disk and $\zeta = 0^+$ means the limit from above (in Weyl-Lewis-Papapetrou coordinates: $n^\rho = 0$, $n^\zeta = \exp(U - \alpha)$). The function σ_P is plotted in Figure 3 for several values of the parameter

$$\mu = 2\Omega^2 e^{-2V_0} \rho_0. \tag{89}$$

The ultrarelativistic distributions (f), (g) have a maximum density close to the rim of the disk. For the critical value $\mu = \mu_0 = 4.62966\ldots$ (μ_0 is a zero of a transcendental equation) the disk-like behaviour breaks down and the exterior solution becomes the extreme Kerr solution. A further illustration to this phenomenon is given in Figure 4, where we have plotted the connection between the total mass M and the angular momentum $\mathcal{J}$ of the solution (G is Newton's gravitational constant and c the velocity of light). There is a phase transition between 'normal' matter and Black Hole at $GM^2 = c\mathcal{J}$ (extreme Kerr state). Figure 5 shows the velocities of co-rotating (v_+) and counter-rotating (v_-) test particles at the rim of the disk. Because of the dragging effect, counter-rotation is impossible in the ultrarelativistic region. Finally, Figure 6 demonstrates the beginning and formation of ergospheres.

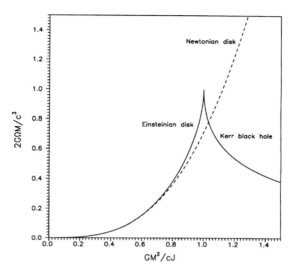

Figure 4. *Rigidly rotating axisymmetric dust in General Relativity (solid line) compared with the Newtonian case (dashed line). The latter is a good approximation for $GM^2/cJ \ll 1$ only. For $GM^2/cJ > 1$ the only possible state is the Kerr black hole where Ω denotes the 'angular velocity of the horizon'. The limit $GM^2/cJ \to \infty$ leads to the Schwarzschild solution ($\Omega = 0$).*

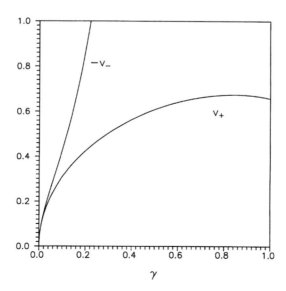

Figure 5. *The velocities $v_\pm$ of test particles on orbits at the rim of the disk ($\gamma = 1 - e^{V_0}$).*

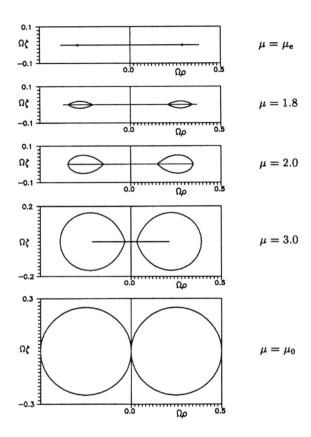

Figure 6. *The formation of ergospheres for characteristic values of μ ($\mu_e = 1.7\ldots$).*

References

Alekseev G A, 1980, *Zh Eksper Teoret Fiz Pis'ma* **32** 301.
Alekseev G A, 1987, in *Trudy Matem Inst Steklova* **17** 6 211, English translation in *Proceedings of the Steklov Institute of Mathematics* **3** 215.
Alekseev G A, 1992, in *International Workshop* NEEDS-*92 Dubna: Nonlinear Evolution Equations and Dynamical Systems*, eds Makhankov V, Puzynin I and Pashaev O (World Scientific, Singapore).
Backer D C, Kulkarni S R, Heiles C, Davis M M, and Goss W M, 1982, *Nature* **300** 615.
Bardeen J M and Wagoner R V, 1969, *Astrophys J* **158** L65.
Bardeen J M and Wagoner R V, 1971, *Astrophys J* **167** 359.
Belinski V A and Zakharov V E, 1978, *Zh Eksper Teoret Fiz* **75** 195.

Harrison B K, 1978, *Phys Rev Lett* **41** 119.
Hauser I and Ernst F J, 1979, *Phys Rev D* **20** 1783.
Hauser I and Ernst F J, 1979, *Phys Rev D* **20** 362.
Hauser I and Ernst F J, 1980, *J Math Phys* **21** 1418.
Herlt E, 1978, *Gen Rel Grav* **9** 711.
Hoenselaers C, Kinnersley W, and Xanthopoulos B C, 1979, *Phys Rev Lett* **42** 481.
Kramer D, 1988, *Astron Nachr* **309** 267.
Kundt W and Trümper M, 1966, *Z Phys* **192** 419.
Lewis T, 1932, *Proc R Soc* **A136** 176.
Lindblom L, 1976, *Astrophys J* **208** 873.
Maison D, 1978, *Phys Rev Lett* **41** 521.
Neugebauer G and Herlt E, 1984, *Classical Quantum Grav* **1** 695.
Neugebauer G and Herold H, 1992, Relativistic Gravity Research, in *Lecture Notes in Physics* **410** 305, eds Ehlers J and Schäfer G (Springer, Berlin).
Neugebauer G and Meinel R, 1993, *Astrophys J* **414** L 97.
Neugebauer G and Meinel R, 1994, *Phys Rev Lett* **73** 2166.
Neugebauer G and Meinel R, 1995, *Phys Rev Lett* **75** 3046.
Neugebauer G, 1979, *J Phys A: Math Gen* **12** L67.
Neugebauer G, 1980, *J Phys A: Math Gen* **13** L19.
Papapetrou A, 1966, *Ann Inst H Poincaré* **A4** 83.
Weyl H, 1917, *Ann Phys, Leipzig* **54** 117.

A Guide to Basic Exact Solutions

Zoltán Perjés

Theory Department
KFKI, RMKI Budapest

1 Introduction

In these lectures we attempt to survey the exact solutions of Einstein's gravitational equations

$$R_{ab} - \tfrac{1}{2}g_{ab}R = -kT_{ab} \ . \tag{1}$$

The task of finding a solution for the components g_{ab} of the Lorentzian metric for a given material source with energy-momentum tensor T_{ab} and gravitational constant k is daunting. The Einstein tensor on the left hand side contains the Ricci tensor, which in turn is constructed from the Riemann curvature,

$$R_{ab} = R_{ar\ b}^{\ r}, \qquad R = R_a^a. \tag{2}$$

The Riemann tensor satisfies the Ricci identity for the covariant derivatives of a vector field v,

$$(\nabla_a \nabla_b - \nabla_b \nabla_a)v^c = R_{ar\ b}^{\ c}v^r. \tag{3}$$

Yet, by now, the supply of exact solutions is ample and the space allotted here confines us to only some basic and most widely studied classes of metrics. More on the subject, including the method of inverse scattering, boundary value problems and rotating objects is discussed in Neugebauer's article in these Proceedings. A detailed review of exact solutions, treasured by many relativists, is to be found in Kramer *et al.* (1980).

Exact solutions have helped the understanding of properties of relativistic states as well as the geometry of General Relativity. There is no universal can opener at hand for solving Einstein's equations. Many tools exist for particular cases, each glittering with its own particular pride: a metric obtained by that particular method. However, looking back at the history of exact solutions, the best working rule appears to be that exact solutions have been distributed by the Lord, apparently whimsically (as the oil fields to nations of the Earth), to individuals deemed by Him most needy or deserving.

2 Spherical Symmetry

In 1916, soon after the discovery of general relativity, the German astronomer Karl Schwarzschild presented the first exact solution of the gravitational equations at a meeting of the Prussian Academy. He achieved this remarkable feat by simplifying the highly complicated gravitational equations of Einstein for the case of spherical symmetry. The Schwarzschild solution describes, quite faithfully, the external gravitational field of most large celestial bodies, including the Earth and the Sun, and has yielded to us the notion of a black hole.

A space-time is spherically symmetric if it admits an $O(3)$ group of isometries. There are two distinct ways in which spherical symmetry is realized in a space-time. The three generators (Killing vectors) of the group may lie in independent spatial directions, in which case the space is homogeneous, or else they may be tangent to 2-spheres. Many cosmological models (such as the Robertson-Walker universes) are built on the Copernican principle of homogeneity of space, and thus they utilise spherical symmetry of the first kind. Models of the external field of astrophysical bodies, on the other hand, exclude a central region filled with the matter source, and bounded by a 2-sphere. These models realize spherical symmetry by a family of nested 2-spheres filling the space-time in a crooked fashion such that a net curvature arises. It is this architecture of space-time that Schwarzschild employed for simplifying the field equations. He arrived at the form of the metric

$$ds^2 = F dt^2 - G(dx^2 + dy^2 + dz^2) - H(x\, dx + y\, dy + z\, dz)^2 \qquad (4)$$

where the functions F, G and H depend on the radius $r = \sqrt{x^2 + y^2 + z^2}$. This is the most general metric with the properties that it is (i) static and (ii) invariant under orthogonal transformations of x, y and z about the origin. Transforming to the polar form by $x = r\sin\theta\cos\phi$, $y = r\sin\theta\sin\phi$, $z = r\cos\theta$, he obtained

$$ds^2 = F dt^2 - (G + Hr^2)dr^2 - Gr^2(d\theta^2 + \sin^2\theta\, d\phi^2). \qquad (5)$$

For large values of r, the limits are $F, G \to 1$ and $H \to 0$. With the above, the solution of the field equations $R_{ab} = 0$ is

$$F = 1 - \frac{2m}{r}, \qquad G = 1, \qquad H = \frac{1}{r^2}\frac{1-F}{F}. \qquad (6)$$

The expressions for G and H remain valid for many other spherically symmetric space-times.

The Schwarzschild space-time is not only important physically, but is the unique spherically symmetric solution of the vacuum gravitational equations. The famous Birkhoff theorem guarantees that this is the only solution of the vacuum equations when having spherical symmetry of the second kind. An important consequence of the Birkhoff theorem is that spherical vacuum space-times are necessarily static. When a tired spherically symmetric star suffers gravitational collapse it may not emit gravitational waves, the external field being described by Schwarzschild's solution.

The Schwarzschild metric cannot be used, in its original form, to follow the gravitational collapse to the final stage involving the black hole. The reason is that the coordinate patch is faulty at the essential loci of space-time belonging to the black hole and its boundary. The shortcomings of the Schwarzschild coordinates have been observed by Finkelstein and Eddington while tracing the geodesic paths of radial light rays for which $\theta =$ constant, $\phi =$ constant and

$$ds^2 \equiv \left(1 - \frac{2m}{r}\right) dt^2 - \frac{1}{1 - \frac{2m}{r}} dr^2 = 0 \tag{7}$$

so that the two sets of rays have the slopes

$$dt = \pm \frac{1}{1 - \frac{2m}{r}} dr. \tag{8}$$

These authors have remedied the situation (Figure 1) by introducing the advanced time coordinate

$$v = t + r^* \qquad \text{where} \qquad r^* = \int \frac{dr}{F} \tag{9}$$

but still left out some space-time regions covered by

$$w = t - r^*. \tag{10}$$

It was Kruskal who finally achieved the description of the full black hole environment by a single coordinate patch. He sought and found a coordinate system (x', t') in which the speed of light is always unity (see Figure 2):

$$x' = \tfrac{1}{2}(v' - w'), \qquad t' = \tfrac{1}{2}(v' + w'), \qquad v' = \exp(\frac{v}{4m}), \qquad w' = -\exp(-\frac{w}{4m}). \tag{11}$$

The external gravitational and electromagnetic field of a spherically symmetric, charged body is described by the two-parameter family of Reissner-Nordstrøm space-times:

$$F = 1 - \frac{2m}{r} + \frac{e^2}{r^2}, \qquad A_t = \frac{e}{r} \tag{12}$$

where $A = A_t \frac{\partial}{\partial t}$ is the four-potential. Alternatively, when the electric charge parameter e is not larger in magnitude than the mass m, the Reissner-Nordstrøm solution describes the space-time region of a charged black hole. The extended coordinate patches (Figures 3,4) are given again by (10) and (9). Coordinates (x^0, x^1) which cover the whole of the maximally extended Reissner-Nordstrøm manifold have been found by Klösch and Strobl (1996):

$$\begin{aligned}
ds^2 &= 2dx^0 dx^1 - U(dx^1)^2 - r^2(d\theta^2 + \sin^2\theta d\phi^2) \\
r &= f(x^1)x^0 + h(x^1), \qquad U = \frac{2}{f} r_{,1} + \frac{F}{f^2}, \qquad h = m - \sqrt{m^2 - e^2} \cos x^1 \\
f &= e^{-4}\left\{\sqrt{m^2 - e^2}(2m^2 - e^2) + 2m(m^2 - e^2)\cos x^1\right\}\sin x^1.
\end{aligned} \tag{13}$$

When Birkhoff's theorem is generalized to account for the presence of electromagnetism, some further insight is needed to establish the uniqueness of the Reissner-Nordstrøm space-time. There exist space-times for which the radius r of the nested

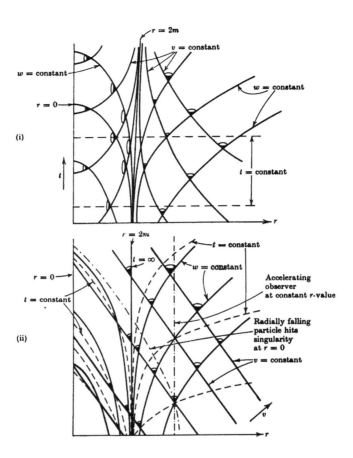

Figure 1. *The radial null geodesics of the Schwarzschild space-time (i) in (t,r) coordinates, (ii) in (v,r) coordinates. (from Hawking and Ellis (1973))*

2-spheres cannot be chosen as a coordinate because r is constant. The corresponding exact solutions are the Robinson-Bertotti metrics (Robinson 1959, Bertotti 1959). The latter, as it turns out, do not represent a separate class of metrics but they describe the region of infinite throat of the extreme ($e=m$) Reissner-Nordstrøm black hole.

Einstein's equations, modified with a cosmological constant term

$$R_{ab} - \tfrac{1}{2}g_{ab}R + g_{ab}\Lambda = -kT_{ab}, \tag{14}$$

have the spherically symmetric solution

$$F = 1 - \frac{2m}{r} + \frac{e^2}{r^2} - \frac{1}{3}\Lambda r^2 \tag{15}$$

describing the external field of the source in a de Sitter space-time. The Reissner-Nordstrøm-de Sitter metric can be transformed to cosmological coordinates (t', r'), as

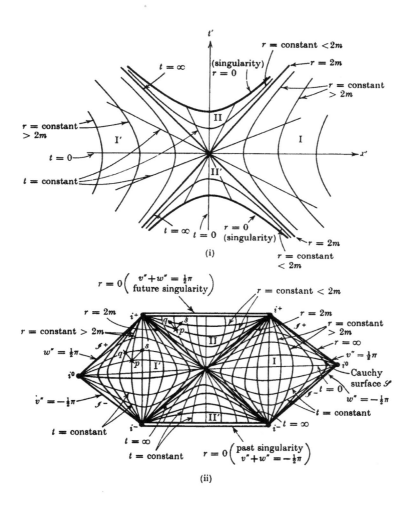

Figure 2. *Schwarzschild space-time in coordinates such that radial null geodesics run at ±45°, with asymptotically flat regions I and I'. (i) The Kruskal diagram, (ii) the conformal Penrose diagram with null infinity (from Hawking and Ellis (1973))*

is the case with the de Sitter space-time:

$$e^{Ht'} r' = r - m, \qquad t' = t - \int \frac{Hr^2}{(r-m)F} dr \qquad (16)$$

where the Hubble constant is defined by $H = \sqrt{\Lambda/3}$. The metric becomes

$$ds^2 = \Omega^{-2}(dt')^2 - e^{2Ht'}\Omega^2[(dr')^2 + (r')^2(d\theta^2 + \sin^2\theta\, d\phi^2)], \qquad \Omega = 1 + e^{-Ht'}\frac{m}{r}. \qquad (17)$$

These are 'contracting' coordinates because the exponential factor in Ω decreases with time t'. There is a horizon (the de Sitter horizon) at $t' = -\infty$. The maximal analytic

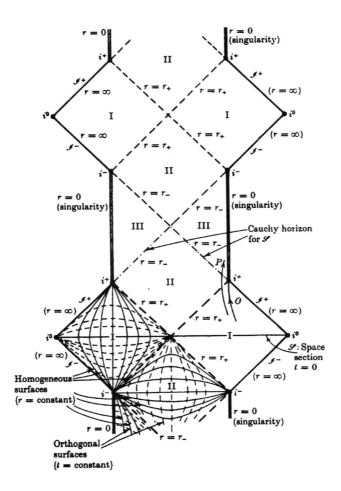

Figure 3. *The Reissner-Nordstrøm space-time for $e^2 < m^2$. (from Hawking and Ellis (1973))*

extension has been given by Brill (1995), noting that the region beyond the horizon can be covered by the 'expanding' coordinates (r', t'), but with opposite sign of H. The structure of the extension depends on the values of the parameters m, Λ and e. When $e^2 = m^2$ a generic black hole exists for the 'undermassive' case $4m|H| < 1$ (beyond that, there is a naked singularity). The extended space-time consists of alternating blocks of expanding and contracting regions (cf. Figure 5).

In a recent work Kastor and Traschen (1993) have succeeded in superimposing several black holes with extremal charge in a de Sitter background space-time. Their

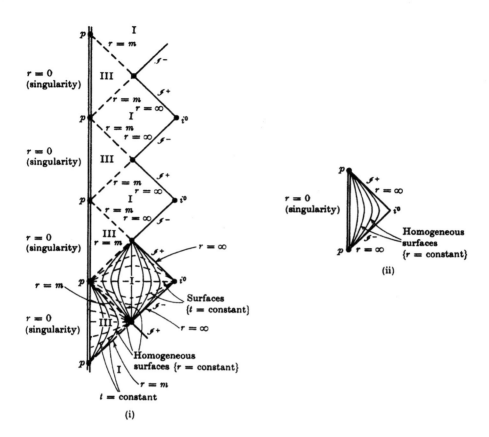

Figure 4. The Reissner-Nordstrøm space-time for (i) $e^2 = m^2$ and (ii) $e^2 > m^2$. (from Hawking and Ellis (1973))

solutions represent colliding black holes, each of mass m_i:

$$\Omega = 1 + e^{-Ht'} \sum_i \frac{m_i}{r_i} \quad \text{where} \quad r_i = \sqrt{(x-x_i)^2 + (y-y_i)^2 + (z-z_i)^2}\ . \tag{18}$$

As the spherical symmetry is lost, the global structure can no longer be represented by two-dimensional conformal diagrams. In the special case of two black holes of equal mass there is still an axis of symmetry on which the black holes lie. The conformal diagram of this space-time is shown on Figure 6, for the overmassive case. The points near $r = \infty$ represent an S^2, whereas the points near $r = 0$ represent *two* S^2's.

Interior solutions with spherical symmetry have been fitted to the Schwarzschild metric along a boundary sphere S^2. Singularity-free configurations exist even for the simple matter source of an incompressible homogeneous perfect fluid,

$$F = \frac{1}{4}\left(3\sqrt{1 - \frac{r_1^2}{R^2}} - \sqrt{1 - \frac{r^2}{R^2}}\right)^2, \quad G = 1, \quad H = \frac{1}{R^2 - r^2} \tag{19}$$

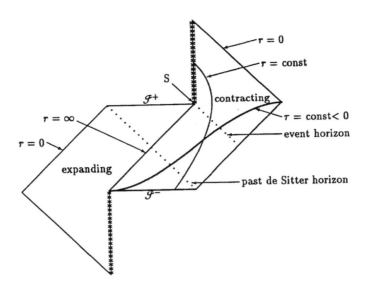

Figure 5. *The Reissner-Nordstrøm-de Sitter space-time for $e^2 = m^2$ and $m < 1/4|H|$. (from Brill (1995))*

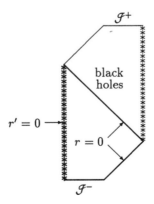

Figure 6. *Conformal diagram of history of axis from one black hole to 'infinity' for a two-black-hole contracting KT geometry. The 'de Sitter background' is the region on the left (from Brill (1995))*

with pressure
$$p = \frac{3}{kR^2} \frac{\sqrt{R^2 - r^2} - \sqrt{R^2 - r_1^2}}{3\sqrt{R^2 - r_1^2} - \sqrt{R^2 - r^2}} \tag{20}$$

where $R^2 = \frac{3}{k\rho}$, ρ is the density and r_1 the radius of the matter. If the metric is real we must have $r_1^2 < R^2$ which imposes an upper limit on the mass $m = \frac{4\pi}{3}\rho r_1^3$.

These and more sophisticated spherically symmetric solutions have frequently been used for modeling relativistic stars.

3 Axial Symmetry

A more realistic description of the fields of relativistic bodies takes into account the deviations from the idealized spherical shape. One obvious idea is to relax the restriction to that of axial symmetry. This gives a good deal of flexibility, and may be justified by the fact that most astrophysical bodies are oblate due to rotation about an axis. Although the further restriction of cylindrical symmetry (translational symmetry along the axis) would greatly facilitate the mathematical simplicity, this would mostly characterize the idealized fields of infinitely extended cylindrical bodies. We will not impose here the condition of cylindrical symmetry, for the sake of fields with greater physical importance. Even so, we cannot claim to have the best possible exact description of the physics.

It is true that astrophysical bodies are oblate due to rotation, but their axis of symmetry does not coincide with the axis of rotation. Theory cannot cope exactly with this complicated system which is not stationary but is radiative. The fields with axial symmetry belong to idealized bodies with coincident axes. For such sources the field is stationary.

A simple class of axisymmetric fields, those generated by *static* sources, was obtained by Weyl in 1917. For static fields, there is a reflectional symmetry in time, $t \to -t$, as well as in the azimuthal direction, $\phi \to -\phi$, due to lack of rotation. All the components of the metric with one t or ϕ index must vanish in a coordinate system adapted to both of these directions. This is because such a component would flip sign under reflection. For similar reasons, there is no cross-term $g_{t\phi}$ in the metric. Thus the nonvanishing metric components form two 2×2 block diagonal matrices, of which the one with t and ϕ components itself is diagonal. There is no dependence on the t and ϕ coordinates since they are adapted to the directions of the symmetry. Hence the components in the other 2×2 block, depending only on the coordinates, say x^1 and x^2, are really the components of a metric of a two-surface. It is an elementary property of two-surfaces that they are all conformally flat. Therefore we can bring this 2-metric to the form

$$g = \alpha(x^1, x^2)[(dx^1)^2 + (dx^2)^2]. \tag{21}$$

As is easily seen, choosing any pair of conjugate harmonic functions as coordinates preserves this form of the metric. There is a natural choice for this pair of harmonic functions whenever the energy-momentum tensor satisfies the condition

$$T_\phi^\phi + T_t^t = 0. \tag{22}$$

This condition certainly holds for vacuum space-times, but also for stationary axisymmetric electrovacua. By Einstein's equations we then have that the sum of the Ricci tensor components

$$R^\phi_\phi + R^t_t = -g^{-1/2}\left(\frac{\partial^2}{(\partial x^1)^2}\rho + \frac{\partial^2}{(\partial x^2)^2}\rho\right) \tag{23}$$

vanishes. Here g is the determinant of the metric and ρ is the determinant

$$\rho^2 = -g_{\phi\phi}g_{tt} + (g_{\phi t})^2 \tag{24}$$

(for static fields, however, this simplifies since $g_{\phi t} = 0$). Thus we choose ρ and its conjugate harmonic function, to be called z, as the new coordinates. The metric then takes the form

$$ds^2 = e^{2\lambda}dt^2 - e^{-2\lambda}[e^{2\gamma}(d\rho^2 + dz^2) + \rho^2 d\phi^2]\,. \tag{25}$$

The components of the Ricci tensor satisfy the equations

$$\begin{aligned}
\tfrac{1}{2}(R_{11} + R_{22}) &= \gamma_{,11} + \gamma_{,22} - \Delta\lambda + (\lambda_{,1})^2 + (\lambda_{,2})^2 \\
\tfrac{1}{2}(R_{11} - R_{22}) &= (\lambda_{,1})^2 - (\lambda_{,2})^2 - \frac{\gamma_{,1}}{\rho} \\
R_{12} &= 2\lambda_{,1}\lambda_{,2} - \frac{\gamma_{,2}}{\rho} \\
R^\phi_\phi - R^t_t &= -2e^{2(\lambda-\gamma)}\Delta\lambda \\
R^\phi_\phi + R^t_t &= 0
\end{aligned} \tag{26}$$

where a comma in the subscript denotes a partial derivative and

$$\Delta = \frac{\partial^2}{\partial\rho^2} + \frac{1}{\rho}\frac{\partial}{\partial\rho} + \frac{\partial^2}{\partial z^2} \tag{27}$$

is the Laplace operator in an Euclidean 3-space expressed in cylindrical coordinates, with ρ as the radius function.

In addition to the boundary conditions at large distances, we prescribe the condition of regularity on the symmetry axis $\rho = 0$. Wherever this holds, the ratio of the

$$\text{circumference} = \int_0^{2\pi}\sqrt{e^{-2\lambda}r^2}\,d\phi = 2\pi r e^{-\lambda} \tag{28}$$

and

$$\text{radius} = \int_0^r \sqrt{e^{-2\lambda}e^{2\gamma}}\,d\rho = e^\gamma e^{-\lambda}r \tag{29}$$

of a small circle is 2π. As is easily seen, this provides the boundary condition

$$\lim_{\rho\to 0}\gamma = 0\,. \tag{30}$$

3.1 Vacuum Space-Times

In vacuum, the components (26) of the Ricci tensor vanish by Einstein's equations. We thus arrive at the amazing fact that at the heart of the static axisymmetric system lies the Laplace equation

$$\Delta\lambda = 0. \tag{31}$$

Given a solution of this linear equation, the remaining metric function γ can be determined by simple line integrals

$$\gamma = \int \rho[((\lambda_{,1})^2 - (\lambda_{,2})^2)d\rho + 2\lambda_{,1}\lambda_{,2}dz] \, . \tag{32}$$

Simple point sources on the z axis have the form

$$\lambda = -\frac{m_1}{r_1} - \frac{m_2}{r_2} - \cdots \tag{33}$$

where the constants $m_1, m_2, \ldots$ describe the masses of the sources and the constants $z_1, z_2, \ldots$ give their locations. The functions in the denominators are the Euclidean distances from each of the sources,

$$r_1^2 = \rho^2 + (z - z_1)^2, \qquad r_2^2 = \rho^2 + (z - z_2)^2, \qquad \ldots \tag{34}$$

For example, the Schwarzschild metric is defined by two point sources on the axis, of equal mass, and equidistant, $z_i = \pm m$, from the origin. In this description, the spherical symmetry is no longer manifest. When we take the potential to be spherically symmetric, we get the Curzon space-time with

$$\lambda = -\frac{2m}{r_1}, \qquad \gamma = -\frac{m^2 \rho^2}{2r_1^4} \, . \tag{35}$$

Although the linear superposition rule (33) works for the potential λ, the nonlinearity shows up in the integral (32) for γ. For the superposition of two Curzon metrics with masses m_1 and m_2 one obtains

$$\gamma = -\frac{m_1^2 \rho^2}{2r_1^4} - \frac{m_2^2 \rho^2}{2r_2^4} + \frac{2m_1 m_2}{(z_1 - z_2)^2} \left\{ \frac{\rho^2 + (z - z_1)(z - z_2)}{r_1 r_2} - 1 \right\} . \tag{36}$$

At this point, one may rightly ask if it is possible to float several mass points along the axis, hanging in the thin vacuum. The answer is no. A fascinating analysis shows that Einstein's gravitational equations have a strong grip on the laws of physics even for the axisymmetric problem. When we insist that the z-axis is regular in the regions outside the sources, the line integral (32) implies that a wire-like singularity must occur between the sources. Further, when there is no material support *between* the sources, wires fixing them to distant objects must be present.

4 Stationary Space-Times

Domains of space-times admitting a time-like Killing vector T are also called *stationary*. When the coordinate system is adapted to the Killing vector, $T = \frac{\partial}{\partial t}$, the metric and all geometric quantities determined by the metric are independent of the coordinate t. Physical fields are stationary if their Lie derivative along T vanishes. We now find the allowable transformations $x^{a'} = x^{a'}(x^b)$ of the adapted coordinates. For such transformations

$$T^{a'} = \delta_0^{a'} = \frac{\partial x^{a'}}{\partial x^b} T^b = \frac{\partial x^{a'}}{\partial t}, \qquad t = x^0 . \tag{37}$$

In a detailed form, this is

$$\frac{\partial t'}{\partial t} = 1 \quad \text{and} \quad \frac{\partial x^{\mu'}}{\partial t} = 0, \quad \mu' = 1, 2, 3. \tag{38}$$

The general solution is

$$\begin{aligned} t' &= t + f(x^\mu) \\ x^{\mu'} &= x^{\mu'}(x^\nu) \quad \mu', \nu = 1, 2, 3. \end{aligned} \tag{39}$$

These are the admissible transformations of the adapted coordinates.

Static space-times are the particular case of stationary space-times with the additional property that the Killing vector T is hypersurface-orthogonal. In this case, one can arrange the adapted coordinate system such that the metric is diagonal. The vector field T is hypersurface-orthogonal (proportional to the gradient of a scalar field) if its vector rotation vanishes:

$$T_b \nabla_d T_c \epsilon^{abcd} = 0. \tag{40}$$

The proof is obvious in one direction: any $T_a = \chi \sigma_{,a}$ satisfies (40). To establish the converse, consider the consequence of (40):

$$T_a \nabla_b T_c + T_c \nabla_a T_b + T_b \nabla_c T_a = 0. \tag{41}$$

Contraction with T^c and use of the Killing equation $\nabla_{(a} T_{b)} = 0$ yields

$$T_a \nabla_b (T^2) - T_b \nabla_a (T^2) + T^2 (\nabla_a T_b - \nabla_b T_a) = 0 \tag{42}$$

where $T^2 = T^a T_a$. Hence

$$\frac{\partial}{\partial x^a} \left(\frac{T_b}{T^2} \right) = \frac{\partial}{\partial x^b} \left(\frac{T_a}{T^2} \right). \tag{43}$$

Therefore there exists a function σ for which

$$\frac{T_a}{T^2} = \frac{\partial \sigma}{\partial x^a} \tag{44}$$

or

$$T_a = T^2 \sigma_{,a}. \tag{45}$$

This completes the proof. We now introduce coordinates adapted to the Killing vector, $T^a = \delta^a_0$. Then, from (45),

$$g_{a0} = T_a = T^2 \sigma_{,a}. \tag{46}$$

By definition,

$$T^2 = g_{ab} \delta^a_0 \delta^b_0 = g_{00} \tag{47}$$

so

$$g_{a0} = g_{00} \sigma_{,a}. \tag{48}$$

Hence, for $a = 0$,

$$\sigma_{,0} = 1 \tag{49}$$

and thus

$$\sigma = t + f(x^\mu) \quad \mu = 1, 2, 3. \tag{50}$$

Let us now take the coordinates

$$t' = \sigma = t + f(x^\mu)$$
$$x^{\mu'} = x^\mu. \qquad (51)$$

This is a special case of (39), and so it preserves the condition $T = \frac{\partial}{\partial t}$. With (48), we obtain in the new coordinate system

$$g_{\mu 0} = 0 \qquad \mu = 1, 2, 3. \qquad (52)$$

Almost all known stationary space-times are also axially symmetric, although some exceptions exist (*e.g.* Israel and Wilson 1972, Perjés 1971). One may argue that the field is stationary if its source is in a stationary state. A stationary lump of matter is either at rest, or in a state of motion preserving its spatial distribution. The latter is achieved, for example, for a body with a symmetry axis by rotating it about this axis.

Following Geroch (1971), the time-like integral curves of the Killing vector T may be considered as the points of a topological space. This space of the Killing trajectories is then made into a differentiable manifold and endowed with a positive-definite metric h related to the space-time by the Lewis-Papapetrou form of the metric:

$$ds^2 = f(dt + \omega_\mu dx^\mu)^2 - f^{-1} h_{\mu\nu} dx^\mu dx^\nu. \qquad (53)$$

Here $f = T^a T_a$ is the norm-square of the Killing vector, and x^μ are coordinates in the three-space of the trajectories.

The essential field equations of the stationary vacuum can be easily memorized in a complex form which is known as the Ernst equation: (see Section 2.3 of Neugebauer's article) "

$$(\operatorname{Re}\mathcal{E})\Delta\mathcal{E} = (\nabla\mathcal{E})^2. \qquad (54)$$

Here, the vector notation refers to the three-space of Killing trajectories and the complex Ernst potential is defined by

$$\mathcal{E} = f + i\psi. \qquad (55)$$

The 'curl-potential' ψ is defined, up to an additive constant, by

$$\nabla_a \psi = 2(\nabla_b T_a)^* T^b. \qquad (56)$$

where an asterisk denotes the dual bivector. The existence of the 'curl-potential' is a consequence of the vacuum Einstein equations.

Introducing the complex 3-vector

$$\mathbf{G} = \frac{\nabla\mathcal{E}}{\mathcal{E} + \overline{\mathcal{E}}}, \qquad (57)$$

the full vacuum equations may be written in the form

$$(\nabla - \mathbf{G} + \overline{\mathbf{G}}) \cdot \mathbf{G} = 0 \qquad (58)$$
$$(\nabla - \mathbf{G} + \overline{\mathbf{G}}) \times \mathbf{G} = 0 \qquad (59)$$
$$R_{\mu\nu} = -G_\mu \overline{G}_\nu - \overline{G}_\mu G_\nu. \qquad (60)$$

Equation (54) contains metric variables, beside $\mathcal{E}$ (unless the system is axisymmetric). How can one handle these field equations? A useful method will be described next.

4.1 Triad Formalism

In analogy with the Newman-Penrose spin coefficient formalism, introduce a real vector l^μ in the tangent 3-space of the manifold of time-like Killing trajectories. Since this space is Riemannian, real vectors are spacelike and we normalize l to unity. However, a complex null vector m still exists, and we complete the triad by adding m and $\overline{m}$:

$$z_o^\mu = l^\mu \qquad z_+^\mu = m^\mu \qquad z_-^\mu = \overline{m}^\mu. \tag{61}$$

As with the Newman-Penrose tetrad, there is a close relationship with spinors (Perjés 1970), but we shall omit here that chapter of the theory.

Conjugation may be expressed in geometric terms as an exchange of the vectors m and $\overline{m}$ while leaving the vector l intact.

The 3-metric has the form

$$(g_{\mathbf{pq}}) = (z_{\mathbf{p}\mu} z_{\mathbf{q}}^\mu) = \begin{pmatrix} 1 & 0 & 0 \\ 0 & 0 & 1 \\ 0 & 1 & 0 \end{pmatrix} \tag{62}$$

where the triad indices $\mathbf{p}, \mathbf{q}, \ldots$ take values from $o, +, -$.

The *Ricci rotation coefficients* are $\gamma_{\mathbf{pqr}} = \nabla_\nu z_{\mathbf{p}\mu} z_{\mathbf{q}}^\mu z_{\mathbf{r}}^\nu$ and will be given the individual notations shown below

qr \ p	$-o$	$+o$	$+-$
o	$\overline{\kappa}$	κ	ϵ
$+$	$\overline{\rho}$	σ	$-\overline{\tau}$
$-$	$\overline{\sigma}$	ρ	τ

The *triad derivatives* acting on scalars are defined by

$$\partial_{\mathbf{p}} = z_{\mathbf{p}}^\mu \partial_\mu$$

and in particular,

$$D = \partial_o \qquad \delta = \partial_+ \qquad \overline{\delta} = \partial_-. \tag{63}$$

The triad derivatives have the commutators

$$D\delta - \delta D = (\overline{\rho} + \epsilon)\delta + \sigma\overline{\delta} + \kappa D \tag{64}$$
$$\delta\overline{\delta} - \overline{\delta}\delta = \overline{\tau}\overline{\delta} - \tau\delta + (\overline{\rho} - \rho)D. \tag{65}$$

The Ricci identities have the form

$$\begin{aligned}
D\sigma - \delta\kappa &= \overline{\tau}\kappa + \kappa^2 + \sigma(\rho + \overline{\rho} + 2\epsilon) - 2\phi_{++} + 2\overline{G}_+ G_+ \\
D\rho - \overline{\delta}\kappa &= \overline{\kappa}\kappa - \kappa\tau + \sigma\overline{\sigma} + \rho^2 - \phi_{oo} + \overline{G}_o G_o \\
D\tau - \overline{\delta}\epsilon &= \overline{\kappa}\rho - (\kappa + \overline{\tau})\sigma + \epsilon(\overline{\kappa} - \tau) + \rho\tau - 2\phi_{o-} + \overline{G}_o G_- + \overline{G}_- G_o \\
\overline{\delta}\sigma - \delta\rho &= 2\sigma\tau + \kappa(\rho - \overline{\rho}) + 2\phi_{o+} - \overline{G}_o G_+ - \overline{G}_+ G_o \\
\delta\tau + \overline{\delta}\overline{\tau} &= \rho\overline{\rho} - \sigma\overline{\sigma} + 2\tau\overline{\tau} - \epsilon(\rho - \overline{\rho}) - 2\phi_{+-} + \phi_{oo} - \overline{G}_o G_o + \overline{G}_- G_+ + \overline{G}_+ G_-
\end{aligned} \tag{66}$$

where
$$\phi_{\text{pq}} = \frac{1}{2} R_{\text{pq}}. \tag{67}$$

We now list the relations between the triad and Newman-Penrose quantities. In the remainder of this subsection, where these quantities simultaneously occur, the 4-dimensional objects are distinguished by a hat. Choosing the gauge

$$\hat{l}^a = \{l^\mu, f^{-1} - (l^\nu \omega_\nu)\} \tag{68}$$
$$\hat{m}^a = \sqrt{|f|}\{m^\mu, -(m^\nu \omega_\nu)\} \tag{69}$$
$$n^a = \frac{1}{2} f \hat{l}^a + T^a \tag{70}$$

the spin coefficients of Newman and Penrose have the decomposition

$$\hat{\kappa} = f^{-1/2}(\kappa - 2G_+)$$
$$\hat{\pi} = \tfrac{1}{2} f^{1/2} \overline{\kappa}$$
$$\hat{\tau} = -\tfrac{1}{2} f^{1/2} \kappa$$
$$\hat{\alpha} = \tfrac{1}{4} f^{1/2}(2\tau + G_- - \overline{G}_-)$$
$$\hat{\beta} = -\tfrac{1}{4} f^{1/2}(2\overline{\tau} + 3G_+ + \overline{G}_+)$$
$$\hat{\gamma} = -\tfrac{1}{8} f(2\epsilon - 3G_o - \overline{G}_o)$$

$$\hat{\epsilon} = \tfrac{1}{4}(2\epsilon + G_o - \overline{G}_o)$$
$$\hat{\rho} = \rho + G_o$$
$$\hat{\sigma} = \sigma$$
$$\hat{\lambda} = \tfrac{1}{2} f \overline{\sigma}$$
$$\hat{\mu} = \tfrac{1}{2} f(\overline{\rho} + G_o)$$
$$\hat{\nu} = \tfrac{1}{4} f^{3/2}(-\overline{\kappa} + 2G_-)$$

and the Weyl curvature spinor is given by

$$\begin{aligned}
\Psi_0 &= 2[\delta G_+ - \sigma G_o + \overline{\tau} G_+ + (2G_+ + \overline{G}_+)G_+] \\
\Psi_1 &= -f^{1/2}[DG_+ - \kappa G_o - \epsilon G_+ + (2G_o + \overline{G}_o)G_+] \\
\Psi_2 &= \tfrac{1}{2} f[DG_o + \overline{\kappa} G_+ + \kappa G_- + (G_o + \overline{G}_o)G_o - 2G_+ G_-] \\
\Psi_3 &= \tfrac{1}{2} f^{3/2}[DG_- - \overline{\kappa} G_o + \epsilon G_- + (2G_o + \overline{G}_o)G_-] \\
\Psi_4 &= \tfrac{1}{2} f^2[\overline{\delta} G_- - \overline{\sigma} G_o + \tau G_- + (2G_- + \overline{G}_-)G_-] .
\end{aligned} \tag{71}$$

In a *vacuum space-time*, the Ricci tensor of the 3-space is determined by Einstein's gravitational equations (60). The corresponding relations in the presence of matter are given in Perjés (1970). The triad components G_p of the gravitational vector satisfy the vacuum field equations

$$\begin{aligned}
DG_o + \overline{\delta} G_+ + \delta G_- &= (\rho + \overline{\rho})G_o - (\kappa - \overline{\tau})G_- - (\overline{\kappa} - \tau)G_+ \\
&\quad +(G_o - \overline{G}_o)G_o + (G_+ - \overline{G}_+)G_- + (G_- - \overline{G}_-)G_+
\end{aligned} \tag{72}$$

$$DG_- - \overline{\delta} G_o = (\rho - \epsilon)G_- + \overline{\sigma} G_+ + \overline{\kappa} G_o - \overline{G}_o G_- + \overline{G}_- G_o \tag{73}$$

$$DG_+ - \delta G_o = \sigma G_- + (\overline{\rho} + \epsilon)G_+ + \kappa G_o - \overline{G}_o G_+ + \overline{G}_+ G_o \tag{74}$$

$$\delta G_- - \overline{\delta} G_+ = (\overline{\rho} - \rho)G_o - \tau G_+ + \overline{\tau} G_- - \overline{G}_+ G_- + \overline{G}_- G_+ \tag{75}$$

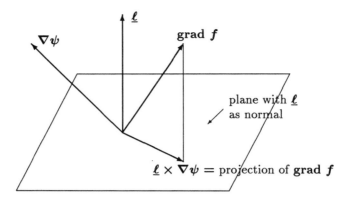

Figure 7. *The condition $G_+ = 0$ fixing the direction of l*

How do we use triads in practice?

(i) Fix the orientation of the triad. For example, set $G_+ = 0$, by which the direction of the vector l^μ is given (Figure 7). Rotations about l, *i.e.* $m \to e^{i\delta}m$, are still possible.

(ii) Find suitable coordinates. For example, one may set $D = \frac{\partial}{\partial r}$.

(iii) Integrate the field equations in some order, *e.g.* getting the r dependencies first.

Here are some applications:

- The vacuum fields with $\kappa = \sigma = 0$ contain the Kerr solution (Perjés 1970).
- The $\kappa = 0$, $\sigma \neq 0$ metrics have been obtained in Kóta and Perjés (1972).
- Some classes of the topological massive fields of Deser, Jackiw and Templeton have been obtained in Hall, Morgan and Perjés (1987).
- The conformastat spaces (see next subsection).

4.2 Conformastat Spaces

One of the many features of the Schwarzschild space-time is that it can be written down in a spatially homogeneous (isotropic) coordinate system (Synge 1960):

$$ds^2 = f dt^2 - f^{-1}\Omega^2(dr^2 + r^2 d\theta^2 + r^2 \sin^2\theta d\phi^2) \tag{76}$$

with

$$f = \left(1 - \frac{m}{2r}\right)^2 \left(1 + \frac{m}{2r}\right)^{-2}, \qquad \Omega = \left(1 - \frac{m}{2r}\right)\left(1 + \frac{m}{2r}\right). \tag{77}$$

That is to say, the space of the time-like Killing trajectories is conformally flat. The question thus naturally arises whether there are any more space-times with the property that the three-space of the time-like Killing vector is conformally flat. Space-times with this property are sometimes called *conformastat*.

The general solution of the vacuum conformastat problem has been given by Perjés (1986). The procedure is rather involved. The first step consists of proving that these spaces have axial symmetry. They contain the three static Levi-Civita metrics (one of them the Schwarzschild metric) shown below. One can then employ the complex Ernst potential and its conjugate as coordinates to show that no further metrics exist in the axisymmetric subclass.

$$g_1 = \left(1 - \frac{2m}{r}\right) dt^2 - \frac{dr^2}{1 - \frac{2m}{r}} - r^2 \left(d\theta^2 + \sin^2\theta \, d\phi^2\right) \quad \text{(Schwarzschild)}$$

$$g_2 = \left(\frac{2m}{z} - 1\right) dt^2 - \frac{dz^2}{\frac{2m}{z} - 1} - z^2(dr^2 + \sinh^2 r \, d\phi^2)$$

$$g_3 = z^{-1} dt^2 - z \, dz^2 - z^2(dr^2 + r^2 \, d\phi^2)$$

4.3 The eigendirections of a Killing vector

Let K^a be a Killing vector:

$$\nabla_{(a} K_{b)} = 0. \tag{78}$$

Then the tensor

$$F_{ab} = \nabla_a K_b = -\nabla_b K_a \tag{79}$$

is skew. One may write, in terms of the symmetric spinor ϕ_{AB},

$$F_{ab} = \epsilon_{A'B'} \phi_{AB} + \epsilon_{AB} \overline{\phi}_{A'B'} \tag{80}$$

where $\epsilon_{AB} = \epsilon_{A'B'} = \begin{pmatrix} 0 & 1 \\ -1 & 0 \end{pmatrix}$. The canonical decomposition

$$\phi_{AB} = \alpha_{(A} \beta_{B)} \tag{81}$$

yields the principal spinors α_A and β_A. The eigenvalue problem of the matrix $\left[\phi_A^B\right]$ reads

$$\phi_A^B \alpha_B = \tfrac{1}{2}(\alpha_A \beta^B + \beta_A \alpha^B)\alpha_B = \lambda \alpha_A \tag{82}$$

with the eigenvalue $\lambda = \tfrac{1}{2}\alpha_B \beta^B$. Thus α_A and β_A are eigenspinors, with the corresponding null eigendirections of the vector K:

$$l_a = \alpha_A \overline{\alpha}_{A'} \qquad n_a = \beta_A \overline{\beta}_{A'} . \tag{83}$$

5 Null Killing Vectors

A null Killing vector K is defined by the relations

$$\nabla_{(a}K_{b)} = 0, \qquad K_a K^a = 0. \tag{84}$$

From $\nabla_b(K^a K_a) = 0$ we get that K is tangent to a family of null geodesics:

$$K^b \nabla_b K_a = 0 \tag{85}$$

and by the Killing equation (84), K has no divergence:

$$\nabla^a K_a = 0. \tag{86}$$

In spinor terms we may write $K_a = o_A \bar{o}_{A'}$. The spinor (80) describing the bivector $\nabla_a K_b$ takes the form

$$\phi_{AB} = o_{(A} \nabla^{A'}_{B)} \bar{o}_{A'} + \bar{o}_{A'} \nabla^{A'}_{(A} o_{B)} . \tag{87}$$

Hence, o_A is an eigenspinor of ϕ_A^B:

$$o^A o^B \phi_{AB} = 0. \tag{88}$$

By setting K to be the vector l of a Newman-Penrose tetrad $(l, m, \bar{m}, n)$, it is easy to see that the null Killing vector field has no shear:

$$\sigma = (\nabla_a K_b) m^a m^b = 0. \tag{89}$$

Thus, by the Goldberg-Sachs theorem, a vacuum space-time containing a null Killing vector is algebraically special. From the Newman-Penrose equation for $D\sigma$ one finds that K is hypersurface-orthogonal.

The vacuum space-times admitting a null Killing vector are all known. Dautcourt (1964) has shown that they form two classes:

1. The *pp-waves*, characterized by $\nabla_a K_b = 0$. These metrics were discovered by Brinkmann in 1923. They have the form

$$ds^2 = 2dudv - 2d\zeta d\bar{\zeta} + 2H du^2 \tag{90}$$

with $H = H(u, \zeta, \bar{\zeta})$. The Weyl curvature is Petrov type N.

In vacuum, we have $H = f(\zeta, u) + \bar{f}(\bar{\zeta}, u)$. The pp-waves may be classified by the group of isometries. The solution of the Killing equation in the non-flat case can be written as

$$\begin{aligned} K = & [ib\zeta + \beta(u)]\frac{\partial}{\partial \zeta} + [-ib\bar{\zeta} + \bar{\beta}(u)]\frac{\partial}{\partial \bar{\zeta}} \\ & + (cu+d)\frac{\partial}{\partial u} + [a(u) + \dot{\beta}(u)\bar{\zeta} + \dot{\bar{\beta}}(u)\zeta - cv]\frac{\partial}{\partial v} \end{aligned} \tag{91}$$

where the constants b, c, d and the function a are real and they satisfy a constraint equation. Hence the isometry group can have 1,2,3,5 or 6 parameters. The superposition of plane waves is governed by the Ernst equation (Griffiths 1993).

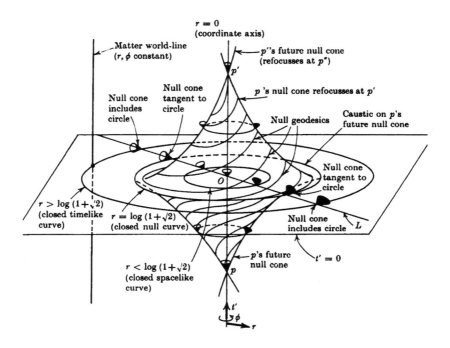

Figure 8. *Gödel's universe (from Hawking and Ellis (1973))*

2. The metric
$$ds^2 = 2xdu(dv + Mdu) - x^{-1/2}(dx^2 + dy^2) \tag{92}$$
where $M = M(u,x,y)$ satisfies $(xM)_{,x} + xM_{,yy} = 0$.

In 1949, Gödel found a space-time filled by a dust flow having the four-velocity $u = (1/\sqrt{2})\omega\frac{\partial}{\partial t}$ and metric
$$ds^2 = \frac{2}{\omega^2}[dt'^2 - dr^2 + (\sinh^4 r - \sinh^2 r)d\phi^2 - 2\sqrt{2}\sinh^2 r\, d\phi dt] - dz^2. \tag{93}$$

The density of the dust is $\rho = \omega^2/2k$. The Gödel metric satisfies the gravitational equations (14) with the cosmological constant $\Lambda = -\omega^2$ and the range of the coordinates is $-\infty < t' < \infty$, $-\infty < r < \infty$, $\phi \in [0, 2\pi)$. These are co-moving coordinates and on the space-time diagram (Figure 8), suppressing the irrelevant z coordinate, the world-line of a dust particle is a straight vertical line. The null geodesics from point p on the symmetry axis through the (arbitrarily chosen) origin O meet again at p'. The light cones tilt in the direction of rotation more and more as one moves away from O in the direction $\frac{\partial}{\partial r}$. As a result, the space-like concentric circles in the plane become closed null curves at the radial distance $r = \log(1 + \sqrt{2})$. Thus this space-time exhibits the worst kind of causality violation and no cosmic time exists, even though the space-time is geodesically complete.

This space-time contains two null Killing vectors and three more isometries.

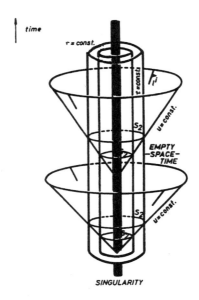

Figure 9. *A Robinson-Trautman space-time*

6 Robinson-Trautman Space-times

Robinson-Trautman (1962) vacuum space-times are characterized by a shear-free and curl-free expanding null geodesic congruence with tangent vector l. Hence these space-times are algebraically special. One may choose coordinates such that the null congruence is normal to the constant u surfaces and the luminosity coordinate is $r = -1/\nabla_a l^a$. Then the intersection of a constant u hypersurface and a constant luminosity hypersurface is a surface $\mathcal{S}$ with metric characterised by $P = P(u, \zeta, \overline{\zeta})$ (Figure 9). The space-time metric takes the form

$$ds^2 = 2drdu - 2r^2 P^{-2} d\zeta d\overline{\zeta} + H du^2 \tag{94}$$

where

$$H = 2P^2 \frac{\partial^2 \ln P}{\partial \zeta \partial \overline{\zeta}} - 2r \frac{\partial \ln P}{\partial u} - \frac{2m}{r} . \tag{95}$$

The Schwarzschild space-time is a particular example of a Robinson-Trautman solution. Truly radiative solutions have been shown to exist and to converge asymptotically to the Schwarzschild metric, but none are known explicitly. Axially symmetric Robinson-Trautman metrics are examples of space-times with a space-like Killing vector. In Robinson coordinates the metric takes the form

$$ds^2 = r^2 \left[\frac{1}{y} \left(dx + \frac{1}{12m} yy_{,xxx} du \right)^2 + y d\phi^2 \right] - 2du dr + \left[\frac{1}{2} y_{,xx} + \frac{r}{12m} (yy_{,xxx})_{,x} + \frac{12m}{r} \right] du^2 \tag{96}$$

and the only field equation left to be satisfied is

$$(1/y)_{,t} = y_{,xxxx} \tag{97}$$

where $t = u/12m$. The Schwarzschild solution reads $y = 1 - x^2$.

7 Einstein-Maxwell Fields

The essential field equations of a *static* electrovacuum (interacting gravitational and electric fields) are

$$\begin{aligned}(\xi^2 + \eta^2 - 1)\Delta\xi &= 2(\xi\nabla\xi + \eta\nabla\eta)\cdot\nabla\xi \\ (\xi^2 + \eta^2 - 1)\Delta\eta &= 2(\xi\nabla\xi + \eta\nabla\eta)\cdot\nabla\eta\end{aligned} \tag{98}$$

where ξ and η are real gravitational and electromagnetic potentials, respectively. For axisymmetric fields, the background metric is flat as is for vacuum, but the linearity of the field equations is here clearly lost. It has been shown (Bonnor 1966) that Equations (98) are mapped to the Ernst equation by making the potential η pure imaginary. This has to be accompanied by a complex extension of the suitable parameter values. For example, writing $\xi = \alpha/\beta, \eta = \gamma/\beta$, one obtains the following static space-time from the Kerr metric:

$$\alpha = x^2 + p^2 - q^2 y^2, \qquad \beta = 2px, \qquad \gamma = 2pqy \tag{99}$$

where x and y are spheroidal coordinates and the parameters p and q are related to the mass and charge by $p = (1/m)\sqrt{m^2 - e^2}$ and $p^2 + q^2 = 1$. From the second Tomimatsu-Sato metric (the rotating Darmois solution) one gets the electrovacuum

$$\begin{aligned}\alpha &= -4p^2q^2y^6 + 8p^2q^2y^4 - 4p^2q^2y^2 + 4p^2x^6 - 8p^2x^4 + 4p^2x^2 + q^4y^8 - 2q^4y^4 + q^4 \\ &\quad -4q^2x^6y^2 + 6q^2x^4y^4 + 2q^2x^4 - 4q^2x^2y^6 + 2q^2y^4 - 2q^2 + x^8 - 2x^4 + 1 \quad (100)\\ \beta &= 4px(q^2x^2y^4 - 2q^2x^2y^2 + q^2x^2 - 2q^2y^6 + 3q^2y^4 - q^2 + x^6 - x^4 - x^2 + 1) \quad (101)\\ \gamma &= 4pqy\left[(2x^2 - y^2 + 1)(x^2 - 1)^2 - q^2(y^2 + 1)(y^2 - 1)^2\right]. \quad (102)\end{aligned}$$

Furthermore, the inclusion of a magnetic field in the system is subject to a theorem according to which the electric and magnetic fields in a static axisymmetric electrovacuum space-time are in a linear relationship (Perjés 1968).

To each static axisymmetric vacuum space-time there exists a solution of Equations (98). These Papapetrou-Majumdar electric counterpart metrics are obtained by assuming that the potential η is a functional of ξ alone. As an example, the electric counterpart of the Schwarzschild space-time is the Reissner-Nordstrøm space-time. The procedure also works for the stationary fields (Ernst 1968).

The field equations of a *stationary* Einstein-Maxwell system (with Lie derivatives of the metric and four-potential A along the time-like vector $K = \frac{\partial}{\partial t}$ vanishing) can be written solely in terms of the complex gravitational and electromagnetic three-vectors

$$\mathbf{G} = \frac{\nabla\mathcal{E} + 2\overline{\Phi}\nabla\Phi}{2(\mathrm{Re}\,\mathcal{E} + |\Phi|^2)} \tag{103}$$

and
$$\mathbf{H} = \frac{\nabla \Phi}{\sqrt{\operatorname{Re} \mathcal{E} + |\Phi|^2}} \qquad (104)$$

where R is the Ricci tensor of the manifold of Killing trajectories and

$$\mathcal{E} = f - |\Phi|^2 + i\psi \qquad (105)$$
$$\Phi = \sqrt{\frac{k}{2}}(A_o + iA') \qquad (106)$$

are the complex Ernst potentials defined in terms of the real scalars A' and ψ,

$$A'_{,\alpha} = \epsilon_{\alpha\beta\gamma}(A^{\alpha;\gamma} - \omega^\gamma A_o^{;\gamma})f \qquad (107)$$
$$\psi_{,\alpha} = \epsilon_{\alpha\beta\gamma}\omega^{\beta;\gamma}f^2 + 2\operatorname{Im}(\Phi\overline{\Phi}_{,\alpha}). \qquad (108)$$

The field equations of the stationary Einstein-Maxwell field are (Perjés 1971)

$$\begin{aligned}
(\nabla - \mathbf{G}) \cdot \mathbf{G} &= \overline{\mathbf{H}} \cdot \mathbf{H} - \overline{\mathbf{G}} \cdot \mathbf{G} \\
\nabla \times \mathbf{G} &= \overline{\mathbf{H}} \times \mathbf{H} - \overline{\mathbf{G}} \times \mathbf{G} \\
(\nabla - \mathbf{G}) \cdot \mathbf{H} &= \tfrac{1}{2}(\mathbf{G} - \overline{\mathbf{G}}) \cdot \mathbf{H} \\
\nabla \times \mathbf{H} &= -\tfrac{1}{2}(\mathbf{G} + \overline{\mathbf{G}}) \times \mathbf{H} \\
R_{\mu\nu} &= -G_\mu \overline{G_\nu} - \overline{G_\mu} G_\nu + H_\mu \overline{H_\nu} + \overline{H_\mu} H_\nu \ .
\end{aligned} \qquad (109)$$

(Note that Einstein's equations can be decomposed similarly with respect to an *arbitrary* non-null congruence of curves (Perjés 1993)). The equilibrium (mass=electric charge) metrics are obtained by assuming that the 3-space is flat (Israel and Wilson 1972, Perjés 1971).

The formulation like (98) is also possible for stationary fields by introducing the complex potentials ξ and η such that $\mathcal{E} = (\xi - 1)/(\xi + 1)$ and $\Phi = \eta/(\xi + 1)$. The beautiful and deep theory of the axisymmetric space-times is discussed in Neugebauer's article in these proceedings.

8 Further reading

The literature of our subject has grown to the extent that the bare list of papers would easily be longer than the length of these notes. An important omission from the exact solutions is the theory of Kerr-Schild space-times containing the algebraically special space-times. As far as techniques are concerned, one should be familiar with the algebraic classification of the curvature and some pertinent theorems (Goldberg-Sachs, Mariot-Robinson), spinors and the Newman-Penrose spin coefficient formalism, conformal techniques and multipole moments.

References

Bertotti B, 1959, *Phys Rev* **116** 1331.
Bonnor W, 1966, *Z Phys* **190** 444.
Brill D R, 1995, preprint gr-qc/9507019
Dautcourt G, 1964, in *Relativistic theories of gravitation* ed Infeld L, Pergamon
Ernst F J, 1968, *Phys Rev* **168** 1415.
Geroch R, 1971, *J Math Phys* **12** 918.
Griffiths J B, 1991, *Colliding plane waves in general relativity*, Clarendon Press
Hall G S, Morgan T and Perjés Z, 1987, *J Math Phys* **19** 1137.
Hawking S W and Ellis G F R, 1973, *The large-scale structure of space-time*, Cambridge University Press
Israel W and Wilson G A, 1972, *J Math Phys* **13** 865.
Kastor D and Traschen J, 1993, *Phys Rev* **D47** 5370.
Klösch T and Strobl T, 1995, *Class Quantum Gravity* **13** 1191.
Kóta J and Perjés Z, 1972, *J Math Phys* **13** 1695.
Kramer D et al. , 1980, *Exact Solutions of Einstein's Field Equations*, VEB Deutscher Verlag
Perjés Z, 1968, *Acta Physica Hungarica* **25** *303*.
Perjés Z, 1971, *Phys Rev Lett* **27** 1668.
Perjés Z, 1970, *J Math Phys* **11** 3383.
Perjés Z, 1986, *Gen Rel and Grav* **18** 511, 531.
Perjés Z, 1993, *Nucl Phys* **B403** 809.
Robinson I, 1959, *Bull Acad Polon Sci,* **7** *351.*
Robinson I and Trautman A, 1962, *Proc R Soc* **A265** 463.
Schwarzschild K, 1916, *Sitzber Preuss Akad Wiss, Physik-Mathematik, Kl* **189** *189.*
Synge J L, 1960, *Relativity: The General Theory*, North-Holland

Relativistic Cosmology

John Wainwright

Department of Applied Mathematics
University of Waterloo

1 Background

These notes contain an introduction to cosmology within the framework of general relativity theory. We discuss the Friedmann-Lemaitre (FL) models, with emphasis on the observational parameters, and illustrate the constraints imposed by observations through the use of a cosmological state space. We also characterise the Friedmann-Lemaitre universes within the general class of cosmological models, and discuss their viability as models of the real universe.

1.1 The scale of the universe

The subject of cosmology, the study of the universe as a whole, involves physical quantities—distances, times, temperatures and densities—which are extreme by terrestrial standards. So we begin by giving a glimpse of these extremes, and indicating the units that it is convenient to use.

Extragalactic distances are measured in megaparsecs where 1Mpc $\approx 3 \times 10^6$ light years. For nearby objects one uses kiloparsecs, where 1Mpc = 1000 kpc. Here are the distances of some well-known extragalactic objects, which play a role in the continuing efforts to establish the cosmological distance scale (see Fukugita *et al.* (1993)):

— the Large Magellanic Cloud (the nearest galaxy) 50 ± 3 kpc

— M81 (the most distant accurate calibrator) 3.63 ± 0.34 Mpc

— the Virgo cluster 14–22 Mpc

— the Coma cluster 75–120 Mpc.

For comparison, the particle horizon, representing the theoretical maximum distance, is at approximately 10^4Mpc , depending on the model and on the value of the Hubble constant.

Cosmological times are measured in gigayears, where 1 Gyr = 10^9 years. The ages of the oldest stars are estimated to be 10–20 Gyr. In contrast, the conjectured 'inflationary epoch' occurs at $t \approx 10^{-35}$ sec, where t is time since the initial singularity.

Temperatures are measured in degrees Kelvin. For example, the current temperature of the cosmic microwave background radiation is $T \approx 2.7$K. In contrast big-bang nucleosynthesis occurs at $T \approx 10^9$K, while the 'inflationary epoch' occurs at $T \approx 10^{27}$ K.

1.2 Observational facts and the standard model

Three observational facts led to the development of the so-called standard model of cosmology:

1. the systematic redshift of light from distant galaxies,

2. the highly isotropic cosmic microwave background radiation (CMB),

3. the uniform primordial abundance (approximately 23% by mass) of helium (^{4}He) in astronomical objects.

The redshifts were initially interpreted as being due to recession of the sources, leading to the idea that the universe is expanding (the expanding universe paradigm; see Harrison (1993)). We refer to Ellis (1989, pages 369–381) and Peebles (1993, pages 78–82) for a discussion of the development of this concept in the 1920's and 1930's. Following an expanding universe back in time leads to the occurrence of a hot dense phase, and subsequently to a breakdown of the laws of classical physics at a spacetime singularity. The first observational evidence for a hot dense phase was provided by the discovery in 1965 of the cosmic microwave background radiation, interpreted as relic radiation from such a phase. Further confirmation was provided by the theory of big-bang nucleosynthesis (BBN), which gave an explanation of the uniform abundance of ^{4}He (and subsequently of the other light isotopes), assuming the existence of an epoch of extreme temperature (approximately 10^{10}K) occurring very soon after the big-bang (about 1 second). We refer to Peebles (1971, pages 125–128, 240–241) for a discussion of the history of the cosmic microwave background and big-bang nucleosynthesis.

When Friedmann, Lemaitre and others first studied expanding cosmological models in the 1920's and 1930's, they assumed that the models were as simple as possible mathematically, namely that the geometry was homogeneous in space and isotropic about each fundamental observer. This assumption, the Cosmological Principle, characterises the so-called *Robertson-Walker geometry* which in conjunction with the Einstein field equations leads to the so-called *Friedmann-Lemaitre*(FL) *cosmologies*. The highly isotropic nature of the cosmic microwave background radiation is interpreted as support for this assumption.

In summary, the principal features of the standard model (see, for example, Peebles et al. (1991), Peebles (1993), pages 3–10) are:

1. Robertson-Walker geometry, incorporating the Cosmological Principle,

2. a hot dense radiation-dominated epoch during which nucleosynthesis produces the observed uniform abundance of the light isotopes, followed by a matter-dominated epoch during which the black body radiation from the hot dense epoch propagates freely,

3. galaxy formation and the growth of structure in their distribution is accounted for by perturbations of the Friedmann-Lemaitre model.

The standard model has by now gained widespread, though by no means universal, acceptance (see, for example, Arp et al. (1990)). Despite its success in giving a coherent explanation of a number of significant features of the universe, there remain a number of puzzles and uncertainties. At present there is no fully satisfactory theory of galaxy formation and the origin of large scale structure in the distribution of galaxies (Peebles 1993, page 9). We refer to Peebles and Silk (1988) and White et al. (1994, pages 327–330) for a survey of the different theories, and to Ostriker (1993) for a critical review of the popular cold dark matter theory. Another uncertainty is the nature of the dark matter, which observations, combined with theoretical considerations, suggest occurs in abundance (see Section 2.3).

It is also important to note that the validity of a Friedmann-Lemaitre model entails the requirement that, when averaged on some suitable scale,

1. the motion of the matter (galaxies) referred to as the Hubble flow, is isotropic, and

2. the matter is distributed homogeneously.

In recent years, however, structure in the distribution of galaxies and deviations from an isotropic Hubble flow have been detected on increasingly large scales, leading to uncertainties about the scale of averaging. These matters are discussed in Section 4.2.

Another difficulty associated with the standard model is the so-called horizon problem: due to the existence of a particle horizon in the standard big-bang Friedmann-Lemaitre models, no causal mechanism could lead to homogeneity on the surface of last scattering (e.g. uniform temperature), and hence this homogeneity has to be postulated rather than explained on physical grounds. In simple terms, the theory of inflation solves this problem by having the universe undergo a brief period of very rapid expansion driven by a scalar field called the inflaton, which effectively eliminates the particle horizon (see, for example, Peebles (1993, page 392)). Some theorists would include inflation as part of the standard model. This author is inclined to adopt Peebles's point of view that inflation is 'an elegant and influential scenario, but since it is difficult to test the idea observationally, it should not be included as part of the standard model' (Peebles 1993, pages xvi-xvii).

1.3 More general cosmologies

The Friedmann-Lemaitre universes are the simplest cosmological models within the framework of general relativity. Their mathematical simplicity is due to the fact that they admit a 6-parameter group G_6 of isometries whose orbits are the hypersurfaces of constant time. More general classes of solutions of the Einstein field equations have been used as models for cosmology in a variety of ways. We list the most important classes below.

1. *Bianchi Models*. These models admit hypersurfaces of homogeneity (the orbits of a 3-parameter group of isometries) but are anisotropic. We refer to MacCallum (1973) for a detailed introduction. They have been used to study the effects of anisotropy on the cosmic microwave background (see, for example, Barrow et al. (1985)) and on big-bang nucleosynthesis (see, for example, Barrow (1984) and Rothman & Matzner (1984)).

2. *Spherically symmetric models*. These models have been used in a variety of ways (MacCallum 1994, pages 192–193), for example, to study the effect on the Hubble constant of a region of underdensity in a Friedmann-Lemaitre model (Wu et al. 1995).

3. G_2 *models*. These models admit two commuting spacelike Killing vector fields, are anisotropic, and admit one degree of spatial inhomogeneity. They have been used in numerical relativity to study the effects of inhomogeneities on big-bang nucleosynthesis (Kurki-Suonio 1989), and on the occurrence of inflation (see, for example, Shinkai and Maeda (1994)).

4. *Szekeres models* (see Goode and Wainwright (1982) for a unified treatment). These models have dust as the source and are known explicitly. Although they admit no Killing fields, they can approximate Friedmann-Lemaitre models arbitrarily closely during part of their evolution.

We refer to MacCallum (1994) for further details and references.

1.4 Overview

In Section 2 we give an introduction to the Friedmann-Lemaitre (FL) universes. We derive the basic equations for these models with an n-component fluid as source, thereby incorporating dust, radiation, and a cosmological constant, and give a general discussion of the associated observational parameters. We also discuss the difficulties in determining these parameters from observations. As regards other references we mention Ellis (1987) and Kolb and Turner (1990, Chapters 1–3) for a detailed introduction, with a somewhat different emphasis, to the Friedmann-Lemaitre universes and the standard model.

Physical systems with a finite number of degrees of freedom, whose evolution in time is described by a system of autonomous differential equations, can be studied qualitatively using the theory of dynamical systems. The state of the system is represented by a point in state space ($\mathbf{R}^n$ or, more generally, a differentiable manifold), and the evolution of the system in time is described by an orbit in state space representing a solution of the system of differential equations (see for example Hirsch (1984), and Wiggins (1990)). Special orbits, the equilibrium points, describe equilibrium states of the physical system. The long-term behaviour of the physical system is described by an attractor (an equilibrium point in the simplest cases), which is an invariant set to which almost all orbits are attracted as time evolves. In Section 3 we introduce a cosmological state space, and use dynamical systems methods to analyse the evolution of various classes of Friedmann-Lemaitre models. An advantage of this approach is that the observations restrict the region of the state space in which the current state of the universe may lie, thereby giving a visual interpretation of the observations. Dynamical systems methods have also been used to analyse Bianchi models and G_2 models, and this section will serve as an introduction to this more general work (see for example Wainwright and Hsu (1989), Hewitt and Wainwright (1990)).

In Section 4, we give a mathematical formulation of the concept of a model universe being close to a Friedmann-Lemaitre universe. We then discuss the observations which potentially determine the extent to which a Friedmann-Lemaitre model gives a valid description of the real universe.

2 Friedmann-Lemaitre models

We begin by giving the basic equations that describe the Friedmann-Lemaitre models (Section 2.1). We then introduce the observational parameters and derive the constraints that follow from the field equations, showing that the models are falsifiable (Section 2.2). Finally we describe the bounds imposed on the parameters by observations (Section 2.3).

2.1 Basic equations

Line-element and field equations

The Cosmological Principle implies that the line-element has the Robertson-Walker form (see, for example, Weinberg (1972), pages 395–404):

$$ds^2 = -dt^2 + S(t)^2 d\Omega^2 \tag{1}$$

where the space sections $t = $ constant are of constant curvature:

$$d\Omega^2 = dr^2 + f(r)^2(d\theta^2 + \sin^2\theta\, d\phi^2) \tag{2}$$

with $f(r) = \sin r,\ r,\ \sinh r$ for $k = +1, 0, -1$. The curvature is positive, zero or negative depending on whether $k = +1, 0$ or -1. The coordinate t measures clock time along the world lines of the fundamental observers, with tangent vector field $\mathbf{u} = \frac{\partial}{\partial t}$.

The function $S(t)$ in the line-element (1) is called the scale function. In the cases $k = \pm 1$ it is uniquely determined by the line-element, while if $k = 0$, $S(t)$ is determined up to a constant multiplicative factor since the r-coordinate can be rescaled.

The symmetry of the line-element and the field equations $G_{ab} = T_{ab}$ imply that the stress-energy tensor has the following algebraic form

$$T_{ab} = (\mu + p)u_a u_b + p\, g_{ab} \tag{3}$$

where $\mu(t)$ is the total energy density, $p(t)$ is the isotropic pressure and the u_a are the covariant components of the vector field **u**.

The evolution of the models is governed by the conservation equation

$$\dot{\mu} = -3\frac{\dot{S}}{S}(\mu + p) \tag{4}$$

the Raychaudhuri equation

$$\frac{\ddot{S}}{S} = -\frac{1}{6}(\mu + 3p) \tag{5}$$

and the Friedmann equation

$$\frac{\dot{S}^2}{S^2} = \frac{1}{3}\mu - \frac{k}{S^2} \tag{6}$$

(see, for example, Weinberg (1972), page 472). We note that (5) and (6) follow from the field equations, and that (4) follows from the contracted Bianchi identities: $T_a{}^b{}_{;b} = 0$. The overdot denotes differentiation with respect to t.

Note that (4) follows from (5) and (6), by differentiating (6). Alternatively, (6) can be viewed as a first integral of (4) and (5). Equations (4)–(6) are valid irrespective of the detailed specification of the source, as described by T_{ab}. These equations are an under-determined system, however, until the pressure p is specified, for example by giving an equation of state $p = p(\mu)$.

The Raychaudhuri and conservation equations give rise to an elementary singularity theorem:

If $\mu + 3p > 0$ for all t, and $\dot{S}(t_0) > 0$, then there exists a time $t_b < t_0$ such that $S(t_b) = 0$ and $\lim_{t \to t_b^+} \mu = +\infty$.

The proof is as follows. Since $\mu + 3p > 0$, (5) implies $\ddot{S} > 0$ for all t, and hence that the graph of $S(t)$ is concave down. Since $\dot{S}(t_0) > 0$, the graph must intersect the t-axis at some time $t_b < t_0$. The second assertion follows from (4), which can be rewritten in the form $(\mu S^2)\dot{} = -(\mu + 3p)S\dot{S}$. The hypotheses imply that μS^2 is monotone increasing into the past and that $\lim_{t \to t_b^+}(\mu S^2) \neq 0$.

Stress-energy content

In the simplest Friedmann-Lemaitre models, the source is assumed to be dust (that is $p = 0$, $\mu > 0$) in the galactic epoch, or radiation ($p = \frac{1}{3}\mu$, $\mu > 0$) in the hot dense epoch. Another possibility is $p = -\mu$, in which case (4) implies $\mu = \Lambda$, a constant. In this case, the source can be interpreted as a cosmological constant (a positive cosmological constant is dynamically equivalent to endowing the vacuum with a non-zero energy density), or as the limiting case of a scalar field $\phi(t)$ with $\dot{\phi} = 0$ (see below). These three cases can be treated simultaneously by writing

$$p = (\gamma - 1)\mu \tag{7}$$

where $\gamma = \frac{4}{3}$ (radiation), $\gamma = 1$ (dust) or $\gamma = 0$ (cosmological constant).

More realistic models can be constructed by using an n-fluid source, with all fluids having the same 4-velocity **u**. The total energy density and pressure in (3) will be

$$\mu = \sum_{A=1}^{n} \mu_A, \qquad p = \sum_{A=1}^{n} p_A \qquad (8)$$

where

$$p_A = (\gamma_A - 1)\mu_A \qquad (9)$$

and the γ_A are constants. We shall assume that the fluids are non-interacting, so that each component separately satisfies that conservation equation (4). There are two cases of interest.

1. A combination of radiation and dust ($\gamma_1 = \frac{4}{3}$, $\gamma_2 = 1$), so that the model will have a radiation-dominated epoch and a matter-dominated epoch,

2. $\gamma_1 = 1, \gamma_2 = 0$, so that the model will describe the effects of a cosmological constant on a dust-filled universe.

A different type of source that is compatible with the stress-energy tensor (3) is a homogeneous massless scalar field (the 'inflaton'), with

$$T_{ab} = \phi_{,a}\phi_{,b} - \left[\frac{1}{2}\phi_{,c}\phi^{,c} + V(\phi)\right] g_{ab} \qquad (10)$$

which is used to describe an inflationary epoch. Note that if $\phi = \phi(t)$ then (10) is of the form (3) with

$$u_a = (-\phi_{,b}\phi^{,b})^{-\frac{1}{2}}\phi_{,a}, \qquad \mu = \frac{1}{2}\dot{\phi}^2 + V(\phi), \qquad p = \frac{1}{2}\dot{\phi}^2 - V(\phi) \qquad (11)$$

(see for example Peebles (1993), pages 395–396).

Redshift and temperature in a Friedmann-Lemaitre model

The redshift parameter z is defined by

$$z = \frac{\lambda_o - \lambda_e}{\lambda_e} \qquad (12)$$

where λ_e and λ_o are the emitted and observed wavelengths respectively. It is a well-known result (see for example, Wald (1984), pages 102–104) that in a Friedmann-Lemaitre model, with scale factor $S(t)$:

$$1 + z = \frac{S(t_o)}{S(t_e)} \qquad (13)$$

where t_e and t_o are the times of emission and observation. In this way, z can be used to label the cosmological timescale, instead of t.

The temperature of the cosmic microwave background radiation also provides a convenient way to label the time scale. For radiation ($\gamma = \frac{4}{3}$), the conservation equation (4) implies

$$\mu_{(\text{rad})} = CS(t)^{-4} \qquad (14)$$

where C is a constant. For black-body radiation $\mu_{(\text{rad})} = aT^4$, where a is Boltzmann's constant. It follows that

$$T(t) = T_0 \left[\frac{S_0}{S(t)}\right] \qquad (15)$$

where S_0 and T_0 are the values of S and T at the present time $t = t_0$. Recent observations from the COBE satellite (see Partridge (1994) for a summary) give

$$T_0 = 2.726 \pm 0.010 \,\text{K}.$$

Special solutions

There are a number of special Friedmann-Lemaitre universes which play an important role in the dynamical systems analysis in Section 3.

1. *The flat Friedmann-Lemaitre universe ($k = 0$).* The line-element is

$$ds^2 = -dt^2 + t^{\frac{4}{3\gamma}}\left[dr^2 + r^2(d\theta^2 + \sin^2\theta\, d\phi^2)\right] \qquad (16)$$

and the energy density and pressure are

$$\mu = \frac{4}{3\gamma^2 t^2}, \qquad p = (\gamma - 1)\mu, \qquad 0 < \gamma \leq 2.$$

In the case of dust ($\gamma = 1$), this solution is the Einstein-de Sitter universe (Einstein and de Sitter (1932)).

2. *The de Sitter universe ($k = 0$)*

The line-element is

$$ds^2 = -dt^2 + e^{\sqrt{\frac{\Lambda}{3}}t}\left[dr^2 + r^2(d\theta^2 + \sin^2\theta\, d\phi^2)\right] \qquad (17)$$

and the energy density and pressure are

$$\mu = \Lambda, \qquad p = -\Lambda$$

where Λ is the cosmological constant.

3. *The Milne universe ($k = -1$)*

The line-element is

$$ds^2 = -dt^2 + t^2\left[dr^2 + \sinh^2 r(d\theta^2 + \sin^2\theta\, d\phi^2)\right] \qquad (18)$$

and $\mu = 0, p = 0$.

4. *The Einstein static universe* ($k = +1$) The line-element is

$$ds^2 = -dt^2 + \ell^2 \left[dr^2 + \sin^2 r(d\theta^2 + \sin^2\theta\, d\phi^2)\right] \qquad (19)$$

where $\ell > 0$ is a constant. The energy density and pressure are

$$\mu = 3\ell^{-2}, \qquad p = -\ell^{-2}$$

(*i.e.* $\gamma = \frac{2}{3}$). One can also interpret the source as consisting of two fluids, with

$$\mu_1 = \left(\frac{2 - 3\gamma_2}{\gamma_1 - \gamma_2}\right)\ell^{-2}, \qquad \mu_2 = \left(\frac{3\gamma_1 - 2}{\gamma_1 - \gamma_2}\right)\ell^{-2}$$

and $p_1 = (\gamma_1 - 1)\mu_1$, $p_2 = (\gamma_2 - 1)\mu_2$, where γ_1 and γ_2 satisfy

$$0 \leq \gamma_2 < \frac{2}{3} < \gamma_1 \leq 2.$$

Einstein's choice was $\gamma_1 = 1$, $\gamma_2 = 0$, so that $\mu_2 = \ell^{-2} = \Lambda$, where Λ is the cosmological constant.

2.2 Theoretical constraints on the observational parameters

In order to analyse the observational properties of the Friedmann-Lemaitre models, it is necessary to introduce the so-called *observational parameters*.

Firstly, the *Hubble variable* H, which measures the rate of expansion of the universe, is defined by

$$H = \frac{\dot{S}}{S}. \qquad (20)$$

The *deceleration parameter* q, which measures whether the expansion is speeding up or slowing down, is defined by

$$q = -\frac{\ddot{S}S}{\dot{S}^2}. \qquad (21)$$

The initial singularity occurs when $S(t) = 0$. We choose the origin of t so that $S(0) = 0$. Then the present day value of t, denoted by t_0, represents the *age of the universe*, which is one of the observational parameters. The present day values of H and q are denoted by $H_0 = H(t_0)$ and $q_0 = q(t_0)$, and H_0 is called the *Hubble constant*.

There is an inequality relating t_0 and H_0 which holds irrespective of the specific nature of the source terms:

$$\text{if} \quad H_0 > 0 \quad \text{and} \quad \mu + 3p > 0 \quad \text{for all } t > 0, \quad \text{then} \quad t_0 H_0 < 1. \qquad (22)$$

The proof is as follows. Since $\mu + 3p > 0$, (5) implies $\ddot{S} > 0$ for all $t > 0$, *i.e.* the graph of $S(t)$ is concave down. The inequality follows by considering the intercept of the tangent line to the graph of $S(t)$ at (t_0, S_0).

We shall refer to (22) as the *age inequality*.

The *density parameter*, which describes the dynamical effects of the total energy density, is defined by

$$\Omega = \frac{\mu}{3H^2}. \tag{23}$$

Both q and Ω are dimensionless, while H has the dimensions of (time)$^{-1}$. It follows from (6) and (12) that

$$\Omega - 1 = \frac{k}{H^2 S^2}. \tag{24}$$

Thus, for an *open* model ($k = -1$), $\Omega < 1$, for a *closed* model ($k = +1$), $\Omega > 1$, and for a *flat* model ($k = 0$), $\Omega = 1$.

For an n-fluid model,

$$\Omega = \sum_{A=1}^{n} \Omega_A \tag{25}$$

where

$$\Omega_A = \frac{\mu_A}{3H^2}. \tag{26}$$

The present day values of the density parameters Ω_A are denoted by $\Omega_{A,0} = \Omega_A(t_0)$.

For an n-fluid model, we thus have the following set of observational parameters

$$\{t_0, H_0, q_0, \Omega_{A,0}\}. \tag{27}$$

The field equations lead to two constraints on these parameters. Firstly, it follows from the Raychaudhuri equation (5), using (8), (9), (21), (25) and (26) that

$$q = \frac{1}{2} \sum_{A=1}^{n} (3\gamma_A - 2)\Omega_A \tag{28}$$

which relates q_0 to $\Omega_{A,0}$.

The second constraint arises in a less direct manner, as follows. For an n-fluid model, the conservation equation (4) may be integrated separately for each fluid, to yield

$$\mu_A = \frac{C_A}{S^{3\gamma_A}}. \tag{29}$$

We introduce a *dimensionless scale factor* y by

$$y = \frac{S}{S_0} \tag{30}$$

where $S_0 = S(t_0)$. By using (8), (20), (25), (26), (28) and (29), the Friedmann equation (6) can be rearranged to read

$$\left(\frac{dy}{dt}\right)^2 = H_0^2 \, F(y, \Omega_{A,0}) \tag{31}$$

where

$$F(y, \Omega_{A,0}) = 1 + \sum_{A=1}^{n} \Omega_{A,0}(y^{2-3\gamma_A} - 1). \tag{32}$$

The differential equation (31), which shows how the matter content determines the scale function and hence the observational properties of the model, is of fundamental importance.

By separating variables in (31) and integrating with $0 \leq t \leq t_0$ and $0 \leq y \leq 1$, we obtain

$$t_0 H_0 = \int_0^1 \frac{1}{\sqrt{F(y, \Omega_{A,0})}} dy \equiv \alpha(\Omega_{A,0}). \tag{33}$$

The content of this equation is that the dimensionless constant $t_0 H_0$, called the *age parameter*, is a function of the dimensionless parameters $\Omega_{A,0}$, which describe the dynamical effects of the source at the present time $t = t_0$. This function, which cannot be written in elementary terms in general, is denoted by α in (33).

In deriving (31), we used the Friedmann equation in the form (24), but evaluated at $t = t_0$, namely,

$$\Omega_0 - 1 = \frac{k}{H_0^2 S_0^2} \tag{34}$$

in order to eliminate k. We note that if $k = \pm 1$, this equation expresses $S_0 = S(t_0)$ in terms of the observational parameters (27), while if $k = 0$ it leads to an additional constraint on the observational parameters, namely $\sum_{A=1}^n \Omega_{A,0} = 1$.

The differential equation (31) also leads to a distance-redshift relation and to an expression for the distance to the particle horizon, which we now derive. Consider a light signal emitted at time $t = t_e$ by a galaxy with worldline $r = r_e$, $\theta, \phi =$ constant, and received by us ($r = 0, t = t_0$). The worldline of the light signal is a past radial null geodesic of the line element (1), and is determined by

$$\frac{dr}{dt} = -\frac{1}{S(t)}.$$

We can rewrite this equation using (30) and (31) in the form

$$\frac{dr}{dt} = -\frac{1}{H_0 S_0} \frac{1}{y\sqrt{F(y, \Omega_{A,0})}}.$$

We separate variables and integrate, with $r = r_e$ to 0 and $y = y_e$ to 1, where $y_e = y(t_e)$. By (12) and (30) $y_e = 1/(1+z)$. Thus the present metric distance of the emitting galaxy, $d = S_0 r_e$, is given as a function of the redshift z (and the $\Omega_{A,0}$):

$$d(z) = H_0^{-1} \int_{1/(1+z)}^1 \frac{1}{y\sqrt{F(y, \Omega_{A,0})}} dy. \tag{35}$$

If $\lim_{z \to +\infty} d(z) = d_0$ exists, then the model will have a *particle horizon* at $t = t_0$, whose metric distance is

$$d_0 = H_0^{-1} \int_0^1 \frac{1}{y\sqrt{F(y, \Omega_{A,0})}} dy \equiv H_0^{-1} \beta(\Omega_{A,0}). \tag{36}$$

A particle horizon (Rindler 1956) divides all fundamental particles into two classes, those that have been observable by the observer $r = 0$ at time $t = t_0$ and those that have not. The constant d_0, as given in (36), is thus regarded as the *size of the observable universe at time* $t = t_0$. In general, d_0 cannot be written as an elementary function of the $\Omega_{A,0}$, and so for convenience this function is denoted by β in (36).

In summary, we have shown that in an n-fluid model the observational parameters (27) satisfy two constraints, namely

$$q_0 = \frac{1}{2}\sum_{A=1}^{n}(3\gamma_A - 2)\Omega_{A,0} \tag{37}$$

as follows from (28), and

$$t_0 = H_0^{-1}\alpha(\Omega_{A,0}) \tag{38}$$

where α is defined by (33). We have also shown that the size of the observable universe is given by

$$d_0 = H_0^{-1}\beta(\Omega_{A,0}) \tag{39}$$

where β is defined by (36), provided that the integral exists. One can interpret (37) and (38) as stating that *the matter content of the universe determines its age and size, up to scaling by the Hubble constant.*

In the case of a single-fluid model with dust ($n = 1, \gamma_1 = 1$), equations (37)–(39) specialize to

$$q_0 = \frac{1}{2}\Omega_0, \quad t_0 = H_0^{-1}\alpha(\Omega_0), \quad d_0 = H_0^{-1}\beta(\Omega_0). \tag{40}$$

In this case the integrals for α and β are elementary:

$$\alpha(\Omega_0) = \begin{cases} (\Omega_0 - 1)^{-3/2}\left[-(\Omega_0-1)^{1/2} + \frac{1}{2}\Omega_0 \cos^{-1}(\frac{2}{\Omega_0}-1)\right] & \Omega_0 > 1 \\ \frac{2}{3} & \Omega_0 = 1 \\ (1-\Omega_0)^{-3/2}\left[+(1-\Omega_0)^{1/2} - \frac{1}{2}\Omega_0 \cosh^{-1}(\frac{2}{\Omega_0}-1)\right] & \Omega_0 < 1 \end{cases} \tag{41}$$

$$\beta(\Omega_0) = \begin{cases} (\Omega_0 - 1)^{-1/2}\cos^{-1}(\frac{2}{\Omega_0}-1) & \Omega_0 > 1 \\ 2 & \Omega_0 = 1 \\ (1-\Omega_0)^{-1/2}\cosh^{-1}(\frac{2}{\Omega_0}-1) & \Omega_0 < 1 \end{cases} \tag{42}$$

(see Weinberg (1972), pages 482–484 and 489, who uses $2q_0$ in place of Ω_0).

One can use these formulae to give estimates of the present age and size of the universe. The Hubble constant is traditionally expressed in units of $\text{km s}^{-1}\text{Mpc}^{-1}$, since the redshift was interpreted as a classical Doppler shift due to a velocity of recession v, giving $v \approx H_0 d$ to first order. To account for the uncertainty in the value of H_0 (see Section 2.3), it is customary to write

$$H_0 = 100h \text{ km s}^{-1}\text{Mpc}^{-1} \tag{43}$$

where h is a dimensionless constant. When using formulae such as (40) for t_0 or d_0, it is convenient to express H_0^{-1} in Gyr or in Mpc:

$$H_0^{-1} \approx \frac{9.8}{h} \text{ Gyr}, \quad H_0^{-1} \approx \frac{3.3}{h}\times 10^3 \text{ Mpc}. \tag{44}$$

These numbers are called the *Hubble time* and *Hubble distance* respectively.

Equations (35)–(39) give the following estimates for the age and size of the universe:

$$t_0 \approx \begin{cases} 6.5h^{-1}\,\text{Gyr} & \text{if } \Omega_0 = 1 \\ 8.8h^{-1}\,\text{Gyr} & \text{if } \Omega_0 = 0.1 \end{cases}$$

$$d_0 \approx \begin{cases} 6.6h^{-1}\times 10^3\,\text{Gyr} & \text{if } \Omega_0 = 1 \\ 12.5h^{-1}\times 10^3\,\text{Gyr} & \text{if } \Omega_0 = 0.1 \end{cases}$$

2.3 Observational constraints

In this section we discuss the constraints that are placed by observations on the parameters $\{t_0, H_0, q_0, \Omega_{A,0}\}$ in (27).

The Hubble constant H_0 and the deceleration parameter q_0

The Hubble constant H_0 and the deceleration parameter q_0 can in principle be determined by measuring the luminosities and redshifts of galaxies. The luminosity distance d_L is defined by

$$d_L = \left(\frac{L}{4\pi \ell}\right)^{1/2}$$

where L is the absolute luminosity (assumed known) and ℓ is the apparent luminosity (measured). For any expanding Friedmann-Lemaitre model, H_0 and q_0 determine to second order the dependence of d_L on the redshift z:

$$d_L = H_0^{-1}[z + \frac{1}{2}(1 - q_0)z^2 + \cdots]$$

(see for example, Weinberg (1972) pages 441–442). For a specific Friedmann-Lemaitre model, one can derive an exact formula of this type (*e.g.* Weinberg (1972), equation (15.3.24), page 485, and Terrell (1977)).

A major difficulty in determining H_0 is that for nearby galaxies ($d \lesssim 3\,\text{Mpc}$), whose distances are reliably known, the peculiar velocities caused by local mass concentrations can be comparable to the Hubble velocity (which is assumed to be isotropic on a sufficiently large scale). On the other hand, for more distant galaxies there are uncertainties in the distance scale. In particular, the distance to the Virgo cluster is not reliably known. Fukugita *et al.* (1993) summarise the two standard methods for calibrating the extra-galactic distance scale, the one leading to a 'low' H_0 ($50 \pm 10\,\text{km s}^{-1}\text{Mpc}^{-1}$) and the other to a 'high' H_0 ($80 \pm 10\,\text{km s}^{-1}\text{Mpc}^{-1}$). Since then, observations of Cepheid variable stars in galaxies in the Virgo cluster by two different groups have led support to the 'high' value (Freedman *et al.* 1994, Pierce *et al.* 1994). However, a recent paper by Sandage and his coworkers (Sandage *et al.* 1994) using supernovae of type Ia as distance indicators, supports the 'low' value.

During the 1960's and 1970's major efforts were made to determine q_0 from redshift observations, but these efforts were discontinued when it was realised that evolutionary effects in galaxies, which are little understood, dominated the observations. A recent estimate is $-1 < q_0 < 1.5$ (Sandage and Tammann 1986).

The age of the universe t_0

One can obtain a lower bound on the age of the universe t_0 by inferring the ages of various astronomical objects. For example, the oldest stars are observed in globular clusters and their ages are estimated to be $15 \pm 3\,\text{Gyr}$ (Peebles et al. 1991, page 773). Fukugita et al. (1993) comment that stellar theories cannot accomodate a universe younger than 13 Gyr, i.e.

$$t_0 > 13\,\text{Gyr} \qquad (45)$$

(see page 312). One can also date lunar and meteoritic rock using long-lived radioactive isotopes such as U^{238}, leading to estimates for the age of the galaxy of 9–16 Gyr (Symbalisty and Schramm 1981). We refer to Sciama (1993, page 59) for further references.

The density parameter Ω_0

Following Sciama (1993, page 61), we distinguish three different types of estimate for Ω_0, labelled Ω_{vis}, Ω_{b}, Ω_{tot}. We refer also to Peebles (1993, page 476) and Rubin (1989, page 99).

1. *The observed baryonic density Ω_{vis}.*

   This quantity represents the visible baryons in galaxies and clusters, and is estimated using mass-to-luminosity ratios. In a recent analysis Persic and Salucci (1992) find $\Omega_{\text{vis}} \approx 0.003$.

2. *The derived baryonic density Ω_{b}*

   This quantity depends on the theory of big-bang nucleosynthesis (BBN), and involves the abundances of the light isotopes D, ^{3}He, ^{4}He and ^{7}Li. The theoretical abundances depend on the ratio $\eta = n_{\text{b}}/n_\gamma$, where n_{b} and n_γ are the baryon and photon number densities. Thus observed abundances lead to bounds on η. The densities n_{b} vary as the universe expands, but η remains constant. The observed present CMB temperature T_0 (Section 2.1) determines the present energy density n_γ (see for example, Sciama 1993). Knowing η, one can successively calculate the present values of n_{b} and μ_{b}. The resulting bounds on Ω_{b} depend on the Hubble constant:

   $$0.009 \leq \Omega_{\text{b}} h^2 \leq 0.02$$

   (Copi et al. 1995, page 192).

3. *The dynamical matter density Ω_{tot}*

   This quantity is estimated by determining how much mass is needed to cause the observed orbital motions in galaxies and clusters. The interpretation is that this quantity represents the total matter density, hence the notation. There seems to be general agreement that $\Omega_{\text{tot}} \geq 0.1$ (Sciama 1993, page 56), and some dynamical estimates on scales $d \gtrsim 30\,h^{-1}\,\text{Mpc}$ give values up to 1 (Peebles 1993, page 476; however, see Bahcall et al. 1995 and White et al. 1993).

The difference in these estimates is striking:

$$\Omega_{\text{vis}} < \Omega_{\text{b}} < \Omega_{\text{tot}}.$$

The fact that $\Omega_{\text{vis}} < \Omega_b$ implies that a significant amount of baryonic matter (perhaps as high as 90%) is *dark*. We refer to Sciama (1993, pages 62–63) for a brief discussion of the possibilities, and to Carr (1994) for further details. Secondly, $\Omega_b < \Omega_{\text{tot}}$ implies the existence of *non-baryonic dark matter*. Since non-baryonic dark matter has not been detected to date, the situation as regards Ω_0 is in a state of uncertainty (Peebles (1993) comments (see page 677) 'It would be difficult to overstate the effect of a laboratory detection of nonbaryonic dark matter would have on cosmology in general ...'). We note that Sasselov and Goldwirth (1995) have re-examined the systematic errors in the observational determination of the primordial helium abundance, which led to an increase in the upper bound for $\Omega_b h^2$ and the possibility that $\Omega_b \approx \Omega_{\text{tot}}$.

Confrontation with observation

We have seen that in a Friedmann-Lemaitre model with dust (and zero cosmological constant), the observational parameters satisfy the restrictions (40). We now discuss the resulting confrontation with observations.

Since q_0 is ill-determined by observation, the constraint $q_0 = \frac{1}{2}\Omega_0$ in (40) does not lead to a confrontation with observation. The age constraint $t_0 H_0 = \alpha(\Omega_0)$, where α is given by (41), does, however, contain the possibility of excluding certain Friedmann-Lemaitre models. We illustrate the situation by considering the implications of the lower bound $t_0 > 13$ Gyr. The age constraint, in conjunction with (44), shows that

$$t_0 > 13\,\text{Gyr} \quad \text{implies} \quad h < 0.75\,\alpha(\Omega_0) \qquad (46)$$

for any Friedmann-Lemaitre dust model, where h is defined by (43). Firstly, the age inequality (22) implies $\alpha(\Omega_0) \leq 1$ for all Ω_0. Thus by equation (46), $t_0 > 13$ Gyr and $H_0 > 75\,\text{km}\,\text{s}^{-1}\text{Mpc}^{-1}$ is incompatible with all Friedmann-Lemaitre dust models. Secondly, suppose $\Omega_0 > 0.1$. By (41), $\alpha(\Omega_0) < 0.9$. Thus, by (46), $t_0 > 13$ Gyr and $H_0 > 68\,\text{km}\,\text{s}^{-1}\text{Mpc}^{-1}$ is incompatible with all Friedmann-Lemaitre dust models with $\Omega_0 > 0.1$. Finally, suppose $\Omega_0 = 1$, so that $\alpha(\Omega_0) = 2/3$ by (41). Thus by (46), $t_0 > 13$ Gyr and $H_0 > 50\,\text{km}\,\text{s}^{-1}\text{Mpc}^{-1}$ is incompatible with the Einstein-de Sitter model. Further implications of the constraints imposed by observations are discussed in the next section.

3 The dynamical systems approach

In this section we will use dynamical systems methods to analyse various classes of Friedmann-Lemaitre models. We show that observations restrict the region of the cosmological state space in which the current state of the universe may lie. This approach has its origins in the work of Stabell and Refsdal (1966) and Gott *et al.* (1974).

3.1 One-fluid models and the ΩH-plane

We introduce a dimensionless time variable τ by

$$\tau = \ln\left(\frac{S}{S_0}\right). \qquad (47)$$

By the definition (20) of H it follows that
$$\frac{dt}{d\tau} = \frac{1}{H}. \tag{48}$$
By differentiating H and using the definition (21) of q, we obtain
$$\frac{dH}{d\tau} = -(1+q)H. \tag{49}$$
Next, we rewrite the conservation equation (4) to obtain an evolution equation for the density parameter Ω, as defined by (23). After using (5), with $\check{S}$ expressed in terms of q, we obtain
$$\frac{d\Omega}{d\tau} = -2q(1-\Omega). \tag{50}$$
Equations (49) and (50) hold for any Friedmann-Lemaitre cosmology. The expression for q, however, depends on the source terms, and for a one-fluid model, (28) gives
$$q = \frac{1}{2}(3\gamma - 2)\Omega. \tag{51}$$
Equations (49)–(51) form an autonomous differential equation (ODE) in $\mathbf{R}^2$ for (Ω, H). We regard a point $(\Omega, H) \in \mathbf{R}^2$ as describing the state of the universe at an instant of time, and the orbits of the autonomous ODE describe the evolution.

The age constraint (40) for a one-fluid model reads $t_0 H_0 = \alpha(\Omega_0)$, where α is given by (41). Since t_0 is arbitrary time, we drop the subscript and rewrite the constraint as
$$H = \frac{1}{t}\alpha(\Omega). \tag{52}$$
This equation describes the curves of constant time ($t = $ constant) in the ΩH-plane. Figure 1 shows typical orbits and curves of constant time in the case of dust ($\gamma = 1$), with the range of the variables, $0 \leq \Omega \leq 2$, $0 \leq H \leq 120$, chosen to accommodate the present epoch. Note that the following three constraints are embodied in Figure 1;
(i) Corresponding to the big-bang at $\tau \to -\infty$
$$\lim_{\tau \to -\infty} \Omega = 1, \qquad \lim_{\tau \to -\infty} H = +\infty.$$
(ii) For open models (*i.e.* $k = -1$, and $\Omega < 1$ by (24)) we must have expansion to infinite size in infinite time.
$$\lim_{\tau \to +\infty} \Omega = 0, \qquad \lim_{\tau \to +\infty} H = 0.$$
(iii) For closed models (*i.e.* $k = +1$, $\Omega > 1$),
$$\lim_{\tau \to \tau_{\max}} \Omega = +\infty, \qquad \lim_{\tau \to \tau_{\max}} H = 0.$$
corresponding to approach to the instant of maximum expansion in a finite time, prior to recollapse.

In order to indicate the observational constraints on the location of the present state of the universe in the ΩH-plane, it is convenient to use a log scale for the Ω-axis, as in Figure 2 (*cf.* Figure 1 on page 544 in Gott *et al.* (1974))

We have drawn the age lower limit $t_0 \geq 12$, the BBN lower bound $\Omega_b h^2 \geq 0.01$ and the Hubble lower bound $H_0 \geq 40$ as illustrative examples of the restrictions that observations can impose on the cosmological state space. The figure illustrates clearly the severe constraints that a 'high' value of H_0 imposes, as discussed in Section 2.3.

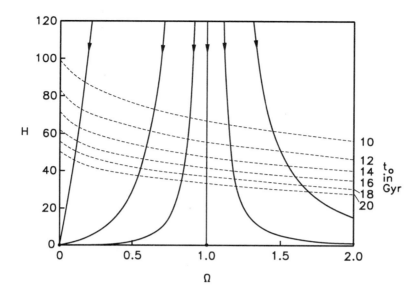

Figure 1. *Evolution of expanding dust* FL *models in the* ΩH-*plane*

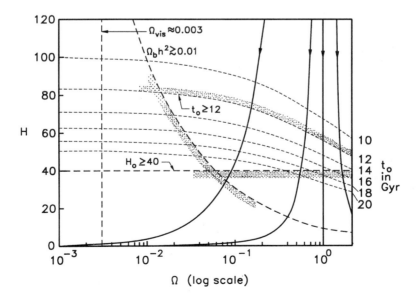

Figure 2. *Observational constraints on expanding* FL *models in the* ΩH-*plane*

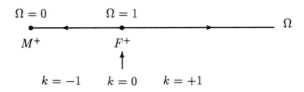

Figure 3. *Dynamics of expanding 1-fluid FL models in the Ω state space*

3.2 The compactified density parameter $\widetilde{\Omega}$

The evolution of the one-fluid Friedmann-Lemaitre models can be shown more simply by using a one-dimensional state space, the Ω-axis. Such a reduction in dimension is possible because in the autonomous ODE for (Ω, H), given by (48)–(51), the H-equation decouples from the Ω-equation:

$$\frac{d\Omega}{d\tau} = -\frac{1}{2}(3\gamma - 2)(1 - \Omega)\Omega. \tag{53}$$

(For $\gamma \neq \frac{2}{3}$) this ODE has two equilibrium points (see Figure 3):

1. $\Omega = 1$, the flat expanding FL model ($k = 0$), which we denote by F^+,

2. $\Omega = 0$, the expanding Milne model ($k = -1$), which we denote by M^+.

The asymptotic behaviour is as follows: all models are past asymptotic to F^+; all open models are future asymptotic to M^+.

The dimensionless age parameter tH, and the deceleration parameter q can be viewed as functions on state space, since, by (51–52) they are determined uniquely by Ω,

$$tH = \alpha(\Omega), \qquad q = \frac{1}{2}(3\gamma - 2)\Omega. \tag{54}$$

These equations provide the confrontation with observation.

The drawback of the preceding Ω-representation is that it does not give a complete description of the evolution of the closed ($k = +1$) models. To circumvent this difficulty, we define the *compactified density parameter* $\widetilde{\Omega}$ by

$$\widetilde{\Omega} = \arctan\left(\frac{\sqrt{3}\,H}{\sqrt{\mu}}\right), \qquad \mu > 0. \tag{55}$$

It follows that

$$-\frac{\pi}{2} < \widetilde{\Omega} < \frac{\pi}{2}, \qquad \text{sign}(\widetilde{\Omega}) = \text{sign}(H),$$

and

$$\Omega = (\tan \widetilde{\Omega})^{-2}, \qquad \widetilde{\Omega} \neq 0. \tag{56}$$

As regards the definition of $\tilde{\Omega}$, note that $H/\sqrt{\mu}$ is bounded at the instant of maximum expansion since $H = 0$, $\mu > 0$ there. In addition, the arctan ensures that $\tilde{\Omega}$ is bounded as $\mu \to 0$ in ever-expanding models.

In order that the evolution equation for $\tilde{\Omega}$ be regular, it is necessary to introduce a new time variable $\tilde{\tau}$ via

$$\frac{dt}{d\tilde{\tau}} = \sqrt{\frac{3}{\mu}} \cos \tilde{\Omega}. \tag{57}$$

In terms of $\tilde{\tau}$, the conservation equation (4), with $p = (\gamma - 1)\mu$, becomes

$$\frac{d\mu}{d\tilde{\tau}} = -3\gamma\mu \sin \tilde{\Omega}. \tag{58}$$

Finally, it follows from (53), (56) and (57) that the evolution equation for $\tilde{\Omega}$ is

$$\frac{d\tilde{\Omega}}{d\tilde{\tau}} = -\frac{1}{4}(3\gamma - 2) \cos 2\tilde{\Omega} \cos \tilde{\Omega}. \tag{59}$$

The autonomous ODE (59) determines the evolution of the single-fluid universes in the state space $-\pi/2 \leq \tilde{\Omega} \leq \pi/2$. The ODEs (57) and (58) play an auxiliary role, in that (58) determines $\mu(\tilde{\tau})$ knowing $\tilde{\Omega}(\tilde{\tau})$, and (57) determines $t(\tilde{\tau})$ knowing $\mu(\tilde{\tau})$ and $\tilde{\Omega}(\tilde{\tau})$. The Hubble parameter H is then determined by (55), knowing $\mu(\tilde{\tau})$ and $\tilde{\Omega}(\tilde{\tau})$.

The equilibrium points of the ODE (59) and their physical interpretation are listed below.

F^+: $\tilde{\Omega} = \pi/4$,   expanding flat FL model $(\Omega = 1, k = 0, H > 0)$
F^-: $\tilde{\Omega} = -\pi/4$,   contracting flat FL model $(\Omega = 1, k = 0, H < 0)$
M^+: $\tilde{\Omega} = \pi/2$,   expanding Milne model $(\Omega = 0, k = -1, H > 0)$
M^-: $\tilde{\Omega} = -\pi/2$,   contracting Milne model $(\Omega = 0, k = -1, H < 0)$

Figure 4 shows the state space $-\pi/2 \leq \tilde{\Omega} \leq \pi/2$; it gives a complete description of the dynamics of the single fluid models—for both expanding and contracting states.

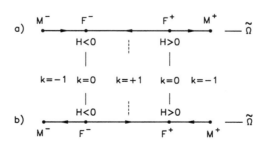

Figure 4. *Dynamics of one-fluid FL models in the $\tilde{\Omega}$ state space. In case (a), $2/3 < \gamma \leq 2$ $(q > 0)$, and in case (b), $0 \leq \gamma < 2/3$ $(q < 0)$.*

3.3 Two-fluid models and the $\tilde{\Omega}\chi$-plane

We now use the compactified density parameter $\tilde{\Omega}$, as defined by (55), to describe the evolution of the 2-fluid Friedmann-Lemaitre models. We assume that

$$\gamma_1 > \gamma_2 \quad \text{and} \quad \gamma_1 > \frac{2}{3}. \tag{60}$$

In order to describe which fluid is dominant dynamically, we use the *transition variable* χ, as in Coley and Wainwright (1992), defined by

$$\chi = \frac{\mu_2 - \mu_1}{\mu_2 + \mu_1}. \tag{61}$$

It follows that the density parameters Ω_1 and Ω_2, as defined by (17), are related to Ω by

$$\Omega_1 = \frac{1}{2}(1-\chi)\Omega, \qquad \Omega_2 = \frac{1}{2}(1+\chi)\Omega. \tag{62}$$

Note that $-1 \leq \chi \leq 1$, and that when $\chi \to -1$, fluid 1 is dominant, and when $\chi \to +1$, fluid 2 is dominant. The expression for the deceleration parameter q is obtained from (28) and (62):

$$q = \frac{3}{4}(\gamma_1 - \gamma_2)(b - \chi)\Omega \tag{63}$$

where

$$b = \frac{3(\gamma_1 + \gamma_2) - 4}{3(\gamma_1 - \gamma_2)}. \tag{64}$$

The restriction (60) implies that $b > -1$.

As in the single fluid case in Section 3.2, we introduce a new time variable $\tilde{\tau}$ by

$$\frac{dt}{d\tilde{\tau}} = \frac{2}{3(\gamma_1 - \gamma_2)}\sqrt{\frac{3}{\mu}} \cos \tilde{\Omega}. \tag{65}$$

Then the conservation equation (4), with (8), (9) and (61) becomes

$$\frac{d\mu}{d\tilde{\tau}} = -\left[\frac{\gamma_1(1-\chi) + \gamma_2(1+\chi)}{\gamma_1 - \gamma_2}\right]\mu \sin \tilde{\Omega}. \tag{66}$$

Finally, the evolution equation for $\tilde{\Omega}$ is obtained from (48), (50), (56), (63) and (65), and the evolution equation for χ is obtained by applying the conservation equation (4) to μ_1 and μ_2 separately. The result is

$$\begin{aligned}\frac{d\tilde{\Omega}}{d\tilde{\tau}} &= -\frac{1}{2}(b-\chi)\cos 2\tilde{\Omega} \cos \tilde{\Omega} \\ \frac{d\chi}{d\tilde{\tau}} &= (1-\chi^2)\sin \tilde{\Omega}.\end{aligned} \tag{67}$$

We note that (65) and (66) play an auxiliary role, analogous to that played by (57) and (58) in the single-fluid case.

The state space is the rectangle given by $-\pi/2 \leq \tilde{\Omega} \leq \pi/2$ and $-1 \leq \chi \leq 1$. The structure of the orbits in the state space (Figure 5) is governed by the following invariant sets of the ODE (67).

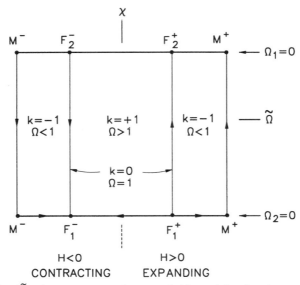

Figure 5. The $(\tilde{\Omega}, \chi)$ state space for two-fluid models showing equilibrium points, invariant sets, and the signs of the Hubble variable H and spatial curvature k.

$\tilde{\Omega} = -\pi/2$	contracting empty models	$(\Omega = 0, H < 0)$
$\chi = -1$	single fluid models	$(\Omega_2 = 0)$
$\chi = +1$	single fluid models	$(\Omega_1 = 0)$
$\tilde{\Omega} = \pi/4$	expanding flat models	$(\Omega = 1, H > 0)$
$\tilde{\Omega} = -\pi/4$	contracting flat models	$(\Omega = 1, H < 0)$
$\tilde{\Omega} = \pi/2$	expanding empty models	$(\Omega = 0, H > 0)$
$\tilde{\Omega} = -\pi/2$	contracting empty models	$(\Omega = 0, H < 0)$

The intersections of these invariant sets give equilibrium points, labelled $F_1^\pm$, $F_2^\pm$, $M^\pm$, in analogy with Figure 4. The subscripts 1,2 indicate which fluid is present. The assumptions (60) determine the direction of increasing time, except on the invariant set $\chi = 1$ ($\Omega_1 = 0$). The invariant set $\chi = -1$ ($\Omega_2 = 0$) coincides with Figure 4a.

There is one additional equilibrium point, given by $\tilde{\Omega} = 0$, $\chi = b$, which, however, lies in the state space if and only if $\gamma_2 \leq 2/3$ ($\gamma \leq 2/3$ is equivalent to $b \leq 1$, as follows from (60) and (63)). This equilibrium point is denoted by E, and corresponds to the two-fluid interpretation of the Einstein static model in Section 2.1 (see (19)).

Figure 6 shows the $(\tilde{\Omega}, \chi)$ state space for the two-fluid models with $\gamma_1 = 1$ and $\gamma_2 = 0$, corresponding to dust and a positive cosmological constant, and gives the complete set of dynamical possibilities for these models, including the contracting models, for which $H < 0$ or equivalently, $\tilde{\Omega} < 0$. The scale factors $S(t)$ for the different orbits (models) in Figure 6 are shown qualitatively in the minigraphs.

This approach complements the usual analysis of the dynamics of the Friedmann-Lemaitre models (see for example, Rindler (1977), pages 235–238), first given by Robertson (1933) and extended by Harrison (1967). Note that the Einstein-de Sitter model and the de Sitter model correspond to the equilibrium points F_1^+ and F_2^+ respectively,

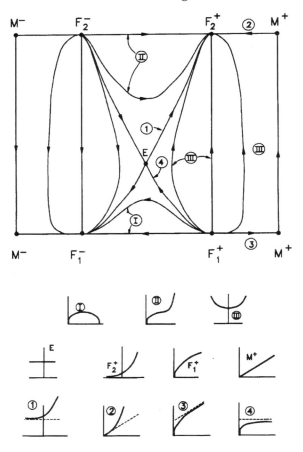

Figure 6. The $(\tilde{\Omega}, \chi)$ state space for two-fluid models with dust and cosmological constant. The minigraphs show the scale factor $S(t)$ for the various orbits.

and the Milne model corresponds to M^+. In addition the so-called Eddington-Lemaître models (Rindler 1977, page 238) are described by the orbit labelled (1), and the Lemaître coasting models are described by orbits labelled (III) which pass close to the Einstein static model E, and hence have a quasi-static phase. Figure 6 thus shows the role played by the Einstein static model in determining the evolution of other models. The fact that E is a saddle point establishes the well-known instability of the Einstein static model, first shown by Eddington. Another feature of interest in that the ODE (67) is invariant under the mapping $(\tilde{\tau}, \tilde{\Omega}) \to (-\tilde{\tau}, -\tilde{\Omega})$. Thus to each model there corresponds a time-reversed model whose orbit is obtained by reflecting in the χ-axis and reversing the direction of time. The graphs of $S(t)$ for the time-reversed models are obtained by reflecting the graphs in the vertical axis.

Stabell and Refsdal (1966) were apparently the first to use dimensionless variables to describe the evolution of Friedmann-Lemaître models with dust and a cosmological constant (they use $\sigma = \Omega/2$ as the 'density parameter'). They used the (q, Ω)-plane

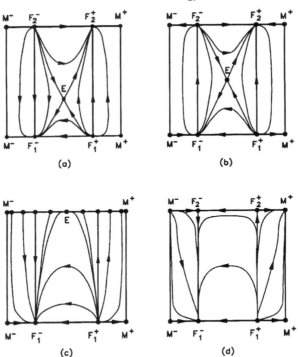

Figure 7. *The (Ω, χ) state space for two-fluid models, whose evolution is governed by (4.20). The four cases (a), (b), (c) and (d) correspond to $b = -1/3, 0, 1, 3$ respectively. The values of γ_1 and γ_2 are given in the text.*

as the state space. (This approach has been generalised by Ehlers and Rindler (1989) to include the effects of radiation—a three-fluid model.) Our figure Figure 6 should be compared with their Figure 4. The drawback of using the (q, Ω)-plane is that the Einstein static model lies 'at infinity', since q and Ω are undefined for that model.

Figure 7 shows the dependence of the dynamics of the two-fluid Friedmann-Lemaitre models on the equation of state parameters, through the constant b given by (64). The four cases shown are (a) $\gamma_1 = 1$, $\gamma_2 = 0$, (b) $\gamma_1 = 4/3$, $\gamma_2 = 0$, (c) $\gamma_1 = 4/3$, $\gamma_2 = 2/3$ and (d) $\gamma_1 = 4/3$, $\gamma_2 = 1$. The main feature is the bifurcation that occurs at $b = 1$ (case (c), $\gamma_2 = 2/3$), when the equilibrium point E leaves the state space, transferring stability from F_2^+ to M_2.

Figure 8 shows the Friedmann-Lemaitre models with dust and cosmological constant (as in Figure 6), illustrating how the $\tilde{\Omega}\chi$-plane can be used to compare the models with observations. In this figure we have illustrated the level curves of the dimensionless age parameter α and of the density parameter Ω_1 for the dust. One can see that a high value of $\alpha = H_0 t_0$ (i.e. $\alpha > 1$), which requires a non-zero cosmological constant, is compatible with a flat Friedmann-Lemaitre model, provided that Ω_1 is not too large (compare Peebles 1984).

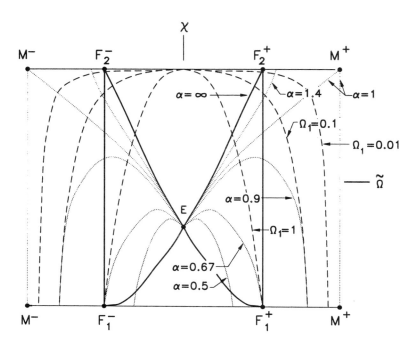

Figure 8. The $(\tilde{\Omega}, \chi)$ state space for the two-fluid models with dust and cosmological constant, showing the level sets $\Omega_1 = $ constant and $\alpha = $ constant, where Ω_1 is the density parameter for the dust, and $\alpha = Ht$ is the dimensionless age parameter.

4 Universes close to a Friedmann-Lemaitre model

A general cosmological model $\{M, g, \mathbf{u}\}$ is specified by giving a manifold M, a Lorenzian metric g and a unit timelike vector field $\mathbf{u}$, which gives the 4-velocity of the family of fundamental observers. We first describe the mathematical differences between Friedmann-Lemaitre models and more general cosmological models, with a view to giving a coordinate independent definition of the concept of a model being 'close to Friedmann-Lemaitre' (Section 4.1). We then discuss specific observations which potentially determine to what extent the Friedmann-Lemaitre models give a valid description of the real universe (Section 4.2).

4.1 Models more general than Friedmann-Lemaitre

The Friedmann-Lemaitre models differ from more general models (see Section 1.3) as regards *kinematic properties*, properties of the *curvature tensor*, and *gradients of physical and geometrical scalars*. The first two properties give a characterisation of the Friedmann-Lemaitre models, while the third provides a necessary condition. We refer to Ellis (1973) and MacCallum (1973) for further details about these concepts and other aspects of general cosmological models.

Kinematic properties

The kinematic quantities associated with a timelike congruence **u** were first introduced by Ehlers in 1961 (see Ehlers 1993 for a reprint of his original article). We refer to Ellis (1973, pages 9–11) for a discussion of these quantities in a cosmological context, and here give a brief review.

A unit timelike vector field **u** determines a projection tensor h_{ab} according to

$$h_{ab} = g_{ab} + u_a u_b$$

which at each point projects into the 3-space orthogonal to **u**. It follows that

$$h_a{}^c h_c{}^b = h_a{}^b, \quad h_a{}^b u_b = 0, \quad h_a{}^a = 3.$$

One can decompose the covariant derivative $u_{a;b}$ into its irreducible parts according to

$$u_{a;b} = \sigma_{ab} + \omega_{ab} + \tfrac{1}{3}\Theta h_{ab} - \dot{u}_a u_b$$

where σ_{ab} is symmetric and tracefree, ω_{ab} is antisymmetric, and $u^a \sigma_{ab} = 0 = u^a \omega_{ab}$. It follows that

$$\begin{aligned}
\Theta &= u^a{}_{;a} \\
\dot{u}_a &= u_{a;b} u^b \\
\sigma_{ab} &= u_{(a;b)} - \tfrac{1}{3}\Theta h_{ab} + \dot{u}_{(a} u_{b)} \\
\omega_{ab} &= u_{[a;b]} + \dot{u}_{[a} u_{b]}.
\end{aligned}$$

The scalar Θ is the (volume) *expansion* and $\dot{u}_a$ is the *acceleration vector*. The tensor σ_{ab} is the *shear tensor* and ω_{ab} is the *vorticity tensor*. These tensors have magnitudes σ and ω defined by

$$\sigma^2 = \frac{1}{2}\sigma_{ab}\sigma^{ab}, \qquad \omega^2 = \frac{1}{2}\omega^{ab}\omega_{ab}.$$

When applied to the fundamental timelike congruence in a cosmological model we shall replace Θ by the Hubble variable H defined by

$$H = \frac{\Theta}{3}.$$

This variable, which is defined in any cosmological model, equals the Hubble variable in the Friedmann-Lemaitre models.

There is a well-known characterisation of the Friedmann-Lemaitre models in terms of the kinematic quantities (see for example Krasinski 1996, section 3.2):

> *A cosmological model which satisfies the Einstein field equations with perfect fluid source is a Friedmann-Lemaitre model if and only if*
>
> $$\sigma_{ab} = 0, \quad \omega_{ab} = 0, \quad \dot{u}_a = 0. \tag{68}$$

The shear tensor enters at first order into the redshift-distance relation, leading to a Hubble constant that is effectively direction dependent. For a dust model ($\dot{u}^a = 0$), one obtains

$$z \approx [H_0 - (\sigma_{ab}e^a e^b)_0] d$$

where e^a is a spacelike unit vector giving the direction of the light signal (see for example Ellis 1973, page 10; Kristian and Sachs 1966, page 393).

The significance of the kinematic quantities also manifests itself through the Raychaudhuri equation, as follows. One can use the Hubble variable to define a scale factor S along the congruence **u**, according to

$$H = \frac{\dot{S}}{S}$$

where $\dot{S} = S_{,a}u^a$. The Raychaudhuri equation then reads

$$\frac{\ddot{S}}{S} = -\frac{1}{6}(\mu + 3p) - \frac{2}{3}\sigma^2 + \frac{2}{3}\omega^2 - \frac{1}{3}\dot{u}^a{}_{;a} \tag{69}$$

(Ellis 1973, page 25), generalizing the Friedmann-Lemaitre form (5) of the equation. Note that the shear contributes a negative term, the vorticity a positive term and the acceleration a term of unknown sign.

One can draw strong conclusions from this equation in the case of non-rotating ($\omega = 0$) dust ($p = 0$) models. Since zero pressure implies $\dot{u}_a = 0$ by the contracted Bianchi identities (Ellis 1973, page 17, equation (47)), (69) reduces to

$$\frac{\ddot{S}}{S} = -\frac{1}{6}\mu - \frac{2}{3}\sigma^2.$$

Since $\ddot{S} < 0$ as in a Friedmann-Lemaitre model (for which $\sigma = 0$), the same argument as in that case implies that a singularity will occur on each fundamental world line at a finite time t_b in the past, and the *the age inequality (22) will hold* (although the time of the singularity and the age may depend on the worldline). Thus, if observations of H_0 and t_0 are found to be incompatible with the age inequality (22) in a Friedmann-Lemaitre model, then they will also be incompatible with the age inequality in any non-rotating dust model, necessitating the introduction of either a cosmological constant or vorticity ($\omega \neq 0$).

Weyl Curvature

In General Relativity, the curvature tensor R_{abcd} is regarded as describing the gravitational field. It is convenient to decompose R_{abcd} into the Weyl and Ricci tensors:

$$R_{abcd} \leftrightarrow \{C_{abcd}, R_{ab}\}$$

where

$$C_{abcd} = R_{abcd} - 2R_{[a[c}g_{d]b]} + \tfrac{1}{3}R\, g_{[a[c}g_{d]b]}.$$

The Ricci tensor couples *algebraically* to the matter distribution via the field equations:

$$R_{ab} = T_{ab} - \tfrac{1}{2}T_c{}^c g_{ab}$$

while the Weyl tensor couples *differentially*, via the Bianchi identities and field equations:
$$C_{abc}{}^d{}_{;d} = T_{c[a;b]} - \tfrac{1}{3}g_{c[a}T_{;b]} \tag{70}$$
(Ellis 1973, page 31). The Weyl tensor can be decomposed into an *electric* part E_{ab} and a *magnetic part* H_{ab} relative to a timelike congruence **u**, in formal analogy with the decomposition of an electromagnetic field, according to
$$E_{ac} = C_{abcd}u^b u^d, \qquad H_{ac} = {}^*C_{abcd}u^b u^d$$
where ${}^*C_{abcd} = \eta_{ab}{}^{st}C_{cdst}$. (Ellis 1973, page 7). These tensors are symmetric and tracefree and satisfy $E_{ab}u^b = 0 = H_{ab}u^b$. It also follows that
$$E_{ab} = 0 = H_{ab} \Leftrightarrow C_{abcd} = 0.$$

It is well-known that a spacetime is conformally flat if and only if the Weyl tensor is zero. It follows that for any Friedmann-Lemaitre model, $C_{abcd} = 0$. Conversely, if the field equations hold and the perfect fluid satisfies an equation of state $p = p(\mu)$, then $C_{abcd} = 0$ implies, using (70), that the model is Friedmann-Lemaitre (this follows from equations (74) in Ellis (1973)).

Thus we have the Weyl tensor characterisation of the Friedmann-Lemaitre models:

A cosmological model which satisfies the Einstein field equations with perfect fluid source and equation of state $p = p(\mu)$ is a Friedmann-Lemaitre model if and only if
$$E_{ab} = 0 \qquad H_{ab} = 0. \tag{71}$$

The electric part E_{ab} is the analogue in General Relativity of the second derivatives of the Newtonian potential, $\phi_{,\alpha\beta}$. It acts as a source term in the shear evolution equation, which for irrotational dust ($\omega_{ab} = 0, \dot{u}_a = 0$) reads
$$h_a{}^c h_b{}^d (\sigma_{cd})\dot{} = -2H\sigma_{ab} - \sigma_a{}^c \sigma_{cb} + \tfrac{2}{3}\sigma^2 h_{ab} - E_{ab} \tag{72}$$
where $(\sigma_{ab})\dot{} = \sigma_{ab;e}u^e$ (Ellis 1973, page 29). It can be shown (Kristian and Sachs 1966) that E_{ab} (but not H_{ab}) enters into the quadratic terms in the redshift-distance relation, and thus redshift observations can in principle place bounds on E_{ab}.

Spatial Gradients

A third way in which the Friedmann-Lemaitre models are special is that the various physical and geometric scalars are constant on the hypersurfaces of constant time. In general models these quantities will have a spatial gradient defined by
$$\widetilde{\nabla}_a f = h_a{}^b \frac{\partial f}{\partial x^b}.$$
The two most important spatial gradients are $\widetilde{\nabla}_a \mu$, the spatial density gradient, and $\widetilde{\nabla}_a H$, the spatial Hubble gradient. In a Friedmann-Lemaitre model we have
$$\widetilde{\nabla}_a \mu = 0, \qquad \widetilde{\nabla}_a H = 0 \tag{73}$$
but these conditions, unlike (68) and (70), do not imply that a model is Friedmann-Lemaitre (*e.g.* (73) is satisfied by the orthogonal Bianchi models).

Definition of 'close to Friedmann-Lemaitre'

We have seen that the kinematic quantities (68), the Weyl curvature (71) and the spatial gradients (73) are zero in a Friedmann-Lemaitre model, and that (68) and (71) characterise these models. It is not sufficient to require that the quantities (68) *or* (71) are close to zero in order to define 'close to Friedmann-Lemaitre', since for example, σ_{ab} small does not imply E_{ab} small ($\dot{\sigma}_{ab}$ in (72) could be large even if σ_{ab} is small).

Furthermore, it is not sufficient to require all the quantities (68), (71) and (73) to be small, since, for example, the shear will tend to zero in *any* ever-expanding model, irrespective of whether the model isotropises. The appropriate quantities are the *dimensionless ratios* formed by normalising these quantities with the Hubble variable H. This leads us to consider scalars such as

$$\frac{\sigma}{H}, \quad \frac{|E_{ab}|}{H^2}, \quad \frac{|\widetilde{\nabla}_a \mu|}{H\mu}$$

where

$$|E_{ab}| = (E_{ab} E^{ab})^{1/2}, \qquad |\widetilde{\nabla}_a \mu| = (g^{ab} \widetilde{\nabla}_a \mu \widetilde{\nabla}_b \mu)^{1/2}.$$

Definition *A cosmological model* $(M, g, \mathbf{u})$ *with perfect fluid source and equation of state* $p = p(\mu)$ *is said to be* close to Friedmann-Lemaitre *in some open set* U *if and only if for some suitably small constant* $\epsilon \ll 1$ *the following inequalities hold in* U:

$$\frac{\sigma}{H} < \epsilon \qquad \frac{\omega}{H} < \epsilon \qquad \frac{|\widetilde{\nabla}_a \mu|}{H\mu} < \epsilon \qquad \frac{|\widetilde{\nabla}_a H|}{H^2} < \epsilon$$

$$\frac{|E_{ab}|}{H^2} < \epsilon \qquad \frac{|H_{ab}|}{H^2} < \epsilon.$$

4.2 Observational constraints; anisotropies and inhomogeneities

The Cosmic Microwave Background Radiation

The COBE Differential Microwave Radiometer compares the intensity of the background radiation (and hence the temperature) in different directions, obtaining bounds on the temperature variation $\Delta T/T$, defined by $\Delta T/T = (T - \overline{T})/\overline{T}$, where $\overline{T}$ is the mean temperature over the sky (Smoot *et al.* 1992). The dominant feature is a dipole variation, with $\Delta T/T \sim 10^{-3}$, which is interpreted as a Doppler effect due to the peculiar velocity of the Local Group (see White *et al.* 1994, page 319 for a summary). When this velocity is accounted for, there remains an anisotropy on angular scales $\gtrsim 10°$ of $\Delta T/T \approx 10^{-5}$ (see Partridge 1994, pages A159-A160 for a summary). This temperature anisotropy can be translated into bounds on the deviation of the universe from a Friedmann-Lemaitre model, for example, the density contrast $\delta\mu/\mu$ for large scale perturbations (*i.e.* larger than the horizon size at the surface of last scattering), and the normalised shear scalar σ/H (large scale anisotropy in the Hubble flow).

(i) *The Sachs-Wolfe effect*

Sachs and Wolfe (1967) considered density perturbations in the Einstein–de Sitter universe, using a particular gauge, and investigated the effect of the perturbations on the temperature of the cosmic microwave background radiation. The deviation involves integrating the null geodesic equations and the resulting simplified formula is

$$\frac{\Delta T}{T} \approx \frac{1}{2}\left(\frac{\delta_0 \mu}{\mu}\right)(H_0 L)^2 \tag{74}$$

where $\delta_0 \mu$ is the amplitude and L is the length scale of the *present day* density perturbations. (see Sachs and Wolfe (1967), page 84, equation [48]; also Peebles (1993), page 501, equation [21.2].) Note that the right side of (74) applies at any time in the Einstein-de Sitter model since $\delta_0\mu/\mu \sim t^{2/3}$ and $HL \sim t^{-1/3}$. Indeed, the formula is usually written in terms of quantities on the surface of last scattering. The Sachs-Wolfe formula has been generalised to perturbations of non-flat multicomponent Friedmann-Lemaitre models (see for example, Panek (1986), page 421, equation [53]). A different analysis, based on the use of observational coordinates, has been given by Stoeger, Ellis and Xu (1994).

(ii) *Temperature anisotropy due to shear and vorticity*

Anisotropy in the Hubble flow, *e.g.* non-zero shear and vorticity, will give rise to anisotropy in the cosmic microwave background radiation temperature distribution. In a Bianchi model, there are two types of anisotropy, a quadrupole or a hotspot, with and without a dipole (Barrow et al. 1985, page 922). The main difficulty in finding the temperature distribution lies in solving the null geodesic equations. In the usual analysis the problem is simplified by considering Bianchi universes which contain the Friedmann-Lemaitre models as special cases (Bianchi types VIIo and I, VIIh and V, and IX), and which are perturbations of a Friedmann-Lemaitre model. By integrating the equations for the null geodesics one can derive the following formula, which gives the angular dependence of the temperature to first order

$$\frac{\Delta T}{T} \approx (p^\alpha u_\alpha)_o - (p^\alpha u_\alpha)_e - \int_{t_e}^{t_o} p^\alpha p^\beta \sigma_{\alpha\beta}\, dt \tag{75}$$

(Collins and Hawking 1973, page 313, equation [3.1], Barrow et al. 1985, page 920, equation [2.18]). Here the p^α are the direction cosines of the null geodesic towards the observer in the unperturbed Friedmann-Lemaitre model, the u_α are the spatial components of the fundamental 4-velocity, and subscripts e and o refer to the time of emission and the present time. The shear contributes to $\Delta T/T_o$ via the integral, and the vorticity through the u_α. An alternative approach, which avoids assuming that the model is close to Friedmann-Lemaitre, relies instead on numerical integration of the null geodesic equations (see for example Bajtlik et al. (1986) for Bianchi type V). Both approaches lead to bounds on σ/H and ω/H at the present time. The details depend on the Bianchi type and on whether reheating occurs (see Bajtlik et al. (1986), page 471, for some explicit bounds).

(iii) *The kinetic theory approach* (Maartens, Ellis and Stoeger 1995)

This approach begins by assuming that not only do we, but all fundamental observers in some spacetime domain, measure the cosmic microwave background radiation to be close to isotropy (a Copernican type of assumption). One then uses a kinetic theory description of the photon distribution to prove that the universe is close to Friedmann-Lemaitre, in the sense of Section 4.1 (Stoeger, Maartens and Ellis 1995). This result generalises the corresponding exact theorem of Ehlers, Geren and Sachs (1968). The cosmic microwave background radiation observations are interpreted as placing bounds on the magnitude of the covariant multipole moments of the temperature anisotropy, in terms of a parameter ϵ. It is also assumed that the derivatives of these quantities are of order ϵ. One then derives bounds, in terms of ϵ, for the dimensionless quantities which characterise closeness to Friedmann-Lemaitre (Section 4.1). In this way the observed low level of anisotropy of the cosmic microwave background radiation, by placing a bound on ϵ, leads in principle to bounds on the deviation of the universe from a Friedmann-Lemaitre model, between the time of last scattering and the present.

Anisotropies in the Hubble flow

In an ideal Friedmann-Lemaitre universe $(M, g, \mathbf{u})$, the cosmic microwave background radiation is isotropic relative to the fundamental observers $\mathbf{u}$, and so is the Hubble flow (*i.e.* the redshift-distance relation is isotropic). In other words, the fundamental congruence $\mathbf{u}$ defines a common reference frame for the cosmic microwave background and for the Hubble flow. In the real universe we expect that galaxies and clusters will have peculiar velocities, caused by local mass concentrations. Thus, in order for the real universe to be described by a Friedmann-Lemaitre model it is necessary that there exists a distance scale beyond which the peculiar velocities are small compared to the Hubble velocity, so that sufficiently distant galaxies and clusters will define a frame of reference (the large-scale Hubble frame) relative to which the large-scale Hubble flow will be isotropic. In principle, redshift observations can be used to determine the velocity $\mathbf{v}_H$ of the Local Group relative to this frame, and the observed dipole moment of the cosmic microwave background radiation determines the velocity $\mathbf{v}_{CMB}$ of the local group relative to it. The requirement that the Hubble frame and the CMB frame coincide in a Friedmann-Lemaitre model leads to the consistency requirement $\mathbf{v}_H = \mathbf{v}_{CMB}$.

Starting with the work of Rubin and others in the 1970's (see Rubin 1977 for a summary), observers have performed surveys of peculiar velocities of increasingly large scales (*e.g.* Dressler *et al.* 1987, Courteau *et al.* 1993, Lauer and Postman 1994). At the present time, it is not clear on what distance scale the observations support the above consistency requirement; although it appears that the distance is at least $50h^{-1}$Mpc, and possibly as large as $180h^{-1}$Mpc (Lauer and Postman 1994, page 418).

Once a large-scale Hubble frame has been established, any small residual (large-scale) anisotropies would lead to bounds on quantities which characterise departures from a Friedmann-Lemaitre model such as σ/H and ω/H (Section 4.1). The necessary theoretical framework was provided in the 1960's by Kristian and Sachs (1966), who gave a detailed analysis of observations in an arbitrary cosmological model $(M, g, \mathbf{u})$ which satisfies the field equations with dust as source. They assumed that all quantities can be expanded in a power series in s, an affine parameter along null geodesics, and derive the following fundamental equation

$$\frac{f_e}{f_o} = 1 - (u_{a;b})_o e^a e^b \, r - \frac{1}{2}(u_{a;bc})_o e^a e^b e^c \, r^2 + \cdots \tag{76}$$

(see page 386, equation [36]), where f_e and f_o are the frequencies of emission and reception, r is the luminosity distance (see page 392, equation [68]), and e^a is a suitably normalized null vector giving the direction of the source. The coefficient of r depends only on the Hubble constant and the shear tensor, while the coefficient of r^2 also depends on the electric part of the Weyl tensor, the vorticity and certain spatial derivatives of the kinematic quantities, called the differential velocity gradient (see pages 393–4, equations [70]-[74]). Their detailed analysis shows the extent to which observations of galaxies can in principle restrict the dimensionless scalars that describe the deviation from a Freidmann-Lemaitre model (Section 4.1).

Inhomogeneities in the distribution of galaxies

In an ideal Friedmann-Lemaitre universe, the matter density is spatially homogeneous. In the real universe, the matter is not distributed homogeneously even on the scale of galaxies, since galaxies are observed to form clusters. In order for the real universe to be described by a Friedmann-Lemaitre model it is thus necessary that there exists a distance scale d Mpc relative to which the mass density is homogeneous, *e.g.* any cube of side d contains the same mass.

Since the late 1970's (see for example Kirshner *et al.* (1981)), astronomers have undertaken extensive redshift surveys of galaxies in order to produce three-dimensional maps of the distribution of galaxies. This work has led to the detection of structure in the distribution of galaxies of unexpectedly large size. We quote from Geller and Huchra (1989): 'The extent of the largest features is limited only by the size of the survey. Voids with a density typically 20% of the mean and with diameters of $5000\,\mathrm{km\,s^{-1}}$ ($50h^{-1}$ Mpc) are present in every survey large enough to contain them. Many galaxies lie in thin sheet-like structures. The largest sheet detected so far is the 'Great Wall' with a minimum extent of $60h^{-1}$ Mpc $\times$ $170h^{-1}$ Mpc.' We refer to Strauss and Willick (1995) (pages 23–26) for a history of redshift surveys.

Indeed, at the present time it is apparently not clear on what scale the matter distribution can be regarded as homogeneous, although the distance is at least $50h^{-1}$ Mpc, and possibly larger than $100h^{-1}$ Mpc. Nevertheless it appears from the literature that most theorists believe that a suitably large scale does exist. Finally we mention a second consistency requirement: the observed peculiar velocity field should be compatible with the observed non-uniformities in the distribution of matter, whose gravitational field is assumed to cause the peculiar velocity (Dekel *et al.* 1993, page 1; Dekel 1994).

5 Concluding remarks

At present, it is clear that there are major uncertainties in using the Friedmann-Lemaitre cosmological models to describe the universe during the galactic epoch. On the first level there are uncertainties relating to the averaging scale, both as regards anisotropy in the Hubble flow and inhomogeneity in the distribution of matter. Clearly, incompatibilities on this level threaten the viability of the whole class of Friedmann-Lemaitre models. On the second level are the uncertainties in determining the observational parameters within the class of Friedmann-Lemaitre models: firstly, uncertainties in the density parameter and the nature of the dark matter, and secondly, uncertainties in the Hubble constant and whether a Friedmann-Lemaitre model with zero cosmological constant is viable. Further accumulation of observational data is needed to resolve these uncertainties.

Acknowledgements

It is a pleasure to thank the members of the Organizing Committee of the Summer School, Roger Clark, Graham Hall, John Pulham and Jim Skea, and Barry Haddow who acted as a steward, for their efforts, before, during and after the meeting. The Financial Support provided by NATO is also gratefully acknowledged. I would also like to thank Bernard Carr, George Ellis, Conrad Hewitt and Claes Uggla for helpful discussions, and Sean Bourdon for his assistance with the analysis and numerical work in Section 3.

References

Arp H C, Burbidge G, Hoyle F, Narlikar J V and Wickramasinghe N C, 1990, 'The extragalactic universe: an alternative view', *Nature* **346**, 807–812.

Bajtlik S, Juszkiewicz R, Prószynński M and Amsterdamski P, 1986, '2.7 K Radiation and the isotropy of the universe', *Ap J*, **300**, 463–473.

Bahcall N A, Lubin L M and Dorman V, 1995, 'Where is the dark matter?', *Ap J* **447**, L81-L85.

Barrow J D, 1984, 'Helium formation in cosmologies with anisotropic curvature', *Mon Not R Astr Soc*, **211**, 221–227.

Barrow J D, Juszkiewicz R and Sonoda D H, 1985, 'Universal rotation: how large can it be?', *Mon Not R Astr Soc* **213**, 917–943.

Carr B, 1994, 'Baryonic dark matter', *Annual Rev. Astron. and Astrophys.* **32**, 531–590.

Coley A A and Wainwright J, 1992, 'Qualitative analysis of two-fluid Bianchi cosmologies', *Class Quantum Grav* **9**, 651–665.

Collins C B and Hawking S W, 1973, 'The rotation and distortion of the universe', *Mon Not R Astr Soc* **162**, 307–320.

Copi C J, Schramm D N and Turner M S, 1995, 'Big-bang nucleosynthesis and the baryon density of the universe', *Science* **267**, 192–199.

Courteau S, Faber S M, Dressler A and Willick J A, 1993, 'Streaming motions in the local universe: evidence for large-scale low amplitude density fluctuations', *Ap J* **42**, L51–L54.

Dekel A, Bertschinger E, Yahil A, Strauss M A, Davis M and Huchra J P, 1993, 'IRAS galaxies versus Potent mass: density fields, biasing, and Ω', *Ap J* **416**, 1–21.

Dekel A, 1994, 'Dynamics of Cosmic Flows', *Annual Rev. Astron. and Astrophys.* **32**, 371–418.

Dressler A, Faber S M, Burstein D, Davies R L, Lynden-Bell D, Terlevich R I and Wegner G, 1987, 'Spectroscopy and photometry of elliptical galaxies: a large-scale streaming motion in the local universe', *Ap J* **313**, L37-L42.

Ehlers J, Geren P and Sachs R K, 1968, 'Isotropic solutions of the Einstein-Louville equations', *J Math Phys* **9**, 1344–1349.

Ehlers J and Rindler W, 1989, 'A phase-space representation of FL universes containing both dust and radiation and the inevitability of a big-bang', *Mon Not R Astr Soc* **238**, 503–521.

Ehlers J, 1993, 'Contributions to the relativistic mechanics of continuous media', *Gen Rel Grav* **25**, 1225–1266.

Einstein A and de Sitter W, 1932, 'On the relation between the expansion and the density of the universe', *Proc Nat Acad Sc (Washington)*, **18**, 213–214.

Ellis G F R, 1973, 'Relativistic cosmology', in Cargese Lectures in Physics, vol. 6, ed Schatzman E, Gordon and Breach.

Ellis G F R, 1987, 'Standard Cosmology', Vth Brazilian School on Cosmology and Gravitation, ed Novello M, World Scientific, pp. 83–151.

Ellis G F R, 1989, 'A history of cosmology 1917–1955', in, *Einstein and the History of General Relativity*, Einstein Study Series, Volume 1, ed Howard D and Stachel J, Birkhauser.

Freedman W et al. , 1994, 'Distance to the Virgo cluster galaxy M100 from Hubble Space Telescope Observations of Cepheids', *Nature* **371**, 757–752.

Fukugita M, Hogan C J and Peebles P J E, 1993, 'The cosmic distance scale and the Hubble constant', *Nature* **366**, 309–312.

Geller M J and Huchra J P, 1989, 'Mapping the Universe', *Science* **246**, 897–903.

Goode S W and Wainwright J, 1982, 'Singularities and evolution of the Szekeres cosmological models', *Phys Rev D* **26**, 3315–3326.

Gott J R, Gunn J E, Schramm D N and Tinsley B M, 1974, 'An Unbound Universe?', *Ap J* **194**, 543–553.

Gunn J E and Tinsley B M, 1975, 'An accelerating universe', *Nature* **257**, 454–457.

Harrison E R, 1967, 'Classification of Uniform Cosmological Models', *Mon Not R Astr Soc* **137**, 69–79.

Harrison E R, 1993, 'The redshift-distance and velocity-distance laws', *Ap J*, **403**, 28–31.

Hewitt C G and Wainwright J, 1990, 'Orthogonally transitive G_2 cosmologies', *Class Quantum Gravity* **7**, 2295–2316.

Hirsch M W, 1984, 'The dynamical systems approach to a differential equations', *Bull Amer Math Soc*, **11**, 1–63.

Kirshner R P, Oemler A, Schechter P and Schectman S A, 1981, 'A million cubic megaparsec void in Boötes?', *Ap J* **248**, L57–L60.

Kolb E W and Turner M S, 1990, 'The Early Universe', (Frontiers in Physics; vol 69), Addison-Wesley.

Krasinski A, 1996, 'Inhomogeneous cosmological models', Cambridge University Press, in press (available as a preprint entitled, 'Physics in an inhomogeneous universe').

Kristian J and Sachs R K, 1966, 'Observations in Cosmology', *Ap J*, **143**, 379–399.

Kurki-Suonio H, 1989, 'Dynamically inhomogeneous cosmic nucleosynthesis', in, *Frontiers in Numerical Relativity*, ed Evans C R, Finn L S and Hobill D W, Cambridge University Press.

Lauer T R and Postman M, 1994, 'The motion of the Local Group with respect to the 15000 kilometer per second Abell cluster inertial frame', *Ap J* **425**, 418–438.

MacCallum M A H, 1973, 'Cosmological models from a geometric point of view', in Cargese Lectures in Physics, vol. 6, ed Schatzman E, Gordon and Breach.

MacCallum M A H, 1994, 'Relativistic Cosmologies', in, *Deterministic Chaos in General Relativity*, ed Hobill D et al. , Plenum Press.

Maartens R, Ellis G F R and Stoeger W, 1995, 'Limits on anisotropy and inhomogeneity from the cosmic background radiation', *Phys Rev D* **51**, 1525–1535.

Maartens R, Ellis G F R and Stoeger W, 1995, 'Improved limits on anisotropy and inhomogeneity from the cosmic background radiation', *Phys Rev D* **51**, 5942–5945.

Ostriker J P, 1993, 'Astronomical tests of the cold dark matter scenario', *Annual Rev Astron and Astrophys* **31**, 689–716.

Panek M, 1986, 'Large-scale microwave background fluctuations: gauge-invariant formalism', *Phys Rev D* **34**, 416–423.

Partridge R B, 1994, 'The cosmic microwave background radiation and cosmology', *Class Quantum Gravity* **11**, A153–A169.

Peebles P J E, 1971, 'Physical Cosmology', Princeton University Press.

Peebles P J E, 1984, 'Tests of cosmological models constrained by inflation', *Ap J* **284**, 439–444.

Peebles P J E, Schramm D N, Turner E L and Kron R G, 1991, 'The case for the relativistic hot big bang cosmology', *Nature* **352**, 769–776.

Peebles P J E, 1993, 'Principles of Physical Cosmology, Princeton University Press.

Peebles P J E and Silk J, 1988, 'A cosmic book', *Nature*, **335**, 601–606.

Persic M and Salucci P, 1992, 'The baryon content of the universe', *Mon Not R Astr Soc* **258**, 14p–18p.

Pierce M J et al., 1994, 'The Hubble Constant and Virgo cluster distance from observations of Cepheid variables' *Nature* **371**, 385–389.

Rindler W, 1956, 'Visual Horizons in World Models', *Mon Not R Astr Soc* **116**, 662–677.

Rindler W, 1977, 'Essential Relativity', Springer-Verlag.

Robertson H P, 1933, 'Relativistic Cosmology', *Rev Mod Phys* **5**, 62–90.

Rothman T and Matzner R, 1984, 'Nucleosynthesis in anisotropic cosmologies revisited', *Phys Rev D*, **30**, 1649–1668.

Rubin V, 1977, 'Is there evidence of anisotropy in the Hubble expansion?', in 'Proceedings of the 8th Texas Symposium on Relativistic Astrophysics', *Annals of the New York Academy of Sciences*, **302**, 408–421.

Rubin V, 1989, 'Weighing the universe: dark matter and missing mass', in Bubbles, voids and bumps in time: the new cosmology, ed Cornell J, Cambridge University Press.

Sachs R K and Wolfe A M, 1967, 'Perturbations of a cosmological model and angular variations of the microwave background', *Ap J* **147**, 73–90.

Sandage A and Tamman G, 1986. In 'Inner Space/Outer Space', ed E.W. Kolb E W et al., University of Chicago Press.

Sandage A, Saha A, Tammann G A, Labhardt L, Schwengeler H, Panagia N and Macchetto F D, 1994, 'The Cepheid distance to NGC 5253: calibration of M(max) for the type Ia supernovae SN1972E and SN1895B', *Ap J* **423**, L13–L17.

Sasselov D and Goldwirth D, 1995, 'A new estimate of the uncertainties in the primordial helium abundance: new bounds on Ω baryons', *Ap J*, **444**, L5–L8.

Sciama D W, 1993, 'Modern Cosmology and the Dark Matter Problem', Cambridge University Press.

Shinkai H-a and Maeda K-i, 1994, 'Generality of inflation in a planar universe', *Phys Rev D* **49**, 6367–6378.

Smoot G F et al., 1992, 'Structure in the COBE differential microwave radiometer first-year maps', *Ap J* **396**, L1–L5.

Stabell R and Refsdal S, 1966, 'Classification of General Relativistic World Models', *Mon Not R Astr Soc* **132**, 379–388.

Stoeger W R, Ellis G F R and Xu C, 1994, 'Observational cosmology VI: the microwave background and the Sachs-Wolfe effect', *Phys Rev D* **49**, 1845–1853.

Stoeger W, Maartens R and Ellis G F R, 1995, 'Proving almost-homogeneity of the universe: as almost Ehlers-Geren-Sachs theorem', *Ap J* **443**, 1–5.

Strauss M A and Willick J A, 1995 'The density and peculiar velocity fields of nearby galaxies', *Physics Reports*.

Symbalisty E M D and Schramm D W, 1981, 'Nucleocosmochronology', *Rep Prog in Phys* **44**, 293–328.

Terrell J, 1977, 'The luminosity distance equation in Friedmann cosmology', *Amer J Physics* **45**, 869–870.

Tinsley B M, 1977, 'The cosmological constant and cosmological change', in Proc of the 8th Texas Symposium on Relativistic Astrophysics, *Annals New York Academy of Sciences*, **302**, 423–437.

Wainwright J and Hsu L, 1989, 'A dynamical systems approach to Bianchi cosmologies: orthogonal models of class A', *Class Quantum Gravity* **6**, 1409–1431.

Weinberg S, 1972, 'Gravitation and Cosmology', John Wiley.

White M, Scott D and Silk J, 1994, 'Anisotropies in the cosmic microwave background', *Annual Rev Astron Astrophys* **32**, 319–370.

White S D M, Navarro J F, Evrard A E and Frenk C S, 1993, 'The baryon content of galaxy clusters: a challenge to cosmological orthodoxy', *Nature* **366**, 429–433.

Wiggins S, 1990, 'Introduction to applied nonlinear dynamical systems and chaos', Springer-Verlag.

Wu X-P, Deng Z, Zou Z, Fang L-Z, and Qin B, 1995, 'On the measurement of the Hubble constant in a local low-density universe', *Ap J* **448**, L65–L68.

Black Holes in Cosmology and Astrophysics

B J Carr

Queen Mary & Westfield College
University of London

1 Introductory Overview

The purpose of these lectures is to review some of the astrophysical and cosmological aspects of black holes. Since the orientation of the School was more mathematical than physical, the discussion will not be very technical but I will give references to more technical articles where appropriate. I include some basic discussion of the mathematical aspects of black holes for completeness but again my treatment will be rather elementary; my purpose is mainly to derive results required in the astrophysical and cosmological discussion and many readers will be able to skip this part. I will say rather little about the quantum aspects of black holes and nothing about their connection with quantum gravity, since these topics are covered by other lecturers, but I will discuss the cosmological consequences of black hole evaporations. Some of the material in these lectures has appeared in my earlier reviews (Carr 1978, 1985, 1993, 1994) but this course provides a useful opportunity to bring it altogether in one place.

The plan of the lectures is as follows: The rest of Section 1 will provide a general overview of the astrophysical and cosmological properties of black holes: how, where and when they form; how many there could be; what effects they might have. Section 2 will review those mathematical features of black holes which are necessary for understanding their astrophysical consequences, focusing primarily on the well-known Schwarzschild, Reissner-Nordstrom and Kerr solutions. Section 3 will consider some astrophysical aspects of black holes, in particular black hole accretion and the evidence this affords for stellar black holes in binaries systems. Section 4 will cover some of their cosmological aspects, with special emphasis on the relevance of black holes to quasars and active galaxies, as well as the dark matter problem and gravitational lensing. Finally Section 5 will discuss the formation and evaporation of primordial black holes. As regards references, I will be fairly thorough where a topic is undergoing current developments

1.1 General Features of Black Holes

One of the most exciting predictions of general relativity theory is that there can exist regions of space-time in which gravity is so strong that nothing, not even light, can ever escape. As shown by the spherically symmetric Schwarzschild solution (Section 2.1), this happens whenever a mass M is concentrated within a radius

$$R_S = \frac{2GM}{c^2} \approx 3\left(\frac{M}{M_\odot}\right) \text{ km}. \tag{1}$$

The density which must be attained for this to happen is

$$\rho_S = \frac{3c^6}{32\pi G^3 M^2} \approx 10^{16}\left(\frac{M}{M_\odot}\right)^{-2} \text{ g cm}^{-3} \tag{2}$$

and, for a solar mass object, this is a hundred times greater than nuclear density. Nevertheless, the discovery of neutron stars (with radius only a few times larger than that required for collapse) forced astrophysicists to realise that such extreme conditions are not necessarily implausible. Indeed our understanding of stellar evolution now suggests that sufficiently massive stars inevitably leave black hole remnants. As we will see, black holes may also arise in other circumstances.

The realisation that black holes could actually exist prompted a renewed interest in their mathematical properties and the last three decades have seen some remarkable developments in this respect. Spherically symmetric gravitational collapse is well understood (Oppenheimer and Snyder 1939) and exhibits two crucial features (Section 2.1). Firstly, an *event horizon* forms when the radius of the object falls below R_S; this is the boundary of the black hole and events inside it can never be seen by an outside observer. Secondly, having formed such an event horizon, the infalling matter collapses to a point of infinite density called a *singularity*; at such a point all known laws of physics break down.

Insight into more general (non-spherical) collapse is provided by four remarkable theorems (Section 2.4). Firstly, the *singularity theorem* shows that a singularity must always occur somewhere once a body gets sufficiently compressed, even though not every part of the collapsed object may encounter it (Hawking and Penrose 1969). It would be embarrassing if this singularity (with its associated impredictability) could influence the outside world, so the second theorem states that an outside observer will always be shielded from its effects by an event horizon (Penrose 1969). This *cosmic censorship* hypothesis has still not been rigorously proved.

The third theorem (which has been rigorously proved) says that, however messy the collapse, a black hole will always settle down to a stationary state which depends only on the mass, angular momentum, and charge of the original object (Israel 1967, Carter 1971, Hawking 1973, Robinson 1974, Mazur 1982). All other information about the object is lost and any irregularities are radiated away as gravitational and electromagnetic radiation (Price 1972). This is called the *No Hair Theorem* and it makes the study of black holes remarkably simple: unlike all other astrophysical objects (which display a wide variety of properties), black holes are described by only three parameters. The stationary solution to which black holes evolve is called the Kerr-Newman solution (Section 2.3) and, in the limit in which the angular momentum and charge go to zero, it just reduces to the Schwarzschild solution.

The fourth theorem shows that the surface area of a black hole never decreases (Hawking 1971a). This is the *Area Theorem* and it implies that black holes can never bifurcate, even though two of them can merge. However, the proof of this theorem applies only in classical theory and it can be violated by quantum effects. This was demonstrated by the discovery of Hawking (1974) that black holes radiate due to quantum effects like a black-body with temperature (Section 5.2)

$$T_{\rm BH} = \frac{\hbar c^3}{8\pi GkM} \approx 10^{-7} \left(\frac{M}{M_\odot}\right)^{-1} {\rm K}. \tag{3}$$

Since the holes lose energy in this way, they must shrink (even though this contradicts the Area Theorem) and eventually evaporate altogether on a timescale

$$\tau \approx 10^{64} \left(\frac{M}{M_\odot}\right)^3 {\rm y} \approx 10^{10} \left(\frac{M}{10^{15}g}\right)^3 {\rm y}. \tag{4}$$

For holes of stellar origin, the temperature given by (3) is tiny and the evaporation timescale is much longer than the age of the Universe, so the classical laws are still effectively valid. However, evaporation could be important for black holes which form in the early Universe.

1.2 How Black Holes Form

MO remnants. The most plausible mechanism for black hole formation invokes the collapse of stars which have completed their nuclear burning (Section 3.1). However, this only happens for stars massive enough to be classified as *Massive Objects* (MOs). Stars smaller than $4\,M_\odot$ evolve into white dwarfs because the collapse of their remnants can be halted by electron degeneracy pressure, while stars in the mass range $4-8\,M_\odot$ may explode due to degenerate carbon ignition (Arnett 1969). Stars that are larger than $8\,M_\odot$ but smaller than about $10^2\,M_\odot$ probably burn stably until they form an iron/nickel core, at which point no more energy can be released by nuclear reactions and so the core collapses (Woosley and Weaver 1986). If the collapse can be halted by neutron degeneracy pressure, a neutron star will form and a reflected hydrodynamic shock then ejects the envelope of the star, giving rise to a type II supernova. If the core is too large, however, it necessarily collapses to a black hole, possibly without envelope ejection. Chemical evolution models suggest that a black hole will result if the initial stellar mass exceeds some critical value $M_{\rm BH}$ in the range $20-50\,M_\odot$ (Schild and Maeder 1985, Maeder 1992).

VMO remnants. Stars larger than $10^2\,M_\odot$ are radiation-dominated and therefore unstable to nuclear-energised pulsations during their hydrogen and helium burning phases (Schwarzschild and Harm 1959) (Section 3.2). It used to be thought that the resulting mass loss would be so rapid as to preclude the existence of such *Very Massive Objects* (VMOs). However, it is now thought that the pulsations will be dissipated as a result of shock formation (Papaloizou 1973) and this could quench the mass loss enough for VMOs to survive for their main-sequence time (which is just a few million years). However, VMOs encounter a serious instability when they commence oxygen-core burning because the temperature attained in this phase is enough to generate electron-positron pairs (Fowler and Hoyle 1964). This instability has the consequence that sufficiently large cores collapse to black holes, while smaller ones explode. Both analytical and numerical calculations indicate that the critical dividing mass for the oxygen core is about $100\,M_\odot$ if there is no rotation (Woosley and Weaver 1982, Ober *et al.* 1983, Bond *et al.* 1984). The critical mass for the initial hydrogen stars depends upon the mass loss in the hydrogen and helium burning phase but would be at least $200\,M_\odot$.

SMO remnants. Stars in the mass range above $10^5\,M_\odot$ are unstable to general relativistic instabilities (Section 3.2). Such *Supermassive Objects* (SMOs) may collapse directly to black holes without any nuclear burning at all, at least if they have zero metallicity and no angular momentum (Fowler 1966). The presence of either metals or rotation may permit SMOs to explode in some mass range above $10^5\,M_\odot$ (Fricke 1973) but sufficiently massive ones will still collapse (Fuller *et al.* 1986). Although the evidence for SMOs is not as conclusive as it is for stellar black holes, one could plausibly envisage their formation through relaxation at the centres of dense star clusters: the stars would be disrupted through collisions and a single supermassive star could then form from the newly released gas. Supermassive holes might also derive from the coalescence of smaller holes (Duncan and Shapiro 1983) or from accretion onto a single hole of more modest mass (Hills 1975). Equation (2) shows that the formation of giant black holes does not involve the same extreme conditions as arise in stellar collapse: an object of $10^9\,M_\odot$ would only have the density of water on falling inside its event horizon.

Primordial remnants. Equation (2) shows that the formation of black holes smaller than a solar mass would require extremely high compression and such conditions are unlikely to arise at the present epoch. They may, however, have arisen naturally in the first few moments of the Big Bang (Zeldovich and Novikov 1967, Hawking 1971b) and this has led to the suggestion that *Primordial Black Holes* (PBHs) may have formed with mass much less than $1\,M_\odot$ (Section 5.1). Such PBHs could have formed either from initial inhomogeneities if the Universe started off 'semi-chaotic' or spontaneously at various cosmological phase transitions (Carr 1993) (Section 5.2). We will see that they are expected to have a size of order the particle horizon at their formation epoch, so they could span an enormous mass range: from 10^{-5}g for those forming at the Planck time to $10^5\,M_\odot$ for those forming at 1s. Although there is no conclusive evidence that PBHs ever formed, they are of great theoretical interest since they are the only holes small enough for quantum effects to be important (Section 5.3).

1.3 Where Black Holes Reside

Galactic Discs. There is good evidence that stellar black holes exist in the disc of our own galaxy. For even though black holes cannot be seen directly, one can still detect their effects on surrounding objects. In particular, one can infer their presence in binary systems, where they may be able to accrete material from their companions and thereby generate X-rays (Section 3.4). There are now at least six convincing binary candidates (Cowley 1992) (Section 3.5). On the other hand, the evidence for VMOs at the present epoch is poor, although some candidates have been proposed (*e.g.* η Carina and SN1961). At one time 30 Doradus seemed to be a good candidate (Cassinelli *et al.* 1981) but Hubble Space Telescope (HST) data has now shown this to be a cluster of stars with more modest mass.

Globular Clusters. Many globular clusters are X-ray sources and—since X-rays are a signature of black hole accretion—this has led several people to propose that VMO-size black holes might reside in their nuclei (Bahcall and Ostriker 1975, Grindlay 1978). Although it is not inconceivable that such black holes could form through stellar relaxation, it is now thought unlikely that central black holes account for the X-ray emission. Neutron stars in binary systems (Joss and Rappaport 1984) probably provide a more plausible explanation, especially in view of the preponderance of millisecond pulsars in globular clusters. However, one might expect globular clusters to contain at least a few stellar black holes in binary systems (Kulkarni *et al.* 1993, Sigurdsson and Hernquist 1993).

Galactic Nuclei. Accretion by supermassive black holes is now regarded as the best explanation for the enormous energy output associated with quasars (Rees 1984) and the holes would need to have a mass of about $10^8 \, M_\odot$ (Section 4.1). Since quasars are thought to be the precursors of galaxies, one might therefore expect many galaxies to contain SMO black holes in their nuclei today (Haehnelt and Rees 1992). The violent activity associated with some galactic nuclei may also indicate the presence of supermassive black holes (Blandford and Rees 1992). In recent years considerable evidence has accumulated to support this view (Section 4.2): the centre of our own galaxy may house a $3 \times 10^6 \, M_\odot$ hole, while more massive galaxies may contain holes as large as $10^9 \, M_\odot$ (Kormendy and Richstone 1995).

Galactic Halos. There is now firm evidence (Ashman 1992) that galaxies have extended dark halos (Section 4.3). There may also be dark matter in galactic discs (Bahcall *et al.* 1992), though this is more contentious. Since darkness is a pronounced property of black holes, they would provide a natural explanation, although several other solutions have been suggested (*e.g.* brown dwarfs or some kind of elementary particle). A variety of constraints (Carr 1994) suggest that only VMO or primordial black holes could contribute appreciably to the halo dark matter; indeed this is the main reason for invoking such holes.

Intergalactic Space. It is also possible (though less likely) that black holes populate intergalactic space. If such holes make up the critical cosmological density required by the inflationary scenario (Guth 1981), they would almost certainly need to be primordial

Name	Mass ($M_\odot$)	Mechanism	Location
SMO		Collapse from GR instability before H-burning	intergalactic space? galactic nuclei
	— 10^5 —		
VMO		Collapse due to pair-instability during O-burning	globular clusters? galactic halos?
	— 200 —		
		Explode due to pair-instability during O-burning, no remnant	
	— 100 —		
MO		Core collapse + envelope ejection after nuclear burning	galactic discs
	— 20-50 —		
		White dwarf or neutron star remnant	
	— 1 —		
PBH		Collapse from primordial density fluctuations	intergalactic space? galactic halos?
	— 10^{15}g —		
		Collapse from phase transitions evaporated by today	
	— 10^{-5}g —		
		Planck mass relics	intergalactic space?

Table 1. *How and where black holes with different masses form*

because the standard cosmological nucleosynthesis explanation of the light element abundances only works if the baryons (from which all non-primordial holes form) have much less than the critical density (Walker *et al.* 1991) (Section 4.3).

1.4 When Black Holes Form

A population of black holes could form at a variety of cosmological epochs, as indicated in Figure 1. The figure also associates a 'probability' with each scenario. The probability estimates are necessarily subjective, since it is difficult to assess the likelihood of any scenario in a field as prone to changing fashions as cosmology. The probabilities are merely supposed to reflect the current consensus; that is if one took a poll, the probabilities quoted might correspond to the fraction of cosmologists who would have credence in each scenario!

Protogalactic holes. The holes most likely to exist are those which derive from ordinary Population I stars (*i.e.* the stars in the discs of spiral galaxies). The number of such holes depends on the uncertain mass M_{BH} above which stars leave black hole remnants, but it would be surprising if there were none of them at all. The probability of 90% indicated in Figure 1 corresponds to the likelihood that the prime candidates (like Cygnus X-1) are black holes. Since galactic discs probably do not form until a fairly recent epoch, Population I holes are assumed to form at around 10^{10}y. Population II stars (those in the spheroidal parts of galaxies which form somewhat earlier) could also

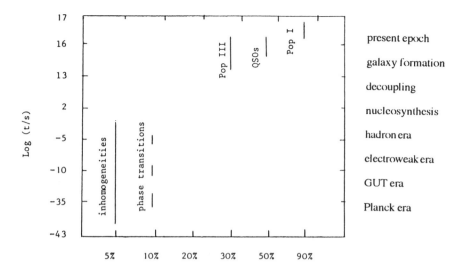

Figure 1. *Probability of black holes forming at various epochs*

leave black hole remnants but probably fewer of them since these stars are less massive. A probability of 50% has been assigned to the hypothesis that supermassive black holes power quasars and active galaxies. If these holes form through relaxation processes at the centres of protogalaxies, they would arise shortly after the protogalaxies themselves. This probably corresponds to a time of order 10^9y, although one cannot exclude the alternative hypothesis that the holes formed *before* the galaxies.

Pregalactic holes. One could expect a generation of stars to form before galaxies providing the density fluctuations surviving at decoupling extend down to subgalactic scales (White and Rees 1978). The remnants of pregalactic stars might in principle comprise dark halos (Carr *et al.* 1984). Such stars are termed 'Population III', although (rather confusingly) this term is also used to describe the stars which generate the first metals. The formation of pregalactic stars is expected if the initial fluctuations were isothermal or if they were adiabatic with the density of the Universe being dominated by 'cold' particles like axions. The time at which such pregalactic stars would form depends on the amplitude of the density fluctuations: it would necessarily exceed the time of decoupling (10^6y) and it would have to precede 10^9y. Various arguments suggest that the first stars may have been bigger than the ones forming today (Tohline 1980, Kashlinsky and Rees 1983)—as required to make a lot of black holes—and perhaps even as large as SMOs (Gnedin and Ostriker 1992, Loeb 1993, Umemura *et al.* 1993). Although the halo stars might in principle be protogalactic, background light constraints preclude this possibility if they are massive enough to leave black holes (Bond *et al.* 1991). The probability of 30% indicated in Figure 1 is supposed to incorporate the likelihood that stars are large enough to leave black holes rather than some other form of remnant (for example, brown dwarfs). If some of the Population III stars were SMOs, they could even produce the holes required to power quasars.

Primordial holes. The probability of black hole formation from inhomogeneities in the early Universe has been estimated as only 10% because the amplitude of the inhomogeneities has to be very finely tuned (Carr 1975) if primordial black holes are to be produced in a sufficient number to be interesting without being overproduced (Section 5.2). The holes would probably need to form before 1s in order for the fluctuations to avoid invalidating the cosmological nucleosynthesis scenario. If the holes form at a phase transition, the situation is less contrived because one does not depend on the prior existence of density fluctuations but one still needs fine-tuning to get an interesting number. Such phase transitions could only occur very early and certainly not after 10^{-5}s (Section 5.2).

1.5 Density of Black Holes

It is convenient to express the density of a population of black holes ρ_B in terms of the critical cosmological density ρ_{crit} which separates models which expand forever from those which recollapse; we will denote this as $\Omega_B = \rho_B/\rho_{crit}$.

Galactic Disc. If one assumes that the initial mass function (IMF) of disc stars has the Salpeter (1955) form [with a number density spectrum $dN/dM \propto M^{-2.35}$] and that a fraction ϕ_B of the mass of all stars larger than M_{BH} ends up as a black hole, then the fraction of the disc density in black hole remnants is $0.2\phi_B(M_{BH}/20\,M_\odot)^{-0.35}$. This corresponds to a density parameter $\Omega_B \sim 0.001$ for reasonable values of ϕ_B and M_{BH}. Although there could be a billion stellar holes in the Galactic disc, they could only comprise a small fraction of its mass.

Galactic Nuclei. There is an indication (Kormendy and Richstone 1995) that every galaxy contains a black hole in its nucleus with a mass of about 0.002 times the mass in the bulge (Section 4.1). The associated cosmological density would then be $\Omega_B \sim 10^{-5}$. Independent constraints on the density parameter of the black holes residing in dead quasars come from background light considerations: if the black holes generate radiation from accreted mass with efficiency ϵ, then the observed energy density in quasar light corresponds to $\Omega_B \sim 10^{-6}(\epsilon/0.1)^{-1}$ (Chokshi and Turner 1982).

Evaporating Black Holes. One expects about 10% of the rest mass of evaporating black holes to go into photons with the temperature given by (3). For PBHs evaporating at the present epoch (with $M \approx 10^{15}$g), this corresponds to an energy of around 100MeV, which is in the γ-ray band. Since the energy density of the γ-ray background is around 10^{-9} in units of the critical density, this implies $\Omega_B < 10^{-8}$ (Page and Hawking 1976). However, PBHs larger than 10^{15}g could contravene this limit, as could the Planck mass relics of evaporating black holes (MacGibbon 1987) if these are stable (Section 5.5).

Dark Matter. If black holes provide the dark matter in galactic halos, they would have a density parameter $\Omega_B \sim 0.1$ (Section 4.3). Such holes would either have to be primordial with a mass between 10^{16}g and $1\,M_\odot$ or the remnants of pregalactic VMOs with mass between $10^2\,M_\odot$ and $10^5\,M_\odot$ (Carr 1994). If intergalactic black holes make

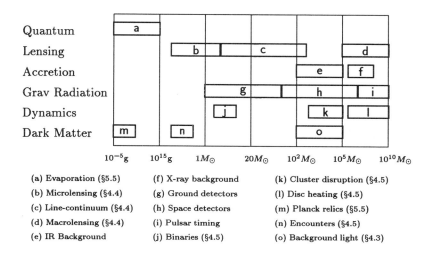

Table 2. *Astrophysical consequences of black holes*

up the critical density required by the inflationary scenario (Guth 1981), they would have $\Omega_B = 1$ (Section 4.3). Such holes would certainly need to be primordial in view of the cosmological nucleosynthesis limits.

1.6 What Black Holes Do

Black holes could have a variety of astrophysical and cosmological consequences, even if their density is too low for them to contribute appreciably to the dark matter. The nature of these consequences depends on their mass and is summarised in Table 2.

Dynamical effects. Large black holes could have important dynamical effects (Section 4.5). For example, black holes comprising the dark matter in galactic halos would contravene observations by puffing up the Galactic disc if larger than $10^6 \, M_\odot$ (Lacey and Ostriker 1985), disrupting globular clusters if larger than $10^5 \, M_\odot$ (Wielen 1985), and sinking into the Galactic nucleus through dynamical friction if larger than $10^4 \, M_\odot$ (Carr and Lacey 1985). These constraints would not apply for black holes residing outside galaxies but black holes in clusters of galaxies would induce unexplained tidal distortions in the visible galaxies if larger than $10^9 \, M_\odot$ (Van den Bergh 1969) and intergalactic ones would induce unacceptably large peculiar galactic motions if larger than $10^{16} \, M_\odot$ (Carr 1978). The dynamical effects of small black holes would be virtually undetectable: even a hole encountering the Earth would have no appreciable effect unless it were bigger that 10^{25}g and such an encounter could occur at most once every 10^5 years.

Lensing effects. Since light bends in a gravitational field, anomalous lensing effects are likely to occur whenever a black hole traverses the line of sight of a more distant source.

Indeed gravitational lensing could permit the detection of black holes over the entire mass range from $10^{-4}\,M_\odot$ to $10^{12}\,M_\odot$ (Section 4.4). Macrolensing constraints associated with the multiple-imaging of distant quasars already exclude a critical density of black holes in the mass ranges $10^7\,M_\odot$ to $10^9\,M_\odot$ (Kassiola et al. 1991) and above $10^{10}\,M_\odot$ (Surdej et al. 1993), while line-continuum effects place an even stronger constraint on those in the mass ranges $10^{-3}\,M_\odot$ to $300\,M_\odot$ (Dalcanton et al. 1994). Microlensing of stars in the Large Magellanic Cloud by objects in our own galactic halo places constraints on black holes in the mass ranges $10^{-7}\,M_\odot$ to $10^2\,M_\odot$ (Paczynski 1986); it may indeed have already been observed, although brown dwarfs rather than black holes may be responsible. Intergalactic black holes with $10^{-3}\,M_\odot$ have also been invoked to explain microlensing of quasars (Hawkins 1993) (Section 4.4).

Black hole accretion. Black holes may generate radiation through accretion. At the present epoch, this may be the chief hallmark of their existence, since the resulting luminosity can be very large for black holes in binary systems or galactic nuclei. Although it has received less attention, accretion by black holes at pregalactic epochs could also be significant: besides generating a significant background radiation density, this could have an important effect on the thermal history of the Universe (Carr 1983a) (e.g. by re-ionizing the intergalactic medium). It has been proposed that the pregalactic accretion of supermassive holes currently in galactic nuclei ($M \sim 10^8\,M_\odot$) could explain the hard X-ray background (Boldt and Leiter 1981, Carr 1983b), whereas the pregalactic accretion of halo VMO holes could produce a significant infrared background (Carr et al. 1983). In any case, pregalactic black holes must have generated a lot of radiation in some waveband.

Gravitational radiation. Black holes should be the most efficient generators of gravitational waves in nature. Most of the radiation would appear as a burst at the time of the original collapse, with the efficiency ε with which radiation energy is generated from the original rest mass depending on the asymmetry of the collapse. To ensure a high value of ε, the collapsing object has to undergo bounces or fragment when rotational effects become important: in the optimal case ε might be as high as 0.1 but it might also be much less. A variety of types of gravitational wave detectors will be operating by the end of the decade (Thorne 1991): bars and ground-based interferometers (like LIGO) could detect the sort of holes formed from stellar collapse, while space-based interferometers (like LISA), Doppler-tracking of interplanetary spacecraft and pulsar timing could detect black holes associated with SMO collapses. If a large number of holes formed at a pregalactic epoch, one would expect their bursts to overlap and form a background of gravitational waves (Bertotti and Carr 1980).

Quantum effects. Although evaporating holes could never have made a significant contribution to the cosmological density, they could still have had important cosmological consequences (Section 5.1). In particular, they may have generated a detectable background of gamma-rays, as well as cosmic ray positrons and antiprotons (MacGibbon and Carr 1991) (Section 5.4). The final explosive phase of evaporation could be particularly dramatic, the energy of a billion megaton bomb being released from a region whose size is only one thousandth of a fermi! Evaporating black holes may conceivably leave stable Planck mass relics, in which case these relics could provide another solution to the dark matter problem (Carr et al. 1994).

In concluding this section, it must be emphasized that we cannot be sure that any of the black holes whose formation mechanisms are indicated in Table 2 actually exist. As indicated by the probabilities in Figure 1, they could exist in theory but there is no guarantee that the conditions required for their formation arose in practice. Nevertheless, even the possibility that black holes could have such a wide variety of effects emphasizes how important it is to study them.

2 Mathematical Aspects of Black Holes

2.1 The Schwarzschild Solution

As is well known, the most general spherically symmetric, static, vacuum, asymptotically flat solution to Einstein's equations is described by the Schwarzschild metric

$$ds^2 = -\left(1 - \frac{2M}{r}\right) dt^2 + \left(1 - \frac{2M}{r}\right)^{-1} dr^2 + r^2 d\Omega^2, \qquad d\Omega^2 = d\theta^2 + \sin^2\theta d\phi^2 \quad (5)$$

where (r, θ, ϕ) are spherical polar coordinates, t is the time measured by a clock at infinity, M is the mass of the central body, and units are chosen such that $G=c=1$ (Schwarzschild 1916). The spherical surface associated with the Schwarzschild radius $r_S = 2M$ is null and corresponds to the black hole's 'event horizon'. Inside this radius, the r and t coordinates change their roles in the sense that the r coordinate becomes timelike and the t coordinate becomes spacelike. The singularity at $r=0$ is a genuine physical singularity, in the sense that the curvature diverges there, but the divergence of g_{rr} at r_S is just a coordinate singularity. This can be seen by introducing the Eddington-Finkelstein coordinate

$$\tilde{t} = t + 2M \ln\left|\frac{r}{2M} - 1\right| \quad (6)$$

in terms of which the metric takes the regular form

$$ds^2 = -\left(1 - \frac{2M}{r}\right) d\tilde{t}^2 + \frac{4M}{r} dr d\tilde{t} + dr^2 \left(1 + \frac{2M}{r}\right) + r^2 d\Omega^2. \quad (7)$$

Radially propagating photons have

$$\frac{d\tilde{t}}{dt} = -1 \quad \text{or} \quad \left(\frac{r + 2M}{r - 2M}\right) \quad (8)$$

so the gradient of the outward light-cone increases as r decreases. For $r < r_S$, it slopes inwards, so that all infalling objects inevitably hit the central singularity. The form of the Schwarzschild solution, together with some of its light-cones, is illustrated in Figure 2. By introducing the Kruskal coordinates

$$u = \pm \left|\frac{r}{2M} - 1\right|^{1/2} e^{r/M} \cosh\left(\frac{t}{4M}\right), \qquad v = \pm \left|\frac{r}{2M} - 1\right|^{1/2} e^{r/M} \sinh\left(\frac{t}{4M}\right) \quad (9)$$

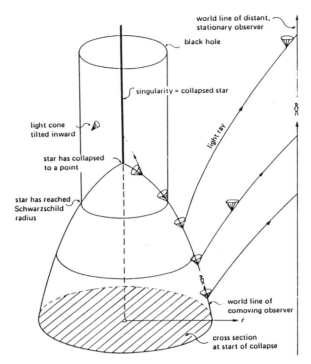

Figure 2. *The Schwarzschild solution in $(r, \tilde{t})$ coordinates*

the metric can be put in the form (Kruskal 1960)

$$ds^2 = \frac{32M^3}{r} e^{-r/2M}(du^2 - dv^2) + r^2 d\Omega^2. \tag{10}$$

All photons now travel at $\pm 45°$ since $du/dv = \pm 1$. This represents the maximal analytic extension of the Schwarzschild solution; as illustrated in Figure 3, it contains a black hole, a white hole (a time-reversed black hole) and two asymptotically flat regions (corresponding to the + and − signs in (9)).

Orbits in the Schwarzschild solution can be analysed by extremising the proper time along spacetime trajectories between fixed points A and B (Shapiro and Teukolsky 1983):

$$\delta \int_A^B L\, d\lambda = 0, \quad L = \left[\left(1 - \frac{2M}{r}\right)\dot{t}^2 - \left(1 - \frac{2M}{r}\right)^{-1}\dot{r}^2 - r^2\dot{\theta}^2 - r^2\sin^2\theta\, \dot{\phi}^2\right]^{1/2}. \tag{11}$$

Here λ parametrises the path and a dot denotes $d/d\lambda$. For timelike orbits, one can take λ to be the proper time τ itself. Since the Lagrangian L is independent of t and ϕ, the Euler-Lagrange equations for equatorial orbits ($\theta = \pi/2$) imply

$$\left(1 - \frac{2M}{r}\right)\dot{t} = E, \quad r^2\dot{\phi} = h \tag{12}$$

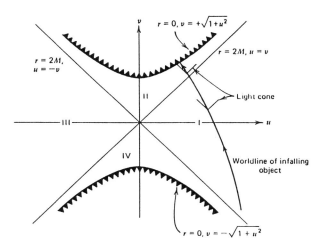

Figure 3. *The Schwarzschild solution in Kruskal (u,v) coordinates*

where E and h are the energy per unit mass at infinity and the angular momentum per unit mass, both of these being constant for a particular orbit. Combining these equations with the identity $L = \varepsilon$, where ε is 1 for timelike orbits and 0 for null orbits, one obtains the energy equation

$$\frac{1}{2}\dot{r}^2 - \varepsilon\frac{M}{r} + \frac{h^2}{2r^2} - \frac{Mh^2}{r^3} = \frac{1}{2}(E^2 - \varepsilon). \tag{13}$$

For timelike orbits one can write this in the form

$$\left(\frac{dr}{d\tau}\right)^2 = E^2 - V(h,r), \quad V(h,r) = \left(1 - \frac{2M}{r}\right)\left(1 + \frac{h^2}{r^2}\right) \tag{14}$$

where the form of the effective potential $V(h,r)$ is indicated in Figure 4. We now consider three applications of the orbit equations.

Radial infall from infinity. In this case, one can take $E=1$ and $h=0$, so (14) implies

$$\frac{dr}{d\tau} = -\left(\frac{2M}{r}\right)^{1/2} \Rightarrow r(\tau) = \left(-\frac{3\tau}{2}\right)^{2/3}(2M)^{1/3} \tag{15}$$

where we have chosen the time origin so that $\tau = 0$ when $r = 0$. Thus in terms of the proper time τ measured by the infalling particle itself, the central singularity is reached at a time

$$\tau(r_S) = \frac{2r_S}{3} \approx 10^{-5}\left(\frac{M}{M_\odot}\right) \text{ s} \tag{16}$$

after passing through the event horizon. From (12) we can also express the trajectory in terms of the t coordinate:

$$\frac{dt}{dr} = \frac{\dot{t}}{\dot{r}} = \frac{r^{3/2}}{(2M)^{1/2}(r - 2M)}$$

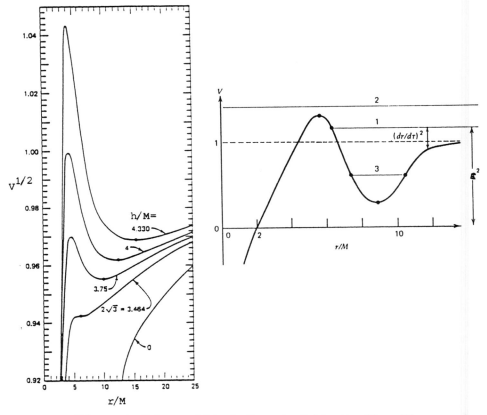

Figure 4. *Schwarzschild effective potential for timelike orbits*

whence

$$t = 2M \left[-\frac{2}{3}\left(\frac{r}{2M}\right)^{3/2} - 2\left(\frac{r}{2M}\right)^{1/2} + \ln\left|\frac{1 + (r/2M)^{1/2}}{1 - (r/2M)^{1/2}}\right| \right]. \quad (17)$$

As $r \to 2M$, $t \to \infty$ and so for the observer at infinity the object never appears to cross the event horizon but merely approaches it asymptotically. This is associated with the fact that a clock at fixed r experiences a gravitational redshift

$$z_g = \frac{dt}{d\tau} - 1 = \left(1 - \frac{2M}{r}\right)^{-1/2} - 1. \quad (18)$$

This also implies that the luminosity of a collapsing star decays exponentially as it approaches the event horizon: the luminosity scales as $(1 + z_g)^{-2}$ (one redshift factor coming from the reduction of the photon energy, the other from the slowing down of the emission rate as measured from infinity) and one can easily show that this falls off as $\exp(-t/4M)$.

Circular orbits. The form of the effective potential given by (14) implies that circular orbits correspond to the condition

$$\frac{dV}{dr} = 0 \quad \Rightarrow \quad h^2 = \frac{Mr^2}{r - 3M}, \quad E^2 = \frac{(r - 2M)^2}{r(r - 3M)}. \quad (19)$$

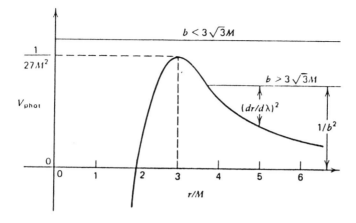

Figure 5. *Schwarzschild effective potential for photon orbits*

For a given value of h, one generally has both a stable orbit at the minimum of V and an unstable orbit at the maximum. As h decreases, the minimum and maximum approach one another until they merge to form an inflection where

$$\frac{d^2V}{dr^2} = 0 \quad \Rightarrow \quad r = 6M, \quad h = 2\sqrt{3}M. \tag{20}$$

This corresponds to the last stable orbit and from (19) the corresponding binding energy is

$$E_{\text{bind}} = 1 - E = 1 - \sqrt{8/9} = 0.057. \tag{21}$$

Photon orbits. In this case, (13) with $\varepsilon = 0$ implies that the orbit equation can be written as

$$\left(\frac{dr}{d\lambda}\right)^2 = \frac{1}{b^2} - V_{\text{phot}}(r), \quad V_{\text{phot}}(r) = \frac{1}{r^2}\left(1 - \frac{2M}{r}\right) \tag{22}$$

where $b = h/E$ is the impact parameter. As illustrated in Figure 5, there is now only an unstable circular orbit and the condition $dV_{\text{phot}}/dr = 0$ implies that this occurs at $r = 3M$.

2.2 The Kerr Solution

The most general axisymmetric, stationary, vacuum, asymptotically flat solution to Einstein's equations has a metric (Kerr 1963) which can be put in the Boyer-Lindquist form:

$$\begin{aligned}ds^2 =& -\left(1 - \frac{2Mr}{\rho}\right)dt^2 - \frac{4aMr\sin^2\theta}{\rho}dt\,d\phi + \frac{\rho}{\Delta}dr^2 + \rho\,d\theta^2 \\ &+ \left(r^2 + a^2 + \frac{2Mra^2\sin^2\theta}{\rho}\right)\sin^2\theta\,d\phi^2\end{aligned} \tag{23}$$

where
$$a = \frac{J}{M}, \qquad \Delta = r^2 - 2Mr + a^2, \qquad \rho = r^2 + a^2\cos^2\theta.$$

This is the Kerr solution and represents the field outside a rotating black hole of mass M and angular momentum J. An event horizon occurs where $\Delta = 0$, which implies

$$r = r_\pm = M \pm (M^2 - a^2)^{1/2} \tag{24}$$

The outer event horizon corresponds to r_+; this reduces to the Schwarzschild radius if $a = 0$ but it decreases as a increases. The outer horizon would disappear altogether for $a > M$, corresponding to an angular momentum $J > J_{\max} = M^2$. However, one can show that centrifugal forces would prevent collapse if J exceeded this, so one would not expect this to occur in practice. There is also an inner event horizon at $r = r_-$, where the radial and time coordinates again interchange their roles. This means that the singularity at $\rho = 0$ is timelike rather than spacelike (unlike the Schwarzschild case). Since $\rho = 0$ corresponds to $r = 0$ and $\theta = \pi/2$, it represents a ring singularity rather than a point singularity and this has the consequence that infalling observers do not necessarily encounter it.

Asymptotically, the Kerr metric takes the form

$$ds^2 = -\left(1 - \frac{2M}{r}\right)dt^2 + \frac{4aM}{r}\sin^2\theta\, dt\, d\phi + \left[1 + O\left(\frac{1}{r}\right)\right](dr^2 + r^2 d\Omega^2) \tag{25}$$

so it resembles the Schwarzschild metric apart from the $dt\, d\phi$ term. The presence of this term is associated with the angular momentum of the black hole and represents the fact that space is being dragged around by the rotation of the source. If one defines a set of stationary observers with $\Omega = d\phi/dt$ being constant, one can show that Ω necessarily lies between $\Omega_{\min}$ and $\Omega_{\max}$ where

$$\Omega_{\max}_{\min} = \frac{-g_{t\phi} \pm (g_{t\phi}^2 - g_{tt}g_{\phi\phi})^{1/2}}{g_{\phi\phi}}. \tag{26}$$

Since $\Omega_{\min} > 0$ for $g_{tt} > 0$, there are no *static* observers for

$$r_+ < r < r_{\text{stat}} = M + (M^2 - a^2\cos^2\theta)^{1/2}. \tag{27}$$

The radius r_{stat} also corresponds to the surface at which the gravitational redshift is infinite. There are no *stationary* observers for

$$g_{t\phi}^2 < g_{tt}g_{\phi\phi} \quad \Rightarrow \quad \Delta < 0 \tag{28}$$

and this just corresponds to values of r between the outer and inner event horizons. Note that the static surface always lies outside the event horizon except at the poles, where they coincide.

Equatorial orbits in the Kerr solution can be analysed in the same way as in the Schwarzschild case except that the Lagrangian now becomes (Shapiro and Teukolsky 1983)

$$L = \left[\left(1 - \frac{2M}{r}\right)\dot{t}^2 + \frac{4aM}{r}\dot{t}\dot\phi - \frac{r^2}{\Delta}\dot{r}^2 - \left(r^2 + a^2 + \frac{2Ma^2}{r}\right)\dot\phi^2\right]^{1/2}. \tag{29}$$

Again L is independent of t and ϕ, so the Euler-Lagrange equations give

$$\dot{t} = \frac{(r^3 + a^2 r + 2Ma^2)E - 2aMh}{r\Delta}, \qquad \dot{\phi} = \frac{(r - 2M)h + 2aME}{r\Delta} \tag{30}$$

and the identity $L = \varepsilon$ gives an energy equation like (14) with

$$V(h, r) = \left(1 - \frac{2M}{r}\right)\left(1 + \frac{h^2}{r^2}\right) + \frac{a^2}{r^2}(1 - E^2) + \frac{2aM}{r^3}. \tag{31}$$

The stable circular orbits have

$$h = \pm \frac{\sqrt{Mr}(r^2 \mp 2a\sqrt{Mr} + a^2)}{r(r^2 - 3Mr \pm 2a\sqrt{Mr})^{1/2}}, \qquad E = \frac{r^2 - 2Mr \pm a\sqrt{Mr}}{r(r^2 - 3Mr \pm 2a\sqrt{Mr})^{1/2}} \tag{32}$$

where + and − refer to co-rotating and counter-rotating orbits, respectively. In both cases there is a last stable orbit and the condition $d^2V/dr^2 = 0$ implies that the associated radius and binding energy are

$$r = \begin{cases} M \\ 9M \end{cases} \quad E = \begin{cases} 1/\sqrt{3} \\ \sqrt{25/27} \end{cases} \Rightarrow E_{\text{bind}} = \begin{cases} 0.42 & \text{co-rotating} \\ 0.04 & \text{counter-rotating} \end{cases} \tag{33}$$

Non-equatorial orbits are more complicated but can be analysed using the fact that there is a Killing tensor $K_{\alpha\beta}$ with the property that $K_{\alpha\beta}U^\alpha U^\beta$ is constant on such orbits.

Astrophysical significance is attached to the fact that orbits inside r_{stat} have $E < 0$. This means that a particle can enter this region and split in two in such a way that one component falls down the hole and the other leaves with more energy than it entered (viz. $E_{\text{in}} = E_{\text{out}} + E_{\text{down}} < E_{\text{out}}$). This is called the Penrose process and the region inside r_{stat} is called the 'ergosphere' since energy can be extracted from it (Penrose 1969). In this situation the energy comes from the rotation of the black hole. The mass of the hole can be written in the form

$$M^2 = M_{\text{irr}}^2 + \frac{J^2}{4M_{\text{irr}}^2} \tag{34}$$

where M_{irr} is the 'irreducible' mass defined by (Christodoulou 1964)

$$M_{\text{irr}} = \frac{1}{\sqrt{2}} M \left[1 + \left(1 - \frac{a^2}{M^2}\right)^{1/2}\right]^{1/2}. \tag{35}$$

Equation (35) implies that the fraction of a rotating black hole's mass which is extractable is

$$\varepsilon = 1 - \frac{M_{\text{irr}}}{M} = 1 - \frac{1}{\sqrt{2}}\left[1 + \left(1 - \frac{a^2}{M^2}\right)^{1/2}\right]^{1/2} \tag{36}$$

and this is at most 29%. Note that (23) implies that the area of the event horizon is

$$A = \iint \sqrt{-g}\, d\theta d\phi = 8\pi M(M + \sqrt{M^2 - a^2}) = 16\pi M_{\text{irr}}^2 \tag{37}$$

so this is directly related to the irreducible mass. The significance of this will become clear shortly.

2.3 Charged Black Hole Solutions

If one adds electric charge Q to a spherically symmetric black hole, one finds a static solution to the Einstein-Maxwell equations called the Reissner-Nordstrom solution (Reissner 1916, Nordstrom 1918). This resembles the Schwarzschild solution except that

$$1 - \frac{2M}{r} \rightarrow 1 - \frac{2M}{r} + \frac{Q^2}{r^2} \qquad (38)$$

so there are now outer and inner event horizons at

$$r = r_\pm = M \pm (M^2 - Q^2)^{1/2}, \qquad (39)$$

the outer horizon reducing to the Schwarzschild radius for $Q=0$. Although there would be no event horizon for $Q > M$, corresponding to $(Q/M)_{\max} = 10^{-18}(e/m_p)$, the electrostatic repulsion would anyway be expected to prevent collapse in this situation. As in the Kerr solution, the singularity at $r = 0$ is timelike rather than spacelike, so not every infalling object need encounter it.

One obtains a charged rotating black hole solution (Newman et al. 1965) by replacing Δ by $\Delta + Q^2$ in the Kerr solution and this gives the Kerr-Newman solution. Asymptotically the magnetic field has the form

$$E_{\hat{r}} = \frac{Q}{r^2}, \quad E_{\hat{\theta}} = E_{\hat{\phi}} = 0, \quad B_{\hat{r}} = \frac{2Qa\cos\theta}{r^3}, \quad B_{\hat{\theta}} = \frac{Qa\sin\theta}{r^3}, \quad B_{\hat{\phi}} = 0. \qquad (40)$$

The hole therefore has electric charge Q and magnetic dipole moment $\mu = Qa$. Its gyromagnetic ratio is $g = 2M\mu/(QJ) = 2$, which is the same as for an electron.

2.4 General Collapse

Although all the solutions discussed above assume special symmetries and one would not expect these symmetries to pertain for realistic collapse, a number of important theorems enable us to understand the more general situation.

Singularity Theorem. This says that a singularity must always form somewhere in the spacetime, in the sense that there are incomplete geodesics, although it is not necessarily a curvature singularity and not every part of the collapsing star need encounter it (Hawking and Penrose 1969). The proof of this theorem assumes the dominant energy condition

$$T^{\alpha\beta}t_\alpha t_\beta \geq 0 \quad \text{for} \quad t^\alpha t_\alpha = -1 \qquad (41)$$

and also that there are no closed timelike curves. Condition (41) requires that the local energy density be non-negative and the local energy flow vector be non-spacelike for all timelike observers.

Cosmic Censorship. This hypothesizes that singularities are always hidden by an event horizon (Penrose 1969), corresponding to the requirement that all physical spacetimes are globally hyperbolic. This theorem has still not been proved for the most general situation. Indeed several counter-examples are known. However, these may merely arise because of a break-down in the physical assumptions. For example, the naked singularity which arises in shell-crossing solutions probably results from the breakdown in the hydrodynamic approximation. However, the situation is less clear in other circumstances. Even when genuine naked singularities do arise, one might hope that the violations of global hyperbolicity would be unstable. For example, it is thought that this is true for the inner event horizon of the Kerr and Reissner-Nordstrom solutions. However, this has not been proved for the general situation.

No Hair Theorem. This says that all black holes settle down to a stationary state described by the Kerr-Newman solution, all perturbations being radiated away as gravitational or electromagnetic waves during the collapse (Price 1972). The proof of this theorem proceeds in six steps, although this is not the sequence in which they were established historically: (i) one first shows that a black hole has spherical topology (Hawking 1973); (ii) one then shows that stationary vacuum black holes are static or axisymmetric (Hawking 1973); (iii) one proves that a static vacuum black hole must be Schwarzschild (Israel 1967); (iv) one then proves that stationary vacuum black holes can only have two parameters, leading to their identification with the Kerr solution (Carter 1971, Robinson 1974); (v) one generalises these results to the electrovacuum case (Mazur 1982); (vi) finally one generalises to other fields (Bekenstein 1972).

Area theorem. This says that the sum of the areas of all the black holes in the Universe can never decrease (Hawking 1971a). This can be proved rigorously providing one assume the dominant energy condition but this may be violated by quantum effects. The area theorem has important implications for how much energy can be extracted from black holes. From equation (37) for the area of a Kerr black hole, if two Kerr holes (rotating in opposite directions) of mass M_i merge to form a single Schwarzschild hole of mass M_f, one necessarily has $M_f > M_i$ and this implies that the fraction of energy extracted, $\varepsilon = (2M_i - M_f)/(2M_i)$, is at most 1/2. On the other hand, if two Schwarzschild holes of mass M_i merge to form a single Schwarzschild hole of mass M_f, one requires $M_f > \sqrt{2}M_i$ and this implies that ε is at most $1 - (1/\sqrt{2}) = 0.29$.

3 Black Holes in Astrophysics

3.1 White Dwarfs, Neutron Stars, Black Holes

In this section, we will use 'order of magnitude' Newtonian arguments to understand the circumstances in which a star can collapse to a black hole. Consider a gas cloud of radius R and mass $M = Nm_p$ (where N is the number of protons) which is contracting quasi-statically. So long as the particles are sufficiently dispersed for degeneracy effects to be negligible, the virial theorem implies that the internal (*i.e.* thermal) energy per particle is of order the gravitational energy per particle and so the temperature T is given by

$$kT \sim \frac{GMm_p}{R} \sim \left(\frac{N}{N_0}\right)^{2/3} \left(\frac{\hbar c}{d}\right). \tag{42}$$

Here d is the interparticle spacing and

$$N_0 \equiv \left(\frac{Gm_p^2}{\hbar c}\right)^{-3/2} \sim 10^{60} \qquad (43)$$

is a dimensionless number (roughly the number of protons in the Sun) which will play a special role in what follows. We infer that the cloud heats up as it collapses. When d gets small enough, the degeneracy pressure of the electrons becomes important and will try to stop the collapse. The exclusion principle requires that each electron has momentum $p = \hbar/d$ and hence energy $\hbar^2/(2m_e d^2)$. Since this contributes to the internal energy per particle, the virial theorem now gives

$$kT \sim \left(\frac{N}{N_0}\right)^{2/3}\left(\frac{\hbar c}{d}\right) - \frac{\hbar^2}{2m_e d^2}. \qquad (44)$$

This shows that T attains a maximum value $T_{\max}$ as d decreases and then falls to zero when the interparticle separation is $d_{\min}$, where

$$kT_{\max} \sim \left(\frac{N}{N_0}\right)^{4/3} m_e c^2, \qquad d_{\min} \sim \left(\frac{N}{N_0}\right)^{-2/3} r_e, \qquad (45)$$

and $r_e = \hbar/m_e c \sim 10^{-10}$cm is the Compton wavelength of the electron. Since T cannot go negative, d cannot fall below $d_{\min}$, so we end up with a cold degeneracy-supported object called a 'white dwarf'. The evolution of T as a function of d is indicated in Figure 6.

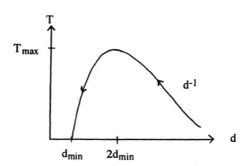

Figure 6. *Temperature evolution of collapsing gas cloud*

If $T_{\max}$ exceeds 10^7K, nuclear reactions can be initiated and the object will pause for a while in its evolution along the $T(d)$ curve as a main-sequence star. This happens for $N > 0.1 N_0$, which corresponds to a mass exceeding around $0.1\,M_\odot$. However, once its nuclear fuel is consumed, collapse will proceed and so the object may still end up as a white dwarf eventually. For $N < 0.1 N_0$, the object will evolve to the degeneracy-supported state directly; such objects are usually referred to as 'brown dwarfs'.

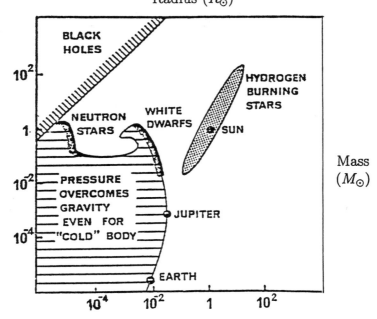

Figure 7. *Mass-radius relation for various astronomical objects*

The preceding analysis fails if the electrons ever become relativistic because the degeneracy energy is then $\hbar c/d$ rather than $\hbar^2/(2m_e d^2)$. Since both terms on the right-hand-side of (44) now go as d^{-1}, there is no longer a value of d for which $T = 0$, so there is no stable end-state. The electrons go relativistic if $kT_{\max} > m_e c^2$ and (45) shows that this applies if M exceeds the critical Chandrasekhar mass

$$M_C \sim N_0 m_p \sim 1 \text{ M}_\odot. \tag{46}$$

A more precise calculation (Chandrasekhar 1931) gives $M_C = 5.6\mu^2 \text{ M}_\odot \approx 1.4 \text{M}_\odot$, where μ is the number of electrons per nucleon (always close to 1/2). The associated size and density are

$$R_C \sim N_0^{1/3} r_e \sim 10^4 \text{ km}, \qquad \rho_C \sim \frac{m_p}{r_e^3} \sim 10^6 \text{ g cm}^{-3}. \tag{47}$$

Equation (45) implies that the mass-radius and mass-density relations for objects smaller than M_C are

$$M = M_C \left(\frac{R}{R_C}\right)^{-3} = M_C \left(\frac{\rho}{\rho_C}\right)^{1/3} \tag{48}$$

and this is indicated in Figure 7.

When ρ goes above 10^{10}g cm^{-3}, the electrons begin to get pushed onto the protons to form neutrons ($e^- + p \to n + \nu_e$), so the nuclei become neutron-rich and the degeneracy pressure of the electrons decreases. Once ρ reaches 10^{14}g cm^{-3} (*i.e.* nuclear density), nearly all the electrons and protons have disappeared, so it is the neutrons rather than the electrons which provide the degeneracy pressure. Objects supported by neutron degeneracy are called 'neutron stars' and their characterisics can be found by replacing r_e by the Compton wavelength of the proton $r_p = \hbar/m_p c \sim 10^{-13}$cm in (47). Thus the upper limit on the neutron star mass is still of order $1\,M_\odot$ but the associated size and density become

$$R_{\rm NS} \sim N_0^{1/3} r_p \sim 10 \text{ km}, \qquad \rho_{\rm NS} \sim \frac{m_p}{r_p^3} \sim 10^{14} \text{ g cm}^{-3}. \tag{49}$$

Note that $R_{\rm NS}$ is close to the Schwarzschild radius (R_S), so a proper analysis would need to allow for general relativistic effects. It should also allow for strong interaction effects since $\rho_{\rm NS}$ corresponds to nuclear density. This makes the maximum mass of a neutron star difficult to determine precisely but it is probably about $3\,M_\odot$ (McClintock 1986).

Objects bigger than $3\,M_\odot$ have no zero temperature equilibrium state and so must ultimately collapse to a black hole. However, the *original* star needs to be much larger than this since stars shed mass at various stages in their evolution. In particular, stars larger than about $8\,M_\odot$ may explode as supernovae after completing their nuclear burning (Weaver et al. 1978). Only stars originally larger than some critical mass $M_{\rm BH}$ are guaranteed to form black holes. The value of $M_{\rm BH}$ is very uncertain but chemical evolution considerations suggest that it is probably in the range 20–50 $M_\odot$ (Schild and Maeder 1985, Maeder 1992).

The locations of black holes, neutron stars, white dwarfs and stars in the (M, R) plane are indicated in Figure 7. For completeness, the location of solids (like the Earth) is also shown. In solids there is a balance between electron degeneracy and chemical bonds rather than gravity. Since the degeneracy energy and chemical energy per particle go as $\hbar^2/(2m_e d^2)$ and e^2/d respectively, the associated density is

$$\rho_0 \sim \frac{e^b m_p m_e^3}{\hbar^6} \sim 1 \text{ g cm}^{-3}. \tag{50}$$

This corresponds to 'atomic density', although we have neglected a weak dependence on chemical composition. The white dwarf density line hits the atomic density line in Figure 7 at a mass and radius

$$M_{\rm max} \sim \alpha^{3/2} M_C \sim 10^{-3} \,M_\odot, \qquad R_{\rm max} \sim \alpha^{-1/2} R_C \sim 10^5 \text{ km} \tag{51}$$

where $\alpha = e^2/\hbar c = 1/137$ is the fine structure constant. Objects more massive than $M_{\rm max}$ (which is roughly the mass of Jupiter) collapse to higher than atomic density but less massive objects stop collapsing at density ρ_0. The radius $R_{\rm max}$ specifies the maximum radius of a cold degeneracy-supported object.

3.2 Very Massive and Supermassive Objects

In order to understand how objects larger than ordinary stars collapse to black holes, one must go beyond a Newtonian analysis. For a relativistic analysis, one uses the fact that the most general spherically symmetric static metric inside a star has the form

$$ds^2 = -e^{2\Phi(r)} dt^2 + \left[1 - \frac{2m(r)}{r}\right]^{-1} dr^2 + r^2 d\Omega^2 \tag{52}$$

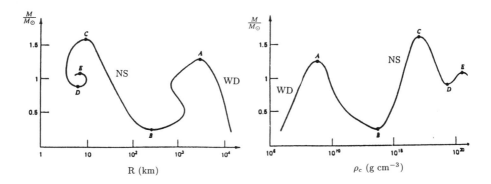

Figure 8. *Mass-radius and mass-density relations for degenerate objects*

where $m(r)$ is the mass within radius r, $\Phi(r)$ is the gravitational potential, and we choose units with $G=c=1$. Outside the star $m(r)$ is constant with the value M, so this goes over to the Schwarzschild metric given by (5). Einstein's equations reduce to the following three relationships (Wald 1984):

$$\frac{dm}{dr} = 4\pi r^2 \rho \qquad (53)$$

$$\frac{d\Phi}{dr} = \frac{m + 4\pi p r^3}{r(r-2m)} \qquad (54)$$

$$\frac{dp}{dr} = -(\rho + p)\frac{d\Phi}{dr} \qquad (55)$$

where ρ is the density and p is the pressure. The first relation is the same as in Newtonian theory. The second and third relations (which are different from Newtonian theory) can be combined to give the equation of hydrostatic support:

$$\frac{dp}{dr} = -(\rho + p)\left[\frac{m + 4\pi p r^3}{r(r-2m)}\right]. \qquad (56)$$

Note that this implies that r must exceed $2m$ everywhere. Given p as a function of ρ, equations (53) and (56) provide two first order ordinary differential equations for $p(r)$ and $m(r)$. By specifying the density at the centre of the star, ρ_c, we can integrate outwards until the pressure falls to zero. This corresponds to the surface of the star and so we can determine the radius $R(\rho_c)$ and total mass $M = M(R) = M(\rho_c)$. Using the appropriate equation of state in the different density regimes gives Figure 8.

Equation (56) has the important feature that p itself contributes to gravity. Sufficiently massive stars are radiation-dominated, in the sense that the radiation pressure p_r exceeds the matter pressure p_m, and such stars are known to be unstable (Schwarzschild and Harm 1959). Since the matter and radiation and pressures are nkT and aT^4 respectively, their ratio is

$$\beta \equiv \frac{p_m}{p_r} \sim \frac{nk}{aT^3} \sim \frac{Nk}{a}\left(\frac{GMm_p}{k}\right)^{-3} \sim 10^3 \left(\frac{N}{N_0}\right)^{-2} \tag{57}$$

where we have used (42) and the relation $a = \pi^2 k^4/(15\hbar^3 c^3)$. Thus stars are unstable for $N > 30N_0$, corresponding to masses exceeding about 60 $M_\odot$. Although this instability would lead to pulsations and hence mass-loss during the main-sequence phase, the existence of such stars is not necessarily precluded. If they did exist, they would have rather interesting properties (Bond et al. 1984). They would all have the Eddington luminosity (i.e. the luminosity for which the radiation drag on the electrons balances the gravitational attraction of the star)

$$L_{\rm ED} = \frac{4\pi cGM}{\sigma_T} \sim 10^5 \left(\frac{M}{M_\odot}\right) L_\odot \tag{58}$$

and a mass-independent main-sequence lifetime

$$\tau_{\rm MS} = \frac{0.007 Mc^2}{L_{\rm ED}} = 0.007\left(\frac{c\sigma_T}{4\pi G}\right) \approx 3\times 10^6 \text{ y}. \tag{59}$$

Here $L_\odot \approx 10^{33}$ erg s^{-1} is the luminosity of the Sun and $\sigma_T \approx 10^{-24}$ cm^2 is the Thomson cross-section for photon-scattering off electrons.

One can understand the instability of stars which are radiation-dominated from a semi-Newtonian analysis in which one neglects the matter pressure. In this situation, the pressure and entropy per particle (s) are given by

$$p = \frac{1}{3}aT^4, \qquad s = \frac{4aT^3}{3n} \tag{60}$$

so the pressure and density are related by

$$p = K\rho^{4/3}, \qquad K \propto s^{4/3} \tag{61}$$

(corresponding to an $n=3$ polytrope). The total energy of the star has the form

$$E = E_{\rm int} + E_{\rm grav} = k_1 KM\rho_c^{1/3} - k_2 GM^{5/3}\rho_c^{1/3} \tag{62}$$

where k_1 and k_2 are constants associated with the $n=3$ polytrope. Since both terms in (62) scale as $\rho_c^{1/3}$, the equilibrium condition $dE/d\rho_c = 0$ just leads to $M \propto s^2$ and does not specify ρ_c. This means that the star can have any central density and so is unstable.

A more precise analysis shows that sufficiently low mass radiation-dominated stars can be stabilised by their plasma content (Shapiro and Teukolsky 1983). The effect of the plasma can be accommodated by writing the $p(\rho)$ relation in the form

$$p = K\rho^\Gamma, \qquad \Gamma = \frac{4}{3} + \frac{\beta}{6} \tag{63}$$

where β is given by (57). Although β is small for radiation-dominated stars, it allows the adiabatic index Γ to be slightly larger than 4/3 and this restores stability. It also adds a small contribution to the internal energy of the star

$$\Delta E_{\rm int} \propto M s^{1/3} \rho_c^{1/3} \ln \rho_c. \tag{64}$$

This means that, for a given value of M and s, the function $E(\rho_c)$ has a minimum and this corresponds to the equilibrium configuration. However, although radiation-dominated stars are stabilised by their plasma content, they are also destabilised by electron-positron pair creation at high temperatures (Fowler and Hoyle 1964) and general relativistic effects at high density (Fowler 1966). The first effect is relevant for VMOs, the second for SMOs.

To understand the first instability, we note that pair-production pushes the effective value of Γ back towards 4/3 (since the electron-positron pairs behave like radiation). The temperature of 10^9K required to produce electrons and positrons will arise during oxygen-burning providing the oxygen core mass exceeds $30\,M_\odot$ and this corresponds to an initial hydrogen mass of around $100\,M_\odot$. Simple energetic arguments show that this instability has two possible consequences (Bond et al. 1984): VMOs with an initial mass below the critical value $M_{\rm crit} \approx 200\,M_\odot$ completely explode (leaving no remnants), while those larger than $M_{\rm crit}$ undergo collapse to a black hole (with no further mass ejection).

To understand the second (purely relativistic) instability, we note that the total energy (excluding rest mass) for a static spherical cloud in general relativity can be written as (Shapiro and Teukolsky 1983)

$$E = \int_0^R \left[\rho \left(1 - \frac{2m}{r}\right)^{1/2} - \rho_0 \right] dV \tag{65}$$

with $\rho = \rho_0(1+u)$. Here ρ_0 is the rest mass density, u is the fraction of the density associated with the internal energy, and the volume element has the Schwarzschild form

$$dV = \left(1 - \frac{2m}{r}\right)^{-1/2} 4\pi r^2 dr. \tag{66}$$

If one expands the expression for E to second order in (m/r), one obtains

$$E = \int_0^R \rho_0 \left[u - \frac{m}{r} - u\left(\frac{m}{r}\right) - \frac{1}{2}\left(\frac{m}{r}\right)^2 \right] dV \tag{67}$$

and comparison with the Newtonian expression for the energy

$$E_{\rm N} = \int_0^R \rho_0 u\, dV - \int_0^M \frac{m'}{r'} dm' \tag{68}$$

with m' and r' defined by $dm' = \rho_0 dV$ and $r' = (3V/4\pi)^{1/3}$, shows that the correction term is

$$\Delta E_{\rm GR} = \int_0^R \rho_0 \left[-u\left(\frac{m}{r}\right) - \frac{1}{2}\left(\frac{m}{r}\right)^2 + \frac{m'}{r'} - \frac{m}{r} \right] dV = -0.9 M^{7/3} \rho_c^{2/3} \tag{69}$$

where the last expression assumes an $n=3$ polytrope.

Combining all the effects discussed above shows that the total energy of a star can be expressed in terms of the mass and central density as (Zeldovich and Novikov 1971)

$$\begin{aligned} E &= E_{\rm int} + E_{\rm grav} + \Delta E_{\rm int} + \Delta E_{\rm GR} \\ &= A M s^{4/3} \rho_c^{1/3} - B M^{5/3} \rho_c^{1/3} + C M s^{1/3} \rho_c^{1/3} \ln \rho_c - D M^{7/3} \rho_c^{2/3} \end{aligned} \tag{70}$$

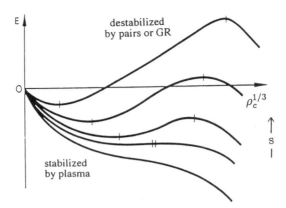

Figure 9. *Energy of a spherical cloud of fixed mass as a function of central density and entropy*

where the constants A, B, C and D are all positive. If we fixed M, the dependence of E upon ρ_c is indicated in Figure 9, each curve corresponding to a particular value of s. For large enough values of s, there is both a minimum and a maximum. The position of the minimum corresponds to the equilibrium configuration and specifies ρ_c as a function of s. However, as s decreases, the minimum turns into an inflection when $d^2 E/d\rho_c^2 = 0$ and this occurs at a critical central density

$$\rho_{\rm crit} \approx 2\times 10^{18} \left(\frac{M}{M_\odot}\right)^{-7/2} {\rm g~cm^{-3}}. \tag{71}$$

For smaller values of s, there is no equilibrium configuration, so the star goes unstable once ρ_c exceeds this. Note that the star is much larger than its gravitational radius at the onset of instability since (2) implies

$$\left(\frac{GM}{Rc^2}\right)_{\rm crit} \approx 0.6 \left(\frac{M}{M_\odot}\right)^{-1/2} \ll 1. \tag{72}$$

Also the temperature at the onset of instability is

$$T_{\rm crit} \approx 2\times 10^{13} \left(\frac{M}{M_\odot}\right)^{-1} {\rm K} \tag{73}$$

which shows that nuclear burning and pair-production are unimportant for M greater than $10^5~M_\odot$; this may be taken to specify the lower mass limit for SMOs. The binding energy at the onset of instability is

$$E_{\rm crit} = -DM^{7/3}\rho_{\rm crit}^{2/3} \approx -4\times 10^{54} {\rm ~ergs} \tag{74}$$

(independent of M) and this corresponds to the maximum energy which can be radiated away prior to collapse. Note that the other sources of instability (associated with lower mass stars) can be analysed in a similar way. Figure 10, from Zeldovich and Novikov

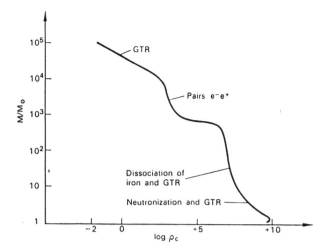

Figure 10. *Mass of a spherical cloud as a function of central density at which instability occurs*

(1971), indicates the value of ρ_c when instability sets in and its cause for different mass ranges.

The evolution of a supermassive star can be regarded as a quasi-static progression through equilibrium states with ρ_c increasing until the critical value indicated by (71) is attained. The time to reach this state is

$$t_{\rm coll} \sim \frac{|E_{\rm crit}|}{L_{\rm ED}} \sim 10^3 \left(\frac{M}{10^6\,M_\odot}\right)^{-1} {\rm y} \qquad (75)$$

which is short but longer than a dynamical time. Thereafter the instability sets in and the collapse proceeds on a dynamical timescale

$$t_{\rm dyn} \sim (G\rho_{\rm crit})^{-1/2} \sim 10^3 \left(\frac{M}{10^6\,M_\odot}\right)^{7/4} {\rm s}. \qquad (76)$$

Note however that SMOs with non-zero metallicity may explode for small enough values of M (Fricke 1973). There is no quasi-static phase for $M > 10^8\,M_\odot$ since $t_{\rm dyn} > t_{\rm coll}$ in this case.

3.3 The Gravitational Collapse of Star Clusters

The most natural way to produce a supermassive star is as the endpoint of evolution of a cluster of smaller stars (Binney and Tremaine 1987). To understand this, consider a cluster of radius R, containing N stars of mass m and radius r. Then the virial theorem implies that the velocity dispersion of the stars and the time they take to cross the cluster are

$$\sigma \sim \left(\frac{GNm}{R}\right)^{1/2}, \qquad t_{\rm cross} \sim \left(\frac{GNm}{R^3}\right)^{-1/2}. \qquad (77)$$

One can show that a typical star will experience a gravitational encounter which deflects its trajectory appreciably on a timescale

$$t_{\text{relax}} \sim \left(\frac{N}{\log N}\right) t_{\text{cross}} \tag{78}$$

and this is also the timescale on which the cluster relaxes into an isothermal distribution with a Maxwellian velocity distribution. When a cluster is relaxed, the fraction of stars with velocity in the range $(V, V + dV)$ is

$$f(V)\, dV \sim \left(\frac{V}{\sigma}\right) \exp\left(-\frac{3V^2}{2\sigma^2}\right) d\left(\frac{V}{\sigma}\right) \tag{79}$$

and this implies that some fraction of the stars can escape from the cluster because their speed exceeds the escape velocity

$$V_{\text{esc}} = \left(\frac{2GNm}{R}\right)^{1/2} \approx 2\sigma. \tag{80}$$

One infers that the fraction of stars escaping at any time is

$$f_{\text{esc}} = \int_{2\sigma}^{\infty} f(V)\, dV = 0.007 \tag{81}$$

so the cluster will evaporate at a rate

$$\frac{dN}{dt} = -\frac{N}{t_{\text{evap}}}, \qquad t_{\text{evap}} \sim f_{\text{esc}}^{-1}\, t_{\text{relax}}. \tag{82}$$

This implies that the cluster will develop an isothermal core, together with a halo of high energy stars (with eccentric orbits) which gain energy via two-body encounters as they pass through the core and eventually escape altogether.

One can calculate the evolution of the core by noting that its total energy (which goes like $N\sigma^2 \sim N^2/R$ from the virial theorem) must remain approximately constant because the escaping stars have nearly zero energy. This implies $N \propto R^{1/2}$, so that the number of stars in the core, the radius of the core and the stellar density there evolve according to

$$N = N_o\left(1 - \frac{t}{t_o}\right)^{2/7}, \qquad R = R_o\left(1 - \frac{t}{t_o}\right)^{4/3}, \qquad n = n_o\left(1 - \frac{t}{t_o}\right)^{-10/7} \tag{83}$$

where a subscript 'o' indicates the initial configuration and $t_o = (2/7)t_{\text{evap}(o)}$. Thus the core is shrinking in terms of size and number of stars but its density is increasing. This process is termed 'gravithermal catastrophe' and is indicated by the arrows in Figure 11 for a galaxy and globular cluster. Of course, equations (83) will not pertain indefinitely: collisions will become important once the collision timescale $t_{\text{coll}} = (n\pi r^2 V)^{-1}$ becomes comparable to t_{evap} and this occurs when N has got down to a value

$$N_{\text{crit}} \approx 10^3 \left(\frac{r}{R_\odot}\right) \left(\frac{N_o}{10^5}\right)^2 \left(\frac{R_o}{10\ \text{pc}}\right)^{-1} \tag{84}$$

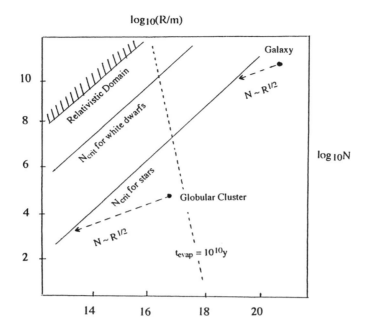

Figure 11. *Evolution of star cluster*

where we have normalised to values appropriate for globular clusters. This is just the value of N where the $N \propto R^{1/2}$ evolution line in Figure 11 hits the $t_{coll} = t_{evap}$ line. We note that the value of N_{crit} depends critically on the value of r and it is much smaller for neutron stars or white dwarfs than ordinary stars. If $N_{crit} \gg 1$, as expected for stellar components, the cluster will evolve to a supermassive star formed from the gas released by the collisions of all the original stars, and this SMO should then itself evolve to a black hole for the reasons discussed in Section 3.2. If $N_{crit} \ll 1$, collisions are never important and the cluster evolves either until it forms a tightly bound binary (which may then coalesce through gravitational radiation losses) or until it approaches its gravitational radius and becomes relativistic. In all cases, one expects a supermassive black hole to be the final product.

3.4 Black Hole Accretion

Even though a black hole cannot be seen, it can still be detected through its gravitational effects. In particular, it will tend to accrete any gas in its vicinity and this gas will heat up as it falls towards the hole, thereby generating radiation. In most situations, the material captured by the hole will have some angular momentum, so one expects it to form an accretion disc with each particle being in a Keplerian circular orbit around the hole (Novikov and Thorne 1973). Because the Keplerian velocity scales with distance from the hole as $r^{-1/2}$, there will be friction between particles in neighbouring orbits and the consequent dissipation will result in the gas losing energy and thus gradually spiraling inwards at a rate determined by the viscosity. The timescale for this will usually be much larger than the orbital time, so one can regard the gas as being in the sort of quasi-static circular orbit described in Section 2. When the gas reaches the last stable orbit, it can no longer follow a circular orbit, so it plunges into the hole.

The same dissipation which allows the gas to spiral inwards also results in its being heated up and, providing cooling is sufficiently efficient, this heat can be radiated away (Shapiro and Teukolsky 1983). If the gas at radius $r(\gg r_S)$ drifts inwards with speed V_r, the heat radiated per unit area is simply the rate at which it loses potential energy and in a Newtonian analysis this is

$$Q(r) = \sigma \frac{d}{dr}\left(-\frac{GM}{r}\right) V_r = \frac{GM\sigma V_r}{r^2} \qquad (85)$$

where σ is the gas density integrated over the disc thickness. Since the mass flux is

$$\dot{M} = 2\pi r \sigma V_r \qquad (86)$$

flux conservation implies

$$\sigma V_r \propto r^{-1} \quad \Rightarrow \quad Q(r) \propto r^{-3} \quad \Rightarrow \quad T(r) \propto r^{-3/4} \qquad (87)$$

providing the gas radiates as a black-body, so the temperature rises as one approaches the hole. The total luminosity of the disc is

$$L = \int_{r_{\min}} Q(r) 2\pi r\, dr = \frac{GM\dot{M}}{r_{\min}} = \varepsilon \dot{M} c^2, \qquad \varepsilon = \frac{E_{\text{bind}}}{c^2} \qquad (88)$$

where $r_{\min}$ is the radius of the last stable orbit. For a relativistic analysis one replaces E_{bind} by the relativistic expression for the gravitational binding energy, so ε is given by (21) for a non-rotating hole and (33) for a maximally-rotating one. The associated black-body temperature is

$$T_{\max} \approx \left(\frac{\varepsilon \dot{M} c}{a r_{\min}^2}\right)^{1/4} \approx 10^7 \left(\frac{\dot{M}}{10^{-9}\, M_\odot y^{-1}}\right)^{1/4} \left(\frac{M}{M_\odot}\right)^{-1/2} \left(\frac{\varepsilon}{0.1}\right)^{1/4} \text{K} \qquad (89)$$

where we have normalised M and $\dot{M}$ to the sort of values which arise in astrophysical situations. This is typically in the X-ray range, so one of the hallmarks of an accreting black hole would be X-ray emission. We stress that the radiation does not come from inside the hole but from the region just outside it.

For an accreting body with a solid surface, like a white dwarf or a neutron star, the gas never gets in as far as the last stable orbit, so $r_{\min}$ is replaced by the radius of the body R and a Newtonian analysis suffices to show that the efficiency parameter is

$$\varepsilon = \frac{GM}{c^2 R} \sim \begin{cases} 10^{-4} & (\text{WD}) \\ 10^{-1} & (\text{NS}) \end{cases} \quad \Rightarrow \quad T_{\max} \propto \varepsilon^{3/4} \sim \begin{cases} 10^{-3} & (\text{WD}) \\ 10^{-1} & (\text{NS}) \end{cases} \qquad (90)$$

Hence $T_{\max}$ is of order 10^5K for a white dwarf or 10^7K for a neutron star (corresponding to ultraviolet and X-ray emission respectively). Since both black holes and neutron stars emit X-rays, one must use more subtle signatures to distinguish between them: for example, one would expect the X-rays from neutron stars (but not black holes) to vary periodically.

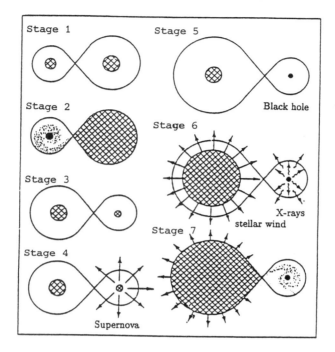

Figure 12. *Evolution of binary system with black hole*

3.5 Stellar Black Holes in Binary Systems

An isolated black hole in the interstellar medium would be hard to detect since the gas density is too low for accretion to make it very luminous. On the other hand, about 30% of stars are in binary systems and one would have a much better chance of detecting a black hole with a companion (Trimble 1991). This is partly because the companion could provide a source of accretable material and partly because the presence of the hole would perturb its orbit: one would expect its light to exhibit a Doppler shift with some period P and velocity amplitude $V_{\rm dop}$. If the mass of the visible star and the black hole are M_V and M_B respectively, and if the inclination of the plane of the orbit to the observer's line of sight is i, then the mass function is given by

$$\mathcal{M} = \frac{M_B^3 (\sin i)^3}{(M_B + M_V)^2} = \frac{P V_{\rm dop}^3}{2\pi G}. \tag{91}$$

All the quantities on the right-hand-side can be measured observationally, so one can determine the mass M_B providing one has information about M_V (*e.g.* from spectroscopic information) and i. Thus one way to search for black holes is to look for binary systems in which one of the stars is invisible but larger than the maximum mass for a white dwarf or neutron star. The evidence would be compounded by the presence of X-rays.

Figure 12, taken from Van den Heuvel (1974), illustrates the complicated evolutionary sequence whereby a binary system evolves into a black hole or neutron star in orbit around a visible star. X-rays are only produced during phases 6 and 7, so the fraction of binary black holes detectable at any time will be small. One can anticipate the number of detectable black holes in the Galaxy by noting that the time between Galactic supernovae is about 30 years, so there should be around 10^9 collapsed cores. Since the observed stellar mass function implies that neutron star remnants should be roughly 10 times as numerous as black hole remnants, there should be about 10^8 black holes. However, the number of X-ray emitting black holes in binaries should only be about 10^3, this being smaller by a factor which represents both the fraction of black holes in binaries and the fraction of time for which X-rays are emitted.

The first black hole candidate was Cygnus X1, which was discovered as an X-ray source by the UHURU satellite in 1970. The visible companion is a blue supergiant whose mass is estimated to be $33\,M_\odot$. Its orbital characteristics ($P = 5.6$ days, $\mathcal{M} = 0.24\,M_\odot$) suggest that the mass of the X-ray source is $16\,M_\odot$ and it is hard to see how something this large could be anything other than a black hole without being visible (Cowley 1992). This conclusion is strengthened by the characteristics of the X-rays. Their luminosity is $10^4 L_\odot$, which from (88) could be explained by a plausible accretion rate of around $10^{-9}\,M_\odot\,y^{-1}$, but they also exhibit rapid (non-periodic) fluctuations in intensity on timescales of order 10^{-2}s. This suggests that the object must be smaller than $10^{-3} R_\odot$ (the distance light can travel in 10^{-2}s), again precluding an ordinary star.

Various other black hole candidates now exist (Cowley 1992): LMC-X3 (an X-ray source in the Large Magellanic Cloud) is thought to be a $9\,M_\odot$ hole in orbit around a $6\,M_\odot$ companion, A0620-00 (Monoceros) a $13\,M_\odot$ hole in orbit around a $0.7\,M_\odot$ companion, GS 2023+33 (V404 Cygni) a $12\,M_\odot$ hole in orbit around a $1.2\,M_\odot$ companion (Casares et al. 1992), and GS 1124-68 (Nova Muscae) a $6\,M_\odot$ hole in orbit around a $0.7\,M_\odot$ companion (Remillard et al. 1992). The last three candidates are 'transient' X-ray sources associated with low mass companions, whereas the first one and Cygnus X1 are 'persistent' sources associated with high mass companions. The persistent sources are likely to be much rarer (Stella et al. 1994).

4 Black Holes in Cosmology

4.1 Black Holes, Quasars, Active Galaxies

Quasars are objects with around 10^3 times the luminosity of the Galaxy (assuming their redshifts are cosmological) but with a size (as inferred from the timescale of their luminosity variations) of order that of the solar system. There is now little doubt that they are associated with the nuclei of galaxies—in some cases one can even resolve the galaxy—and the fact that their comoving density peaks at a redshift of around 2 (Hartwick and Schade 1990) suggests that they are associated with *young* galaxies. Although the comoving number density of quasars is only about 1% that of galaxies, the fraction of galaxies which pass through a quasar phase could be more than that if this phase is of short duration (Haehnelt and Rees 1993).

Three types of model have been proposed to explain the energetic output of quasars (Blandford 1987): (i) the 'supernovae' model assumes that the luminosity is generated by the explosions of stars in a dense star cluster; (ii) the 'spinar' model assumes that

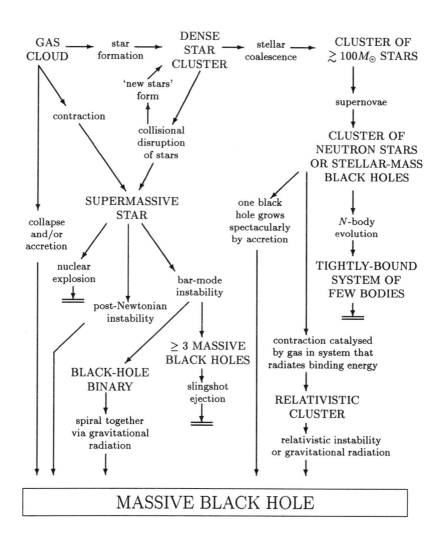

Figure 13. *Evolution to massive black hole in galactic nucleus*

one has a rotating (possibly magnetized) supermassive star which radiates either by undergoing periodic explosions or by synchroton emission; (iii) the 'black hole' model assumes that one has a central supermassive black hole which radiates by accreting gas and stars. Most theorists now favour the black hole model: as illustrated in Figure 13, taken from Rees (1984), even if a galactic nucleus contains a dense cluster or a supermassive star for some period, these objects should eventually evolve into a supermassive black hole. Also it is now clear that quasars just represent the most extreme form of a range of phenomena involving active galactic nuclei (AGN): since black hole accre-

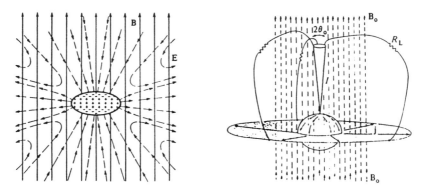

Figure 14. *Blandford-Znajek mechanism*

tion affords the most efficient means of generating radiation, it would seem natural to associate quasars with this final phase.

In discussing the accretion model, to which we henceforth confine attention, we will exploit some of the mathematical results discussed in Section 2. The black hole hypothesis has various key features (Blandford 1987).

• Even without accretion (36) shows that up to 29% of the energy of a rotating black hole can be extracted. In practice, the most plausible way of achieving this is via the Blandford-Znajek mechanism. This depends upon the fact that a rotating black hole acts like a spinning conductor (Blandford and Znajek 1977). As illustrated in Figure 14, adapted from Thorne et al. (1976), if an isolated hole spins with angular velocity parallel to an external magnetic field, an electric field and apparent quadrupolar surface charge density will be induced, so there will be a potential difference between the pole and the equator. If the pole and equator are connected, a current will flow and power will be dissipated. This power derives from the spin of the black hole.

• We saw in Section 3.4 that maximum luminosity generated by disc accretion is $L = \varepsilon \dot{M} c^2$ where ε lies between 0.06 and 0.4, depending on the rotation of the hole. The nature of the accretion disc depend on how the accretion rate compares to the accretion rate associated with the Eddington luminosity (Rees 1984), which from (58) is given by

$$\dot{M}_{\rm ED} = \frac{L_{\rm ED}}{\varepsilon c^2} \sim \left(\frac{M}{10^8 \, M_\odot}\right) M_\odot \, y^{-1}. \qquad (92)$$

For $\dot{M} \ll \dot{M}_{\rm ED}$, one may get an ion-supported torus and this is the situation which is likely to pertain for radio galaxies. For $\dot{M} \sim \dot{M}_{\rm ED}$, one expects to get a standard thin disc with a characteristic 'blue bump' in the spectrum and this is probably the situation for Seyfert galaxies. For $\dot{M} \gg \dot{M}_{\rm ED}$, one expects to get a radiation-supported torus and this is probably the situation for quasars.

• The presence of an accretion torus permits the formation of 'jets' and these may provide the radio beams required to power double-radio sources. In Figure 15, taken from Blandford (1987), the contours of constant effective potential are modelled by using a Newtonian potential $\Phi = -m/(r - 2m)$ and the specific angular momentum is

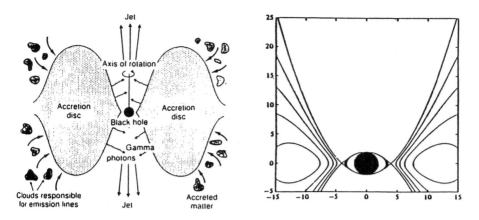

Figure 15. *Production of jets by accretion torus*

chosen to have the value $h = 4M$, so that the contour passing through the cusp has zero binding energy. Gas can fill up the region occupied by the contours outside $r = 4M$ and then pour through the cusp onto the black hole. The zero energy surfaces define a pair of funnels which may be responsible for chanelling some of the outflow into jets. Gamma-rays emitted by the inner face of the disc create electron-positron pairs which then multiply by cascading and accelerate to near the speed of light.

• Lens-Thirring precession may cause the inner part of the accretion disc to become aligned with the black hole spin and this may serve as a 'gyroscope' which maintains the directions of the jets (Bardeen and Petterson 1975). Otherwise it would be hard to explain why the jets in double-radio sources are straight. Where jets are wiggly, this may be because there are two black holes in orbit around each other (Begelman *et al.* 1980). Since the motions are relativistic, one may sometimes appear to get 'superluminal' velocities in the core of a quasar or radio source. This 'illusion' can arise whenever a source is moving near to the speed of light in a direction close to the observer's line of sight.

How large would the black hole powering a quasar need to be? Since accretion can never generate radiation at a rate exceeding the Eddington luminosity given by (58), the observed luminosities of quasars (at least $10^{45} \text{erg s}^{-1}$) imply that their holes must be bigger than $10^7 \, M_\odot$. On the other hand, detailed modelling of the line-widths of the emission lines suggests that they must be smaller than $10^9 \, M_\odot$. It would therefore seem that the holes powering quasars should have a mass of order $10^8 \, M_\odot$ (Blandford and Rees 1992). Somewhat smaller holes might be associated with less extreme forms of nuclear activity. For example, Seyferts might be associated with $10^6 \, M_\odot$ black holes. Indeed one could hypothesize that *every* galaxy has a black hole at its centre, with the larger ones residing in elliptical galaxies (since these are associated with quasars and radio sources) and the smaller ones in spiral galaxies (since these are associated with Seyferts). The black holes would be fed by gas and stars when the galaxies were young but their luminosity would decrease (as observed) once all the 'food' was consumed and this would explain why the number of quasars declines at late epochs.

4.2 Evidence for Black Holes in Galactic Nuclei

We have seen that there are good reasons for anticipating that many galaxies should have massive black holes in their nuclei today, even though they may now be quiescent. In the last few decades much effort has gone into seeking evidence for such black holes, evidence coming from both dynamical studies of the central regions and measurements of the brightness profile there. The first method depends on the fact that the gravitational field of a central black hole would induce an anomalously high velocity in the nearby stars or gas, leading to a high dynamical mass-to-light ratio M/L in the central regions. The second method depends upon the fact that the black hole should induce an enhancement in the density of stars in its neighbourhood, thereby producing a cusp in the light profile. Although early attempts to search for black holes in galactic nuclei used this second method, it turns out that brightness profile is not a very good indicator since the galaxies which are believed to have the largest black holes often have the lowest central surface brightness. Below we therefore focus on the dynamical method.

If the stars near the centre of a galaxy have density ν, rotation velocity V, and velocity dispersions σ_r, σ_θ, σ_ϕ in the radial, latitudinal and azimuthal directions (all of these depending on the galactocentric radius r), then the mass within radius r is

$$M(r) = \frac{V^2 r}{G} + \frac{\sigma_r^2 r}{G}\left[-\frac{d\ln\nu}{d\ln r} - \frac{d\ln\sigma_r^2}{d\ln r} - \left(1 - \frac{\sigma_\theta^2}{\sigma_r^2}\right) - \left(1 - \frac{\sigma_\phi^2}{\sigma_r^2}\right)\right]. \tag{93}$$

Evidence for a central *Massive Dark Object* (MDO) arises whenever $M(r)$ exceeds the mass in visible form. In spiral galaxies, which are rotationally supported, the first term on the right-hand-side dominates and the method is unambiguous. With elliptical galaxies, however, the second term dominates and the mass estimate is more difficult because it depends on whether the velocity dispersion is isotropic or anisotropic (which is uncertain). The mass could be much reduced in the anisotropic case (with $\sigma_\theta \ll \sigma_r$ and $\sigma_\phi \ll \sigma_r$), since the term in square brackets might then be much less than 1. Although the MDO would not necessarily have to be a black hole, one can often argue that alternative candidates (*e.g.* dense star clusters) are implausible on dynamical grounds.

Kormendy and Richstone (1995) have reviewed the evidence that at least eight nearby galaxies have central MDOs. The galaxies have to be nearby in order to resolve the innermost part of their cores and much of the recent progress in this area can be attributed to the high resolution of the Hubble Space Telescope. We discuss the candidates below in order of decreasing certainty. In all cases, the mass turns out to be in the range 10^6–$10^{9.5}$ $M_\odot$, which is what one would expect from the models for quasars and active galactic nuclei. The data suggests that 20% of galaxies contain an MDO and its mass is always about 0.002 times the mass of the galactic bulge.

- The spiral galaxy M31 (our nearest neighbour) is probably the best candidate since V, σ and M/L have all been measured very precisely in this case. The central region is rotating very rapidly (there may even be a double nucleus) and M/L increases very steeply within the central 2 arcsecs. The data is shown in Figure 16, taken from Kormendy and Richstone (1995), and indicates the presence of a black hole with a mass of $M_{\rm BH} = 3 \times 10^7$ $M_\odot$.

- The edge-on S0 galaxy NGC 3115 (at a distance of 8.4 Mpc) is the second best

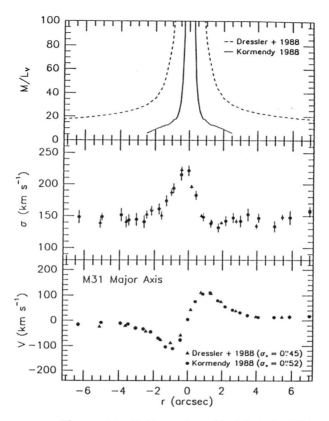

Figure 16. *Evidence for a black hole in M31*

candidate. Again M/L increases dramatically within the central region and this provides evidence for an MDO with mass $M_{BH} = 1 \times 10^9 \, M_\odot$. However, the conclusion that this is a black hole is not as definite as for M31.

- The dwarf elliptical M32 (at a distance of 0.7Mpc) is probably the most studied case, having been observed and modelled by five independent groups. Successful models require that the MDO have the relatively modest mass of $2 \times 10^6 \, M_\odot$, although again one cannot be sure that this is a black hole.

- The Sombrero galaxy NGC 4594 (at a distance of 9.2Mpc) may contain an MDO of mass $5 \times 10^8 \, M_\odot$. However, the M/L profile is less certain in this case since the models have explored fewer degrees of freedom.

- The nucleus of our own galaxy can only be studied in the infrared and radio bands because of the obscuring dust. The rotation curve obtained from gas and different types of stars gives the mass profile shown in Figure 17, taken from Genzel *et al.* (1994), and suggests an MDO of mass $2 \times 10^6 \, M_\odot$.

- The normal elliptical NGC 3377 (at a distance of 10Mpc) may have an MDO of mass $8 \times 10^7 \, M_\odot$ but this assumes that the velocity dispersion is isotropic and this may

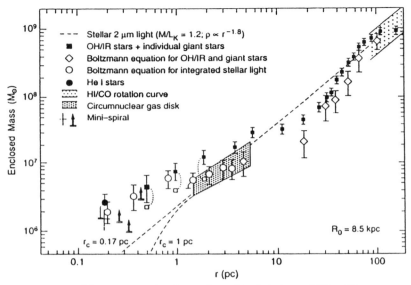

Figure 17. *Evidence for a black hole in Milky Way*

not be the case. The evidence here is therefore less secure.

• The giant elliptical M87 (at a distance of 15Mpc) was the first galaxy postulated to have a central black hole (Young et al. 1978) and the presence of a jet supports this suggestion. The putative mass is 3×10^9 $M_\odot$ (Harms et al. 1994) but the arguments are still not conclusive. The evidence was originally associated with the presence of a light-cusp but we have noted that this feature is not convincing. The evidence from the dynamical studies of stars is also ambiguous because of the uncertainty in the anisotropy. Perhaps the best evidence comes from studying the gas disc since Hubble Space Telescope data shows that the central ionized gas distribution is elongated perpendicular to the jet.

• The spiral galaxy NGC 4258 (at a distance of 7Mpc) is an interesting case because it contains a modest AGN as well as a jet. The central mass appears to be 4×10^7 $M_\odot$. In this case, the rotation curve in the centre has been mapped very precisely by the VLBA telescope down to a distance of 0.2pc (using the radio emission of maser sources) and this confirms that the rotation velocity profile is Keplerian (Miyoshi et al. 1995).

Another interesting recent candidate is MCG-6-30-15. In this case, X-ray observations of iron Ka emission give a velocity width of 10^5 km s^{-1} ($0.3c$) and this indicates that one is probing down to a distance of only 10 times the Schwarzschild radius (10^5 times smaller than the distance which can be probed in the optical or radio). The line profile is particularly interesting since it appears to be compatible with a face-on accretion disc (Tanaka et al. 1995).

4.3 Black Holes and Dark Matter

Equation (77) shows that a gravitationally bound system of mass M and radius R has a characteristic velocity $V \approx (GM/R)^{1/2}$ and a dark matter problem arises whenever the mass inferred from the measured values of V and R exceeds the mass in visible form. Evidence for dark matter has been claimed in four different contexts (Carr 1994).

• There may be *local* dark matter in the Galactic disc with a mass comparable to that in visible form ($M_{\text{dark}} \sim M_{\text{vis}}$); in this case, R is associated with the thickness of the disc of $\sim$ 300pc and V with the vertical velocity dispersion of the stars of $\sim$ 20km s^{-1}.

• There may be dark matter in the *halos* of galaxies with a mass which depends on the extent of the halos ($M_{\text{dark}} \sim 10 M_{\text{vis}}(R_h/100\text{kpc})$ where R_h is the typical halo radius); V is associated with the rotation velocity of the stars for spirals and the velocity dispersion for ellipticals, both being of order 100km s^{-1} and roughly independent of radius.

• There may be dark matter associated with *clusters* of galaxies ($M_{\text{dark}} \sim 10 M_{\text{vis}}$); in this case, R characterises the size of the cluster $\sim$ 10Mpc and V the velocity dispersion of the galaxies of $\sim 10^3$ km s^{-1}.

• In the inflationary scenario, there may also be smoothly distributed *background* dark matter, required in order that the total cosmological density have the critical value which separates ever-expanding models from recollapsing ones ($M_{\text{dark}} \sim 100 M_{\text{vis}}$); in this case, one can interpret V as the speed of light and R as the Hubble radius of around 6000Mpc.

A key question is whether these various forms of dark matter are baryonic or non-baryonic. The main argument for both baryonic and non-baryonic dark matter comes from Big Bang nucleosynthesis. This is because the success of the standard picture in explaining the primordial light element abundances only applies if the baryon density parameter Ω_b lies in the range (Walker *et al.* 1991)

$$0.010 h^{-2} < \Omega_b < 0.015 h^{-2} \tag{94}$$

where h is the Hubble parameter in units of 100km s^{-1}Mpc^{-1} (somewhere between 0.5 and 1). The upper limit implies that Ω_b is well below 1, which suggests that no baryonic candidate could provide the critical density required in the inflationary scenario (Guth 1981). This conclusion also applies if one invokes inhomogeneous nucleosynthesis since one requires $\Omega_b < 0.09 h^{-2}$ even in this case (Mathews *et al.* 1993). The standard scenario therefore assumes that the total density parameter is 1, with only the fraction given by (94) being baryonic. On the other hand, the value of Ω_b allowed by (94) almost certainly exceeds the density of visible baryons Ω_v, a careful inventory by Persic and Salucci (1992) showing that the density in galaxies and cluster gas is about 0.003 for reasonable values of h. Thus it seems that one needs both non-baryonic and baryonic dark matter.

Although one cannot associate a baryon number with a black hole, one can associate one with its progenitor. Since stellar progenitors are made of ordinary baryons, this means that any pregalactic or protogalactic black holes must have a density less than Ω_b. In principle, such holes could certainly provide the dark matter in galactic discs:

even if all discs have the 50% dark component envisaged for the Galaxy (Bahcall et al. 1992), this only corresponds to $\Omega_d \approx 0.001$. Black holes might also provide the dark matter in galactic halos: if the Milky Way is typical, the density associated with halos would be $\Omega_h \approx 0.03 h^{-1}(R_h/100\text{kpc})$, so (94) implies that all the dark matter in halos could be baryonic providing $R_h < 50 h^{-1}\text{kpc}$ and this is marginally possible (Fich and Tremaine 1991). On the other hand, the cluster dark matter, which has a density $\Omega_c \approx 0.1$–0.2, could only be baryonic if one invoked inhomogeneous nucleosynthesis and the background dark matter (if it exists) would definitely need to be non-baryonic. Therefore if the last two dark matter problems are attributed to black holes, they would need to be primordial in the sense that they formed when the density of the Universe was dominated by its radiation content. Indeed, as shown in Section 5, they would probably have to form before cosmological nucleosynthesis at 1s.

It should be stressed that black holes are not the only dark matter candidates. The non-baryonic dark matter required to make up the critical density might also be in the form of 'Weakly Interacting Massive Particles' or 'WIMPs'. Although none of these particles has yet been detected, their existence is certainly no more speculative than that of primordial black holes. Nor need the baryonic dark matter comprise black holes. The discrepancy between Ω_b and Ω_v could be also resolved if the missing baryons were in a hot intergalactic medium (Barcons et al. 1991) or intergalactic Lyman-α clouds (Rees 1986) or a population of dwarf galaxies (Bristow and Phillipps 1994) or low surface brightness galaxies (McCaugh 1994). None of these are included in the Persic-Salucci estimate of Ω_v. Even if the missing baryons are in halo objects (corresponding to what is termed the 'Massive Compact Halo Object' or 'MACHO' scenario), they may comprise brown dwarfs (Kerins and Carr 1994) or cold molecular clouds (Pfenniger et al. 1994) rather than black holes and the disc dark matter (if it exists) may comprise white dwarfs. Despite these reservations, one cannot strictly exclude any of the four dark matter problems being attributable to black holes, so we now consider possible consequences of this.

4.4 Gravitational Lensing Effects

One of the most useful signatures of black holes (or other compact objects) is their gravitational lensing effects. Indeed it is remarkable that lensing could permit their detection over the entire mass range $10^{-7}\,M_\odot$ to $10^{12}\,M_\odot$. All sorts of astronomical objects can serve as lenses (Blandford and Narayan 1992) but the crucial advantage of compact objects is that they are small and spherically symmetric and this makes their effects very clean. There are two distinct lensing effects and these probe different but nearly overlapping mass ranges: macrolensing (the multiple-imaging of a source) can be used to search for objects larger than $10^5\,M_\odot$, while microlensing (modifications to the intensity of a source) can be used for objects smaller than $10^3\,M_\odot$. The current constraints on the density parameter of compact objects in various mass ranges are brought together in Figure 18.

Macrolensing by Compact Objects. If one has a population of compact objects with mass M and density parameter Ω_C, then the probability of one of them multiply-imaging a source at redshift $z \sim 1$ and the separation between the images are given

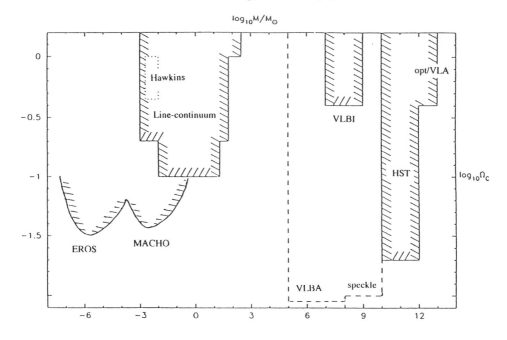

Figure 18. *Lensing constraints on the density parameter for compact objects*

by

$$P \approx (0.1 - 0.2)\,\Omega_C, \qquad \theta \approx 6\times 10^{-6}(M/\,\mathrm{M_\odot})^{1/2} h^{1/2} \text{ arcsec} \qquad (95)$$

(Press and Gunn 1973). One can therefore use upper limits on the frequency of macrolensing for different image separations to constrain Ω_C as a function of M (Nemiroff 1989). In particular, for quasars, VLA data imply $\Omega_C(10^{11}-10^{13}\,\mathrm{M_\odot}) < 0.4$ (Hewitt 1986) and Hubble Space Telescope data imply $\Omega_C(10^{10}-10^{12}\,\mathrm{M_\odot}) < 0.02$ (Surdej et al. 1993). To probe smaller scales, one can use high resolution radio sources: Kassiola et al. (1991) have invoked lack of lensing in forty VLBI objects to infer that $\Omega_C(10^7-10^9\,\mathrm{M_\odot}) < 0.4$. Future observations could strengthen these constraints: as indicated by the broken lines in Figure 18, speckle interferometry could push the value of $\Omega_C(10^8-10^{10}\,\mathrm{M_\odot})$ down to 0.01, while VLBA could push $\Omega_C(10^5-10^8\,\mathrm{M_\odot})$ down to 0.001 (Surdej et al. 1993).

Microlensing in Macrolensed Quasars. Even if a lens is too small to produce resolvable multiple-imaging of a source, it may still induce detectable intensity variations. In particular, one can look for microlensing in quasars which are already macrolensed. This possibility arises because, if a galaxy is suitably positioned to image-double a quasar, then there is also a high probability that an individual halo object will traverse the line of sight of one of the images (Gott 1981) and this will give intensity fluctuations in one but not both images. The effect would be observable for objects bigger than $10^{-4}\,\mathrm{M_\odot}$ but the timescale of the fluctuations, being of order $40(M/\,\mathrm{M_\odot})^{1/2}$y, would only show up over a reasonable time for $M < 0.1\,\mathrm{M_\odot}$. There is already evidence of this effect for the quasar 2237+0305 (Irwin et al. 1989, Corrigan et al. 1991), the observed timescale for the variation in the luminosity of one of the images indicating a mass below $0.1\,M_\odot$ (Webster et al. 1991).

Effect of Microlensing on Quasar Luminosity. More dramatic but rather controversial evidence for the microlensing of quasars comes from Hawkins (1993), who has been monitoring 300 quasars in the redshift range 1-3 over the last 17 years using a wide-field Schmidt camera. He finds quasi-sinusoidal variations with an amplitude of 0.5 magnitudes on a timescale 5y and attributes this to lenses with mass $\sim 10^{-3}\,M_\odot$. The crucial point is that the timescale decreases with increasing redshift, which is the opposite to what one would expect for intrinsic variations. The timescale also increases with the luminosity of the quasar and he explains this by noting that the variability timescale should scale with the size of the accretion disc (which should itself correlate with luminosity). A rather striking feature of Hawkins' claim is that he requires the density of the lenses to be close to critical (in order that the sources are transited continuously), so he has to invoke primordial black holes.

Line-Continuum Effects for Quasars. In some circumstances, only part of a quasar may be microlensed. In particular, the line and continuum fluxes may be affected differently because they may come from regions which act as extended and point-like sources respectively. (For a lens at a cosmological distance the Einstein radius is $0.05(M/M_\odot)^{1/2}h$pc, whereas the size of the optical continuum and line regions are of order 10^{-4}pc and 1pc respectively.) This effect can show up in statistical studies of many quasars: one would expect the characteristic equivalent width of quasar emission lines to decrease as one goes to higher redshift because there would be an increasing probability of having an intervening lens. Recently Dalcanton et al. (1994) have compared the equivalent widths for a high and low redshift sample of quasars and find no difference. They infer the following limits:

$$\Omega_C(0.001 - 60\,M_\odot) < 0.2, \quad \Omega_C(60 - 300\,M_\odot) < 1, \quad \Omega_C(0.01 - 20\,M_\odot) < 0.1 \quad (96)$$

The mass limits come from the fact that the amplification of even the continuum region would be unimportant for $M < 0.001\,M_\odot$, while the amplification of the line regions would be important (cancelling the effect) for $M > 20\,M_\odot$ if $\Omega_C = 0.1$ or $M > 60\,M_\odot$ if $\Omega_C = 0.2$ or $M > 300\,M_\odot$ if $\Omega_C = 1$. These limits are indicated in Figure 18 and are marginally incompatible with Hawkins' claim that $\Omega_C(10^{-3}\,M_\odot) \sim 1$.

Microlensing of Stars by Halo Objects in our own Galaxy. Attempts to detect microlensing by objects in our own halo by looking for intensity variations in stars in the Magellanic Clouds and the Galactic Bulge have now been underway for several years and have already met with success. In this case, the timescale for the variation is $P \approx 0.2(M/M_\odot)^{1/2}$y, so one can seek lenses over the mass range $10^{-7} - 10^2\,M_\odot$, but the probability of an individual star being lensed is only $\tau \sim 10^{-6}$, so one has to look at many stars for a long time (Paczynski 1986). The likely event rate is

$$\Gamma \sim N\tau P^{-1} \sim (M/M_\odot)^{-1/2}\,\text{y}^{-1}$$

where $N \sim 10^6$ is the number of stars. Thus small masses give frequent short-duration events, while large masses give rare long-duration events. The key feature of these microlensing events is that the light-curve is time-symmetric and achromatic and this may allow them to be distinguished from intrinsic stellar variations (Griest 1991). Three groups are involved and each now claims to have detected lensing events. The American group (MACHO) has used a dedicated telescope to study 10^7 stars in red and blue light in the Large Magellanic Cloud (LMC), the Small Magellanic Cloud (SMC) and the Galactic bulge. They currently have 3 LMC events (with durations of order a month) and around 45 bulge events (Alcock et al. 1993). The timescale for the LMC events suggests that the halo objects have a mass of around $0.1\,\mathrm{M}_\odot$ but the frequency (although larger than that expected from ordinary stars) is only about a fifth that anticipated if the halo consists entirely of such objects. The French group (EROS) has been studying stars in the LMC and is seeking both 1–100 day events (corresponding to $10^{-4} - 1\,\mathrm{M}_\odot$ lenses) with digitised red and blue Schmidt plates and 1 hour to 3 day events (corresponding to $10^{-7} - 10^{-3}\,\mathrm{M}_\odot$) with charged coupled devices (CCD) (Auborg et al. 1993). The CCD searches have given no results, which implies a limit $\Omega_C(10^{-7} - 10^{-3}\,\mathrm{M}_\odot) < 0.1$, but analysis of 3×10^6 stars on the Schmidt plates yields two events, each with duration of about two months. The Polish collaboration (OGLE) has looked at 7×10^5 stars in the Galactic bulge and has claimed 11 events (Udalski et al. 1993). The most plausible LMC lensing objects are brown dwarfs but they might also be primordial black holes.

4.5 Dynamical Constraints

A variety of constraints can be placed on black holes in different sites by considering their dynamical effects (Carr and Sakellariadou 1996). The limits are summarised as upper limits on the density parameter $\Omega_\mathrm{B}(M)$ for black holes of mass M in Figure 19, where the disc, halo and cluster densities are assumed to be 0.001, 0.1 and 0.2, respectively.

Disc Heating by Halo Holes. As halo objects traverse the Galactic disc, they will impart energy to the stars there. This will lead to a gradual puffing up of the disc, with older stars being heated more than younger ones. Indeed Lacey and Ostriker (1985) have argued that black holes of around $10^6\,\mathrm{M}_\odot$ could generate the observed disc-puffing providing the number density of the holes n satisfies $nM^2 \approx 3\times10^4\,\mathrm{M}_\odot^2\mathrm{pc}^{-3}$; combining this with the local halo density $\rho_h = nM \approx 0.01\,\mathrm{M}_\odot\mathrm{pc}^{-3}$ gives $M = 2\times10^6\,\mathrm{M}_\odot$. More recent data is probably inconsistent with this picture and heating by spiral density waves or giant molecular clouds is now usually invoked (Lacey 1991). Nevertheless, one can still use the Lacey-Ostriker argument to place an upper limit on the density in halo objects of mass M (Carr et al. 1984):

$$\Omega_\mathrm{B} < \Omega_h \min[1, (M/M_\mathrm{heat})^{-1}], \qquad M_\mathrm{heat} = 3\times10^6 (t_g/10^{10}\,\mathrm{y})^{-1}\,\mathrm{M}_\odot \qquad (97)$$

where t_g is the age of the Galaxy. Otherwise the disc would be more puffed up than observed.

Disruption of Stellar Clusters by Halo Objects. Another type of dynamical effect associated with halo objects would be their influence on bound groups of stars (e.g. globular clusters). Every time a halo object passes near a star cluster, the object's tidal field heats up the cluster and thereby reduces its binding energy. If the object is sufficiently large it will disrupt the cluster in a single fly-by. For smaller objects

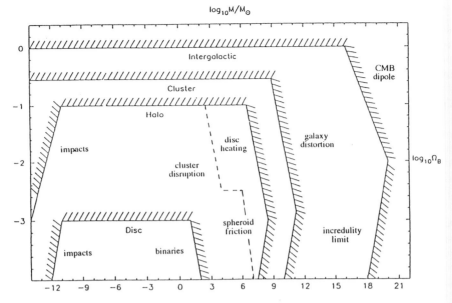

Figure 19. *Dynamical constraints on the density parameter for black holes located in the Galactic disc, the Galactic halo, clusters of galaxies and the intergalactic medium.*

the disruption will be gradual, requiring the cumulative effect of many traversals. By comparing the expected disruption time for clusters of mass m_c and radius r_c with the cluster lifetime t_L, one finds that the local density of halo holes of mass M must satisfy (Sakellariadou 1984, Wielen 1985, Carr and Sakellariadou 1996)

$$\rho_B < \begin{cases} m_c V/(GMt_L r_c) & \text{for} \quad M < M_c(V/V_c) \\ (m_c/Gt_L^2 r_c^3)^{1/2} & \text{for} \quad M_c(V/V_c) < M < M_c(V/V_c)^3 \end{cases} \quad (98)$$

where $V \approx 300 \mathrm{km\,s^{-1}}$ is the characteristic speed of the holes. Any lower limit on t_L therefore places an upper limit on ρ_B. The crucial point is that the limit is independent of M in the single-encounter regime, so it bottoms out at density of order $(\rho_c/Gt_L^2)^{1/2}$. If one applies this argument to globular clusters, for which $m_c = 10^5\,\mathrm{M_\odot}$, $r_c = 10\mathrm{pc}$ and $t_L \approx 10^{10}\mathrm{y}$, one finds that the upper limit on ρ_B is comparable to the actual halo density, which suggests that halo objects might *determine* the characteristics of globular clusters. Numerical calculations for the disruption of globular clusters by Moore (1993) confirm the qualitative features indicated above but he infers a stronger upper limit of $10^3\,\mathrm{M_\odot}$ since his clusters are more diffuse. The line corresponding to Moore's result is shown broken in Figure 19.

Effect of Dynamical Friction on Halo Objects. Another important dynamical effect is that halo objects will tend to lose energy to lighter objects and consequently drift towards the Galactic nucleus (Chandrasekhar 1964). In particular, one can show that halo objects will be dragged into the nucleus by the dynamical friction of the

Spheroid stars from within a Galactocentric radius (Carr and Lacey 1987)

$$R_{\text{df}} \approx (M/10^6 \, M_\odot)^{2/3}(t_g/10^{10} \, \text{y})^{2/3} \, \text{kpc}. \tag{99}$$

The total mass dragged into the Galactic nucleus therefore exceeds the observational upper limit of $3 \times 10^6 \, M_\odot$ unless

$$\Omega_B < \Omega_h \min[1, (M/M_{\text{df}})^{-2}], \qquad M_{\text{df}} = 3 \times 10^4 (a/2 \, \text{kpc})(t_g/10^{10} \, \text{y})^{-1} \, M_\odot \tag{100}$$

where a is the halo core radius. This is stronger than the disc-heating limit but there is an important caveat here since, once more than two black holes have accumulated in the centre of the Galaxy, they may be ejected via the 'slingshot' mechanism (Hut and Rees 1992). Because limit (100) is not completely firm, it is only shown broken in Figure 19.

Constraints on Dark Objects Outside Halos. Bahcall et al. (1985) have argued that the *disc* dark matter could not comprise objects larger than $2 \, M_\odot$ else they would disrupt the wide binaries observed by Latham et al. (1984). However, this has been disputed by Wasserman and Weinberg (1987) on the grounds that there is no sharp cut-off in the distribution of binary separations above 0.1pc. Dynamical constraints are much weaker for dark objects in *clusters* of galaxies but one still gets an interesting constraint from upper limits on the fraction of galaxies with unexplained tidal distortions. Van den Bergh (1969) applied this argument to the Virgo cluster and inferred that black holes binding the cluster could not be bigger than $10^9 \, M_\odot$. The form of both this and the wide binary limit can be inferred from (98). The most interesting dynamical constraint on *intergalactic* black holes comes from the fact that each galaxy should have a peculiar velocity due to its gravitational interaction with the nearest one (Carr 1978). If the holes were smoothly distributed and had a number density n, one would expect every galaxy to have a peculiar velocity of order $GMn^{2/3}t_g$. Since the cosmic microwave background (CMB) dipole anisotropy shows that the peculiar velocity of our own Galaxy is only 600km s^{-1}, one infers a limit

$$\Omega_B < (M/10^{16} \, M_\odot)^{-1/2}. \tag{101}$$

Various other limits are shown in Figure 19, including the *lower* limits (at the bottom right) which correspond to there being at least one black hole of mass M within each site (the incredulity limit) and the constraints (on the left) associated with upper limit on the frequency of impacts by interstellar comets (Hills 1986).

5 Primordial Black Holes and Quantum Effects

5.1 Historical Introduction

It was first pointed out by Zeldovich and Novikov (1967) and Hawking (1971b) that black holes could have formed in the early Universe as a result of density inhomogeneities. Indeed we saw in Section 1.1 that this is the only time when black holes smaller than a solar mass could form. In order to collapse against the background pressure, overdense regions would need to have a size comparable to the particle horizon at maximum expansion. On the other hand, they could not be much bigger than this, else they would be a separate closed universe rather than part of our universe. PBHs forming at time t would therefore need to have of order the horizon mass

$$M_H(t) \approx \frac{c^3 t}{G} \approx 10^{15} \left(\frac{t}{10^{-23} \text{ s}}\right) g. \qquad (102)$$

PBHs could thus span an enormous mass range: those formed at the Planck time (10^{-43}s) would have the Planck mass (10^{-5}g), whereas those formed at 1s would be as large as 10^5 $M_\odot$, comparable to the mass of the holes thought to reside in galactic nuclei.

For a while the existence of PBHs seemed unlikely since Zeldovich and Novikov (1967) had pointed out that they might be expected to grow catastrophically. This is because a simple Newtonian argument suggests that, in a radiation-dominated universe, the mass of a black hole should evolve according to

$$M(t) = M_H(t) \left[1 + \frac{t}{t_o}\left(\frac{M_H(t_o)}{M_o} - 1\right)\right]^{-1} \approx \begin{cases} M_o & \text{for } M_o \ll M_H(t_o) \\ M_H(t) & \text{for } M_o \approx M_H(t_o) \end{cases} \qquad (103)$$

where M_0 is the mass of the black hole at some initial time t_0. This implies that holes much smaller than the horizon cannot grow much at all, whereas those of size comparable to the horizon could could continue to grow at the same rate as it ($M \sim t$) throughout the radiation era. Since we have seen that a PBH *must* be of order the horizon size at formation, this suggests that all PBHs could grow to have a mass of order 10^{15} $M_\odot$ (the horizon mass at the end of the radiation era). We saw in Section 4.4 and Section 4.5 that there are strong observational limits on how many such giant black holes the Universe could contain, so the implication seemed to be that very few PBHs ever existed.

The Zeldovich-Novikov argument was questionable since it neglected the cosmological expansion and this would presumably hinder the black hole growth. Indeed the notion that PBHs could grow at the same rate as the particle horizon was disproved by Carr and Hawking (1974), who demonstrated that there is no spherically symmetric similarity solution which represents a black hole attached to an exact Friedmann model via a sound-wave. Since a PBH must therefore soon become much smaller than the horizon, at which stage cosmological effects become unimportant and (103) *does* pertain, one concludes that PBHs cannot grow very much at all.

The realisation that small PBHs might exist after all prompted Hawking to study their quantum properties. This led to his famous discovery (Hawking 1974) that black holes radiate thermally and have a finite lifetime (*cf.* (3) and (4)). Despite the conceptual importance of this result, it was bad news for PBH enthusiasts. For since PBHs with a mass of 10^{15}g, which evaporate at the present epoch, would have a temperature of order 100MeV, the observational limit on the γ-ray background density at 100MeV immediately implied that the density of such holes could not exceed 10^{-8} times the critical density (Page and Hawking 1976). Not only did this render PBHs unlikely dark matter candidates, it also implied that there was little chance of detecting black hole explosions at the present epoch (Porter and Weekes 1979).

Despite this negative conclusion, it was realised that PBH evaporations could still have interesting cosmological consequences and the next five years saw a spate of papers focusing on these. In particular, people were interested in whether PBH evaporations could generate the microwave background (Zeldovich and Starobinsky 1976) or modify the standard cosmological nucleosynthesis scenario (Novikov et al. 1979) or account for the cosmic baryon asymmetry (Barrow 1980). On the observational front, people were interested in whether PBH evaporations could account for the unexpectedly high fraction of antiprotons in cosmic rays (Kiraly et al. 1981, Turner 1982) or the interstellar electron/positron spectrum (Carr 1976) or the annihilation-line radiation coming from the Galactic centre (Okeke and Rees 1980). Renewed efforts were also made to look for black hole explosions after the realisation that—due to the interstellar magnetic field—these might appear as radio rather than γ-ray bursts (Rees 1977).

In the 1980s attention turned to several new formation mechanisms for PBHs. It was realised that PBHs might form very naturally if the equation of state of the Universe was ever soft (Khlopov and Polnarev 1980) or if there was a cosmological phase transition leading to bubble collisons (Hawking et al. 1982). The formation of PBHs during an inflationary era (Naselsky and Polnarev 1985) or at the quark-hadron era (Crawford and Schramm 1982) also received attention and, more recently, people have considered the formation of PBHs through the collapse of cosmic strings (Polnarev and Zembovicz 1988, Hawking 1989). All these scenarios are constrained by the quantum effects of the resulting black holes.

In the last decade work on the cosmological consequences of PBH evaporations has been revitalised by calculations of MacGibbon, who realised that the usual assumption that particles are emitted with a black-body spectrum as soon as the temperature of the hole exceeds their rest mass is too simplistic. If one adopts the conventional view that all particles are composed of a small number of fundamental point-like constituents (quarks and leptons), it would seem natural to assume that it is these fundamental particles rather than the composite ones which are emitted directly once the temperature goes above the QCD confinement scale of 250MeV. MacGibbon therefore envisages a black hole as emitting relativistic quark and gluon jets which subsequently fragment into the stable leptons and hadrons (i.e. photons, neutrinos, gravitons, electrons, positrons, protons and antiprotons). On the basis of both experimental and Monte Carlo simulations one now has a good understanding of how such jets fragment. It is therefore straightforward in principle to convolve the thermal emission spectrum of the quarks and gluons with the jet fragmentation function to obtain the final particle spectra (MacGibbon and Webber 1990, MacGibbon 1991, MacGibbon and Carr 1991). As discussed in Section 5.3, the results of such a calculation are very different from a simple direct emission calculation.

Recently attention has turned to the issue of Planck mass relics. It is usually assumed that PBHs evaporate completely but several people have argued that evaporation could terminate when the black hole gets down to the Planck mass (Bowick et al. 1988, Coleman et al. 1991). In this case, one could end up with stable Planck mass relics and such relics might in principle have the critical density (MacGibbon 1987, Barrow et al. 1992, Carr et al. 1994).

5.2 The Formation of Primordial Black Holes

Inhomogeneities with hard equation of state. If the PBHs form directly from primordial density perturbations, then the fraction of the Universe undergoing collapse at any epoch is just determined by the root-mean-square amplitude ε of the fluctuations entering the horizon at that epoch and the equation of state $p = \gamma\rho$ $(0 < \gamma < 1)$. One usually expects a radiation equation of state with $\gamma = 1/3$ in the early Universe. In order to collapse against the pressure, an overdense region must be larger than the Jeans length at maximum expansion and this is just $\sqrt{\gamma}$ times the horizon size. This implies that the density fluctuation must exceed γ at the horizon epoch, so—providing the fluctuations have a Gaussian distribution and are spherically symmetric—one can infer that the fraction of regions of mass M which collapse is (Carr 1975)

$$\beta(M) \sim \varepsilon(M) \exp\left[-\frac{\gamma^2}{2\varepsilon(M)^2}\right] \quad (104)$$

where $\varepsilon(M)$ is the value of ε when the horizon mass is M. The PBHs can have an extended mass spectrum only if the fluctuations are scale-invariant (*i.e.* with ε independent of M) and, in this case, the PBH mass spectrum is given by (Carr 1975)

$$\frac{dN}{dM} = (\alpha - 2)\left(\frac{M}{M_*}\right)^{-\alpha} M_*^{-2} \Omega_{\text{PBH}} \rho_{\text{crit}} \quad (105)$$

where $M_* \approx 10^{15}$g is the current lower cut-off in the mass spectrum due to evaporations, Ω_{PBH} is the total density of the non-evaporated PBHs in units of the critical density (which itself depends on β) and the exponent α is determined by the equation of state:

$$\alpha = \frac{2(1+2\gamma)}{1+\gamma}. \quad (106)$$

If one has a radiation equation of state ($\gamma = 1/3$), then $\alpha = 5/2$. This means that the integrated PBH mass density falls off as $M^{-1/2}$, so most of the PBH density is contained in the smallest ones. If $\varepsilon(M)$ decreases with M, then the spectrum falls off exponentially with M and PBHs can form around the Planck time ($t_{\text{pl}} \sim 10^{-43}$s) if at all; if $\varepsilon(M)$ increases with M the spectrum rises exponentially with M and PBHs would form very prolifically at large scales but the cosmic microwave background anisotropies would then be larger than observed. Fortunately, most scenarios for the origin of the cosmological density fluctuations *do* predict that ε is scale-invariant, so (105) represents the most likely mass spectrum.

Inhomogeneities with soft equation of state. The pressure may be reduced for a while ($\gamma \ll 1$) if the Universe's mass is ever channelled into particles which are massive enough to be non-relativistic (Khlopov and Polnarev 1980). In this case, the effect of pressure in stopping collapse is unimportant and the probability of PBH formation depends upon the fraction of regions which are sufficiently spherical to undergo collapse; this can be shown to be (Polnarev and Khlopov 1981)

$$\beta \approx 0.02\varepsilon^{13/2}. \quad (107)$$

Most of the holes will have a mass smaller than the horizon mass at formation by a factor $\varepsilon^{3/2}$, so the period for which the equation of state is soft directly specifies their mass range. In this case, the value of β is not as sensitive to ε as in (104).

Inflationary period. In the standard inflationary scenario, the amplitude of the density fluctuations increases logarithmically with mass and the normalisation required to explain large-scale structure would then preclude the fluctuations being large enough to give PBHs on a smaller scale. One way around this would be to invoke a 'double inflation' scenario, in which there is a second period of inflation associated with larger fluctuations over some range of scales (Khlopov et al. 1985). Another possibility is to invoke a non-standard inflationary scenario in which the fluctuations have a stronger than logarithmic dependence on mass-scale. If PBH formation is to occur at all, one needs the fluctuations to decrease with increasing mass and—in the chaotic inflation scenario—it turns out that this is only possible if the scalar field is accelerating sufficiently fast. This means that one must violate the usual slow-roll friction-dominated assumptions (Carr and Lidsey 1993). For example, one can generate fluctuations which decrease as a power of mass ($\varepsilon \propto M^{-\alpha}$) if the potential $V(\phi)$ has terms which involve powers of $\sec\phi$ (Gilbert 1995). As discussed in Section 5.5, the COBE quadrupole anisotropy measurement implies that one needs $\alpha=0.08$ if PBH formation is to be interesting.

Bubble collisions. Even if the Universe starts off perfectly smooth, bubbles of broken symmetry might arise at a spontaneously broken symmetry epoch and it has been suggested that PBHs could form as a result of bubble collisions (Hawking et al. 1982, La and Stenhardt 1989). However, this happens only if the bubble formation rate is finely tuned: if it is too large, the entire Universe undergoes the phase transition immediately; if it is too small, the bubbles could never collide. The holes should have a mass of order the horizon mass at the phase transition, so PBHs forming at the Grand Unification epoch (10^{-35}s) would have a mass of order 10^3g, whereas those forming at the electroweak unification epoch (10^{-10}s) would have a mass of 10^{28}g. Only a phase transition before 10^{-23}s would be relevant in the context of evaporating PBHs.

Collapse of cosmic loops. A typical cosmic loop will be larger than its Schwarzschild radius by the inverse of the factor $G\mu$ which represents the mass per unit length. In the favoured scenario, $G\mu$ is of order 10^{-6} but Hawking (1989) and Polnarev and Zemboricz (1988) have shown that there is still a small probability that a cosmic loop will get into a configuration in which every dimension lies within its Schwarzschild radius. Hawking estimates this to be

$$\beta \approx (G\mu)^{-1}(G\mu x)^{2x-2} \tag{108}$$

where x is the ratio of the loop length to the correlation scale. If one takes x to be 3, Ω_{PBH} exceeds 1 for $G\mu > 10^{-7}$, so he argues that one overproduces PBHs. However, Ω_{PBH} is very sensitive to x and a slight reduction would give a rather interesting value. Note that spectrum (105) still applies since the holes are forming at every epoch.

In all these scenarios, the value of Ω_{PBH} associated with PBHs which form at a redshift z or time t is related to β by

$$\Omega_{\text{PBH}} = \beta\Omega_R(1+z) \approx 10^6 \beta \left(\frac{t}{s}\right)^{-1/2} \tag{109}$$

where $\Omega_R \sim 10^{-4}$ is the density of the microwave background. Since t is very small, the constraint $\Omega_{\rm PBH} < 1$ implies that β must be tiny over all mass ranges. This is because the radiation density scales as $(1+z)^4$, whereas the PBH density scales as $(1+z)^3$. If the PBHs form at a phase transition, then they have a very narrow mass spectrum and t is just the time of the transition. If they have a continuous mass spectrum, then the dominant contribution to $\Omega_{\rm PBH}$ comes from the holes evaporating at the present epoch. These form at $t \sim 10^{-23}$s and so (109) implies $\beta \sim 10^{-17}\Omega_{\rm PBH}$.

5.3 Evaporation of Primordial Black Holes

A black hole of mass M will emit particles in the energy range $(Q, Q+dQ)$ at a rate (Hawking 1975)

$$d\dot{N} = \frac{\Gamma dQ}{2\pi\hbar} \left[\exp(\frac{Q}{T}) \pm 1 \right]^{-1} \quad (110)$$

where T is the black hole temperature, Γ is the absorption probability and the + and − signs refer to fermions and bosons respectively. This assumes that the hole has no charge or angular momentum. This is a reasonable assumption since charge and angular momentum will also be lost through quantum emission but on a shorter timescale that the mass (Page 1977). Γ goes roughly like $Q^2 T^{-2}$, though it also depends on the spin of the particle and decreases with increasing spin, so a black hole radiates roughly like a black-body. The temperature is given by (3) and can be expressed as

$$T \approx 10^{26} \left(\frac{M}{g}\right)^{-1} {\rm K} \approx \left(\frac{M}{10^{13}g}\right)^{-1} {\rm GeV}. \quad (111)$$

This means that it loses mass at a rate

$$\dot{M} = -5 \times 10^{25} M^{-2} f(M) \text{ g s}^{-1} \quad (112)$$

where the factor $f(M)$ depends on the number of particle species which are light enough to be emitted by a hole of mass M, so the lifetime is

$$\tau(M) = 6 \times 10^{-27} f(M)^{-1} M^3 \text{ s}. \quad (113)$$

The factor f is normalised to be 1 for holes larger than 10^{17}g and such holes are only able to emit 'massless' particles like photons, neutrinos and gravitons. Holes in the mass range 10^{15}g $< M < 10^{17}$g are also able to emit electrons, while those in the range 10^{14}g $< M < 10^{15}$g emit muons which subsequently decay into electrons and neutrinos. The latter range includes, in particular, the critical mass for which τ equals the age of the Universe. This can be shown to be (MacGibbon and Webber 1990)

$$M_* = 4.4 \times 10^{14} h^{-0.3} \text{ g} \quad (114)$$

where we have assumed that the total density parameter is 1.

Once M falls below 10^{14}g, the hole can also begin to emit hadrons. However, hadrons are composite particles made up of quarks held together by gluons. For temperatures exceeding the QCD confinement scale of $\Lambda_{\rm QCD} = 250 - 300$GeV, one would therefore expect these fundamental particles to be emitted rather than composite particles. Only pions would be light enough to be emitted below Λ_{QCD}. Since there are 12 quark degrees of freedom per flavour and 16 gluon degrees of freedom, one would also expect the emission rate (i.e. the value of f) to increase dramatically once the QCD temperature is reached.

The physics of quark and gluon emission from black holes is simplified by a number of factors. Firstly, since the spectrum peaks at an energy of about $5kT$, (111) implies that most of the emitted particles have a wavelength $\lambda = 2.5M$ (in units with $G = c = 1$), so the particles have a size comparable to the hole. Secondly, one can show that the time between emissions is $\Delta\tau = 20\lambda$, which means that short range interactions between successively emitted particles can be neglected. Thirdly, the condition $T > \Lambda_{\text{QCD}}$ implies that $\Delta\tau$ is much less than $\Lambda_{\text{QCD}}^{-1} \approx 10^{-13}$cm (the characteristic strong interaction range) and this means that the particles are also unaffected by gluon interactions. The implication of these three conditions is that one can regard the black hole as emitting quark and gluon jets of the kind produced in collider events. The jets will decay into hadrons over a distance which is always much larger than M, so gravitational effects can be neglected. The hadrons will themselves decay into protons, antiprotons, electrons, positrons, neutrinos and photons on an even longer timescale.

To find the final spectra of stable particles emitted from a black hole, one must convolve the Hawking emission spectrum given by (110) with the jet fragmentation function. This gives

$$\frac{d\dot{N}_x}{dE} = \sum_j \int_{Q=0}^{Q=\infty} \frac{\Gamma_j(Q,T)}{2\pi\hbar} \left[\exp\left(\frac{Q}{T}\right) \pm 1\right]^{-1} \frac{dg_{jx}(Q,E)}{dE} dQ. \quad (115)$$

Here x and j label the final particle and the directly emitted particle, respectively, and the last factor—the fragmentation function—specifies the number of final particles with energy in the range $(E, E+dE)$ generated by a jet of energy Q. For hadrons this can be represented by

$$\frac{dg_{jh}}{dE} = \frac{1}{E}\left(1 - \frac{E}{Q}\right)^{2m-1} \theta(E - km_h c^2) \quad (116)$$

where m_h is the hadron mass, k is a constant of order 1, and m is 1 for mesons and 2 for baryons. The fragmentation function therefore has an upper cut-off at Q, a lower cut-off and peak around m_h, and an E^{-1} Bremsstrahlung tail in between. By examining the dominant contribution to the Q integral, one obtains

$$\frac{d\dot{N}}{dE} \sim \begin{cases} E^2 \exp(-E/T) & \text{for } E \gg T & (Q \sim E) \\ E^{-1} & \text{for } T > E \gg m_h & (Q \sim T) \\ dg/dE & \text{for } E \sim m_h \ll T & (Q \sim m_h) \end{cases} \quad (117)$$

where the terms in parentheses indicate the value of Q which dominates. This explains the qualitative form of the instantaneous emission spectrum show in Figure 20 for a $T=1$GeV black hole (MacGibbon and Webber 1990). The direct emission just corresponds to the small bumps on the right. All the particle spectra show a peak at 100MeV due to pion decays; the electrons and neutrinos also have peaks at 1MeV due to neutron decays.

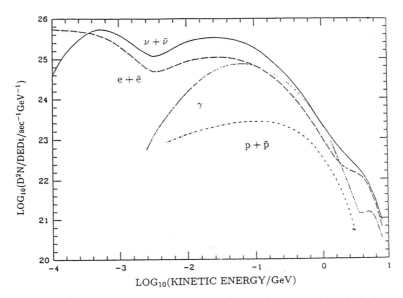

Figure 20. *Instantaneous emission from a 1 GeV black hole*

5.4 Cosmic Rays from Primordial Black Holes

In order to determine the present day background spectrum of particles generated by PBH evaporations, we must first integrate over the lifetime of each hole of mass M and then over the PBH mass spectrum (MacGibbon 1991). In doing so, we must allow for the fact that smaller holes will evaporate at an earlier cosmological epoch, so the particles they generate will be redshifted in energy by the present epoch. If the holes are uniformly distributed throughout the Universe, the background spectra should have the form indicated in Figure 21. All the spectra have rather similar shapes: an E^{-3} fall-off for $E > 100$ MeV due to the final phases of evaporation at the present epoch and an E^{-1} tail for $E < 100$ MeV due to the fragmentation of jets produced at the present and earlier epochs. Note that the E^{-1} tail masks any effect associated with the PBH mass spectrum (Carr 1976) for reasonable values of the spectral index α.

The situation is more complicated if the PBHs evaporating at the present epoch are clustered inside our own Galactic halo (as is most likely). In this case, any charged particles emitted after the epoch of galaxy formation will have their flux enhanced relative to the photon spectra by a factor ζ which depends upon the halo concentration factor and the time for which particles are trapped inside the halo by the Galactic magnetic field. Assuming for simplicity that the particles are uniformly distributed throughout a halo of radius R_h, one finds

$$\zeta = \left(\frac{\tau_{\text{leak}}}{t_g}\right)\left(\frac{\rho_h}{\rho_{\text{crit}}}\right) \approx 10^6 h^2 \left(\frac{\tau_{\text{leak}}}{t_g}\right)\left(\frac{R_h}{10 \text{ kpc}}\right)^{-2}. \qquad (118)$$

The ratio of the leakage time τ_{leak} to the age of the Galaxy t_g is rather uncertain and also energy-dependent. At 100 MeV we take τ_{leak} to be about 10^7y for electrons or positrons ($\zeta \sim 10^3$) and 10^8y for protons or antiprotons ($\zeta \sim 10^4$). The postgalactic

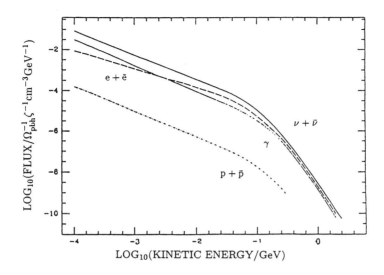

Figure 21. *Spectrum of particles from uniformly distributed PBHs*

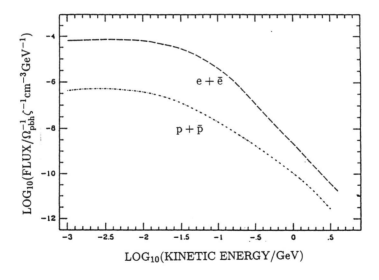

Figure 22. *Spectrum of charged particles from PBHs in our halo*

contribution of charged particles is shown in Figure 22 and comes from just a narrow range of masses below M_*.

For comparison with the observed cosmic ray spectra (MacGibbon and Carr 1991), one needs to determine the amplitude of the spectra at 100MeV. This is because the observed fluxes all have slopes between E^{-2} and E^{-3}, so the strongest constraints come from measurements at 100MeV. The amplitudes all scale with $\Omega_{\rm PBH}$ and are found to be

$$\frac{dF}{dE} = \begin{cases} 1.5\times 10^{-5} h^2 \Omega_{\rm PBH} \text{ GeV}^{-1}\text{cm}^{-3} & (\gamma) \\ 9.5\times 10^{-3} h^2 \Omega_{\rm PBH}(\zeta/10^3) \text{ GeV}^{-1}\text{cm}^{-3} & (e^+, e^-) \\ 4.5\times 10^{-4} h^2 \Omega_{\rm PBH}(\zeta/10^4) \text{ GeV}^{-1}\text{cm}^{-3} & (p, \bar{p}) \end{cases} \quad (119)$$

Comparison with the observed γ-ray background spectrum (Fichtel et al. 1975) places a firm upper limit $\Omega_{\rm PBH} \leq 8\times 10^{-9} h^{-2}$ and suggests that PBH emission may even be the dominant contribution above 50MeV. An interesting feature of the observed electron and positron spectra is that they have comparable fluxes at 100MeV, even though electrons are more numerous at higher energies. This feature is unexplained in most cosmic ray models but it is a natural consequence of the PBH scenario since electrons and positrons are emitted in equal numbers. To explain the observed interstellar positron flux at 300MeV (Ramaty and Westergaard 1976), one requires

$$\Omega_{\rm PBH} \simeq 2\times 10^{-8} \left(\frac{\eta_{\rm leak}}{10^7 \text{ y}}\right)^{-1} \left(\frac{R_h}{10 \text{ kpc}}\right)^{-2}. \quad (120)$$

Since the ratio of antiprotons to protons in cosmic rays is around 10^{-4} over the energy range 100MeV − 10GeV, whereas PBHs should produce them in equal numbers, PBHs could only contribute appreciably to the antiprotons. It is usually assumed that antiproton cosmic rays are secondary particles, produced by spallation of the interstellar medium by primary cosmic rays. However, there is an indication of a primary antiproton contribution around 100MeV at the 10^{-5} level (Streitmatter et al. 1990) and this could be generated by PBH evaporations for

$$\Omega_{\rm PBH} \simeq (0.6 - 4)\times 10^{-9} \left(\frac{\eta_{\rm leak}}{10^8 \text{ y}}\right)^{-1} \left(\frac{R_h}{10 \text{ kpc}}\right)^{-2}. \quad (121)$$

Both (120) and (121) are compatible with the γ-ray limit for reasonable values of $\eta_{\rm leak}$ and R_h, so in principle PBHs could contribute to all three cosmic ray backgrounds.

5.5 Constraints on Primordial Black Holes

Even if PBHs are not the source of cosmic rays, it is interesting to examine the constraints on the fraction of the early Universe going into them (Novikov et al. 1979). The first such constraint derives from the fact that any PBHs which survive today must certainly have less than the critical density. In the standard radiation-dominated model for the early universe, (109) then implies that the fraction of the universe going into PBHs of mass M must satisfy

$$\beta(M) < 10^{-17} \left(\frac{M}{10^{15}\text{g}}\right)^{1/2}, \qquad M > 10^{15}\text{g}. \quad (122)$$

In fact, the constraints are much stronger just below 10^{15}g since we have seen that the 100MeV γ-ray background measurements imply that 10^{15}g PBHs could have at most 10^{-8} times the critical density, so the exponent 17 in (122) becomes 25. For PBHs which

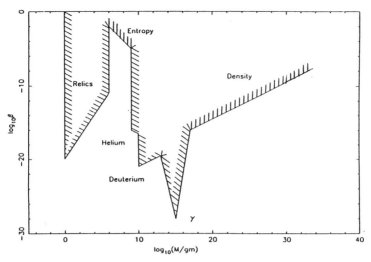

Figure 23. *Constraints on $\beta(M)$*

have evaporated completely, there are also limits on $\beta(M)$ associated with entropy production (Zeldovich and Starobinsky 1976)

$$\beta(M) < 10^{-8} \left(\frac{M}{10^{11}\text{g}}\right)^{-1}, \quad M < 10^{11}\text{g}, \tag{123}$$

distortion of the microwave background (Naselsky 1978)

$$\beta(M) < 10^{-18} \left(\frac{M}{10^{11}\text{g}}\right)^{-1}, \quad 10^{11}\text{g} < M < 10^{13}\text{g}, \tag{124}$$

and cosmological nucleosynthesis constraints

$$\beta(M) < \begin{cases} 10^{-15}(\frac{M}{10^9\text{g}})^{-1}, & 10^9\text{g} < M < 10^{13}\text{g} \\ 10^{-21}(\frac{M}{10^{10}\text{g}})^{1/2}, & M > 10^{10}\text{g} \\ 10^{-16}(\frac{M}{10^9\text{g}})^{-1/2}, & 10^9\text{g} < M < 10^{10}\text{g} \end{cases} \tag{125}$$

The last three limits are associated with the increase of the background photon-to-baryon ratio by PBH photons emitted after nucleosynthesis (Miyama and Sato 1978), photodissociation of deuterium by such photons (Lindley 1980), and modification of the neutron-to-proton ratio by PBH nucleons emitted before nucleosynthesis (Rothman and Matzner 1981). There is also a limit for Planck mass relics (Carr *et al.* 1994)

$$\beta(M) < 10^{-27} \left(\frac{M}{10^{-5}\text{g}}\right)^{3/2}, \quad M < 10^6\text{g} \tag{126}$$

although this only pertains if such relics are stable, which is uncertain.

These limits on $\beta(M)$ are summarised in Figure 23 and the corresponding limits on $\varepsilon(M)$ in the standard radiation-dominated scenario, obtained from (104), are shown

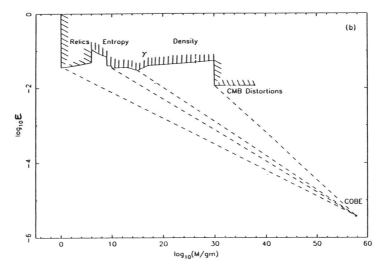

Figure 24. *Constraints on $\varepsilon(M)$*

in Figure 24. Also shown is the constraint on ε from the CMB quadrupole anisotropy measurement and the constraint implied by the lack of spectral distortion in the CMB (Hu *et al.* 1994). If we assume that the horizon-scale fluctuations have a power-law form, $\varepsilon \propto M^{-\alpha}$, then the COBE quadrupole observation (which specifies ε on a scale of 10^{22} M$_\odot$) implies that one needs $\alpha = 0.08$ if PBH formation at 10^{15}g is to be interesting. The COBE constraint on scales above 10° requires $0.1 > \alpha > -0.07$, which just allows this. In any case, the important point is that an analysis of PBH formation places very interesting constraints on the form of the density fluctuations since COBE and PBHs together limit the fluctuations over 57 decades of mass.

Glossary of Acronyms

AGN	Active Galactic Nuclei
CCD	Charged Coupled Device
CMB	Cosmic Microwave Background
COBE	COsmic Background Explorer
EROS	Experience de Recherche d'Objects Sombres
HST	Hubble Space Telescope
IMF	Initial Mass Function
LIGO	Laser Interferometer Gravity Wave Observatory
LISA	Laser Interferometer Space Antenna
LMC	Large Magellanic Cloud
MACHO	MAssive Compact Halo Object
MDO	Massive Dark Object
MO	Massive Object
OGLE	Optical Gravitational Lensing Experiment
PBH	Primordial Black Hole
QCD	Quantum-Chromo-Dynamics

SMC	Small Magellanic Cloud
SMO	Super-Massive Object
VLA	Very Large Array
VLBA	Very Large Baseline Array
VLBI	Very Long Baseline Interferometer
VMO	Very Massive Object
WIMP	Weakly Interacting Massive Particle

References

Alcock C et al., 1993, *Nature* **365** 621.
Arnett W D, 1969, *Astrophys Space Sci* **5** 180.
Ashman K M, 1992, *Pub Astron Soc Pac* **247** 662.
Auborg E et al., 1993, *Nature* **365** 623.
Bahcall J N and Ostriker J P, 1975, *Nature* **256** 23.
Bahcall J N, Hut P and Tremaine S, 1985, *Ap J* **290** 15.
Bahcall J N, Flynn C and Gould A, 1992, *Ap J* **389** 234.
Barcons X, Fabian A C and Rees M J, 1991, *Nature* **350** 685.
Bardeen J M and Petterson J A, 1975, *Ap J Lett* **195** L65.
Barrow J D, 1980, *MNRAS* **192** 427.
Barrow J D, Copeland E J and Liddle A R, 1992, *Phys Rev D* **46** 645.
Begelman M C, Blandford R D and Rees M J, 1980, *Nature* **287** 307.
Bekenstein J, 1972, *Phys Rev D* **5** 1239.
Bertotti B and Carr B J, 1980, *Ap J* **236** 1000.
Binney J and Tremaine S, 1987, *Galactic Dynamics* (Princeton University Press).
Blandford R D, 1987, in *300 Years of Gravitation*, ed S W Hawking, W Israel, p 277 (CUP).
Blandford R D and Znajek R L, 1977, *MNRAS* **179** 433.
Blandford R D and Narayan R, 1992, *Ann Rev Astron Astrophys* **30** 311.
Blandford R D and Rees M J, 1992, in *Testing the AGN Paradigm*, ed S S Holt, S G Neff and C M Urry, p 3 (American Institute of Physics).
Boldt E and Leiter D, 1981, *Nature* **290** 483.
Bond J R, Arnett W D and Carr B J, 1984, *Ap J* **280** 825.
Bond J R, Carr B J and Hogan C J, 1991, *Ap J* **367** 420.
Bowick M J et al., 1988, *Phys Rev Lett* **61** 2823.
Bristow P and Phillipps S, 1994, *MNRAS* **267** 13.
Carr B J, 1975, *Ap J* **201** 1.
Carr B J, 1976, *Ap J* **206** 8.
Carr B J, 1978, *Comm Astrophys* **7** 161.
Carr B J, 1983a, *MNRAS* **194** 639.
Carr B J, 1983b, *Nature* **284** 326.
Carr B J, 1985, in *Observational and Theoretical Aspects of Relativistic Astrophysics and Cosmology*, ed J L Sanz and L J Goicoechea, p 1 (World Scientific).
Carr B J, 1993, in *The Renaissance of General Relativity and Cosmology*, ed G Ellis, A Lanza and J Miller, p 258 (Cambridge University Press).
Carr B J, 1994, *Ann Rev Astron Astrophys* **32** 531.
Carr B J and Hawking S W, 1974, *MNRAS* **168** 399.
Carr B J and Lacey C G, 1987, *Ap J* **316** 23.
Carr B J and Lidsey J E, 1993, *Phys Rev D* **48** 543.

Carr B J and Sakellariadou M, 1996, Preprint.
Carr B J, McDowell J and Sato H, 1983, *Nature* **306** 666.
Carr B J, Bond J R and Arnett W D, 1984, *Ap J* **277** 445.
Carr B J, Gilbert J H and Lidsey J E, 1994, *Phys Rev D* **50** 4853.
Carter B, 1971, *Phys Rev Lett* **26** 331.
Casares J, Charles P and Naylor T, 1992, *Nature* **355** 614.
Cassinelli J P, Mathis J S and Savage B D, 1981, *Science* **212** 1497.
Chandrasekhar S, 1931, *Ap J* **74** 81.
Chandrasekhar S, 1964, *Ap J* **140** 417.
Chokshi A and Turner E L, 1992, *MNRAS* **259** 421.
Christodoulou D, 1970, *Phys Rev Lett* **25** 1596.
Coleman S, Preskill J and Wilczek F, 1991, *Mod Phys Lett A* **6** 1631.
Corrigan R et al. , 1991, *Astron J* **102** 34.
Cowley A P, 1992, *Ann Rev Astron Astrophys* **30** 287.
Crawford M and Schramm D N, 1982, *Nature* **298** 538.
Dalcanton J, Canizares C, Granados A, Steidel C C and Stocke J T, 1994, *Ap J* **424** 550.
Duncan M J and Shapiro D L, 1983, *Ap J* **268** 565.
Fich M and Tremaine S, 1991, *Ann Rev Astron Astrophys* **29** 409.
Fichtel C E et al. , 1975, *Ap J* **198** 163.
Fowler W, 1966, *Ap J* **144** 180.
Fowler W and Hoyle F, 1964, *Ap J Supp* **9** 201.
Fricke K J, 1973, *Ap J* **183** 941.
Fuller G M, Woosley S E and Weaver T A, 1986, *Ap J* **307** 675.
Genzel R, Hollenbach D and Townes C H, 1994, *Rep Prog Phys* **57** 417.
Gilbert J H, 1995, *Phys Rev D* **52** 5486.
Gnedin N Yu and Ostriker J P, 1992, *Ap J* **400** 1.
Gott J R, 1981, *Ap J* **243** 140.
Griest K, 1991, *Ap J* **366** 412.
Grindlay J E, 1978, *Ap J* **221** 234.
Guth A H, 1981, *Phys Rev D* **23** 347.
Haehnelt M G and Rees M J, 1992, *MNRAS* **263** 168.
Harms R J et al. , 1994, *Ap J Lett* **435** L35.
Hartwick F D A and Schade D, 1990, *Ann Rev Astron Astrophys* **28** 437.
Hawking S W, 1971a, *Phys Rev Lett* **26** 1344.
Hawking S W, 1971b, *MNRAS* **152** 75.
Hawking S W, 1973, in *Black Holes*, ed B DeWitt and C DeWitt (Gordon and Breach).
Hawking S W, 1974, *Nature* **248** 30.
Hawking S W, 1975, *Comm Math Phys* **43** 199.
Hawking S W, 1989, *Phys Lett B* **231** 237.
Hawking S W and Penrose R, 1969, *Proc R Soc Lond A* **314** 529.
Hawking S W, Moss I and Stewart J, 1982, *Phys Rev D* **26** 2681.
Hawkins M R S, 1993, *Nature* **366** 242.
Hewitt J N, 1986, PhD thesis (Massachussets Institute of Technology).
Hills J G, 1975, *Nature* **254** 294.
Hills J G, 1986, *Astron.J* **92** 595.
Hu W, Scott D and Silk J, 1994, *Ap J Lett* **430** L5.
Hut P and Rees M J, 1992, *MNRAS* **259** 27P.
Irwin M J et al. , 1989, *Astron J* **98** 1989.
Israel W, 1967, *Phys Rev* **164** 1776.
Joss P C and Rappaport S A, 1984, *Astron Astrophys* **22** 537.

Kashlinsky A and Rees M J, 1983, *MNRAS* **205** 955.
Kassiola A, Kovner I and Blandford R D, 1991, *Ap J* **381** 6.
Kerins E and Carr B J, 1994, *MNRAS* **266** 775.
Kerr R P, 1963, *Phys Rev Lett* **11** 237.
Kiraly P et al. , 1981, *Nature* **293** 120.
Khlopov M Yu and Polnarev A G, 1980, *Phys Lett B* **97** 383.
Khlopov M Yu, Malomed B E and Zeldovich Ya B, 1985, *MNRAS* **215** 575.
Kormendy J and Richstone D, 1995, *Ann Rev Astron Astrophys* **33** 581.
Kruskal M D, 1960, *Phys Rev* **119** 1743.
Kulkarni S R, Hut P and McMillan S, 1993, *Nature* **364** 421.
La D and Steinhardt P J, 1989, *Phys Lett B* **220** 375.
Lacey C G, 1991, in *Dynamics of Disk Galaxies*, ed B Sundelius p 257.
Lacey C G and Ostriker J P, 1985, *Ap J* **299** 633.
Lacy J H, Townes C H and Hollenbach D J, 1983, *Ap J* **262** 120.
Latham D W et al. , 1984, *Ap J Lett* **281** L41.
Lindley D, 1980, *MNRAS* **196** 317.
Loeb A, 1993, *Ap J* , **403** 542.
MacGibbon J H, 1987, *Nature* **329** 308.
MacGibbon J H, 1991, *Phys Rev D* **44** 376.
MacGibbon J H and Webber B R, 1990, *Phys Rev D* **41** 3052.
MacGibbon J H. and Carr B J, 1991, *Ap J* **371** 447.
Maeder A, 1992, *Astron Astrophys* **264** 105.
Mathews G J, Schramm D N and Meyer B S, 1993, *Ap J* **404** 476.
Mazur P O, 1982, *J Phys A* **15** 3173.
McClintock J E, 1986, in *The Physics of Accretion onto Compact Objects*, ed K O Mason, M G Watson and N E White, p 211 (Springer-Verlag).
McGaugh S, 1994, *Nature* **367** 538.
Miyama S and Sato K, 1978, *Prog Theor Phys* **59** 1012.
Miyoshi M et al. , 1995, *Nature* **373** 127.
Moore B, 1993, *Ap J Lett* **413** L93.
Naselsky P D, 1978, *Sov Astron Lett* **4** 209.,
Naselsky P D and Polnarev A G, 1985, *Sov Astron* **29** 487.
Nemiroff R J, 1989, *Ap J* **341** 579.
Newman E T et al. , 1965, *J Math Phys* **6** 918.
Nordstrom G, 1918, *Proc Kon Ned Akad Wet* **20** 1238.
Novikov I D, Thorne K S, 1973, in *Black Holes*, ed B DeWitt and C DeWitt (Gordon & Breach).
Novikov I D, Polnarev A G, Starobinsky A A, Zeldovich Ya B, 1979, *Astron Astrophys* **80** 104.
Ober W W, El Eid M F and Fricke K L, 1983, *Astron Astrophys* **119** 61.
Okeke P N and Rees M J, 1980, *Astron Astrophys* **81** 263.
Oppenheimer J R and Snyder H, 1939, *Phys Rev* **56** 455.
Paczynski B, 1986, *Ap J* 304, 1.
Page D N, 1977, *Phys Rev D* **16** 2402.
Page D N, and Hawking, S.W. 1976, *Ap J* **206** 1.
Papaloizou J, 1973, *MNRAS* **162** 169.
Penrose R, 1969, *Nuovo Cimento* **1** 252.
Persic M and Salucci P, 1992, *MNRAS* **258** 14P.
Pfenniger D, Combes F and Martinet L, 1994, *Astron Astrophys* **285** 79.
Polnarev A G and Khlopov M Yu, 1981, *Astron Zh* **58** 706.

Polnarev A G and Zemboricz R, 1988, *Phys Rev D* **43** 1106.
Porter N A and Weekes T C, 1979, *Nature* **277** 199.
Press W H and Gunn J E, 1973, *Ap J* **185** 397.
Price R H, 1972, *Phys Rev D* **5** 2439.
Ramaty R and Westergaard N J, 1976, *Astrophys Sp Sci* **45** 143.
Rees M J, 1977, *Nature* **266** 333.
Rees M J, 1984, *Ann Rev Astron Astrophys* **22** 471.
Rees M J, 1986, *MNRAS* **218** 25P.
Reissner H, 1916, *Ann Physik* **50** 106.
Remillard R A, McClintock J E and Bailyn C D, 1992, *Ap J Lett* **399** L145
Robinson D C, 1974, *Phys Rev D* **10** 458.
Rothman T and Matzner R, 1981, *Astrophys Space Sci* **5** 229.
Sakellariadou M, 1984, MSc thesis (Cambridge).
Salpeter E E, 1955, *Ap J* **121** 161.
Schild H and Maeder A, 1985, *Astron Astrophys Lett* **143** L7.
Schwarzschild K, 1916, *Sitzber Deut Akad Wiss Berlin, Kl Math Phys Tech* 424.
Schwarzschild M and Harm R, 1959, *Ap J* **243** 1.
Shapiro S L and Teukolsky S A, 1983, *Black Holes, White Dwarfs and Neutron Stars* (Wiley).
Sigurdsson S and Hernquist L, 1993, *Nature* **364** 423.
Stella L, Israel G L, Mereghetti S and Ricci D, 1994, in *Proceedings of 7th Marcel Grossmann Meeting on General Relativity*.
Streitmatter R E et al., 1990, *Bull Am Phys Soc* **35** 1066.
Surdej J et al., 1993, *Astron J* **105** 2064.
Tanaka Y et al., 1995, *Nature* **375** 659.
Thorne K S, 1991, in *Recent Advances in General Relativity*, ed A Janis, J Porter (Birkhauser, Boston).
Thorne K S, Price R H, MacDonald A, 1976, *Black Holes: The Membrane Paradigm* (Yale UP).
Tohline J E, 1980, *Ap J* **239** 417.
Trimble V, 1991, *Contemp Phys* **32** 103.
Turner M S, 1982, *Nature* **297** 379.
Udalski A et al., 1993, *Acta Astron* **43** 289.
Umemura M, Loeb A and Turner E L, 1993, *Ap J* Lett., **419** 459.
Van den Bergh S, 1969, *Nature* **224** 891.
Van den Heuvel E P J, 1974, *Ap J Lett* **198** L109.
Young P J et al., 1978, *Ap J* **221** 721.
Wald R M, 1984, *General Relativity* (University of Chicago Press).
Walker T et al., 1991, *Ap J* **376** 51.
Wasserman I and Weinberg M D, 1987, *Ap J* **312** 390.
Weaver T A, Zimmerman G B and Woosley S E, 1978, *Ap J*, **225** 1021.
Webster R L et al., 1991, *Astron J* **102** 1939.
White S D M and Rees M J, 1978, *MNRAS* **183** 341.
Wielen R, 1985, in *Dynamics of Star Clusters*, ed J Goodwin and P Hut, p 449 (Reidel).
Woosley S E and Weaver T A, 1982, in *Supernovae: A Survey of Current Research*, ed M J Rees and R J Stoneham, p 79 (Reidel: Dordrecht).
Woosley S E and Weaver T A, 1986, *Rev Astron Astrophys* **24** 205.
Zeldovich Ya B and Novikov I D, 1967, *Sov Astron A J* **10** 602.
Zeldovich Ya B and Novikov I D, 1971, *Relativistic Astrophysics: Stars and Relativity* (University of Chicago).
Zeldovich Ya B and Starobinsky A A, 1976, *JETP Lett* **24** 571.

Sources of Gravitational Waves

José Alberto Lobo

Departament de Física Fonamental
Universitat de Barcelona, Spain

1 Introduction

Newton's theory of the gravitational interaction constitutes a major landmark in the evolution of scientific thought in Europe. In his magnum opus, *Principia Mathematica Philosophiæ Naturalis*, Newton laid the foundations of Theoretical Mechanics and put forward his revolutionary ideas about the motion of masses under the influence of their mutual gravitational pull. Such ideas were to remain unchallenged for over two centuries, the theory being crowned with success almost every time it was applied to a new problem. In particular, celestial mechanics was to become the paradigm of exact science.

In modern notation, the Newtonian gravitational interaction is described by a scalar potential, $\phi(\mathbf{x}, t)$, satisfying Poisson's equation

$$\nabla^2 \phi(\mathbf{x}, t) = -4\pi G \rho(\mathbf{x}, t) \tag{1}$$

where $G = 6.672 \times 10^{-11} \, \mathrm{m}^3/\mathrm{kg\, sec}^2$ is a fundamental constant and $\rho(\mathbf{x}, t)$ is the density of matter, which acts as the source of gravitational forces. Remarkably, Equation (1) contains no time derivatives, which implies that changes in $\rho(\mathbf{x}, t)$ are instantly felt by $\phi(\mathbf{x}, t)$, no matter how far from the source.

On the other hand, the physics of electric and magnetic phenomena culminated towards the early 1870's in another major breakthrough in the history of science: Maxwell's *Treatise of Electricity and Magnetism*. In this book, Maxwell revealed the deep connections between electricity and magnetism, and developed a theory whereby both kind of phenomena are included within a unified body of doctrine. Most relevant for our purposes here, Maxwell's theory contains the concept of field. This is a new and powerful concept which is very well suited to describe, amongst many other things, such phenomena as the radiation of electromagnetic waves, like radio waves and light. In modern notation, the electromagnetic interaction is described by a vector potential $A_\mu(\mathbf{x}, t)$ in the four dimensional continuum of space and time, satisfying Maxwell's

equations
$$\Box A_\mu(\mathbf{x},t) = -j_\mu(\mathbf{x},t) \qquad (2)$$

where $j_\mu(\mathbf{x},t)$ contains the sources of electromagnetic fields in the form of electric currents and densities. (Here and elsewhere greek indices run from 0 to 3; the index 0 is used for the time coordinate, and latin indices $i, j, \ldots$ are used for the space coordinates). Unlike (1), Equation 2 does contain time derivatives of the unknown $A_\mu(\mathbf{x},t)$, since $\Box \equiv c^{-2}\,\partial^2/\partial t^2 - \nabla^2$, where $c = 2.9979 \times 10^8$ m/sec is the speed of light in vacuum. As it turns out, a consequence of (2) is that changes in $j_\mu(\mathbf{x},t)$ are not felt instantly by $A_\mu(\mathbf{x},t)$, but rather with a time delay of r/c, with r the distance between the field point and the place where the changes occurred. This time delay led physicists to the idea of causal interaction: a perturbation in the source travels across the intervening space at a finite speed, and is thus felt remotely only after a well determined amount of time has elapsed. Einstein's theory of Special Relativity of 1905 definitively settled this idea thanks to the proof that no signal can possibly propagate faster than light.

Back to gravitation. Why should it constitute an exception to such a general principle? If we are not, on epistemological grounds, prepared to accept that exception then we need a modification of Newton's theory. After a number of attempts in this direction the solution came in the form of a remarkably beautiful theory: Einstein's General Relativity of 1915. Einstein's deep insight into the nature of gravitational phenomena led him to formulate his Principle of Equivalence, whereupon he was further led to conclude that gravitation shows as changes in the *geometry* of space-time, the changes being linked to the distribution of mass and energy. Since the geometric properties of a manifold, space-time in this case, are fully characterised by a symmetric tensor $g_{\mu\nu}(x)$, this tensor is the appropriate mathematical tool to describe gravitation. Einstein's equations

$$R_{\mu\nu} - \frac{1}{2}R\,g_{\mu\nu} = -\frac{8\pi G}{c^4}T_{\mu\nu} \qquad (3)$$

specify how the energy-momentum density, $T_{\mu\nu}$, actually affects $g_{\mu\nu}$. Here $R_{\mu\nu}$ and R are the so called Ricci tensor and curvature scalar, respectively, and are somewhat complicated combinations of first and second order derivatives of the $g_{\mu\nu}$. These equations are certainly compatible with the causality requirements alluded to above, so that their solutions are, in some suitable sense, of the retarded type also. General Relativity thus predicts the existence of gravitational radiation, *i.e.* gravitational fields which propagate at a finite speed away from time-varying distributions of mass-energy, very much like electromagnetic waves do from time-varying charge distributions. Also, in curious analogy with electromagnetic waves, they were predicted before experimental evidence of their existence had been found. Unlike radio waves, though, gravitational waves (GW) have been waiting for (and still await) direct experimental observation for almost 80 years.

In these lectures, I would like to convey to the reader a flavour of the current orthodoxy in gravitational wave generation physics. Amongst other things, I hope this will help him understand the reasons for the unprecedented time lag between theory and experiment in this field, and also to correctly assess the formidable experimental challenge posed by gravitational wave detection, on which Norna Robertson will expand in detail in her article. I will first review the theory of linear waves, their properties and their sources, as expressed by the renowned quadrupole formulae. I will then go on to describe the physics and observational data of the binary pulsar PSR B1913+16, given the outstanding importance of this system for gravitational wave physics. After that, I will present a classified list of what we currently believe are the most likely sources of gravitational waves to be looked for. Finally, very general notions of data filtering techniques will be outlined, which will serve to illustrate the importance of source properties research.

2 Linear theory of Gravitational Waves

Einstein's equations (3) have a complication due to their non-linearity in the unknowns $g_{\mu\nu}(x)$. As it turns out, and as we shall confirm later, gravitational waves are very weak on any forseeable scale in the Earth's surface, so one is fully justified in taking the linear approximation to (3) in the study of gravitational waves. What this precisely means will now be explained.

In the absence of gravitation, space-time is flat, so that we may take

$$g_{\mu\nu}(x) = \eta_{\mu\nu}$$

where $\eta_{\mu\nu} = \text{diag}(-1, 1, 1, 1)$ is the Lorentz-Minkowski metric of Special Relativity. In the presence of a weak field $g_{\mu\nu}(x)$ will differ little from this, so we write

$$g_{\mu\nu}(x) = \eta_{\mu\nu} + h_{\mu\nu}(x) \qquad |h_{\mu\nu}(x)| \ll 1 \qquad (4)$$

and we shall henceforth only be interested in calculations to first order in the small quantities $h_{\mu\nu}(x)$. To this end, we first write Einstein's equations (3) in their linearised form. For this, a formula to evaluate its left hand side is required in the first place. I shall use Weinberg's (Weinberg 1972). It so happens that the following redefinition of the field variables is expedient for first order calculations:

$$\bar{h}_{\mu\nu} \equiv h_{\mu\nu} - \frac{1}{2} h \, \eta_{\mu\nu} \qquad h \equiv \eta^{\mu\nu} h_{\mu\nu} \qquad (5)$$

This can be easily inverted to give

$$h_{\mu\nu} = \bar{h}_{\mu\nu} - \frac{1}{2} \bar{h} \, \eta_{\mu\nu} \qquad \bar{h} \equiv \eta^{\mu\nu} \bar{h}_{\mu\nu} = -h \qquad (6)$$

The linearised field equations can be seen to be

$$\Box \bar{h}_{\mu\nu} - \bar{h}_{\mu,\sigma\nu}^{\sigma} - \bar{h}_{\nu,\sigma\mu}^{\sigma} + \bar{h}^{\rho\sigma}{}_{,\rho\sigma}\, \eta_{\mu\nu} = -\frac{16\pi G}{c^4} T_{\mu\nu} \qquad (7)$$

where indices are raised and lowered with $\eta_{\mu\nu}$ to this order of approximation, *i.e.* $h_\mu{}^\nu \equiv \eta^{\nu\sigma} h_{\mu\sigma}$, *etc.*, and commas stand for partial derivatives. Equation (7) has a rather odd appearance as a wave equation—it would look nicer should the last three terms in its left hand side drop out. We are fortunate that this can actually be accomplished by the following procedure.

The Equivalence Principle implies that gravitational effects can be traced to geometry in space-time. Geometry is expressed in terms of coordinates, of course, but the *choice* of them remains arbitrary. We are thus entitled to play at will with the four functions $x'^\mu(x)$ which define a coordinate set x'^μ in terms of a pre-existing one x^μ. We can certainly benefit from this, and choose those coordinates which most simplify our equations. We must keep in mind at this point, however, that we have already used up some of the available freedom we initially had since, in writing Equation (4), we are implicitly assuming that our coordinates are 'nearly' Cartesian. But we can still make changes of order h and maintain that property, so we are really not so very much constrained. Let, for example,

$$x'^\mu = x^\mu + \epsilon^\mu(x) \tag{8}$$

where $\epsilon^\mu(x)$ are four arbitrary functions, only subject to the condition that they be of the same order of magnitude as $h_{\mu\nu}$. It is straightforward to prove that

$$h'_{\mu\nu}(x') = h_{\mu\nu}(x) - \epsilon_{\mu,\nu}(x) - \epsilon_{\nu,\mu}(x) \tag{9}$$

up to higher order terms. We can now choose the four functions $\epsilon_\mu(x)$ so as to ensure that the four conditions

$$\frac{\partial}{\partial x'^\nu} \bar{h}'_{\mu\nu}(x') = 0 \tag{10}$$

be satisfied. In these new coordinates the linearised equations (7) thus simplify to

$$\Box \bar{h}_{\mu\nu} = -\frac{16\pi G}{c^4} T_{\mu\nu} \tag{11}$$

where accents on h's have now been dropped, as the names of the variables are immaterial. I will shortly come back to this fundamental equation, but let me first go a bit further into the consequences of coordinate choice. It follows from (9) that

$$\bar{h}'^{\mu\nu}{}_{,\nu} = \bar{h}^{\mu\nu}{}_{,\nu} - \Box \epsilon^\mu \tag{12}$$

This means that conditions (10) are preserved if a further coordinate change of the type (8) is applied, subject to the condition that $\Box \epsilon^\mu = 0$, *i.e.* that the four ϵ^μ be *harmonic* functions.

Summing up, this is what we can do before we exhaust our freedom to choose the coordinates in order to accomplish the highest possible simplification of our equations: given $h_{\mu\nu}$, make a coordinate change of the form (8) with

$$\epsilon^\mu(x) = \epsilon_1^\mu(x) + \epsilon_2^\mu(x) \tag{13}$$

and such that

$$\Box \epsilon_1^\mu = \bar{h}^{\mu\nu}{}_{,\nu} \qquad \Box \epsilon_2^\mu = 0. \tag{14}$$

The first of these imposes conditions (10) upon $\bar{h}_{\mu\nu}$ and ensures that (11) holds, while the second can be used to impose further constraints. All in all, we are free to use up to eight functions $\epsilon_1^\mu(x)$ and $\epsilon_2^\mu(x)$ which result in an equal number of constraints on the h's without affecting their information content, as the constraints are derived from mere coordinate changes. Since there were 10 independent h's in the first place (recall that $h_{\mu\nu} = h_{\nu\mu}$), our argument shows that weak gravitational waves only possess two real degrees of freedom ($10 - 8 = 2$). There is a remarkable parallelism between the

	Field	Gauge symmetry	Gauge fixing	Degrees of freedom
e.m. field	A_μ	$A_\mu \to A_\mu - \phi_{,\mu}$	$\phi = \phi_1 + \phi_2$ $\Box\phi_1 = A^\mu{}_{,\mu}$ $\Box\phi_2 = 0$	$4 - 2 = 2$
GW field	$h_{\mu\nu}$	$h_{\mu\nu} \to h_{\mu\nu} - \epsilon_{\{\mu,\nu\}}$	$\epsilon^\mu = \epsilon_1^\mu + \epsilon_2^\mu$ $\Box\epsilon_1^\mu = \bar{h}^{\mu\nu}{}_{,\nu}$ $\Box\epsilon_2^\mu = 0$	$10 - 8 = 2$

Table 1. *Parallelism between electromagnetic and gravitational wave gauge symmetries.*

reduction process just described and the one found in Electromagnetism due to gauge invariance of Maxwell's equations (Barut 1980). It is summarised in Table 1.

Remarkably, the final number of degrees of freedom is 2 in both cases, indicating that there are only two polarisation states for either kind of waves. We take a closer look now.

3 Plane waves and polarisation states

Let us first consider vacuum solutions to (11). This is a very important case to study since the eventual detection of gravitational waves will take place away from its sources, of course. The vacuum Einstein equations read

$$\Box \bar{h}_{\mu\nu} = 0 \qquad (15)$$

with the supplementary 'gauge' conditions

$$\bar{h}^{\mu\nu}{}_{,\nu} = 0. \qquad (16)$$

A solution to these equations which is useful far from the sources can be written as a superposition of plane waves. If we take the direction of propagation as the z-axis, we have

$$\bar{h}_{\mu\nu}(\mathbf{x}, t) = \int_{-\infty}^{\infty} H_{\mu\nu}(\omega) \exp\left[-i\omega\left(t - \frac{z}{c}\right)\right] d\omega \qquad (17)$$

with the additional constraints that

$$H_{\mu\nu}(\omega)\, k^\nu = 0 \qquad (18)$$

k^ν being the wave four-vector $(\omega,0,0,\omega)$. These derive from the divergence condition (10). As mentioned earlier, we can still reduce $H_{\mu\nu}(\omega)$ by a further coordinate change. We want to choose it of the type (8) with a harmonic

$$\epsilon_2^\mu(\mathbf{x}, t) = \int_{-\infty}^{\infty} A^\mu(\omega) \exp[-i\omega(t - z/c)]\, d\omega \qquad (19)$$

where the free values $A^\mu(\omega)$ are chosen so that

$$H_{\mu 0} = 0 \qquad \eta^{\mu\nu} H_{\mu\nu} = 0. \qquad (20)$$

It looks as if (20) imposes *five* conditions, one too many for the four available components of the A_μ. Note, however, that $H_{\mu 0} = 0$ is in fact only three, since $H_{i0} = 0$ implies $H_{00} = 0$, too, as a consequence of (18). We thus need a fourth, and take it as the tracelessness condition. After all these reductions, $H_{\mu\nu}(\omega)$ finally simplifies to

$$H_{\mu\nu}(\omega) = \begin{pmatrix} 0 & 0 & 0 & 0 \\ 0 & H_{11}(\omega) & H_{12}(\omega) & 0 \\ 0 & H_{12}(\omega) & -H_{11}(\omega) & 0 \\ 0 & 0 & 0 & 0 \end{pmatrix} \qquad (21)$$

or, returning to (17),

$$h_{\mu\nu}(\mathbf{x},t) = \begin{pmatrix} 0 & 0 & 0 & 0 \\ 0 & h_+(\mathbf{x},t) & h_\times(\mathbf{x},t) & 0 \\ 0 & h_\times(\mathbf{x},t) & -h_+(\mathbf{x},t) & 0 \\ 0 & 0 & 0 & 0 \end{pmatrix} \qquad (22)$$

where the top bar on $h_{\mu\nu}$ has been dropped, as $\bar{h}_{\mu\nu}$ is traceless. It is often said that the form (22) corresponds to the *transverse-traceless* (or TT) gauge, for obvious reasons. The only two degrees of freedom left for the gravitational radiation field are therefore $h_+(\mathbf{x},t)$ and $h_\times(\mathbf{x},t)$, nicely displayed in the TT gauge, as we see. The notation of subindices '+' and '×' also has an interesting meaning, as will now be briefly sketched.

True gravitational fields are distinguished from fictitious or reference frame acceleration fields by the presence of genuine tidal forces. In the language of differential geometry, these are expressed in terms of the Riemann tensor $R_{\mu\nu\rho\sigma}(x)$. The non-vanishing of this four index tensor is the indication that true gravitational fields do exist. The Riemann tensor is thus a measurable quantity, and can be determined by measuring the relative displacements of two nearby test masses, which are otherwise in free fall in the gravitational field. This is so because such relative motion is governed by purely tidal forces. In the jargon of differential geometry, the relative four-vector ξ^μ is called the geodesic deviation, and satisfies the equations

$$\frac{D^2 \xi^\mu}{d\tau^2} = R^\mu{}_{\nu\rho\sigma} u^\nu \xi^\rho u^\sigma. \qquad (23)$$

Here, u^ν is the four-velocity of one of the particles, τ its proper time parameter, and D stands for *covariant* derivative (see Weinberg (1972) for more details).

It is not difficult to find approximate solutions to (23) for the above wave field: let the two particles lie on a straight line defined by the spherical angles (θ,φ) in the reference frame in which the gravitational wave is given by (22)—the TT frame. Let their spatial distance be $\ell(t)$, and let the value of this distance be ℓ_0 in the absence of gravitational waves. If the significant wavelengths in $h_{\mu\nu}$ are much larger than ℓ_0, and $|h_{\mu\nu}| \ll 1$, then it can be easily proved that

$$\ell(t) \simeq \ell_0 \left[1 + \tfrac{1}{2} \eta(t) \right] \qquad (24)$$

where

$$\eta(t) = [h_+(t) \cos 2\varphi + h_\times(t) \sin 2\varphi] \sin^2 \theta \,.df \qquad (25)$$

The space dependence of h_+ and $h_\times$ has been dropped, on the understanding that it corresponds to the mid point between the two test particles.

The last two equations show that tracking of the relative motion between two particles directly provides measurement of the gravitational wave amplitudes, and also show that the relative displacements of the test masses is of order h, *i.e.* $\delta\ell/\ell \simeq h$. This is the right place to mention that current gravitational wave detectors are designed to operate on the basis of this principle, and that we need therefore to assess what the order of magnitude of $h_{\mu\nu}$ is in realistic cases, in order to make the appropriate sensitivity design.

Equations (24) and (25) are also useful in characterising the physical meaning of the amplitudes h_+ and $h_\times$, *i.e.* of the polarisation states of the GW. Imagine having test particles as beads on an ideally flexible thread of circular shape, in diametrically opposite pairs, lying in a plane perpendicular to the incoming gravitational wave direction ($\theta=\pi/2$ in (25)). We have the following extreme cases:

$$\begin{cases} h_\times(t) = 0 & \Rightarrow \quad \delta\ell/\ell = \tfrac{1}{2} h_+(t) \cos 2\varphi \\ h_+(t) = 0 & \Rightarrow \quad \delta\ell/\ell = \tfrac{1}{2} h_\times(t) \sin 2\varphi \end{cases} \quad (26)$$

The beads' motion appears modulated by $\cos 2\varphi$ or $\sin 2\varphi$ according to their location on the circumference given by φ. Note that the modulation factors are offset by an angle of 45 degrees. If the incoming gravitational wave happens to be periodic then the motions of the beads take place between the positions pictorially displayed in Figure 1, according to whether $h_\times$ is zero ('+' polarisation) or h_+ is zero ('×' polarisation). The figure also justifies the somewhat fancy notation '+' and '×' for the modes. Let me finally stress that no effect can be possibly seen if the test particles are aligned with the GW's incidence direction ($\theta=0$ or π), *cf.* (25). This is why we say that gravitational waves are *transverse*.

4 The quadrupole formulas

We now come to the discussion of how gravitational waves are generated. In other words, we want to find the solution to Equations (11). A formal expression can be readily written down for it:

$$\bar{h}_{\mu\nu}(\mathbf{x},t) = \frac{4G}{c^4} \int T_{\mu\nu}\left(\mathbf{x}', t - \frac{|\mathbf{x}-\mathbf{x}'|}{c}\right) \frac{d^3 x'}{|\mathbf{x}-\mathbf{x}'|} \quad (27)$$

As expected, we see that this solution is of the retarded type, consistent with causality requirements. Equation (27) is too general and it is more useful to consider some particular cases. First of all, we shall consider the wave zone, *i.e.* far from the source, where 'far' means that $|\mathbf{x}|$ is much larger than the source's dimensions, or $|\mathbf{x}| \gg |\mathbf{x}'|$ for all $\mathbf{x}'$ in the above integral. Then

$$\bar{h}_{\mu\nu}(\mathbf{x},t) = \frac{4G}{c^4 r} \int T_{\mu\nu}(\mathbf{x}', t - r/c)\, d^3 x' \quad \text{(wave zone)} \quad (28)$$

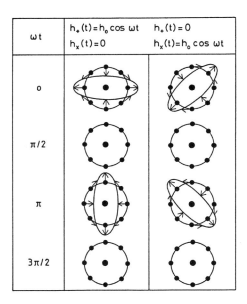

Figure 1. *The polarisation modes of linear gravitational waves, reflected in the motion of test particles which are beads arranged in a circular layout before the gravitational wave hits them. The wave is assumed to be periodic, and so therefore is their motion.*

where $r \equiv |\mathbf{x} - \mathbf{x}'| \simeq |\mathbf{x}|$, taking as coordinate origin a point in the source. It may now be recalled that, in the wave zone, we only need to worry about the TT part of $h_{\mu\nu}$ (the wave zone is a vacuum region), so there are far too many components given in (28), and simplifications are both possible and desirable. We note first of all that all components $h_{\mu 0}$ are zero in the TT gauge (see (22) above) so we concentrate on the ij components only. If a further assumption is made that velocities within the source are small compared to c, a reasonable hypothesis in most cases of interest, then relatively simple algebraic manipulations permit us to write integrals of the T_{ij} components of the energy-momentum tensor in terms of those of T_{00} only. I quote the result (see *e.g.* Landau and Lifshitz 1985):

$$h_{ij}(\mathbf{x},t) = \frac{2G}{c^4 r} \ddot{Q}_{ij}(t - r/c) \qquad (29)$$

where

$$Q_{ij}(t) \equiv \frac{1}{c^2} \int \left(x_i x_j - \tfrac{1}{3} |\mathbf{x}|^2 \delta_{ij} \right) T_{00}(\mathbf{x},t) \, d^3 x' \qquad (30)$$

is the quadrupole moment of the source, and overdots stand for time derivatives. Of course, only the TT part of $\ddot{Q}_{ij}$ contributes to observable gravitational waves. $T_{00}(\mathbf{x}, t)$ is the density of matter $\rho(\mathbf{x}, t)$ multiplied by c^2. Equations (29) and (30) thus show that gravitational waves are generated in distributions of mass which undergo accelerations —uniform motions are not radiative, as expected.

If a source radiates gravitational waves it then consequently looses energy to that radiation. In the quadrupole approximation, the energy loss rate can be seen to be given by

$$\frac{dE}{dt} = -\frac{G}{5c^5} \dddot{Q}_{ij} \dddot{Q}_{ij} \qquad (31)$$

where summation over the repeated indices ij is understood.

The reader may now wonder why gravitational waves have not been seen in spite of the permanent display of accelerated motions in our everyday environment. The answer to this will be most clearly illustrated with an example.

Imagine that a mass M, attached to the free end of a spring of natural angular frequency ω, is set to oscillate with amplitude A. Simple calculations based on the above formulae give

$$h_+(z,t) = \frac{2GM}{c^2 z} \frac{\omega^2 A^2}{c^2} \cos 2\omega(t - z/c) \qquad \langle \dot{A} \rangle = \frac{8}{15} \frac{2GM}{c^2 A} \left(\frac{\omega A}{c}\right)^4 c \qquad (32)$$

where I have made the assumption that the oscillations take place along the y–axis, and that the observation point lies in the wave zone a distance z from the source, and on the z–axis. The amplitude $h_\times(z,t)$ is seen to vanish. In the second equation angled brackets mean average over one period. We can substitute rather exaggerated values for the parameters in (32), $M = 1$ ton, $A = 1$ metre, $\omega/2\pi = 1$kHz, say, and still get the following amazing figures:

$$h_+(z,t) \simeq 10^{-39} \left(\frac{150 \text{ km}}{z}\right) \qquad \langle \dot{A} \rangle \simeq 5 \times 10^{-35} \text{ m/sec} \qquad (33)$$

Here, I have taken a standard z value equal to a wavelength in order to get into the wave zone (note, incidentally, that gravitational waves are produced at *twice* the source's frequency). These numbers are a clear indication of how absolutely hopeless it is to even dream of detecting the gravitational radiation generated by such a device as the one described. Both the following alternatives are equally ridiculous: either to measure the relative distance of two test masses in the wave zone to one part in 10^{39}, or to sense the source's amplitude decay $\dot{A}$, whose accumulated decrease would only amount to about 3×10^{-17} of its original value (1 metre) over the whole age of the Universe...

These numbers suggest that we take a look back into Equations (32) for a more general understanding of what is happening. We recast the first of them in the form

$$h_+(z,t) = \frac{R}{z} \frac{v^2}{c^2} \cos 2\omega(t - z/c) \qquad (34)$$

where $R \equiv 2GM/c^2$ and $v \equiv \omega A$. These parameters are the Schwarzschild radius and the maximum speed of the oscillating mass, respectively. Equation (34) nicely displays now the reason why we have found such small figures in (33) upon substitution of numerical values: two very small quantities conspire to multiply one another in (34) to make up for the amplitude h_+, namely, the ratio of the mass's Schwarzschild radius, R, to several times the radiation wavelength, z, *and* the ratio of the source's (typical) velocity, v, to the speed of light, c. Now, inspection of (29) and (30) immediately shows that an order of magnitude estimation of h is given by the product $(R/r) \times v^2/c^2$ for *any* source, not just the one considered in our example—provided, of course, that the quadrupole formulae are applicable. Which fact, in turn, shows how absolutely hopeless it is to think of a realistic earth-based source of gravitational waves. Recall for example that the Earth's Schwarzschild radius is only about 4.5 millimetres, and that any thinkable motions of large masses are extremely small compared to c.

We are thus led to turn our eyes into the outer Universe, where phenomena involving enormous masses and velocities may result in large amounts of energy being converted into gravitational waves. Places where such things happen, however, are remote, and so signals coming from them will be severely damped on arrival here, simply due to distance. I come to the discussion of these matters in the following three sections.

5 Gravitationally bound binary systems

Another interesting example of a gravitational wave source is a system formed by two point masses revolving around each other in a Keplerian orbit. The characteristics of the gravitational radiation emitted by such a system, as well as its *back action* effects, were studied long ago by Peters and Mathews (Peters 1964). For reasons which will soon become clear, I shall only focus on the latter effects. These are of course due to the loss of energy to gravitational waves, and consist in a series of secular changes in the orbital parameters: the masses are seen to follow an in-spiralling motion with a revolution period which gets progressively shorter as the semimajor axis shrinks, the masses speed up, and the orbital shape gets more circular. Specifically, the period, T, and eccentricity, e, are found to *decay* according to

$$\langle \dot{T} \rangle = -\frac{48\pi}{5\sqrt{2}} \left(\frac{2GM}{c^2 a} \right)^{5/2} \frac{\mu}{M} (1-e^2)^{-7/2} \left(1 + \tfrac{73}{24} e^2 + \tfrac{37}{96} e^4 \right) \tag{35}$$

and

$$\langle \dot{e} \rangle = -\frac{38}{15} \left(\frac{2GM}{c^2 a} \right)^{3} \frac{\mu}{M} \frac{a}{c} e (1-e^2)^{-5/2} \left(1 + \tfrac{121}{304} e^2 \right) \tag{36}$$

where a is the semimajor axis, $M \equiv m_1 + m_2$ and $\mu \equiv m_1 m_2 / M$, with m_1, m_2 the masses of the components of the binary, respectively. As before, angled brackets mean period averages. It is again interesting to see what numbers come from these formulae when numerical values are substituted in them. Take for example the Earth-Sun system: the Earth's orbit is nearly circular, so we do not worry about eccentricity. The period decay is found to be

$$\langle \dot{T} \rangle_{\text{Earth-Sun}} \simeq 6 \times 10^{-25} \text{ years/year} \tag{37}$$

If one thinks that the age of the Universe is some 10^{10} years, the above is a reassuring result: the possibility that we all vanish from the Earth's surface like roast chicken, due to inexorable gravitational bremsstrahlung in the Sun's atmosphere, can definitely be discounted. But let us go a bit further.

In the winter of 1967-68 Professor Anthony Hewish and his PhD student Jocelyn Bell, working at the Mullard radio observatory near Cambridge, announced the discovery of what was to become known as a *pulsar*. A pulsar is a pulsating radio source with an amazingly precise pulse emission rate, even better than the best atomic clocks.

It is believed to be a rotating neutron star surrounded by a magnetosphere, where huge magnetic fields reside. Not surprisingly, this outstanding discovery immediately resulted in a 'pulsar hunting fever', as a new and promising window for the exploration of the sky had opened. A major landmark in such 'hunting' happened in 1974, when Rusell Hulse and Joseph Taylor reported the finding of a binary pulsar. This is a binary system one of whose members is a pulsar, while the companion is an invisible star. The existence of the latter is established by analysing the periodic delays and advances in pulse arrival times, which are attributed to the pulsar getting further and closer to the observer, respectively, as it moves in its orbit around the invisible companion. This pulsar is now catalogued PSR B1913+16, and was found with the 305 metre diameter radio telescope at Arecibo, Puerto Rico.

The unprecedented precision with which the details of the orbital motion of the binary can be established has made of this system an extraordinary testbed for gravitational physics, in particular certain aspects of gravitational wave physics. The methods and techniques whereby valuable information is retrieved from raw pulsar data are remarkable, and a good example, in my opinion, of thorough scientific work. I devote the next section to give the reader a flavour of this beautiful piece of research.

6 The binary pulsar

Let me first of all describe briefly the kind of observational data generated by the pulsar, and how these data are fed into a theoretical framework in order to draw conclusions of scientific value from their analysis.

As mentioned earlier, pulsars are extremely regular radio 'lighthouses': they emit radio pulses of a specific profile shape at impressively even time intervals. (It is important to stress that pulsar emission is *not* an intermittent emission. Rather, it is a continuous emission of a directional beam, which sweeps the observer's location once per cycle of rotation of the pulsar–very much a lighthouse effect. This being clear, I will maintain the terminology 'pulse emission' for brevity). The actual times of arrival (TOAs) of the pulses to the telescope are *not* evenly spaced, however, as recorded with a local clock, because

a) the Earth moves

b) the pulsar itself moves around its companion

c) the pulsar shows a (small) spin-down effect

d) the interstellar medium causes dissipative delays in signal travelling times.

Let us call τ the pulsar's proper time, and let τ_N be the time of emission of the N-th pulse. Then

$$N = \nu \tau_N + \tfrac{1}{2} \dot\nu \tau_N^2 + \cdots \tag{38}$$

where ν is the pulsar's proper frequency, and $\dot\nu$ its time derivative. For PSR B1913+16, $\nu = 16.94053930\ldots$ Hz and $\dot\nu = -2.47 \times 10^{-15}$ sec^{-2}. The latter is a very small value, and higher derivatives are undetectably small—which facts account for the stability of pulse emission rates. The idea is now to relate the actually measured TOAs to the four effects listed above, on the assumption that (38) holds, in order to quantify those effects. This is accomplished through an equation giving the relationship between the pulsar's proper time and the observer time, t, say. In this equation one should include any conceivable causes of asynchronism between the pulsar and the observer clocks. Such an equation is (Taylor 1993)

$$\begin{aligned}\tau = \ & t - t_0 - D/\nu^2 \\ & + \Delta_{R\odot} + \Delta_{E\odot} - \Delta_{S\odot} \\ & - \Delta_R - \Delta_E - \Delta_S - \Delta_A\end{aligned} \quad (39)$$

which I have split into three lines in order to distinctly classify the different effects. In the first line, t_0 is simply a nominal equivalent TOA, and D/ν^2 accounts for the effect (d) in the above list (Smith 1979). In the second line, time offsets due to orbital motion of the earth around the Sun are considered: $\Delta_{R\odot}$ stands for the difference in light travel time from the solar system barycentre (SSB) to the observatory ('Roemer' lag), $\Delta_{E\odot}$ includes gravitational redshifts caused by all bodies in the solar system but the Earth itself, and proper time corrections due to Earth velocity relative to the SSB ('Einstein' lag), and $\Delta_{S\odot}$ is the 'Shapiro' lag, which is a further delay occurring as a consequence of the pulsar signal passing near the surface of the Sun. Finally, in the third line we have the effects caused by the pulsar motion in its own orbit around the invisible companion: Δ_R, Δ_E, Δ_S are the same effects just described, but applied to the binary system, and Δ_A is an aberration effect correction due to pulsar spin.

The corrections in the first two lines are well under control, as they depend on well understood and accurately modelled physics, so pulse TOA tracking is potentially a source of information about the binary system motions. Of course, the value this information may have strongly depends on the model of dynamics one makes for the system. Current levels of timing accuracy are in the range of 20 μsec, which places very demanding requirements on theory. As a matter of fact, not even a Schwarzschild orbital motion model is sufficient, let alone the Keplerian model of section 5. The best solution to date is due to Damour and Deruelle (Damour 1987), and it depends on the analysis of a more fundamental problem, that is, to solve the motion of two point masses in the dynamical gravitational fields they create. This is by no means a simple problem, and a solution to it can only be found perturbatively. In particular, gravitational radiation damping appears in Damour and Deruelle's treatment of the two body dynamics as a natural consequence of such dynamics, thus in a conceptually different context than that considered in section 4 above—remember that there it was purely a radiation field effect. Remarkably, though, Damour and Deruelle reproduce Equations (35) and (36) to the corresponding order of approximation.

Given the theory, Damour and Deruelle's in this case, the parameters it depends on are estimated on the basis of a statistical fit of the recorded TOAs to Equation (39). Such parameters are classified into 'Keplerian' and 'post-Keplerian', the latter being there simply because the former are not constant. A detailed discussion of these matters is not the subject of this lecture, but there is certainly one post-Keplerian parameter in the model which is of our concern here: it is the system's period decay, $\dot{T}$, which the theory associates to the energy lost by the binary to radiation. This period decay is of course a secular effect, so precision in its determination improves as data accumulate over time. Data have now been steadily recorded for 20 years. If one compares the the measured $\dot{T}$ with that predicted by Einstein's General Relativity Theory, one finds the remarkable result that (Taylor and Weisberg 1989)

$$\frac{\dot{T}_{\text{obs}}}{\dot{T}_{\text{GR}}} = 1.010 \pm 0.011 \tag{40}$$

which thus confirms the theory in this particular respect to a 1% level of accuracy at present, and further improvement is naturally expected in the future. See also Clifford Will's article for a more recent update.

It is some people's view that the binary pulsar analysis can only produce 'indirect' evidence of the existence of gravitational waves. Without going into a deep discussion of what 'evidence' means in science, I think that opinion is somewhat scathing. More precisely, I look upon the binary pulsar data as true evidence of *one* aspect of gravitational wave physics, so I find inappropriate the adjective 'indirect' for that evidence. It is however true that pulsar observations are currently insufficient to reveal important gravitational wave parameters, such as amplitudes and polarisation states of the waves.

The search for the latter is, and has been for more than three decades, actively pursued by a number of scientists committed to the design, building and running of gravitational wave detectors. Their work has not been crowned by success to date, which gives an idea of the formidable difficulties posed by detection. For example, the gravitational wave amplitude of the binary pulsar signal is about 10^{-23} at the Earth surface, and it comes at a main frequency of near 10^{-4}Hz (the orbital period is 27906.98 seconds (Taylor and Weisberg 1989) and the frequency is *twice* the inverse of this period), so very sensitive antennae are needed to sense this. In fact, current state of the art detectors are far from the required sensitivity but, curiously, the reason is not so much the 10^{-23} amplitude as the fact that the signal comes at too low a frequency: current, earth-based, detectors are generally very noisy below 100Hz, a long way from PSR B1913+16 frequency. Although several other binary pulsars have been discovered, their orbital periods fall also within the one day range, and so the same difficulties also persist with them.

We must again turn our eyes into other possible sources of gravitational waves, which will emit at higher frequencies, in the vicinity of 1kHz, say. To these I come next.

7 Astrophysical sources of gravitational waves

An important lesson we learn from the binary pulsar is that it is absolutely critical to have an accurate theoretical model of the dynamics of the system if we wish to draw reliable conclusions from the observational data, and in particular those concerned with its gravitational wave generation properties. It is hard to imagine a simpler, or cleaner, source than the binary pulsar (it is a two mass point system!), yet its theoretical description is a remarkably complex problem.

We now want to look into other possible sources, which will therefore be with high probability much more complicated than the binary. So, can we possibly model a signal coming from one such source with sufficient accuracy that it will be eventually identified? Currently, the answer is a somewhat pessimistic 'not really'. Although the literature is rich in models of different astrophysical objects, detailed knowledge of them is hardly available, particularly if General Relativity effects are relevant to their dynamics. What people tend to do is to make simplified assumptions which enable them to at least calculate order of magnitude estimates, a truly fundamental piece of information for detector builders. Further details can at present only be obtained by by massive numerical calculations programmed in large computers.

I will devote this section to giving a list of some of the possible sources of gravitational waves, along with their estimated amplitude on arrival in Earth, and their believed event rate. I will however generally omit discussions on possible waveforms and other signal fine structure details. To this end, it will be expedient to adopt a general classification scheme, where sources are divided into short duration signals, or bursts, long duration signals, and stochastic signals. I will make no explicit reference to original work in this section. The interested reader is referred to Thorne's comprehensive bibliography in (Thorne 1987).

7.1 Short signals

These are, as can be guessed, produced in explosions and other cataclysmic events. The qualifier 'short' is understood to mean something from a few seconds to a small fraction of a second, perhaps down to the millisecond range. I include as instances in this category supernovae and coalescing binaries.

Supernovae

These are probably amongst the most powerful sources of gravitational radiation known to astronomers. Their physics is however very difficult to model, more specifically those parts relating to gravitational wave emission from them. In order to emit gravitational waves, a supernova must have some degree of non-sphericity, because it is an exact result in General Relativity that spherically symmetric distributions of mass/energy only produce static gravitational fields, even if they themselves are non-stationary. Very roughly speaking, a supernova can be estimated to produce a final matter rebound in a time scale of a few milliseconds, during which time length 10^{52}ergs are converted into radiant energy. These figures are only an indication of scale and can vary depending on whether the supernova is type I, or type II, and other factors. A quantitative estimate yields

$$h_{\rm SN} \simeq 10^{-20} \left(\frac{\Delta E}{M_\odot c^2}\right) \left(\frac{\Delta t}{1 \text{ msec}}\right) \left(\frac{r}{15 \text{ Mpc}}\right) \qquad (41)$$

where ΔE is the amount of energy radiated away, and Δt the rebound time. For $\Delta E = 10^{52}$ergs, corresponding to 1% of a solar mass, we see that $h_{\rm SN} \simeq 10^{-21}$ when $r = 15$Mpc, *i.e.* a typical Virgo cluster distance. If the same event should happen in our galaxy then $h_{\rm SN}$ would be about a thousand times stronger. Millisecond explosions are interesting candidates for earth-based detectors, because their frequency spectra stretch up to the kHz region.

Supernova event rates are estimated to be one or two type I and one or two type II per galaxy every forty years or so. This means that several per month can in principle be seen out to the Virgo cluster, and this number goes up as r^3 as we get further and further away. The amplitude of the events, however, progressively weakens as r^{-1}.

Coalescing binaries

Equation (35) above shows that the period decay rate in a binary gets stronger as the orbit shrinks, since it is proportional to $a^{-5/2}$. One can also readily prove that the energy emission rate too is stronger as the orbit shrinks—it is actually proportional to a^{-5}. Although these figures are based on an oversimplified Keplerian model for the binary, they strongly suggest that such a system may eventually become an intense source of gravitational waves as their members approach coalescence. Also, the system will continue to be clean before tidal disruption, mass accretion and other complications enter the scenario. For a typical neutron star (NS) binary, such as PSR B1913+16, these effects are only expected to set in after the orbital period has gone beyond some 800Hz (Clark and Eardley 1977), and it would take such a system about 3 seconds to sweep in frequency from 100 to 800Hz. This is why coalescing binary signals are classified as short. Also, final coalescence is expected to produce a sudden and violent outburst of radiation.

The coalescing binary waveform is a modified sinusoid with increasing amplitude and frequency as coalescence is approached, see Figure 2b. For a typical NS binary again, a few hundred oscillations are expected from 100Hz to coalescence. An estimate of the signal amplitude is

$$h_{\text{CB}} \simeq 3 \times 10^{-23} \left(\frac{\mathcal{M}}{M_\odot}\right)^{-5/3} \left(\frac{\nu}{100 \text{ Hz}}\right)^{2/3} \left(\frac{r}{100 \text{ Mpc}}\right)^{-1} \qquad (42)$$

where $\mathcal{M}^{5/3} \equiv \mu M^{2/3}$. $\mathcal{M}$ is often named the mass parameter of the binary, and its value is $1.2 M_\odot$ for PSR B1913+16, which is also a typical value for NS binaries. I have taken a standard of distance equal to 100Mpc in (42) in order to embrace a large volume of space. The binary pulsar is much closer to us (only 7.1kpc) but it is also far from coalescence. As we see, the coalescing binary signal is weak, but the fact that its waveform can be modelled with some accuracy makes this an interesting signal to search for.

Now, how many coalescing binaries do we expect? This is a difficult question to answer, as we have no observational evidence of any so far. We can however estimate the coalescence time for the potential progenitor we know best, the binary pulsar: it is of the order of 10^9 years. This is a long time indeed, but it is reassuringly shorter than the age of the Universe, which means that it is not impossible for other binary systems to have formed that long ago, which are now reaching coalescence, or that coalesced 10^8 years back, about 30Mpc away from here, and we may now see. This, and other considerations, lead Thorne to conjecture that about 3 coalescing binary events per year should occur out to 100Mpc (Thorne 1987).

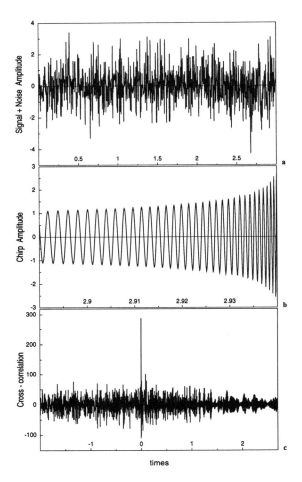

Figure 2. *A sketch of the matched filter working. In* **a**, *signal and noise are superimposed.* **b** *shows a zoom-in of the signal (chirp) buried in the noisy data: note the time scale, which is arbitrary but correctly matched. Note also the impossibility of sighting any traces of it in the noise, whose variance is 2.5 times the signal's mean amplitude. In* **c** *the matched filter output is plotted: a prominent peak sticks out at $\tau = 0$, when the template exactly hits the chirp. The signal-to-noise ratio approaches 2.5.*

7.2 Long signals

Long signals are produced by steady sources of radiation. We expect these to be largely monochromatic, their frequency being modulated by Doppler shifts due to relative motions between emitter and receiver, plus perhaps other internal effects. The binary pulsar would be included in this category were it not for the fact, already stressed, that its current frequency is too low. If one wants to think of higher frequency, long duration, signals, the first candidate which springs to mind is a pulsar, since it is a rapidly rotating mass: the rotation period of a pulsar is typically from 10^{-1} to 10^{-2} seconds, the fastest catalogued to date (PSR 1937+214) having a period of 1.558 milliseconds. In order to emit gravitational waves, however, a pulsar ought to have some asymmetry since otherwise it would not radiate, as already stressed in Section 7.1. It is very difficult to estimate those asymmetries, so very rough numbers can only be advanced. An optimistic one is

$$h_{\text{pulsar}} \simeq 10^{-25} \tag{43}$$

Several hundred galactic pulsars are currently catalogued, so the number of sources is lavish—though their signal is weak.

Chandrasekhar, and later Friedman and Schutz, conjectured that certain instabilities in rotating stars result in the emission of periodic gravitational waves. Estimates of the amplitudes of such waves strongly depend on the star's equation of state, viscosities and other complications. A typical event of this kind could be seen with

$$h_{\text{CFS}} \simeq 10^{-26} \tag{44}$$

Another interesting effect which results in the emission of monochromatic 'Wagoner waves' emitted as a consequence of the accretion of matter onto a rotating star, for example in a binary system. An estimate for a typical event is

$$h_{\text{Wag}} \simeq 10^{-28}. \tag{45}$$

The event rate for the last two kinds of events is quite uncertain.

We see that the amplitudes of long signals are generally very small. This looks discouraging in view of the the already weak short-signal estimates, but we must remember that long signals can be observed and recorded over long periods of time, and therefore signal-to-noise ratios are larger for these signals, as much longer integration times are available for them.

7.3 Stochastic signals

Just as there is an electromagnetic microwave radiation background of cosmological origin filling intergalactic space, it is not hard to imagine that a similar, in some sense, background of gravitational waves should also exist. There is of course no direct observational evidence of such background, but there definitely are arguments and calculations which support the conjecture of its eventual presence in the Universe. You will find more details about this in Bernard Carr's article.

Naturally, the cosmic gravitational wave background is classified as stochastic, since it will be the result of the cooperative effect of a variety of different sources randomly distributed across space, e.g., binary systems, black holes, phase transitions, cosmic strings, primordial gravitational waves, etc. This being the case, it is interesting to mention that if we want to measure such a signal we need a minimum of two detectors. This is because, the signal being random, it will be indistinguishable from detector noise—unless the latter is very small compared to the signal, which we obviously do not expect. Cross-correlation of two detectors' output is therefore the only way one can separate a random process common to the two data series from two independent, therefore uncorrelated, noise processes.

8 The matched filter

Let me close this brief review with a few basic considerations about data filtering and signal dig-out in a real detector.

As we have seen, any conceivable gravitational waves bathing the Earth ought to have a very weak effect on a detector. To reach requisite detector sensitivity has taken the endeavours of two generations of scientists, yet no sufficiently satisfactory record of an event has been reported to date. To gain in detector sensitivity, a battle has to be fought in two fronts: one needs to build extremely quiet systems, *i.e.* devices with a very low level of noise, and one also needs to design an adequate data analysis strategy, whereby any valuable information possibly buried in the noisy data be extracted with maximum efficiency. I devote the next few paragraphs to this second part of the problem. Further details will be found in Norna Robertson's article.

Let $y(t)$ be the detector's readout: it will be the superposition of a random signal, or noise, $n(t)$, and the true signal, $h(t)$, say:

$$y(t) = n(t) + h(t) \tag{46}$$

If we could directly subtract out the noise then data analysis would be marvelously straightforward: $h(t) = y(t) - n(t)$. Of course this is a silly idea, as $n(t)$ is random and cannot therefore be predicted. In order to separate out $h(t)$ from $y(t)$ we must filter the data, *i.e.* to process them in some intelligent way. If a linear filter is used, then its output will have the generic form

$$C(\tau) = \int_{t_1}^{t_2} y(t)\, k(t+\tau)\, dt \tag{47}$$

where (t_1, t_2) is the stretch of data analysed, and $k(t)$ is a strategically chosen function—the subject of much elegant theoretical and practical research, (originally in radar and communications problems, see Helstrom 1968). The theory tells us which, in a precisely defined sense, is the optimum filter to use in (47): it is called the *matched filter*, and its design requires prior knowledge of both the signal to be searched in the data *and* the noise spectral density. If the latter is constant over the detector's bandwidth, corresponding to white, or uncorrelated, noise, then $k(t)$ is seen to be precisely a template of the signal, *i.e.*

$$k(t) = h(t) \quad \text{(white noise)}. \tag{48}$$

This can be qualitatively understood as follows. Split the filter output into

$$C(\tau) = \int_{t_1}^{t_2} n(t)\, h(t+\tau)\, dt + \int_{t_1}^{t_2} h(t)\, h(t+\tau)\, dt \tag{49}$$

If the noise is white, then the first term is a random background, whereas the second, while completely deterministic, will in general be small due to poor overlapping of the signal $h(t)$ with its offset template $h(t+\tau)$. Nevertheless, for $\tau = 0$ the template exactly matches the signal, so that there is an important buildup in the second integral in (49), resulting in a prominent peak in $C(\tau)$ which will stick above the noisy background if the signal-to-noise ratio is sufficiently high. Figure 2 graphically illustrates the situation:

while no traces of any signal can possibly be 'eyeballed' in the first graph, the matched filter output shows a remarkably clear peak, which happens when the signal is exactly matched by the τ-moving template.

This is an example of how real data analysis proceeds: in a long stretch of data, a signal is searched for by sliding down the time series a template of it: if a sufficiently prominent peak is found then a signal is said to have been recorded. If not then the signal is said to be absent. Of course the concept of 'sufficiently prominent' requires a suitable threshold prescription, and this in turn depends on the specific statistical criteria set up during the analysis. I will skip these matters here, but the reader can consult the classical literature if he or she is further interested (Helstrom 1968).

A problem which will be found in practice is that the information about the signal one wants to detect will usually be incomplete. We may for instance be interested in looking for a coalescing binary signal, like the one in Figure 2b, but we may ignore *a priori* the actual values of the parameters it depends on. Imagine, in this example, that we know there is a chirp—this is the name commonly given to the coalescing binary waveform—in the data, but do not know its frequency sweep rate. If we use a template in our filter which is incorrectly matched to the signal we will not be able to tell it from the background noise. In a situation like this, a bank of templates is needed, each element being run across the data in parallel or cascade, and a decision on the signal being made on the basis of the maximum height peak comparatively found in the different $C(\tau)$'s corresponding to the respective filters. This, of course, requires suitable algorithms and computer programmes—by no means trivial to design.

The discussion so far has been made on the assumption that detector noise is white. This is unfortunately an oversimplification of what happens in real life (Krolak *et al.* 1993). For example, a major problem one has to face when noise is not uncorrelated is the estimation of its spectral density. But there are many others of both a conceptual and an engineering nature; so many in fact as to fill another series of lectures like this for a brief introduction to its intricacies.

9 Conclusion

In the last section I have tried to introduce you to a subject which is somewhat a diversion from 'pure' gravitational wave physics, but which we cannot escape to come across if we want to see gravitational waves. Data analysis techniques should not be seriously looked upon as mere practicalities. On the contrary, I think there is here an important lesson we must again learn: theoretical work on source properties, however assisted by observational and/or empirical evidence of all sorts, is of paramount importance in gravitational wave searches. We had already come to this conclusion in relation to the binary pulsar data analysis, and it now recurs. To put it in a simplified, but perhaps pedagogical, fashion, we could listen to the matched filter's declaration of principles: "Tell me what you looking for and I'll find it for you. Omit that, and I'm lost..."

Acknowledgements

I want to express gratitude to the organisers of SUSSP46 for inviting me to give these lectures—an excellent opportunity to put some order into otherwise chaotically spread material in one's files. Especially, I am indebted with Jim Skea for his prompt attention and patience with all my queries, both silly and sillier, since the earliest stages of the programme. Thanks are also due to John Pulham for his thorough and patient help in the editing of these notes.

References

Barut A O, 1980, *Electrodynamics and Classical Theory of Fields and Particles* (Dover, New York).
Clark J P A and Eardley D M, 1977, *Ap J* **215** 311.
Damour T, 1987, The Problem of Motion in Newtonian and Einsteinian Gravitation, in *Three hundred years of Gravitation*, eds Hawking S W and Israel W (CUP, Cambridge).
Helstrom C W, 1968, *Statistical Theory of Signal Detection* (Pergamon, Oxford).
Krolak A, Lobo J A, and Meers B J, 1993, *Phys Rev* **D 48** 3451.
Landau L D, and Lifshitz E M, 1985, *The Classical Theory of Fields* (Pergamon, Oxford).
Peters P C, 1964, *Phys Rev* **136** B1224.
Smith F G, 1979, *Pulsars* (CUP, Cambridge).
Taylor J H, 1993, *Classical Quantum Grav* **10** S157.
Taylor J H and Weisberg, 1989, *Ap J* **345** 434.
Thorne K S, 1987, Gravitational Radiation, in *Three hundred years of Gravitation*, eds Hawking S W and Israel W (CUP, Cambridge).
Weinberg S, 1972, *Gravitation and Cosmology* (Wiley & Sons, New York).

Detection of Gravitational Waves

Norna A Robertson

Department of Physics and Astronomy
University of Glasgow

1 Introduction

The prediction of the existence of gravitational waves dates back to the publication of Einstein's General Theory of Relativity in 1915. However, the first attempts to detect such waves came many years later with the pioneering experiments of Joseph Weber in the 1960's using resonant bar detectors (Weber 1969). It is generally accepted that the signals he reported were not due to gravitational waves. Nevertheless, Weber's work stimulated this field of research, and since that time there has been an ongoing research effort to develop detectors of sufficient sensitivity to allow detection of these waves from astrophysical sources.

The effect of a gravitational wave of amplitude h is to produce a strain in space given by $\Delta L/L = h/2$. The magnitude of the problem facing researchers in this area can be appreciated when one realises that theorists predict that the strongest sources might cause strains at the earth of the order of 10^{-18}. This would mean that if one were monitoring the separation of two free test masses one metre apart, the change in separation would be 10^{-18}m. However events of this magnitude would be very rare, and so for a reasonable 'event rate' one should aim for a strain sensitivity of 10^{-21} to 10^{-22}. Such figures should give the reader a feeling for the enormity of the experimental challenge facing those developing gravitational wave detectors!

2 Methods of detection

All methods of detection rely in some way on being able to measure the strain in space, h, but may differ in sensitivity level, frequency band of operation and bandwidth of detection. The major division of frequency band of operation is between ground based detectors, which should operate in the few hertz to few kilohertz region, and space based detectors which will be sensitive at much lower frequencies (for example 10^{-1} to 10^{-4}Hz for the LISA project, see 2.2). In addition astronomical observations such as those of pulsar arrival times can be used to search for signals in the 10^{-7} to 10^{-9}Hz region, and observations of the anisotropy of the cosmic background radiation can yield information in the 10^{-17} to 10^{-19}Hz region. Thus a large spectrum of possible sources can be covered by these different techniques. The types of sources which are likely to produce waves at different frequencies is addressed in the article by Alberto Lobo in this volume. Some of the major techniques for detection, and the results to date, are presented below.

2.1 Resonant bar detectors

The original type of detector as pioneered by Weber consisted of a large (approximately one tonne) aluminium bar suspended around its circumference and hanging in a vacuum system to isolate it from external disturbances. Bonded to the bar around its centre were piezoelectric transducers. These devices produce an electrical signal when subject to a strain. The effect of a gravitational wave incident on such a detector is to produce a mechanical strain in the bar, which in principle can be detected by monitoring the amplified electrical signals from the transducers. A bar does not initially resemble our canonical two free test masses whose separation is monitored to look for strains in space. However, for its fundamental longitudinal mode of vibration it can be thought of as two masses joined together by a spring. The gravitational wave sets the bar into oscillation and hence produces a varying strain in the transducers. Aluminium was chosen as the material because this metal has an intrinsic high quality factor. This implies that once excited, the bar continues in oscillation for a relatively long time, which effectively increases the sensitivity of the detector.

The fundamental limitation to its sensitivity comes from the thermal motion of the atoms in the bar, which produces an average background displacement noise level against which one looks for the gravitational wave signal. Spurious random events such as sudden local disturbances could look like a signal. To reduce the likelihood of misinterpreting such signals, two or more such bars separated in space would be operated in coincidence.

After Weber announced his detection of signals which he interpreted as gravitational waves, many groups set up similar detectors, but no one else detected any signals, despite working at sensitivities as good as and ultimately better than his original detectors. The best upper limit for such room temperature bars was around 10^{-17} for bursts at around 1kHz (Billing 1975).

The obvious way to improve the sensitivity of such detectors is to cool them, and the next generation of detectors were cooled by liquid helium to 4 K. The operation of these cooled detectors has required development of suitable low noise superconducting transducers such as for example SQUIDS. Low temperature detectors have been developed by groups at Stanford, Louisiana State University, Rome and Perth, Western Australia. The current best RMS sensitivity has been achieved by LSU and Rome at a level of $h \sim 6 \times 10^{-19}$ for kHz band bursts (Solomonson et al. 1978, Astone et al. 1993).

Future plans for such detectors include cooling them further to temperatures of order 50mK. This could lead to sensitivity $h \sim 10^{-19}$ in the near-term future, with an eventual aim of achieving $h \sim 10^{-20}$. However this requires continued improvement in transducer noise performance. Further increase in sensitivity could come from increasing the mass of the 'bar' and the 'number' of them, and one way of doing this which is receiving some attention is to use a spherical antenna. The increased mass reduces the effect of thermal noise, and the shape has 5 quadrupole modes, thus giving a system with essentially 5 detectors in one object. In fact the shape might not be truly spherical but, for example, a 'truncated icosahedral' gravitational antenna, hence the name TIGA given to such a device. All the major bar groups, as well as a group from Leiden, have formed a study group to pursue the idea of spherical detectors.

Resonant bar detectors by their very nature tend to be inherently narrow band devices. By that we mean that they are sensitive to a small band of frequencies of the order of a few hertz around their natural fundamental resonant frequency, which is dictated by the speed of sound in the material and their size. Methods to widen the bandwidth which could be promising have been proposed, but they are not likely to achieve the potential bandwidth of the laser interferometric detectors to be discussed in the next section.

2.2 Laser interferometric detectors

Ground-Based

A laser interferometric detector can be thought of in its simplest form as a Michelson interferometer whose mirrors are suspended as pendulums so that they act as free test masses at frequencies above their pendulum resonant frequency (see Figure 1). The first such device was investigated by Forward and his group in the 1970's (Forward 1978), and a detailed consideration of the possible limiting factors to the performance of such detectors was carried out by Weiss (1972). Since then several groups have developed prototype interferometric detectors, at Glasgow, Garching, Caltech, MIT and Tokyo. There are also proposals and projects to build larger versions of these prototypes in several places around the world (discussed later).

A brief revision of the mode of operation of a Michelson interferometer is instructive at this point. Monochromatic light (*e.g.* from a laser) is split into two orthogonal beams via a beamsplitter. Each beam traverses a path length L and is reflected by a mirror back to the splitter. Here the beams are recombined and the output can be monitored on a photodiode. The output observed will depend on the relative path difference between the two arms; *e.g.* if the effective path difference is zero or an integer number of wavelengths, constructive interference is observed. A gravitational wave incident on such a detector will in general induce a change in the difference in arm lengths, which will be seen as a change in the output at the photodiode. In fact the layout of a

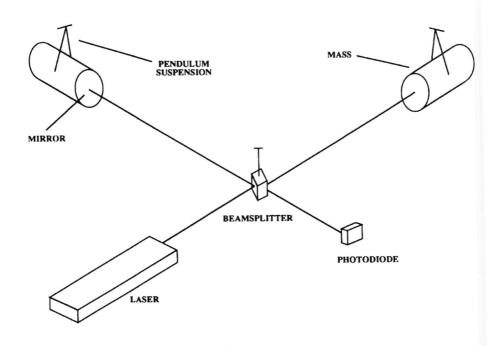

Figure 1. *Schematic diagram illustrating the basic principle of a gravitational wave detector using a Michelson interferometer*

Michelson interferometer is nicely matched to the quadrupole nature of the gravitational wave. For example a wave of amplitude $\Delta L/L$, suitably polarised in the directions of the two arms, will cause the effective arm length of one to increase by ΔL at the same time as the other decreases by ΔL, and thus taking the difference gives a signal twice that due to the change in one arm alone.

There are further practical advantages in the use of an interferometer. Firstly by making a differential measurement it avoids making an absolute measurement of distance, which would require an extremely stable wavelength light source. The measurement technique reduces the requirements of stability by a factor of typically about 100. Another advantage of this type of detector is that its sensitivity can be increased simply by increasing the length of the arms (at least to a certain limit given by the wavelength of the gravitational wave, discussed later). A more sensitive detector for a given arm length can in principle be obtained by the use of optical delay lines, or Fabry-Perot cavities in each arm, which effectively increases the optical path length of the device. Further, the sensitivity is inherently wideband, so that such a detector can be operated over a wide range of frequencies.

Thus there appear to be several advantages to this type of detector, but there are many practical problems which require to be overcome in producing a working detector. However great improvements from the first such detector have been made. The RMS strain sensitivity noise levels which have been achieved in the kHz region by the detectors at Glasgow (10m arm length Fabry-Perot), Garching (30m delay line), and Caltech (40m Fabry-Perot) are in the range from a few times 10^{-18} to a few times 10^{-19} for wideband bursts (Robertson et al. 1995, Shoemaker et al. 1988, Savage 1994). Most of the effort with these detectors has been directed at investigating their noise performance and developing the techniques to improve their sensitivity. However the Glasgow and Garching groups did carry out a 100 hour coincidence experiment in 1989 which demonstrated that such interferometers can function as working instruments with a high duty cycle. Research within the interferometer groups is now being directed towards developing the techniques for extending such detectors to scale of order of a kilometre, which should have the potential to reach the sensitivity levels required to start making astrophysical observations.

The operation of such detectors, and the limiting noise sources will be looked at in more detail in Section 3.

Space-Based

Space-borne laser interferometric detectors offer the opportunity to explore the low frequency part of the gravitational wave spectrum. Such detectors are free from the irreducible effects of gravity gradients on the Earth, and arm lengths of the order of 10^6km can be utilised. Of course putting such a detector into space would not be without its own technical difficulties. However such a mission has been proposed to ESA and is under active consideration at present as a possible post Horizon 2000 cornerstone mission. This is LISA (Laser Interferometer Space Antenna) (Danzmann et al. 1993), a proposal to put up to 6 spacecraft into heliocentric orbit, two at each corner of an equilateral triangle of side 5×10^6km, with the whole assembly being positioned approximately 20° behind the Earth (see Figure 2). This arrangement allows up to three semi-independent interferometers to run, each consisting of two central and two far spacecraft, and it provides for some redundancy if a spacecraft fails. The laser on each central spacecraft is beamed to the respective end spacecraft, where local lasers are phase-locked to the incoming beam and then transmitted back to the centre. By comparing the phases of the incoming beams to the central lasers the changes in path length of the two arms can be deduced. LISA is expected to reach a sensitivity level in a year's observing, allowing for a signal to noise ratio of 5, of around $h \sim 10^{-23}$ from approximately 10^{-3} to 10^{-2} Hz and have useful sensitivity between 10^{-4} and 10^{-1}Hz. The limiting noise sources at the low frequency end are likely to be random forces producing spurious accelerations of the test masses and, at the high frequency end, photon shot noise in the laser light used. The designed operating lifetime would be at least two years. There are many technical details which must be addressed for the successful operation of such a project, and these are the subject of studies at present underway. This is a long term project, with a potential launch date of 2015. However it promises to open up a different window of the spectrum of gravitational waves, and thus is very much complementary to the ground-based detectors.

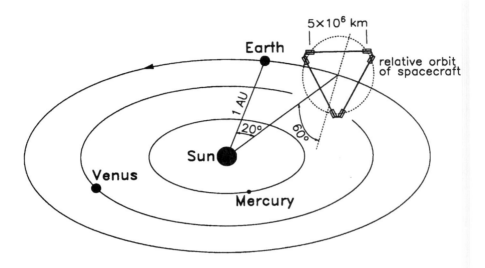

Figure 2. *Schematic diagram of the proposed* LISA *experiment in space*

2.3 Doppler tracking of spacecraft

A gravitational wave will cause small changes in the relative separation of the Earth and an interplanetary spacecraft, which could in principle be detected by looking for the resultant Doppler shift in the frequency of the radio signals used to track the spacecraft. Searches for such signals have been carried out for several years as part of the NASA Planetary programme, and the current best upper limit at the one sigma level is $h \sim 2 \times 10^{-15}$ at 10^{-3}Hz (Bertotti *et al.* 1995). The limits to sensitivity are given by the stability of the master clock used, and by propagation fluctuations in the interplanetary medium. The CASSINI mission to Saturn, scheduled for the turn of the century, offers an opportunity to improve the sensitivity by about a factor of 10, by using more precise clocks, and higher and multiple tracking frequencies to reduce the effects of the interplanetary medium.

2.4 Pulsar timing

One can consider monitoring the changes in separation between the earth and a pulsar by looking at the arrival times of the radio pulses. This is somewhat similar to the Doppler tracking technique, except that it is the pulsar itself which is the reference clock. Also because of the much larger distance involved, the frequencies to which this technique is sensitive to are in the range 10^{-7} to 10^{-9}Hz. For example, observations of the millisecond pulsars PSR B1937+21 and PSR B1855+09 have set an upper limit to the gravitational wave background at a frequency of around 10^{-8} Hz. This limit is about 6×10^{-8} times the energy density per logarithmic frequency interval required to close the Universe, at the 2σ level (Kaspi et al. 1994).

2.5 Indirect detection of gravitational waves

No discussion about the detection of gravitational waves would be complete without the mention of the first indirect detection made from the observations of the binary pulsar PSR 1913+16. This system of two neutron stars, one of which is a pulsar, was discovered by Hulse and Taylor in 1974, and has been carefully monitored now for approximately 20 years. The orbital period of the system is decreasing as the two stars move closer together, and the change in period agrees to high precision with the predictions of General Relativity, where energy is being radiated from the system in the form of gravitational waves (Taylor and Weisberg 1982,1989). For the discovery of binary pulsars, and for the confirmation of Einstein's prediction of the existence of gravitational waves, Hulse and Taylor were awarded the Nobel Prize in Physics in 1993. This indirect evidence for these waves, and the recognition of its importance, has undoubtedly given a boost to the efforts of physicists worldwide involved in the development of gravitational wave detectors.

3 A case study: ground-based interferometric detectors

The basic idea of a Michelson interferometer has already been introduced. We shall now look in some more detail at noise sources which can limit the sensitivity of such a detector, and consider how the effects of these sources can be minimised.

3.1 Photon noise

One of the fundamental sources of noise in an interferometer is the statistical fluctuation in the number of photons detected at the output, known as photon shot noise or photon noise. Consider N photons in time τ, and a phase difference ϕ between the light from the two arms of the interferometer. One can show that $\Delta N \Delta \phi \sim 1$, (this comes from the Uncertainty Principle relationship $\Delta E \Delta t \sim \hbar$). Further, since $\phi = (4\pi/\lambda) \times$ (difference in length of the arms) we can derive the following relationship for the limiting sensitivity expressed as an equivalent gravitational wave amplitude, where Δx is the fluctuation in arm length difference, making use of the relationship $\Delta N = \sqrt{N}$ for the uncertainty associated with an average count of N photons.

$$h \approx \frac{\Delta x}{L} \approx \frac{1}{L}\frac{\lambda}{4\pi}\frac{1}{\sqrt{N}} \approx \frac{1}{L}\left(\frac{\hbar c \lambda}{8\pi I_0 \tau}\right)^{1/2} \quad (1)$$

Here I_0 is the input power, λ is the wavelength of the light, and $\hbar$ and c have their usual meanings. We should include a factor ε, the quantum efficiency of the photodiode, and we can reexpress the relationship considering detection over a bandwidth Δf where $\Delta f = 1/2\tau$ and get

$$h \approx \frac{1}{L}\left(\frac{\hbar c \lambda \Delta f}{4\pi I_0 \varepsilon}\right)^{1/2} \qquad (2)$$

This relationship applies when the time spent by the light in the arms is less than the timescale of the gravitational wave signals. From this equation it can be seen that the sensitivity increases with increasing arm length L. As mentioned before, the effective optical path can be greatly increased from the physical arm length by including either optical delay lines or Fabry-Perot cavities in the arms. With the availability of ultra-high reflectivity, low loss mirrors which have been developed for the laser gyro industry it is possible to choose the effective number of bounces in the arms of the interferometer such that the light storage time matches the period of the waves being looked for. In that case the above equation no longer applies. The sensitivity limited by photon noise becomes independent of the effective arm length, and is given by

$$h \approx \left(\frac{\hbar \pi \lambda f^2 \Delta f}{c I_0 \varepsilon}\right)^{1/2} \qquad (3)$$

For a burst signal at 1kHz with bandwidth 500Hz, using argon laser light of $\lambda = 514.5$nm, and with $\varepsilon = 50\%$, this limit becomes

$$h \approx \frac{2.4 \times 10^{-20}}{I_0^{1/2}} \qquad (4)$$

Thus it can be seen that to reach a level of sensitivity of around 10^{-22} requires 50kW of light! Commercial argon lasers can give at most of order 10 W. A method of increasing the effective light power, suggested by Drever (1983), and known as *power recycling* was stimulated by the fact that very low loss mirrors are available. The storage time limit can be achieved on a long baseline detector with a relatively modest number of effective bounces in the arms. It can be shown that the best photon noise limited sensitivity is obtained when the output photodetector observes an interference minimum. In this situation, with the use of very low loss mirrors, one finds that most of the input light will be directed back towards the laser. An additional mirror introduced in front of the laser can be used to reflect most of this light back into the interferometer, with its position controlled in such a way that the phase of the returned light adds coherently to the original laser light. In effect this mirror plus the combined mirrors of the interferometer looks like one large cavity resonating at the frequency of the light, f_{light}. Large power recycling factors should be achievable, limited by the losses such as scattering at the mirrors. For example in the proposed GEO 600 detector, a power recycling factor of $\times 2000$ is planned. There is however a practical limit to the amount of light which one can allow to be built up inside the interferometer. This is due to thermal distortion effects which can lead to instability in the cavities. And there is a further more fundamental effect with very high light power – that of fluctuating radiation pressure on the mirrors. In fact a fundamental limit is reached where one minimises photon noise and radiation pressure noise with respect to the laser power, and this limit corresponds to that imposed by the Heisenberg Uncertainty Principle. It should be pointed out here that for the laser powers which are currently being considered for use in long baseline detectors, this limit is not reached, but its presence cannot ultimately be ignored.

Further improvement in photon noise can be achieved by the technique of *signal recycling* (Meers, 1988). This technique makes use of another partially reflecting mirror, this time placed at the output port of the interferometer. Its purpose is to make the interferometer into a cavity resonant at the frequencies $f_{\text{light}} + f_{\text{sig}}$ or $f_{\text{light}} - f_{\text{sig}}$, where f_{sig} is the signal frequency. The effect is to improve the sensitivity of the detector for a particular signal frequency. The position of the mirror controls the f_{sig} which is resonant and the reflectivity controls the bandwidth. Such a technique will be particularly useful for looking for continuous signals of a certain frequency such as from pulsars.

At this point it is useful to pause and consider the following. Suppose one wishes to achieve photon noise limited sensitivity in the storage time limit optimised for 1kHz signals, *i.e.* wavelengths of 3×10^5m. One would choose an effective optical path length of one half of this wavelength, 1.5×10^5m, so that the total time spent by the light in one arm is equal to half the gravitational wave period. This could be achieved with a 30m armlength and 5000 beams (about 5000 reflections, which is easily achievable with good mirrors), or equally with 3 km armlength and 50 beams. Which does one choose? The answer lies in considering the other noise sources which are likely to limit the performance of such detectors.

3.2 Seismic noise

There is an ever present background level of seismic motion, which, if not allowed for in the design of detectors, would completely swamp any gravitational wave signal. To appreciate the magnitude of the problem, consider a few figures. The typical RMS spectrum of seismic noise at a quiet site, far from man-made sources of noise, can be approximated to $10^{-7}/f^2$ m/$\sqrt{\text{Hz}}$ (*i.e.* in a one hertz bandwidth) at frequencies, f, above 1Hz. Consider for example the goal of achieving a strain sensitivity of 10^{-22} for a burst source of frequency 100Hz and bandwidth of 50Hz. Now assume the detector consists of 4 mirrors, whose motions due to seismic noise are considered to be uncorrelated. With no isolation present, the RMS noise introduced into the relative displacement of the two arms would be $2 \times 10^{-7}/f^2$ times $\sqrt{50} = 1.4 \times 10^{-10}$m. For a detector of arm length $L = 3$ km, this would lead to a sensitivity limit due to seismic noise $h \sim x/L \sim 5 \times 10^{-14}$. To achieve 10^{-22} requires an isolation factor of 5×10^8. If instead an arm length of 30m were used, the isolation needed would be 100 times more than even this figure!

We can immediately conclude two things. Firstly using the longer arm length helps us, since this source of noise produces a displacement, whereas the gravitational wave signal is a strain in space. This argument will be true for any noise source which gives an effective displacement of the mirrors. Secondly the isolation required, even in a long baseline detector, is a formidable figure. And it should be pointed out that the situation gets worse towards lower frequencies, not only because of the shape of the spectrum of the seismic noise itself, but also because the isolation techniques used are typically frequency dependent, with decreasing isolation towards lower frequencies. It is fair to say that the extent to which these detectors can be operated at the lower end of the frequency spectrum will depend crucially on the seismic isolation achievable.

Simple Pendulum Suspension

A pendulum suspension is probably the simplest isolation system one can envisage. It is not difficult to show that above the resonant frequency f_0 of the suspension, and assuming it has a relatively high quality factor Q, the horizontal transmissibility, or ratio of horizontal motion of the suspended mass to that of the point of suspension, is given by f_0^2/f^2. Thus for example with a 25 cm length pendulum for which f_0 is 1Hz, the isolation at 100Hz would be 10^4. An obvious way to increase the isolation is to cascade two or more of such suspensions, where one finds that the resulting transmissibility from top to bottom is given by the product of the individual stages, above the highest resonant frequency of the system. This idea is being used by the VIRGO project (see below) where a seven-stage pendulum of overall length approximately 7m is being implemented.

Acoustic Isolation Stacks

Another method of isolation, which has been applied with success in gravitational wave detectors both of bar and interferometric design, is the use of isolation stacks consisting of alternating layers of a heavy metal and a soft elastic material such as rubber. These stacks are normally situated between the 'ground' and the point of suspension of the pendulum. Each stage, consisting of a layer of metal on rubber, behaves like a mass on a spring, and it can be shown that the isolation property has a similar form to a simple pendulum, though normally the quality factors are much smaller, which may lead to isolation proportional to $1/f$, and not $1/f^2$, per stage.

The above isolation systems may be described as passive systems. One may also consider active systems, where the performance of a passive isolator is enhanced with the use of electronic feedback. For example, by sensing the relative position of a suspended mass and its point of suspension and feeding back a suitable signal to the point of suspension it is possible to reduce a pendulum's natural resonant frequency, thus increasing its isolation capability.

It might be thought that the principal isolation necessary for gravitational wave detectors, in which horizontal motions are sensed, is against horizontal motions of the ground. However due to effects of cross-coupling in the fairly complicated suspension systems, vertical and tilt motions of the ground must also be isolated against. Fortunately systems such as stacks can provide isolation in more than one dimension.

Using a suitable combination of simple passive isolation systems such as stacks plus single or double pendulums, the required isolation for long baseline detectors should be achieved at 100 Hz. However for good performance at lower frequencies, a system such as the VIRGO multi-stage pendulum or the implementation of an active system will be required.

3.3 Thermal noise

We have seen above that in resonant bar detectors, thermal noise effects are one of the most important sources of noise, and that such detectors are cooled to reduce the effects. Thermal noise is also very important in interferometric detectors, but with careful design, the need to cool can be avoided. However, it may be the limiting factor in sensitivity around the 100Hz region, with seismic noise dominating at lower frequencies and photon noise at higher frequencies.

There are several important sources of thermal noise, namely the pendulum modes of the suspended test masses, the violin modes of the suspension wires, and the internal modes of the test masses themselves. A thermal energy $k_b T$ is associated with each mode of oscillation. However this is integrated over all frequencies. To find out the effective displacement produced as a function of frequency, we need to know the spectrum of the noise. A general model for damping in a harmonic oscillator can be represented using a complex form of Hooke's Law

$$F = -k[1 + i\phi(\omega)]\, x \quad (5)$$

where ϕ is the phase lag between the applied force F and the corresponding displacement x. The phase ϕ is related to the quality factor Q by $Q = 1/\phi(\omega_0)$ where ω_0 is the resonant angular frequency. Applying the fluctuation-dissipation theorem, one can find the power spectral density of thermal motion. This is given by

$$\tilde{x}^2(\omega) = \frac{4 k_b T \omega_0^2 \phi(\omega)}{\omega m \left[(\omega_0^2 - \omega^2)^2 + \omega_0^4 \phi(\omega)^2 \right]} \quad (6)$$

where $\omega_0^2 = k/m$.

When this equation is considered both below resonance and above resonance, it can be seen that the magnitude and shape of the curve is determined by the factor ϕ. In fact low thermal noise away from resonance is achieved by small ϕ, or correspondingly high Q. Thus the conclusion is that one wishes to maximise the quality factor Q of all relevant modes of oscillation, and also work at frequencies away from the resonances where possible. The latter requirement can be satisfied for the pendulum modes and the internal modes of the masses, but the violin modes will fall within the working band of such detectors, and at those frequencies there will be sharp spikes in the sensitivity curve which will require to be notched out electronically.

To predict the actual level of noise due to thermal effects requires knowledge of the magnitude of ϕ as a function of frequency. Conventionally it has in the past been assumed that ϕ is proportional to frequency when sensitivity curves were being estimated. But for many materials there is strong evidence for ϕ being independent of frequency, and this has led to a recent reassessment of the thermal noise levels. This is a very active area of investigation at present.

In summary therefore one wishes to choose materials both for the suspension and the test masses themselves with low ϕ. Silica is a good material for the test masses, having intrinsic ϕ's in the region 10^{-6} to 10^{-7}. Work is in progress to investigate the use of suspensions also made of silica, rather than the conventionally used steel wires with much lower intrinsic ϕ.

To give a feeling for the sensitivity levels which might be achieved, limited by the above noise sources, the predicted sensitivity for the proposed German-British 600m detector GEO-600 is given in Figure 3. Curve **a** is the photoelectric shot noise for a 4 pass delay line illuminated with 5W of laser light (1.06 μm) into the interferometer and typical mirror losses of 20 ppm per mirror. The system incorporates dual recycling with a power recycling factor of 2000 and the signal recycling set to give relatively wide bandwidth. Curve **b** is the thermal noise in the system. The test masses are each 16 kg, of fused silica of 25 cm diameter and 15 cm thickness, s in the beamsplitter. The quality factors of the internal modes are taken to be 5×10^6 and of the pendulums to be 10^7. Curve **c** represents the limit set by the Heisenberg Uncertainty Principle. Curve **d** represents a likely seismic noise for the sensitive components isolated by 4 layer stacks and suspended as double pendulums.

3.4 Other noise sources, and operation of an interferometer

The above noise sources may be considered among the most important, but other sources of noise cannot be ignored. For example fluctuations in the frequency, power and geometry of the laser light can all lead to limitations in the sensitivity of such detectors, and must be stabilised to a high degree to reduce these effects below the noise sources discussed above. Also the actual operation of such a detector requires many control systems to operate simultaneously and continuously, for example to hold the suspended masses aligned correctly, to stabilise the laser, and to keep cavities on resonance. However it is beyond the scope of these notes to address the details of how this is done, and the reader is directed for more detail to the references.

3.5 Future plans for interferometric detectors

We have seen that two important noise sources, thermal and seismic noise, are such that their effects on sensitivity decrease with increasing physical arm length of the detector. For this reason several groups world-wide have proposed to build long baseline detectors, which have the potential to achieve sensitivity levels which give the opportunity to make observations of astrophysical sources. Two projects are already funded. In the LIGO project (USA) (Abramovici et al. 1992), two detectors each of 4 km armlength are to be built, one sited in Washington State, and the other in Louisiana. The VIRGO project (Bradaschia et al. 1990) is a French/Italian collaboration to build a 3 km detector near Pisa in Italy. GEO-600 (Hough et al. 1994), a more modest 600m detector to be sited near Hannover, has been proposed by a collaboration of the University of Glasgow, University of Cardiff, Max-Planck Institute for Quantum Optics in Garching and the University of Hannover. This project has already received partial funding in Germany, and we await news of funding in Britain. Further proposals have been put forward in Australia and Japan for long baseline detectors. GEO-600 is intended to reach a sensitivity comparable to that of the three funded larger detectors in their first phase of operation so that meaningful coincidence searches can be made with all these detectors. In the long term the longer baseline detectors can more easily reach the sensitivity goal of $h \sim 10^{-22}$. As for the timescale involved, it is hoped that the majority of these detectors should come on line towards the turn of the century.

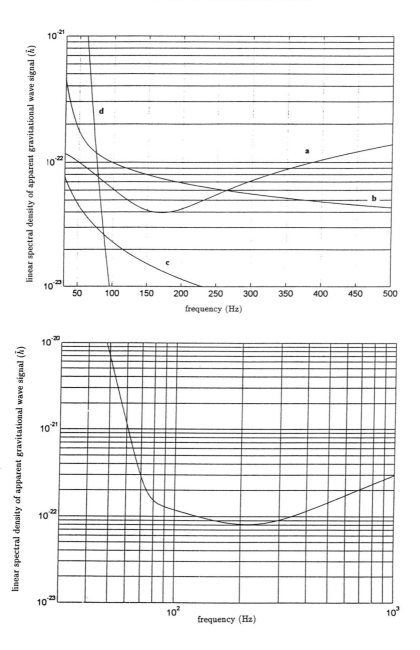

Figure 3. The upper figure shows the expected noise budget of the proposed 600m detector; the y-axis scales are in units of $Hz^{-1/2}$. The curves (a)–(d) are discussed in the text at the top of the opposite page. The lower figure shows the overall noise spectrum of the proposed 600m detector.

4 Some data analysis ideas

Assuming one has several working detectors around the world, how does one extract signals from the noise? Firstly, as was mentioned above, spurious random noise above the mean background level could look like a burst signal in one detector. Such 'events' can be discarded by operating two or more detectors in coincidence. In fact to extract the most information one wishes to use a minimum of three but more effectively four detectors with similar sensitivities. By using the strengths of the signals and the difference in arrival times, one can get information on both the direction to the source and the polarisation of the signal. Obviously data taken at each site must be recorded with very accurate absolute timing. However with the availability of radio timing signals or atomic clocks this is not a problem. Burst type sources can be picked up by monitoring the noise level and looking out for signals which exceed a certain threshold level. Chirp type signals, where the waveform can be predicted, can be searched for in a data stream using matched filter techniques. For continuous signals at a constant frequency, such as from pulsars, signal to noise is greatly increased by taking data over a long time period. This is because the signal, being coherent, increases linearly with time, whereas the noise background, typically random, will increase only as the square root of time. Pulsar signals in principle can be searched for with one detector alone, using the knowledge of the expected frequency available from radio observations. As for a stochastic background of gravitational waves, this can be looked for by cross-correlating the output of two detectors. However for good correlation the detectors should be close together, so that they respond to the same background at the same time. Thus for example GEO-600 and VIRGO should be able to put a better limit on the background than the two LIGO detectors which are 3 times further apart.

The above brief discussion is included to give an idea of the possibilities for detection and analysis. This is a very large topic, and for more details once again the reader is directed to the references given below. However one more point is worth making here. Because of the broadband nature of interferometric detectors, the resulting storage requirements are very large (up to a few terabytes per year of operation). Thus the whole subject of storage exchange and analysis needs to be given due consideration, and commonly agreed procedures within the network of detectors will be of importance.

5 Conclusion

The detection of gravitational waves has been an active area of research for over 25 years. This is a very challenging areas of experimental physics, requiring the development of new and more sensitive methods of making ultra-small displacement measurements. Much progress has already been made, and with the current proposals and plans for more sensitive detectors it now seems likely that within the next decade the first detection will take place, heralding the advent of a new astronomy—gravitational wave astronomy.

Acknowledgements

The author acknowledges with thanks the support of her colleagues in the Gravitational Waves Group at the University of Glasgow.

References

As well as the more detailed references referred to in the text, which are not intended to be comprehensive, the author recommends for consultation the following books and articles, and references therein, many of which she has made use of in writing this contribution.

General References

Thorne K S, 1987, Gravitational Radiation, in *300 years of Gravitation* eds Hawking S W and Israel W (Cambridge University Press)
Blair D G (ed), 1991, *The Detection of Gravitational Waves* (Cambridge University Press)
Saulson P R, 1994, *Fundamentals of Interferometric Gravitational Wave Detectors* (World Scientific)
Schutz B F (ed), 1989, *Gravitational Wave Data Analysis* NATO ASI Series (Kluwer Academic Publishers)
Abramovici A, Bender P, Drever R, Finn L S, Flaminio R, Grischuk L, Johnson W, Kawabe K, Michelson P, Robertson N, Rüdiger A, Sandeman R J, Saulson P, Shoemaker D, Thorn K, Tinto M, Tobar M, Weiss R and Whitcomb S, Gravitational Wave Astrophysics, to appear in *Proceedings of the Snowmass 94 Summer Study on Particle and Nuclear Astrophysics and Cosmology* eds Kolb E W and Peccei R (World Scientific, Singapore)

Text References

Abramovici A, Althouse W E, Drever R W P, Gursel Y, Kawamura S, Raab F J, Shoemaker D, Sievers L, Spero R E, Thorne K S, Vogt R E, Weiss R, Whitcomb S E and Zucker M E, 1992, *Science* **256** 325
Astone et al. 1993, *Phys Rev D* **47** 2
Bertotti B et al. 1995, *Astronomy and Astrophysics* **296** 13
Billing H, Kafka P, Maischberger K, Meyer F, Winkler W, 1975, *Lett al Nuovo Cim* **12** 111
Bradaschia C, Del Fabbro R, Di Virgilio A, Giazotto A, Kautzky H, Montelatici V, Passuello D, Brillet A, Cregut O, Hello P, Man C N, Manh P T, Marraud A, Shoemaker D, Vinet J-Y, Barone F, Di Fiore L, Milano L, Russo G, Solimeno S, Aguirregabiria J R, Bel H, Duruisseau J-P, Le Denmat G, Tourrenc P, Capozzi M, Longo M, Lops M, Pinto I, Rotoli G, Damour T, Bonazzola S, Marck J A, Gourghoulon Y, Holloway L E, Fuligni F, Iafolla V and Natale G, 1990, *Nucl Instr Meth Phys Res* **A289** 518
Danzmann K, Rüdiger A, Schilling R, Winkler W, Hough J, Newton G P, Robertson D I, Robertson N A, Ward H, Bender P, Faller J, Hils D, Stebbins R, Edwards C D, Folkner W, Vincent M, Bernard A, Bertotti B, Brillet A, Man C N, Cruise M, Gray P, Sandford M, Drever R W P, Kose V, Kuhne M, Schutz B F, Weiss R and Welling H, 1993, *Max-Planck-Institut für Quantenoptic* Report MPQ 177

Drever R W P, 1983, Interferometric Detectors for Gravitational Radiation in *Proceedings of the Les Houches* NATO *Advanced Study Institute*, eds Piran T and Derouelle (North Holland Publishing Company, Amsterdam)

Forward R L, 1978, *Phys Rev D* **17** 379

Hough J, Newton G P, Robertson N A, Ward H, Campbell A M, Logan J E, Robertson D I, Strain K A, Danzmann K, Lück H, Rüdiger A, Schilling R, Schrempel M, Winkler W, Bennett J R J, Kose V, Kuhne M, Schutz B F, Nicholson D, Schuttleworth J, Welling H, Aufmuth P, Rinkleff R, Tünnermann A and Wilke B, 1994, GEO-600 — A 600m Laser Interferometric Gravitational Wave Antenna, to be published in *Proceedings of 7th Marcel Grossman Meeting* Stanford, USA

Kaspi V M, Taylor J H, and Ryba M F, 1994, *Astrophysical Journal* **428** 713

Meers B J, 1988, *Phys Rev D* **38** 2317

Robertson D I, Morrison E, Hough J, Killbourn S, Meers B J, Newton G P, Robertson N A, Strain K A and Ward H, 1995, accepted for publication in *Rev Sci Instr*

Savage R L Jr, 1994, Status of the LIGO 40-meter Interferometer, submitted to *Proceedings of 7th Marcel Grossman Meeting* Stanford, USA

Shoemaker D, Schilling R, Schnupp L, Winkler W, Maischberger K and Rudiger A, 1988, *Phys Rev D* **38** 423

Solomonson et al., 1992, in *Proceedings of the Sixth Marcel Grossman Meeting*, eds Sato H and Nakamura A (World Scientific, Singapore)

Taylor J H and Weisberg J M, 1982, *Astrophysical Journal* **253** 908

Taylor J H and Weisberg J M, 1989, *Astrophysical Journal* **345** 434

Weber J, 1969, *Phys Rev Lett* **22** 1320

Weiss R, 1972, *MIT Quart Progress Report no.105*

The Confrontation between General Relativity and Experiment: A 1995 Update

Clifford M Will

McDonnell Center for the Space Sciences
Department of Physics
Washington University in St. Louis

1 Introduction

At the time of the birth of general relativity, experimental confirmation was almost a side issue. Einstein did calculate observable effects of general relativity, such as the deflection of light, which were tested, but compared to the inner consistency and elegance of the theory, he regarded such empirical questions as almost peripheral. But today, experimental gravitation is a major component of the field, characterized by continuing efforts to test the theory's predictions, to search for gravitational imprints of high-energy particle interactions, and to detect gravitational waves from astronomical sources.

The modern history of experimental relativity can be divided roughly into four periods, Genesis, Hibernation, a Golden Era, and an Era of Opportunism. The Genesis (1887–1919) comprises the period of the two great experiments which were the foundation of relativistic physics—the Michelson-Morley experiment and the Eötvös experiment—and the two immediate confirmations of general relativity—the deflection of light and the perihelion advance of Mercury. Following this was a period of Hibernation (1920–1960) during which relatively few experiments were performed to test general relativity, and at the same time the field itself became sterile and stagnant, relegated to the backwaters of physics and astronomy.

But beginning around 1960, astronomical discoveries (quasars, pulsars, cosmic background radiation) and new experiments pushed general relativity to the forefront. Experimental gravitation experienced a Golden Era (1960–1980) during which a systematic, world-wide effort took place to understand the observable predictions of general relativity, to compare and contrast them with the predictions of alternative theories of gravity, and to perform new experiments to test them. The period began with an experiment to confirm the gravitational frequency shift of light (1960) and ended with the reported decrease in the orbital period of the binary pulsar at a rate consistent with the general relativity prediction of gravity-wave energy loss (1979). The results all supported general relativity, and most alternative theories of gravity fell by the wayside (for a popular review, see Will 1993a).

Since 1980, the field has entered what might be termed an Era of Opportunism. Many of the remaining interesting predictions of the theory are extremely small and difficult to check, in some cases requiring further technological development to bring them into detectable range. The sense of a systematic assault on the predictions of general relativity has been supplanted to some extent by an opportunistic approach in which novel and unexpected (and sometimes inexpensive) tests of gravity have arisen from new theoretical ideas or experimental techniques, often from unlikely sources. Examples include the use of laser-cooled atom and ion traps to perform ultra-precise tests of special relativity, and the startling proposal of a 'fifth' force, which led to a host of new tests of gravity at short ranges. Several major ongoing efforts also continue, including the Stanford Gyroscope experiment, and the program to develop sensitive detectors for gravitational radiation observatories. It is also an era when astronomical tests using binary pulsars have become important.

In these lectures, we shall review theoretical frameworks for studying experimental gravitation, summarize the current status of experiments, and attempt to chart the future of the subject. We shall not provide complete references to work done in this field but instead will refer the reader to the appropriate technical review articles and monographs, specifically to *Theory and Experiment in Gravitational Physics* (Will 1993b), hereafter referred to as TEGP. Additional recent reviews in this subject are Will (1987, 1992a), and Damour (1995). Other references will be confined to reviews or monographs on specific topics, and to important recent papers that are not included in TEGP. References to TEGP will be by chapter or section, *e.g.* 'TEGP 8.9'.

2 Tests of the foundations of gravitation theory

2.1 The Einstein equivalence principle

The principle of equivalence has historically played an important role in the development of gravitation theory. Newton regarded this principle as such a cornerstone of mechanics that he devoted the opening paragraph of the *Principia* to it. In 1907, Einstein used the principle as a basic element of general relativity. We now regard the principle of equivalence as the foundation, not of Newtonian gravity or of general relativity, but of the broader idea that spacetime is curved.

One elementary equivalence principle is the kind Newton had in mind when he stated that the property of a body called 'mass' is proportional to the 'weight', and is known as the weak equivalence principle (WEP). An alternative statement of WEP is that the trajectory of a freely falling body (one not acted upon by such forces as electromagnetism and too small to be affected by tidal gravitational forces) is independent of its internal structure and composition. In the simplest case of dropping two different bodies in a gravitational field, WEP states that the bodies fall with the same acceleration.

A more powerful and far-reaching equivalence principle is known as the Einstein equivalence principle (EEP). It states that (i) WEP is valid, (ii) the outcome of any local non-gravitational experiment is independent of the velocity of the freely-falling reference frame in which it is performed, and (iii) the outcome of any local non-gravitational experiment is independent of where and when in the universe it is performed. The second piece of EEP is called local Lorentz invariance (LLI), and the third piece is called local position invariance (LPI).

For example, a measurement of the electric force between two charged bodies is a local non-gravitational experiment; a measurement of the gravitational force between two bodies (Cavendish experiment) is not.

The Einstein equivalence principle is the heart and soul of gravitational theory, for it is possible to argue convincingly that if EEP is valid, then gravitation must be a 'curved spacetime' phenomenon, in other words, the effects of gravity must be equivalent to the effects of living in a curved spacetime. As a consequence of this argument, the only theories of gravity that can embody EEP are those that satisfy the postulates of 'metric theories of gravity', which are (i) spacetime is endowed with a symmetric metric, (ii) the trajectories of freely falling bodies are geodesics of that metric, and (iii) in local freely falling reference frames, the non-gravitational laws of physics are those written in the language of special relativity. The argument that leads to this conclusion simply notes that, if EEP is valid, then in local freely falling frames, the laws governing experiments must be independent of the velocity of the frame (local Lorentz invariance), with constant values for the various atomic constants (in order to be independent of location). The only laws we know of that fulfill this are those that are compatible with special relativity, such as Maxwell's equations of electromagnetism. Furthermore, in local freely falling frames, test bodies appear to be unaccelerated, in other words they move on straight lines; but such 'locally straight' lines simply correspond to 'geodesics' in a curved spacetime (TEGP 2.3).

General relativity is a metric theory of gravity, but then so are many others, including the Brans-Dicke theory. The nonsymmetric gravitation theory (NGT) of Moffat is not a metric theory. So the notion of curved spacetime is a very general and fundamental one, and therefore it is important to test the various aspects of the Einstein Equivalence Principle thoroughly.

A direct test of WEP is the comparison of the acceleration of two laboratory-sized bodies of different composition in an external gravitational field. If the principle were violated, then the accelerations of different bodies would differ. The simplest way to quantify such possible violations of WEP in a form suitable for comparison with experiment is to suppose that for a body with inertial mass m_I, the passive gravitational mass m_P is no longer equal to m_I, so that in a gravitational field g, the acceleration is given by $m_I a = m_P g$. Now the inertial mass of a typical laboratory body is made up of several types of mass-energy: rest energy, electromagnetic energy, weak-interaction energy, and so on. If one of these forms of energy contributes to m_P differently than it does to m_I, a violation of WEP would result. One could then write

$$m_P = m_I + \sum_A \eta^A E^A/c^2 \qquad (1)$$

where E^A is the internal energy of the body generated by interaction A, and η^A is a dimensionless parameter that measures the strength of the violation of WEP induced by that interaction, and c is the speed of light. A measurement or limit on the fractional difference in acceleration between two bodies then yields a quantity called the 'Eötvös ratio' given by

$$\eta \equiv \frac{2|a_1 - a_2|}{|a_1 + a_2|} = \sum_A \eta^A \left(\frac{E_1^A}{m_1 c^2} - \frac{E_2^A}{m_2 c^2} \right) \qquad (2)$$

where we drop the subscript I from the inertial masses. Thus, experimental limits on η place limits on the WEP-violation parameters η^A.

Many high-precision Eötvös-type experiments have been performed, from the pendulum experiments of Newton, Bessel and Potter, to the classic torsion-balance measurements of Eötvös, Dicke, Braginsky and their collaborators. In the modern torsion-balance experiments, two objects of different composition are connected by a rod or placed on a tray and suspended in a horizontal orientation by a fine wire. If the gravitational acceleration of the bodies differs, there will be a torque induced on the suspension wire, related to the angle between the wire and the direction of the gravitational acceleration **g**. If the entire apparatus is rotated about some direction with angular velocity ω, the torque will be modulated with period $2\pi/\omega$. In the experiments of Eötvös and his collaborators, the wire and **g** were not quite parallel because of the centripetal acceleration on the apparatus due to the Earth's rotation; the apparatus was rotated about the direction of the wire. In the Princeton (Dicke et al.), and Moscow (Braginsky et al.) experiments, **g** was that of the Sun, and the rotation of the Earth provided the modulation of the torque at a period of 24 hr (TEGP 2.4). Beginning in the late 1980s, numerous experiments were carried out primarily to search for a 'fifth-force' (see below), but their null results also constituted tests of WEP. In the 'free-fall Galileo experiment' performed at the University of Colorado, the relative free-fall acceleration of two bodies made of uranium and copper was measured using a laser interferometric technique. The 'Eöt-Wash' experiment carried out at the University of Washington used a sophisticated torsion balance tray to compare the accelerations of beryllium and copper toward the Earth, the Sun and the galaxy (Su et al. 1995). The resulting upper limits on η are summarized in Figure 1 (TEGP 14.1; for a bibliography of experiments, see Fischbach et al. (1992)).

The second ingredient of EEP, local Lorentz invariance, can be said to be tested every time that special relativity is confirmed in the laboratory. However, many such experiments, especially in high-energy physics, are not 'clean' tests, because in many cases it is unlikely that a violation of Lorentz invariance could be distinguished from effects due to the complicated strong and weak interactions.

However, there is one class of experiments that can be interpreted as 'clean', high-precision tests of local Lorentz invariance. These are the 'mass anisotropy' experiments:

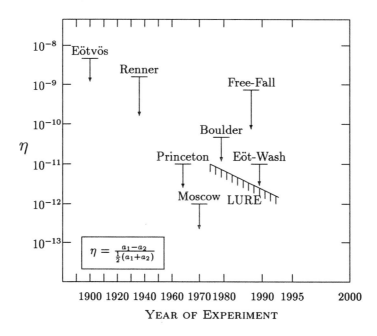

Figure 1. *Selected tests of the weak equivalence principle, showing bounds on η, which measures fractional difference in acceleration of different materials or bodies. Free-fall and 'Eöt-Wash' experiments originally performed to search for fifth force. Hatched line shows current bounds on η for gravitating bodies from lunar laser ranging (LURE).*

the classic versions are the Hughes-Drever experiments, performed in the period 1959-60 independently by Hughes and collaborators at Yale University, and by Drever at Glasgow University (TEGP 2.4(b)). Dramatically improved versions were carried out during the late 1980s using laser-cooled trapped atom techniques (TEGP 14.1). A simple and useful way of interpreting these experiments is to suppose that the electromagnetic interactions suffer a slight violation of Lorentz invariance, through a change in the speed of electromagnetic radiation c relative to the limiting speed of material test particles (c_0, chosen to be unity via a choice of units), in other words, $c \neq 1$ (see below). Such a violation necessarily selects a preferred universal rest frame, presumably that of the cosmic background radiation, through which we are moving at about 300 km/s. Such a Lorentz-non-invariant electromagnetic interaction would cause shifts in the energy levels of atoms and nuclei that depend on the orientation of the quantization axis of the state relative to our universal velocity vector, and on the quantum numbers of the state. The presence or absence of such energy shifts can be examined by measuring the energy of one such state relative to another state that is either unaffected or is affected

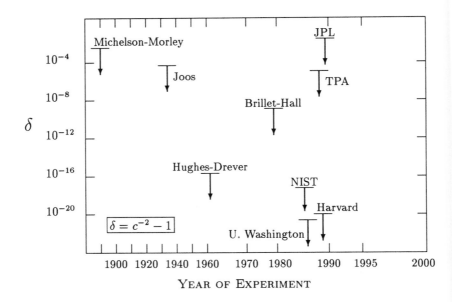

Figure 2. *Selected tests of local Lorentz invariance showing bounds on parameter δ, which measures degree of violation of Lorentz invariance in electromagnetism. Michelson-Morley, Joos, and Brillet-Hall experiments test isotropy of round-trip speed of light, the latter experiment using laser technology. Two-photon absorption (TPA) and JPL experiments test isotropy of light speed using one-way propagation. Remaining four experiments test isotropy of nuclear energy levels. Limits assume speed of Earth of 300 km/s relative to the mean rest frame of the universe.*

differently by the supposed violation. One way is to look for a shifting of the energy levels of states that are ordinarily equally spaced, such as the four $J=3/2$ ground states of the ^{7}Li nucleus in a magnetic field (Drever experiment); another is to compare the levels of a complex nucleus with the atomic hyperfine levels of a hydrogen maser clock. These experiments have all yielded extremely accurate results, quoted as limits on the parameter $\delta \equiv c^{-2} - 1$ in Figure 2. Also included for comparison is the corresponding limit obtained from Michelson-Morley type experiments.

Recent advances in atomic spectroscopy and atomic timekeeping have made it possible to test LLI by checking the isotropy of the speed of light using one-way propagation (as opposed to round-trip propagation, as in the Michelson-Morley experiment). In one experiment, for example, the relative phases of two hydrogen maser clocks at two stations of NASA's Deep Space Tracking Network were compared over five rotations of the Earth by propagating a light signal one-way along an ultrastable fiberoptic link connecting them (see below). Although the bounds from these experiments are not as tight as those from mass-anisotropy experiments, they probe directly the fundamental postulates of special relativity, and thereby of LLI. (TEGP 14.1).

The principle of local position invariance, the third part of EEP, can be tested by the gravitational redshift experiment, the first experimental test of gravitation proposed by Einstein. Despite the fact that Einstein regarded this as a crucial test of general relativity, we now realize that it does not distinguish between general relativity and any other metric theory of gravity, instead is a test only of EEP. A typical gravitational redshift experiment measures the frequency or wavelength shift $Z \equiv \Delta\nu/\nu = -\Delta\lambda/\lambda$ between two identical frequency standards (clocks) placed at rest at different heights in a static gravitational field. If the frequency of a given type of atomic clock is the same when measured in a local, momentarily comoving freely falling frame (Lorentz frame), independent of the location or velocity of that frame, then the comparison of frequencies of two clocks at rest at different locations boils down to a comparison of the velocities of two local Lorentz frames, one at rest with respect to one clock at the moment of emission of its signal, the other at rest with respect to the other clock at the moment of reception of the signal. The frequency shift is then a consequence of the first-order Doppler shift between the frames. The structure of the clock plays no role whatsoever. The result is a shift

$$Z = \Delta U/c^2 \qquad (3)$$

where ΔU is the difference in the Newtonian gravitational potential between the receiver and the emitter. If LPI is not valid, then it turns out that the shift can be written

$$Z = (1+\alpha)\Delta U/c^2 \qquad (4)$$

where the parameter α may depend upon the nature of the clock whose shift is being measured (see TEGP 2.4(c) for details).

The first successful, high-precision redshift measurement was the series of Pound-Rebka-Snider experiments of 1960-1965, that measured the frequency shift of gamma-ray photons from ^{57}Fe as they ascended or descended the Jefferson Physical Laboratory tower at Harvard University. The high accuracy achieved–one percent–was obtained by making use of the Mössbauer effect to produce a narrow resonance line whose shift could be accurately determined. Other experiments since 1960 measured the shift of spectral lines in the Sun's gravitational field and the change in rate of atomic clocks transported aloft on aircraft, rockets and satellites. Figure 3 summarizes the important redshift experiments that have been performed since 1960 (TEGP 2.4(c)).

The most precise experiment to date was the Vessot-Levine rocket redshift experiment that took place in June 1976. A hydrogen-maser clock was flown on a rocket to an altitude of about 10,000 km and its frequency compared to a similar clock on the ground. The experiment took advantage of the masers' frequency stability by monitoring the frequency shift as a function of altitude. A sophisticated data acquisition scheme accurately eliminated all effects of the first-order Doppler shift due to the rocket's motion, while tracking data were used to determine the payload's location and the velocity (to evaluate the potential difference ΔU, and the special relativistic time dilation). Analysis of the data yielded a limit $|\alpha| < 2\times 10^{-4}$.

A 'null' redshift experiment performed in 1978 tested whether the *relative* rates of two different clocks depended upon position. Two hydrogen maser clocks and an ensemble of three superconducting-cavity stabilized oscillator (SCSO) clocks were compared over a 10-day period. During this period, the solar potential U/c^2 changed sinusoidally

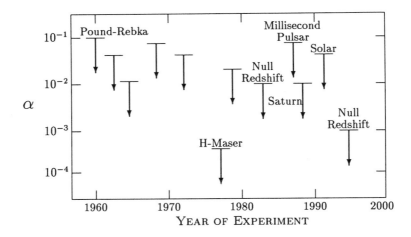

Figure 3. *Selected tests of local position invariance via gravitational redshift experiments, showing bounds on α, which measures degree of deviation of redshift from the formula $\Delta\nu/\nu = \Delta U/c^2$.*

with a 24-hour period by 3×10^{-13} because of the Earth's rotation, and changed linearly at 3×10^{-12} per day because the Earth is 90 degrees from perihelion in April. However, analysis of the data revealed no variations of either type within experimental errors, leading to a limit on the LPI violation parameter $|\alpha^H - \alpha^{SCSO}| < 2\times10^{-2}$. This bound is likely to be improved using more stable frequency standards (Godone et al. 1995, Prestage et al. 1995). The varying gravitational redshift of Earth-bound clocks relative to the highly stable Millisecond Pulsar, caused by the Earth's motion in the solar gravitational field around the Earth-Moon center of mass (amplitude 4000 km), has been measured to about 10 percent, and the redshift of stable oscillator clocks on the Voyager spacecraft caused by Saturn's gravitational field yielded a one percent test. The solar gravitational redshift has been tested to about two percent using infrared oxygen triplet lines at the limb of the Sun, and to one percent using oscillator clocks on the Galileo spacecraft (TEGP 2.4(c) and 14.1(a)).

Modern advances in navigation using Earth-orbiting atomic clocks and accurate time-transfer must routinely take gravitational redshift and time-dilation effects into account. For example, the Global Positioning System (GPS) provides absolute accuracies of around 15 m (even better in its military mode) anywhere on Earth, which corresponds to 50 nanoseconds in time accuracy at all times. Yet the difference in rate between satellite and ground clocks as a result of special and general relativistic effects is a whopping 40 *microseconds* per day. If these effects were not accurately accounted for, GPS would fail to function at its stated accuracy. This represents a welcome practical application of general relativity!

For other tests of local position invariance, including the constancy of the fundamental non-gravitational constants, see TEGP 2.4(c).

2.2 Theoretical frameworks for analyzing EEP

Schiff's conjecture

Because the three parts of the Einstein equivalence principle discussed above are so very different in their empirical consequences, it is tempting to regard them as independent theoretical principles. On the other hand, any complete and self-consistent gravitation theory must possess sufficient mathematical machinery to make predictions for the outcomes of experiments that test each principle, and because there are limits to the number of ways that gravitation can be meshed with the special relativistic laws of physics, one might not be surprised if there were theoretical connections between the three sub-principles. For instance, the same mathematical formalism that produces equations describing the free fall of a hydrogen atom must also produce equations that determine the energy levels of hydrogen in a gravitational field, and thereby the ticking rate of a hydrogen maser clock. Hence a violation of EEP in the fundamental machinery of a theory that manifests itself as a violation of WEP might also be expected to show up as a violation of local position invariance. Around 1960, Schiff conjectured that this kind of connection was a necessary feature of any self-consistent theory of gravity. More precisely, Schiff's conjecture states that *any complete, self-consistent theory of gravity that embodies* WEP *necessarily embodies* EEP. In other words, the validity of WEP alone guarantees the validity of local Lorentz and position invariance, and thereby of EEP.

If Schiff's conjecture is correct, then Eötvös experiments may be seen as the direct empirical foundation for EEP, hence for the interpretation of gravity as a curved-spacetime phenomenon. Of course, a rigorous proof of such a conjecture is impossible (indeed, some special counter-examples are known) yet a number of powerful 'plausibility' arguments can be formulated.

The most general and elegant of these arguments is based upon the assumption of energy conservation. This assumption allows one to perform very simple cyclic gedanken experiments in which the energy at the end of the cycle must equal that at the beginning of the cycle. This approach was pioneered by Dicke, Nordtvedt and Haugan. A system in a quantum state A decays to state B, emitting a quantum of frequency ν. The quantum falls a height H in an external gravitational field and is shifted to frequency ν', while the system in state B falls with acceleration g_B. At the bottom, state A is rebuilt out of state B, the quantum of frequency ν', and the kinetic energy $m_B g_B H$ that state B has gained during its fall. The energy left over must be exactly enough, $m_A g_A H$, to raise state A to its original location. (Here an assumption of local Lorentz invariance permits the inertial masses m_A and m_B to be identified with the total energies of the bodies.) If g_A and g_B depend on that portion of the internal energy of the states that was involved in the quantum transition from A to B according to

$$g_A = g(1 + \alpha E_A/m_A c^2) \qquad g_B = g(1 + \alpha E_B/m_B c^2) \qquad E_A - E_B \equiv h\nu \qquad (5)$$

(violation of WEP) then by conservation of energy, there must be a corresponding violation of LPI in the frequency shift of the form (to lowest order in $h\nu/mc^2$)

$$Z = (\nu' - \nu)/\nu' = (1 + \alpha)gH/c^2 = (1 + \alpha)\Delta U/c^2. \qquad (6)$$

Haugan generalized this approach to include violations of LLI (TEGP 2.5).

Box 1. The $TH\epsilon\mu$ Formalism

1. **Coordinate System and Conventions:**

   $x^0 = t$ = time coordinate associated with the static nature of the static spherically symmetric (SSS) gravitational field; $\mathbf{x} = (x, y, z)$ = isotropic quasi-Cartesian spatial coordinates; spatial vector and gradient operations as in Cartesian space.

2. **Matter and Field Variables:**

   - m_{0a} = rest mass of particle a.
   - e_a = charge of particle a.
   - $x_a^\mu(t)$ = world line of particle a.
   - $v_a^\mu = dx_a^\mu/dt$ = coordinate velocity of particle a.
   - A_μ = electromagnetic vector potential; $\mathbf{E} = \nabla A_0 - \partial \mathbf{A}/\partial t$, $\mathbf{B} = \nabla \times \mathbf{A}$

3. **Gravitational Potential:** $U(\mathbf{x})$

4. **Arbitrary Functions:**

   $T(U)$, $H(U)$, $\epsilon(U)$, $\mu(U)$; EEP is satisfied iff $\epsilon = \mu = (H/T)^{1/2}$ for all U.

5. **Action:**

   $$I = -\sum_a m_{0a} \int (T - Hv_a^2)^{1/2} dt + \sum_a e_a \int A_\mu(x_a^\nu) v_a^\mu dt + (8\pi)^{-1} \int (\epsilon E^2 - \mu^{-1} B^2) d^4x.$$

6. **Non-Metric Parameters:**

   $$\Gamma_0 = -c_0^2 (\partial/\partial U) \ln[\epsilon(T/H)^{1/2}]_0$$
   $$\Lambda_0 = -c_0^2 (\partial/\partial U) \ln[\mu(T/H)^{1/2}]_0$$
   $$\Upsilon_0 = 1 - (TH^{-1}\epsilon\mu)_0$$

   where $c_0 = (T_0/H_0)^{1/2}$ and subscript '0' refers to a chosen point in space. If EEP is satisfied, $\Gamma_0 \equiv \Lambda_0 \equiv \Upsilon_0 \equiv 0$.

The $TH\epsilon\mu$ formalism

The first successful attempt to prove Schiff's conjecture more formally was made by Lightman and Lee. They developed a framework called the $TH\epsilon\mu$ formalism that encompasses all metric theories of gravity and many non-metric theories (Box 1). It restricts attention to the behaviour of charged particles (electromagnetic interactions only) in an external static spherically symmetric (SSS) gravitational field, described by a potential U. It characterizes the motion of the charged particles in the external potential by two arbitrary functions $T(U)$ and $H(U)$, and characterizes the response of electromagnetic fields to the external potential (gravitationally modified Maxwell equations) by two functions $\epsilon(U)$ and $\mu(U)$. The forms of T, H, ϵ and μ vary from theory to theory, but every metric theory satisfies

$$\epsilon = \mu = (H/T)^{1/2} \tag{7}$$

for all U. This consequence follows from the action of electrodynamics with a 'minimal' or metric coupling:

$$\begin{aligned} I &= -\sum_a m_{0a} \int (g_{\mu\nu} v_a^\mu v_a^\nu)^{1/2} dt + \sum_a e_a \int A_\mu(x_a^\nu) v_a^\mu dt \\ &- \frac{1}{16\pi} \int \sqrt{-g} g^{\mu\alpha} g^{\nu\beta} F_{\mu\nu} F_{\alpha\beta} d^4x \end{aligned} \tag{8}$$

where the variables are defined in Box 1, and where $F_{\mu\nu} \equiv A_{\nu,\mu} - A_{\mu,\nu}$. By identifying $g_{00} = T$ and $g_{ij} = H\delta_{ij}$ in a SSS field, $F_{i0} = E_i$ and $F_{ij} = \epsilon_{ijk} B_k$, one obtains Equation 7.

Conversely, every theory within this class that satisfies Equation 7 can have its electrodynamic equations cast into 'metric' form. Lightman and Lee then calculated explicitly the rate of fall of a 'test' body made up of interacting charged particles, and found that the rate was independent of the internal electromagnetic structure of the body (WEP) if and only if Equation 7 was satisfied. In other words WEP → EEP and Schiff's conjecture was verified, at least within the restrictions built into the formalism.

Certain combinations of the functions T, H, ϵ and μ reflect different aspects of EEP. For instance, position or U-dependence of either of the combinations $\epsilon(T/H)^{1/2}$ and $\mu(T/H)^{1/2}$ signals violations of LPI, the first combination playing the role of the locally measured electric charge or fine structure constant. The 'non-metric parameters' Γ_0 and Λ_0 (Box 1) are measures of such violations of EEP. Similarly, if the parameter $\Upsilon_0 \equiv 1 - (TH^{-1}\epsilon\mu)_0$ is non-zero anywhere, then violations of LLI will occur. This parameter is related to the difference between the speed of light, c, and the limiting speed of material test particles, c_o, given by

$$c = (\epsilon_0 \mu_0)^{-1/2} \qquad c_o = (T_0/H_0)^{1/2}. \tag{9}$$

In many applications, by suitable definition of units, c_0 can be set equal to unity. (compare with Figure 2). If EEP is valid, $\Gamma_0 \equiv \Lambda_0 \equiv \Upsilon_0 = 0$ everywhere.

The rate of fall of a composite spherical test body of electromagnetically interacting particles then has the form

$$\mathbf{a} = (m_P/m)\nabla U \tag{10}$$

$$m_P/m = 1 + (E_B^{ES}/Mc_0^2)[2\Gamma_0 - \frac{8}{3}\Upsilon_0] + (E_B^{MS}/Mc_0^2)[2\Lambda_0 - \frac{4}{3}\Upsilon_0] + \ldots \tag{11}$$

where E_B^{ES} and E_B^{MS} are the electrostatic and magnetostatic binding energies of the body, given by

$$E_B^{ES} = -\frac{1}{4} T_0^{1/2} H_0^{-1} \epsilon_0^{-1} \left\langle \sum_{ab} e_a e_b r_{ab}^{-1} \right\rangle \tag{12}$$

$$E_B^{MS} = -\frac{1}{8} T_0^{1/2} H_0^{-1} \mu_0 \left\langle \sum_{ab} e_a e_b r_{ab}^{-1} [\mathbf{v}_a \cdot \mathbf{v}_b + (\mathbf{v}_a \cdot \mathbf{n}_{ab})(\mathbf{v}_b \cdot \mathbf{n}_{ab})] \right\rangle \tag{13}$$

where $r_{ab} = |\mathbf{x}_a - \mathbf{x}_b|$, $\mathbf{n}_{ab} = (\mathbf{x}_a - \mathbf{x}_b)/r_{ab}$, and the angle brackets denote an expectation value of the enclosed operator for the system's internal state. Eötvös experiments place limits on the WEP-violating terms in Equation 11, and ultimately place limits on the non-metric parameters $|\Gamma_0| < 2\times10^{-10}$ and $|\Lambda_0| < 3\times10^{-6}$. (We set $\Upsilon_0 = 0$ because of very tight constraints on it from tests of LLI–see below.) These limits are sufficiently tight to rule out a number of non-metric theories of gravity thought previously to be viable (TEGP 2.6(f)).

The $TH\epsilon\mu$ formalism also yields a gravitationally modified Dirac equation that can be used to determine the gravitational redshift experienced by a variety of atomic clocks. For the redshift parameter α (Equation 4), the results are (TEGP 2.6(c))

$$\alpha = \begin{cases} -3\Gamma_0 + \Lambda_0 & \text{hydrogen hyperfine transition, H} - \text{Maser clock} \\ -\frac{1}{2}(3\Gamma_0 + \Lambda_0) & \text{electromagnetic mode in cavity, SCSO clock} \\ -2\Gamma_0 & \text{phonon mode in solid, principal transition in hydrogen.} \end{cases} \quad (14)$$

The redshift is the standard one ($\alpha = 0$), independently of the nature of the clock if and only if $\Gamma_0 \equiv \Lambda_0 \equiv 0$. Thus the Vessot-Levine rocket redshift experiment sets a limit on the parameter combination $|3\Gamma_0 - \Lambda_0|$ (Figure 3); the null-redshift experiment comparing hydrogen-maser and SCSO clocks sets a limit on $|\alpha_H - \alpha_{SCSO}| = \frac{3}{2}|\Gamma_0 - \Lambda_0|$.

The c^2 formalism

The $TH\epsilon\mu$ formalism can also be applied to tests of local Lorentz invariance, but in this context it can be simplified. Since most such tests do not concern themselves with the spatial variation of the functions T, H, ϵ, and μ, but rather with observations made in moving frames, we can treat them as spatial constants. Then by rescaling the time and space coordinates, the charges and the electromagnetic fields, we can put the action in Box 1 into the form (TEGP 2.6(a)).

$$I = -\sum_a m_{0a} \int (1-v_a^2)^{1/2} dt + \sum_a e_a \int A_\mu(x_a^\nu) v_a^\mu dt + (8\pi)^{-1} \int (E^2 - c^2 B^2) d^4x \quad (15)$$

where $c^2 \equiv H_0/T_0\epsilon_0\mu_0 = (1-\Upsilon_0)^{-1}$. This amounts to using units in which the limiting speed c_o of massive test particles is unity, and the speed of light is c. If $c \neq 1$, LLI is violated; furthermore, the form of the action above must be assumed to be valid only in some preferred universal rest frame. The natural candidate for such a frame is the rest frame of the microwave background.

The electrodynamical equations which follow from Equation 15 yield the behavior of rods and clocks, just as in the full $TH\epsilon\mu$ formalism. For example, the length of a rod moving through the rest frame in a direction parallel to its length will be observed by a rest observer to be contracted relative to an identical rod perpendicular to the motion by a factor $1 - V^2/2 + O(V^4)$. Notice that c does not appear in this expression. The energy and momentum of an electromagnetically bound body which moves with velocity **V** relative to the rest frame are given by

$$\begin{aligned} E &= M_R + \frac{1}{2}M_R V^2 + \frac{1}{2}\delta M_I^{ij} V^i V^j \\ P^i &= M_R V^i + \delta M_I^{ij} V^j \end{aligned} \quad (16)$$

where $M_R = M_0 - E_B^{ES}$, M_0 is the sum of the particle rest masses, E_B^{ES} is the electrostatic binding energy of the system (Equation 12 with $T_0^{1/2} H_0 \epsilon_0^{-1} = 1$), and

$$\delta M_I^{ij} = -2(\frac{1}{c^2} - 1)[\frac{4}{3}E_B^{ES}\delta^{ij} + \tilde{E}_B^{ESij}] \tag{17}$$

where

$$\tilde{E}_B^{ESij} = -\frac{1}{4}\left\langle \sum_{ab} e_a e_b r_{ab}^{-1}\left((n_{ab}^i n_{ab}^j - \frac{1}{3}\delta^{ij})\right)\right\rangle. \tag{18}$$

Note that $(c^{-2} - 1)$ corresponds to the parameter δ plotted in Figure 2.

The electrodynamics given by Equation 15 can also be quantized, so that we may treat the interaction of photons with atoms via perturbation theory. The energy of a photon is $\hbar$ times its frequency ω, while its momentum is $\hbar\omega/c$. Using this approach, one finds that the difference in round trip travel times of light along the two arms of the interferometer in the Michelson-Morley experiment is given by $L_0(v^2/c)(c^{-2} - 1)$. The experimental null result then leads to the bound on $(c^{-2} - 1)$ shown on Figure 2. Similarly the anisotropy in energy levels is clearly illustrated by the tensorial term in Equation 18; by evaluating $\tilde{E}_B^{ESij}$ for each nucleus in the various Hughes-Drever-type experiments and comparing with the experimental limits on energy differences, one obtains the extremely tight bounds also shown on Figure 2.

The behavior of moving atomic clocks can also be analysed in detail, and bounds on $(c^{-2}-1)$ can be placed using results from tests of time dilation and of the propagation of light. In some cases, it is advantageous to combine the c^2 framework with a 'kinematical' viewpoint that treats a general class of boost transformations between moving frames. Such kinematical approaches have been discussed by Robertson, Mansouri and Sexl, and Will (see Will 1992b).

For example, in the 'JPL' experiment, in which the phases of two hydrogen masers connected by a fiberoptic link were compared as a function of the Earth's orientation, the predicted phase difference as a function of direction is, to first order in $\mathbf{V}$, the velocity of the Earth through the cosmic background,

$$\Delta\phi/\tilde{\phi} \approx -\frac{4}{3}(1 - c^2)(\mathbf{V} \cdot \mathbf{n} - \mathbf{V} \cdot \mathbf{n}_0) \tag{19}$$

where $\tilde{\phi} = 2\pi\nu L$, ν is the maser frequency, $L = 21$ km is the baseline, and where $\mathbf{n}$ and $\mathbf{n}_0$ are unit vectors along the direction of propagation of the light, at a given time, and at the initial time of the experiment, respectively. The observed limit on a diurnal variation in the relative phase resulted in the bound $|c^{-2}-1| < 3\times10^{-4}$. Tighter bounds were obtained from a 'two-photon absorption' (TPA) experiment, and a 1960s series of 'Mössbauer-rotor' experiments, which tested the isotropy of time dilation between a gamma ray emitter on the rim of a rotating disk and an absorber placed at the center (Will 1992b).

3 Tests of post-Newtonian gravity

3.1 Metric theories of gravity: the strong equivalence principle

Universal coupling and the metric postulates

The overwhelming empirical evidence supporting the Einstein equivalence principle, discussed in the previous section, has convinced many theorists that only metric theories of gravity have a hope of being completely viable. Therefore for the remainder of these lectures, we shall turn our attention exclusively to metric theories of gravity, which assume that (i) there exists a symmetric metric, (ii) test bodies follow geodesics of the metric, and (iii) in local Lorentz frames, the non-gravitational laws of physics are those of special relativity.

The property that all non-gravitational fields should couple in the same manner to a single gravitational field is sometimes called 'universal coupling'. Because of it, one can discuss the metric as a property of spacetime itself rather than as a field over spacetime. This is because its properties may be measured and studied using a variety of different experimental devices, composed of different non-gravitational fields and particles, and, because of universal coupling, the results will be independent of the device. Thus, for instance, the proper time between two events is a characteristic of spacetime and of the location of the events, not of the clocks used to measure it.

Consequently, if EEP is valid, the non-gravitational laws of physics may be formulated by taking their special relativistic forms in terms of the Minkowski metric η and simply 'going over' to new forms in terms of the curved spacetime metric g, using the mathematics of differential geometry. The details of this 'going over' can be found in standard textbooks (Misner, et al. 1973, Weinberg 1972, TEGP 3.2)

The strong equivalence principle

In any metric theory of gravity, matter and non-gravitational fields respond only to the spacetime metric g. In principle, however, there could exist other gravitational fields besides the metric, such as scalar fields, vector fields, and so on. If matter does not couple to these fields what can their role in gravitation theory be? Their role must be that of mediating the manner in which matter and non-gravitational fields generate gravitational fields and produce the metric; once determined, however, the metric alone acts back on the matter in the manner prescribed by EEP.

What distinguishes one metric theory from another, therefore, is the number and kind of gravitational fields it contains in addition to the metric, and the equations that determine the structure and evolution of these fields. From this viewpoint, one can divide all metric theories of gravity into two fundamental classes: 'purely dynamical' and 'prior-geometric'.

By 'purely dynamical metric theory' we mean any metric theory whose gravitational fields have their structure and evolution determined by coupled partial differential field equations. In other words, the behavior of each field is influenced to some extent by a coupling to at least one of the other fields in the theory. By 'prior geometric' theory, we mean any metric theory that contains 'absolute elements', fields or equations whose structure and evolution are given *a priori*, and are independent of the structure and evolution of the other fields of the theory. These 'absolute elements' typically include flat background metrics η, cosmic time coordinates t, algebraic relationships among otherwise dynamical fields, such as $g_{\mu\nu} = h_{\mu\nu}+k_\mu k_\nu$, where $h_{\mu\nu}$ and k_μ may be dynamical fields.

General relativity is a purely dynamical theory since it contains only one gravitational field, the metric itself, and its structure and evolution are governed by partial differential equations (Einstein's equations). Brans-Dicke theory and its generalizations are purely dynamical theories; the field equation for the metric involves the scalar field (as well as the matter as source), and that for the scalar field involves the metric. Rosen's bimetric theory is a prior-geometric theory: it has a non-dynamical, Riemann-flat background metric, η, and the field equations for the physical metric **g** involve η.

By discussing metric theories of gravity from this broad point of view, it is possible to draw some general conclusions about the nature of gravity in different metric theories, conclusions that are reminiscent of the Einstein equivalence principle, but that are subsumed under the name 'strong equivalence principle'.

Consider a local, freely falling frame in any metric theory of gravity. Let this frame be small enough that inhomogeneities in the external gravitational fields can be neglected throughout its volume. On the other hand, let the frame be large enough to encompass a system of gravitating matter and its associated gravitational fields. The system could be a star, a black hole, the solar system or a Cavendish experiment. Call this frame a 'quasi-local Lorentz frame' . To determine the behavior of the system we must calculate the metric. The computation proceeds in two stages. First we determine the external behavior of the metric and gravitational fields, thereby establishing boundary values for the fields generated by the local system, at a boundary of the quasi-local frame 'far' from the local system. Second, we solve for the fields generated by the local system. But because the metric is coupled directly or indirectly to the other fields of the theory, its structure and evolution will be influenced by those fields, and in particular by the boundary values taken on by those fields far from the local system. This will be true even if we work in a coordinate system in which the asymptotic form of $g_{\mu\nu}$ in the boundary region between the local system and the external world is that of the Minkowski metric. Thus the gravitational environment in which the local gravitating system resides can influence the metric generated by the local system via the boundary values of the auxiliary fields. Consequently, the results of local gravitational experiments may depend on the location and velocity of the frame relative to the external environment. Of course, local *non*-gravitational experiments are unaffected since the gravitational fields they generate are assumed to be negligible, and since those experiments couple only to the metric, whose form can always be made locally Minkowskian at a given spacetime event. Local gravitational experiments might include Cavendish experiments, measurement of the acceleration of massive self-gravitating bodies, studies of the structure of stars and planets, or analyses of the periods of 'gravitational clocks'. We can now make several statements about different kinds of metric theories.

(i) A theory which contains only the metric **g** yields local gravitational physics which is independent of the location and velocity of the local system. This follows from the fact that the only field coupling the local system to the environment is **g**, and it is always possible to find a coordinate system in which **g** takes the Minkowski form at the boundary between the local system and the external environment. Thus the asymptotic values of $g_{\mu\nu}$ are constants independent of location, and are asymptotically Lorentz invariant, thus independent of velocity. General relativity is an example of such a theory.

(ii) A theory which contains the metric **g** and dynamical scalar fields φ_A yields local gravitational physics which may depend on the location of the frame but which is independent of the velocity of the frame. This follows from the asymptotic Lorentz invariance of the Minkowski metric and of the scalar fields, but now the asymptotic values of the scalar fields may depend on the location of the frame. An example is Brans-Dicke theory, where the asymptotic scalar field determines the effective value of the gravitational constant, which can thus vary as φ varies.

(iii) A theory which contains the metric **g** and additional dynamical vector or tensor fields or prior-geometric fields yields local gravitational physics which may have both location and velocity-dependent effects.

These ideas can be summarized in the strong equivalence principle (SEP), which states that (i) WEP is valid for self-gravitating bodies as well as for test bodies, (ii) the outcome of any local test experiment is independent of the velocity of the (freely falling) apparatus, and (iii) the outcome of any local test experiment is independent of where and when in the universe it is performed. The distinction between SEP and EEP is the inclusion of bodies with self-gravitational interactions (planets, stars) and of experiments involving gravitational forces (Cavendish experiments, gravimeter measurements). Note that SEP contains EEP as the special case in which local gravitational forces are ignored.

The above discussion of the coupling of auxiliary fields to local gravitating systems indicates that if SEP is strictly valid, there must be one and only one gravitational field in the universe, the metric **g**. These arguments are only suggestive however, and no rigorous proof of this statement is available at present. Empirically it has been found that every metric theory other than general relativity introduces auxiliary gravitational fields, either dynamical or prior geometric, and thus predicts violations of SEP at some level (here we ignore quantum-theory inspired modifications to general relativity involving 'R^2' terms). General relativity seems to be the only metric theory that embodies SEP completely. This lends some credence to the conjecture SEP $\rightarrow$ General Relativity. In Sec. 3.6, we shall discuss experimental evidence for the validity of SEP.

3.2 The parametrized post-Newtonian formalism

Despite the possible existence of long-range gravitational fields in addition to the metric in various metric theories of gravity, the postulates of those theories demand that matter and non-gravitational fields be completely oblivious to them. The only gravitational field that enters the equations of motion is the metric **g**. The role of the other fields that a theory may contain can only be that of helping to generate the spacetime curvature associated with the metric. Matter may create these fields, and they plus the matter may generate the metric, but they cannot act back directly on the matter. Matter responds only to the metric.

Parameter	What it measures relative to general relativity	General relativity	Semi-conservative theories	Fully-conservative theories
γ	Amount of space-curvature produced by unit rest mass	1	γ	γ
β	Amount of 'nonlinearity' in the superposition law for gravity	1	β	β
ξ	Preferred-location effects	0	ξ	ξ
α_1		0	α_1	0
α_2	Preferred-frame effects	0	α_2	0
α_3		0	0	0
α_3		0	0	0
ζ_1		0	0	0
ζ_2	Violation of conservation of total momentum	0	0	0
ζ_3		0	0	0
ζ_4		0	0	0

Table 1. *The PPN Parameters and their significance. (Note that α_3 has been included twice to show that it is a measure of two effects.)*

Thus the metric and the equations of motion for matter become the primary entities for calculating observable effects, and all that distinguishes one metric theory from another is the particular way in which matter and possibly other gravitational fields generate the metric.

The comparison of metric theories of gravity with each other and with experiment becomes particularly simple when one takes the slow-motion, weak-field limit. This approximation, known as the post-Newtonian limit, is sufficiently accurate to encompass most solar-system tests that can be performed in the foreseeable future. It turns out that, in this limit, the spacetime metric **g** predicted by nearly every metric theory of gravity has the same structure. It can be written as an expansion about the Minkowski metric ($\eta_{\mu\nu} = \text{diag}(-1,1,1,1)$) in terms of dimensionless gravitational potentials of varying degrees of smallness. These potentials are constructed from the matter variables (Box 2) in imitation of the Newtonian gravitational potential

$$U(\mathbf{x},t) \equiv \int \rho(\mathbf{x}',t)|\mathbf{x}-\mathbf{x}'|^{-1}d^3x'. \quad (20)$$

The 'order of smallness' is determined according to the rules $U \sim v^2 \sim \Pi \sim p/\rho \sim O(2)$, $v^i \sim |d/dt|/|d/dx| \sim O(1)$, and so on. A consistent post-Newtonian limit requires determination of g_{00} correct through $O(4)$, g_{0i} through $O(3)$ and g_{ij} through $O(2)$ (for details see TEGP 4.1). The only way that one metric theory differs from another is in the numerical values of the coefficients that appear in front of the metric potentials. The parametrized post-Newtonian (PPN) formalism inserts parameters in place of these coefficients, parameters whose values depend on the theory under study. In the current version of the PPN formalism, summarized in Box 2, ten parameters are used, chosen in such a manner that they measure or indicate general properties of metric theories of gravity Table 1. The parameters γ and β are the usual Eddington-Robertson-Schiff parameters used to describe the 'classical' tests of general relativity; ξ is non-zero in any theory of gravity that predicts preferred-location effects such as a galaxy-induced anisotropy in the local gravitational constant G_L (also called 'Whitehead' effects); α_1, α_2, α_3 measure whether or not the theory predicts post-Newtonian preferred-frame effects; α_3, ζ_1, ζ_2, ζ_3, ζ_4 measure whether or not the theory predicts violations of global conservation laws for total momentum. In Table 1 we show the values these parameters take (i) in general relativity, (ii) in any theory of gravity that possesses conservation laws for total momentum, called 'semi-conservative' (any theory that is based on an invariant action principle is semi-conservative, and (iii) in any theory that in addition possesses six global conservation laws for angular momentum, called 'fully conservative' (such theories automatically predict no post-Newtonian preferred-frame effects). Semi-conservative theories have five free PPN parameters (γ, β, ξ, α_1, α_2) while fully conservative theories have three (γ, β, ξ).

The PPN formalism was pioneered by Kenneth Nordtvedt, who studied the post-Newtonian metric of a system of gravitating point masses, extending earlier work by Eddington, Robertson and Schiff (TEGP 4.2). A general and unified version of the PPN formalism was developed by Will and Nordtvedt. The canonical version, with conventions altered to be more in accord with standard textbooks such as MTW, is discussed in detail in TEGP, Chapter 4. Other versions of the PPN formalism have been developed to deal with point masses with charge, fluid with anisotropic stresses, bodies with strong internal gravity, and post-post-Newtonian effects (TEGP 4.2, 14.2).

Box 2. The Parametrized Post-Newtonian Formalism

1. **Coordinate System:** The framework uses a nearly globally Lorentz coordinate system in which the coordinates are (t, x^1, x^2, x^3). Three-dimensional, Euclidean vector notation is used throughout. All coordinate arbitrariness ('gauge freedom') has been removed by specialization of the coordinates to the standard PPN gauge (TEGP 4.2). Units are chosen so that $G = c = 1$, where G is the physically measured Newtonian constant far from the solar system.

2. **Matter Variables:**

    - ρ = density of rest mass as measured in a local freely falling frame momentarily comoving with the gravitating matter.
    - $v^i = (dx^i/dt)$ = coordinate velocity of the matter.
    - w^i = coordinate velocity of PPN coordinate system relative to the mean rest-frame of the universe.
    - p = pressure as measured in a local freely falling frame momentarily comoving with the matter.
    - Π = internal energy per unit rest mass. It includes all forms of non-rest-mass, non-gravitational energy, e.g. energy of compression and thermal energy.

3. PPN **Parameters:**

$\gamma, \beta, \xi, \alpha_1, \alpha_2, \alpha_3, \zeta_1, \zeta_2, \zeta_3, \zeta_4$.

4. **Metric:**

$$\begin{aligned}
g_{00} &= -1 + 2U - 2\beta U^2 - 2\xi\Phi_W + (2\gamma + 2 + \alpha_3 + \zeta_1 - 2\xi)\Phi_1 \\
&\quad + 2(3\gamma - 2\beta + 1 + \zeta_2 + \xi)\Phi_2 + 2(1 + \zeta_3)\Phi_3 + 2(3\gamma + 3\zeta_4 - 2\xi)\Phi_4 \\
&\quad - (\zeta_1 - 2\xi)\mathcal{A} - (\alpha_1 - \alpha_2 - \alpha_3)w^2 U - \alpha_2 w^i w^j U_{ij} + (2\alpha_3 - \alpha_1)w^i V_i \\
g_{0i} &= -\frac{1}{2}(4\gamma + 3 + \alpha_1 - \alpha_2 + \zeta_1 - 2\xi)V_i - \frac{1}{2}(1 + \alpha_2 - \zeta_1 + 2\xi)W_i \\
&\quad - \frac{1}{2}(\alpha_1 - 2\alpha_2)w^i U - \alpha_2 w^j U_{ij} \\
g_{ij} &= (1 + 2\gamma U)\delta_{ij}
\end{aligned}$$

5. **Metric Potentials:**

$$U = \int \frac{\rho'}{|\mathbf{x} - \mathbf{x}'|} d^3x', \quad U_{ij} = \int \frac{\rho'(x-x')_i(x-x')_j}{|\mathbf{x}-\mathbf{x}'|^3} d^3x'$$

$$\Phi_W = \int \frac{\rho'\rho''(\mathbf{x}-\mathbf{x}')}{|\mathbf{x}-\mathbf{x}'|^3} \cdot \left(\frac{\mathbf{x}'-\mathbf{x}''}{|\mathbf{x}-\mathbf{x}''|} - \frac{\mathbf{x}-\mathbf{x}''}{|\mathbf{x}'-\mathbf{x}''|}\right) d^3x' d^3x''$$

$$\mathcal{A} = \int \frac{\rho'[\mathbf{v}' \cdot (\mathbf{x}-\mathbf{x}')]^2}{|\mathbf{x}-\mathbf{x}'|^3} d^3x', \quad \Phi_1 = \int \frac{\rho' v'^2}{|\mathbf{x}-\mathbf{x}'|} d^3x'$$

$$\Phi_2 = \int \frac{\rho' U'}{|\mathbf{x}-\mathbf{x}'|} d^3x', \quad \Phi_3 = \int \frac{\rho' \Pi'}{|\mathbf{x}-\mathbf{x}'|} d^3x', \quad \Phi_4 = \int \frac{p'}{|\mathbf{x}-\mathbf{x}'|} d^3x'$$

$$V_i = \int \frac{\rho' v'_i}{|\mathbf{x}-\mathbf{x}'|} d^3x', \quad W_i = \int \frac{\rho'[\mathbf{v}' \cdot (\mathbf{x}-\mathbf{x}')](x-x')_i}{|\mathbf{x}-\mathbf{x}'|^3} d^3x'$$

6. **Stress-Energy Tensor** (perfect fluid)

$$\begin{aligned}
T^{00} &= \rho(1 + \Pi + v^2 + 2U) \\
T^{0i} &= \rho v^i(1 + \Pi + v^2 + 2U + p/\rho) \\
T^{ij} &= \rho v^i v^j(1 + \Pi + v^2 + 2U + p/\rho) + p\delta^{ij}(1 - 2\gamma U)
\end{aligned}$$

7. **Equations of Motion**

- Stressed Matter, $T^{\mu\nu}{}_{;\nu} = 0$
- Test Bodies, $d^2x^\mu/d\lambda^2 + \Gamma^\mu{}_{\nu\lambda}(dx^\nu/d\lambda)(dx^\lambda/d\lambda) = 0$
- Maxwell's Equations, $F^{\mu\nu}{}_{;\nu} = 4\pi J^\mu \quad F_{\mu\nu} = A_{\nu;\mu} - A_{\mu;\nu}$

3.3 Competing theories of gravity

One of the important applications of the PPN formalism is the comparison and classification of alternative metric theories of gravity. The population of viable theories has fluctuated over the years as new effects and tests have been discovered, largely through the use of the PPN framework, which eliminated many theories thought previously to be viable. The theory population has also fluctuated as new, potentially viable theories have been invented.

In these lectures, we shall focus on general relativity and the general class of scalar-tensor modifications of it, of which the Jordan-Fierz-Brans-Dicke theory (Brans-Dicke, for short) is the classic example. The reasons are several-fold:

- A full compendium of alternative theories is given in TEGP, Chapter 5.

- Many alternative metric theories developed during the 1970s and 1980s could be viewed as 'straw-man' theories, invented to prove that such theories exist or to illustrate particular properties. Few of these could be regarded as well-motivated theories from the point of view, say, of field theory or particle physics. Examples are the vector-tensor theories studied by Will, Nordtvedt and Hellings.

- A number of theories fall into the class of 'prior-geometric' theories, with absolute elements such as a flat background metric in addition to the physical metric. Most of these theories predict 'preferred-frame' effects, that have been tightly constrained by observations (see below). An example is Rosen's bimetric theory.

- A large number of alternative theories of gravity predict gravitational-wave emission substantially different from that of GR, in strong disagreement with observations of the binary pulsar (see below).

- Scalar-tensor modifications of GR have recently become very popular in cosmological model building and in unification schemes, such as string theory.

General relativity

The metric **g** is the sole dynamical field and the theory contains no arbitrary functions or parameters, apart from the value of the Newtonian coupling constant G, which is measurable in laboratory experiments. Throughout these lectures, we ignore the cosmological constant λ. Although λ has significance for quantum field theory, quantum gravity, and cosmology, on the scale of the solar-system or of stellar systems, its effects are negligible, for values of λ corresponding to a cosmological closure density.

The field equations of GR are derivable from an invariant action principle $\delta I = 0$, where

$$I = (16\pi G)^{-1} \int R(-g)^{1/2} d^4x + I_m(\psi_m, g_{\mu\nu}) \tag{21}$$

where R is the Ricci scalar, and I_m is the matter action, which depends on matter fields ψ_m universally coupled to the metric **g**. By varying the action with respect to $g_{\mu\nu}$, we obtain the field equations

$$G_{\mu\nu} \equiv R_{\mu\nu} - \frac{1}{2}g_{\mu\nu}R = 8\pi G T_{\mu\nu} \tag{22}$$

where $T_{\mu\nu}$ is the matter energy-momentum tensor. General covariance of the matter action implies the equations of motion $T^{\mu\nu}{}_{;\nu} = 0$; varying I_m with respect to ψ_M yields the matter field equations. By virtue of the *absence* of prior-geometric elements, the equations of motion are also a consequence of the field equations via the Bianchi identities $G^{\mu\nu}{}_{;\nu} = 0$.

Theory	Arbitrary Functions or Constants	Cosmic Matching Parameters	PPN parameters				
			γ	β	ξ	α_1	α_2
General Relativity	none	none	1	1	0	0	0
Scalar-Tensor							
Brans-Dicke	ω	ϕ_0	$\frac{1+\omega}{2+\omega}$	1	0	0	0
General	$A(\varphi), V(\varphi)$	φ_0	$\frac{1+\omega}{2+\omega}$	$1+\Lambda$	0	0	0
Rosen's Bimetric	none	c_0, c_1	1	1	0	0	$\frac{c_0}{c_1}-1$

Table 2. *Metric theories and their PPN parameter values. (Note that $\alpha_3 = \zeta_i = 0$ for all cases).*

The general procedure for deriving the post-Newtonian limit is spelled out in TEGP 5.1, and is described in detail for GR in TEGP 5.2. The PPN parameters values are listed in Table 2.

Scalar-tensor theories

These theories contain the metric **g**, a scalar field φ, a potential function $V(\varphi)$, and a coupling function $A(\varphi)$ (generalizations to more than one scalar field have also been carried out (Damour and Esposito-Farèse 1992)). For some purposes, the action is conveniently written in a non-metric representation, sometimes denoted the 'Einstein frame', in which the gravitational action looks exactly like that of GR:

$$I = (16\pi G)^{-1} \int [R_* - 2g_*^{\mu\nu}\partial_\mu\varphi\partial_\nu\varphi - V(\varphi)](-g_*)^{1/2}d^4x + I_m(\psi_m, A^2(\varphi)g_{\mu\nu}^*) \quad (23)$$

where $R_* \equiv g_*^{\mu\nu} R_{\mu\nu}^*$ is the Ricci scalar of the 'Einstein' metric $g_{\mu\nu}^*$. This representation is a 'non-metric' one because the matter fields ψ_m couple to a combination of φ and $g_{\mu\nu}^*$. Despite appearances, however, it is a metric theory, because it can be put into a metric representation by identifying the 'physical metric'

$$g_{\mu\nu} \equiv A^2(\varphi)g_{\mu\nu}^* . \quad (24)$$

The action can then be rewritten in the metric form

$$I = (16\pi G)^{-1} \int [\phi R - \phi^{-1}\omega(\phi)g^{\mu\nu}\partial_\mu\phi\partial_\nu\phi - \phi^2 V](-g)^{1/2}d^4x + I_m(\psi_m, g_{\mu\nu}) \quad (25)$$

where

$$\phi \equiv A(\varphi)^{-2}$$
$$3 + 2\omega(\phi) \equiv \alpha(\varphi)^{-2}$$
$$\alpha(\varphi) \equiv d(\ln A(\varphi))/d\varphi . \quad (26)$$

The Einstein frame is useful for discussing general characteristics of such theories, and in some cosmological applications, while the metric representation is most useful for calculating observable effects. The field equations, post-Newtonian limit and PPN parameters are discussed in TEGP 5.3, and the values of the PPN parameters are listed in Table 2.

The parameters that enter the post-Newtonian limit are

$$\omega \equiv \omega(\phi_0) \qquad \Lambda \equiv [(d\omega/d\phi)(3+2\omega)^{-2}(4+2\omega)^{-1}]_{\phi_0} \qquad (27)$$

where ϕ_0 is the value of ϕ today far from the system being studied, as determined by appropriate cosmological boundary conditions. The following formula is also useful: $1/(2+\omega) = 2\alpha_0^2/(1+\alpha_0^2)$. In Brans-Dicke theory ($\omega(\phi)$ = constant), the larger the value of ω, the smaller the effects of the scalar field, and in the limit $\omega \to \infty$ ($\alpha_0 \to 0$), the theory becomes indistinguishable from general relativity in all its predictions. In more general theories, the function $\omega(\phi)$ could have the property that for the present value of the scalar field ϕ_0, ω is very large, and Λ is very small (theory almost identical to general relativity today), but that for past or future values of ϕ, ω and Λ could take on values that would lead to significant differences in cosmological models. Indeed, Damour and Nordtvedt have shown that in such general scalar-tensor theories, GR is a natural 'attractor': regardless of how different the theory may be from GR in the early universe (apart from special cases), cosmological evolution naturally drives the fields toward small values of the function α, thence to large ω. Estimates of the expected relic deviations from GR today in such theories depend on the cosmological model, but range from 10^{-5} to a few times 10^{-7} for $1-\gamma$ (Damour and Nordtvedt 1993a,b).

Scalar fields coupled to gravity or matter are also ubiquitous in particle-physics-inspired models of unification, such as string theory. In some models, the coupling to matter may lead to violations of WEP, which are tested by Eötvös-type experiments. In many models the scalar field is massive; if the Compton wavelength is of macroscopic scale, its effects are those of a 'fifth force'. Only if the theory can be cast as a metric theory with a scalar field of infinite range or of range long compared to the scale of the system in question (solar system) can the PPN framework be strictly applied. If the mass of the scalar field is sufficiently large that its range is microscopic, then, on solar-system scales, the scalar field is suppressed, and the theory is essentially equivalent to general relativity. This is the case, for example in the 'oscillating-G' models of Accetta, Steinhardt and Will (see Steinhardt and Will 1995), in which the potential function $V(\varphi)$ contains both quadratic (mass) and quartic (self-interaction) terms, causing the scalar field to oscillate (the initial amplitude of oscillation is provided by an inflationary epoch); high-frequency oscillations in the 'effective' Newtonian constant $G_{\text{eff}} \equiv G/\phi = GA(\varphi)^2$ then result. The energy density in the oscillating scalar field can be enough to provide a cosmological closure density without resorting to dark matter, yet the value of ω today is so large that the theory's local predictions are experimentally indistinguishable from GR.

3.4 Tests of the parameter γ

With the PPN formalism in hand, we are now ready to confront gravitation theories with the results of solar-system experiments. In this section we focus on tests of the

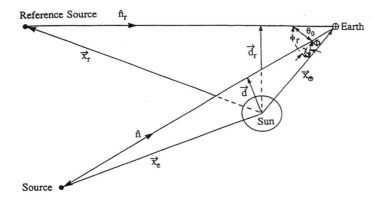

Figure 4. *Geometry of light deflection measurements.*

parameter γ, consisting of the deflection of light and the time delay of light.

The deflection of light

A light ray (or photon) which passes the Sun at a distance d is deflected by an angle

$$\delta\theta = \frac{1}{2}(1+\gamma)(4m_\odot/d)[(1+\cos\Phi)/2] \tag{28}$$

(TEGP 7.1) where $m_\odot$ is the mass of the Sun and Φ is the angle between the Earth-Sun line and the incoming direction of the photon (Figure 4). For a grazing ray, $d \approx d_\odot$, $\Phi \approx 0$, and

$$\delta\theta \approx \frac{1}{2}(1+\gamma)1.''75 \tag{29}$$

independent of the frequency of light. Another, more useful expression gives the change in the relative angular separation between an observed source of light and a nearby reference source as both rays pass near the Sun:

$$\delta\theta = \frac{1}{2}(1+\gamma)\left[-\frac{4m_\odot}{d}\cos\chi + \frac{4m_\odot}{d_r}\left(\frac{1+\cos\Phi_r}{2}\right)\right] \tag{30}$$

where d and d_r are the distances of closest approach of the source and reference rays respectively, Φ_r is the angular separation between the Sun and the reference source, and χ is the angle between the Sun-source and the Sun-reference directions, projected on the plane of the sky (Figure 4). Thus, for example, the relative angular separation between the two sources may vary if the line of sight of one of them passes near the Sun ($d \sim R_\odot$, $d_r \gg d$, χ varying with time).

It is interesting to note that the classic derivations of the deflection of light that use only the principle of equivalence or the corpuscular theory of light yield only the

'1/2' part of the coefficient in front of the expression in Equation 28. But the result of these calculations is the deflection of light relative to local straight lines, as defined for example by rigid rods; however, because of space curvature around the Sun, determined by the PPN parameter γ, local straight lines are bent relative to asymptotic straight lines far from the Sun by just enough to yield the remaining factor '$\gamma/2$'. The first factor '1/2' holds in any metric theory, the second '$\gamma/2$' varies from theory to theory. Thus, calculations that purport to derive the full deflection using the equivalence principle alone are incorrect.

The prediction of the full bending of light by the Sun was one of the great successes of Einstein's general relativity. Eddington's confirmation of the bending of optical starlight observed during a solar eclipse in the first days following World War I helped make Einstein famous. However, the experiments of Eddington and his co-workers had only 30 percent accuracy, and succeeding experiments were not much better: the results were scattered between one half and twice the Einstein value (Figure 5), and the accuracies were low.

However, the development of very-long-baseline radio interferometry (VLBI) produced greatly improved determinations of the deflection of light. These techniques now have the capability of measuring angular separations and changes in angles as small as 100 microarcseconds. Early measurements took advantage of a series of heavenly coincidences: each year, groups of strong quasistellar radio sources pass very close to the Sun (as seen from the Earth), including the group 3C273, 3C279, and 3C48, and the group 0111+02, 0119+11 and 0116+08. As the Earth moves in its orbit, changing the lines of sight of the quasars relative to the Sun, the angular separation $\delta\theta$ between pairs of quasars varies (Equation 30). The time variation in the quantities d, d_r, χ and Φ_r in Equation 30 is determined using an accurate ephemeris for the Earth and initial directions for the quasars, and the resulting prediction for $\delta\theta$ as a function of time is used as a basis for a least-squares fit of the measured $\delta\theta$, with one of the fitted parameters being the coefficient $\frac{1}{2}(1+\gamma)$. A number of measurements of this kind over the period 1969–1975 yielded an accurate determination of the coefficient $\frac{1}{2}(1+\gamma)$ which has the value unity in general relativity. Their results are shown in Figure 5.

A recent series of transcontinental and intercontinental VLBI quasar and radio galaxy observations made primarily to monitor the Earth's rotation ('VLBI' in Figure 5) was sensitive to the deflection of light over almost the entire celestial sphere (at 90° from the Sun, the deflection is still 4 milliarcseconds). The value obtained from the data was $\frac{1}{2}(1+\gamma) = 1.000 \pm 0.001$. A VLBI measurement of the deflection of light by Jupiter was reported; the predicted deflection of about 300 microarcseconds was seen with about 50 percent accuracy.

The time delay of light

A radar signal sent across the solar system past the Sun to a planet or satellite and returned to the Earth suffers an additional non-Newtonian delay in its round-trip travel time, given by (see Figure 4)

$$\delta t = 2(1+\gamma)m_\odot \ln[(r_\oplus + \mathbf{x}_\oplus \cdot \mathbf{n})(r_e - \mathbf{x}_e \cdot \mathbf{n})/d^2] \tag{31}$$

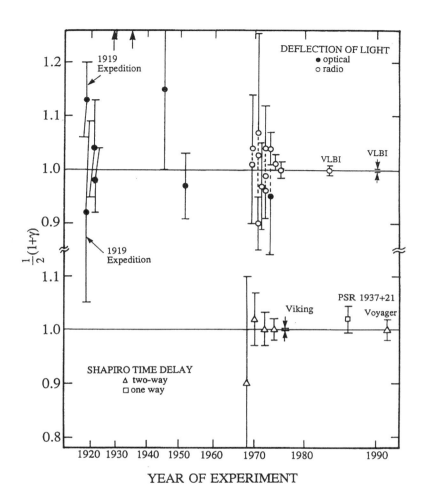

Figure 5. *Measurements of the coefficient $(1+\gamma)/2$ from light deflection and time delay measurements. General relativity value is unity. Arrows denote anomalously large values from 1929 and 1936 eclipse expeditions. Shapiro time-delay measurements using Viking spacecraft and* VLBI *light deflection measurements yielded agreement with general relativity to 0.1 percent.*

(TEGP 7.2). For a ray which passes close to the Sun,

$$\delta t \approx \frac{1}{2}(1+\gamma)[240 - 20\ln(d^2/r)] \ \mu s \tag{32}$$

where d is the distance of closest approach of the ray in solar radii, and r is the distance of the planet or satellite from the Sun, in astronomical units.

In the two decades following Irwin Shapiro's 1964 discovery of this effect as a theoretical consequence of general relativity, several high-precision measurements were made using radar ranging to targets passing through superior conjunction. Since one does not have access to a 'Newtonian' signal against which to compare the round-trip travel time of the observed signal, it is necessary to do a differential measurement of the variations in round-trip travel times as the target passes through superior conjunction, and to look for the logarithmic behavior of Equation 32. In order to do this accurately however, one must take into account the variations in round-trip travel time due to the orbital motion of the target relative to the Earth. This is done by using radar-ranging (and possibly other) data on the target taken when it is far from superior conjunction (*i.e.*, when the time-delay term is negligible) to determine an accurate ephemeris for the target, using the ephemeris to predict the PPN coordinate trajectory $x_e(t)$ near superior conjunction, then combining that trajectory with the trajectory of the Earth $x_\oplus(t)$ to determine the Newtonian round-trip time and the logarithmic term in Equation 32. The resulting predicted round-trip travel times in terms of the unknown coefficient $\frac{1}{2}(1+\gamma)$ are then fit to the measured travel times using the method of least-squares, and an estimate obtained for $\frac{1}{2}(1+\gamma)$.

The targets employed included planets, such as Mercury or Venus, used as a passive reflectors of the radar signals ('passive radar'), and artificial satellites, such as Mariners 6 and 7, and Voyager 2, and the Viking Mars landers and orbiters, used as active retransmitters of the radar signals ('active radar').

The results for the coefficient $\frac{1}{2}(1+\gamma)$ of all radar time-delay measurements performed to date (including a measurement of the one-way time delay of signals from the millisecond pulsar PSR 1937+21) are shown in Figure 5 (see TEGP 7.2 for discussion and references). The Viking experiment resulted in a 0.1 percent measurement.

From the results of light-deflection and time-delay experiments, we can conclude that the coefficient $\frac{1}{2}(1+\gamma)$ must be within at most 0.1 percent of unity. Scalar-tensor theories must have $\omega > 500$ to be compatible with this constraint.

3.5 The perihelion shift of Mercury

The explanation of the anomalous perihelion shift of Mercury's orbit was another of the triumphs of general relativity. This had been an unsolved problem in celestial mechanics for over half a century, since the announcement by Le Verrier in 1859 that, after the perturbing effects of the planets on Mercury's orbit had been accounted for, and after the effect of the precession of the equinoxes on the astronomical coordinate system had been subtracted, there remained in the data an unexplained advance in the perihelion of Mercury. The modern value for this discrepancy is 43 arcseconds per century. A number of *ad hoc* proposals were made in an attempt to account for this excess, including, among others, the existence of new planet Vulcan near the Sun, a ring of planetoids, a solar quadrupole moment and a deviation from the inverse-square law of gravitation but none was successful. General relativity accounted for the anomalous shift in a natural way without disturbing the agreement with other planetary observations.

The predicted advance, $\Delta\tilde{\omega}$, per orbit, including both relativistic PPN contributions and the Newtonian contribution resulting from a possible solar quadrupole moment, is

Parameter	Effect	Value or Limit	Remarks
γ	time delay	1.000 ± 0.002	Viking ranging
	light deflection	1.000 ± 0.002	VLBI
β	perihelion shift	1.000 ± 0.003	$J_2 = 10^{-7}$ from helioseismology
	Nordtvedt effect	1.000 ± 0.001	$\eta = 4\beta - \gamma - 3$ assumed
ξ	Earth tides	$< 10^{-3}$	gravimeter data
α_1	orbital preferred-frame effects	$< 4 \times 10^{-4}$	Mars-Mercury data
α_2	solar spin precession	$< 4 \times 10^{-7}$	alignment of solar equator and ecliptic
α_3	pulsar acceleration	$< 4 \times 10^{-16}$	pulsar $\dot{P}$ statistics
η	Nordtvedt effect	$< 10^{-3}$	lunar laser ranging
ζ_1	–	< 0.02	combined PPN bounds
ζ_2	binary acceleration	$< 4 \times 10^{-5}$	$\ddot{P}_p$ for PSR 1913+16
ζ_3	Newton's 3rd law	$< 10^{-8}$	Lunar acceleration
ζ_4	–	–	–

Table 3. *Current limits on the PPN parameters. (The parameter η for the Nordvedt effect is defined in Equation 35).*

given by

$$\Delta\tilde{\omega} = (6\pi m/p)[\frac{1}{3}(2 + 2\gamma - \beta) + \frac{1}{6}(2\alpha_1 - \alpha_2 + \alpha_3 + 2\zeta_2)\mu/m + J_2(R^2/2mp)] \quad (33)$$

where $m \equiv m_1 + m_2$ and $\mu \equiv m_1 m_2/m$ are the total mass and reduced mass of the two-body system respectively; $p \equiv a(1-e^2)$ is the semi-latus rectum of the orbit, with a the semi-major axis and e the eccentricity; R is the mean radius of the oblate body; and J_2 is a dimensionless measure of its quadrupole moment, given by $J_2 = (C-A)/m_1 R^2$, where C and A are the moments of inertia about the body's rotation and equatorial axes, respectively (for details of the derivation see TEGP 7.3). We have ignored preferred-frame and galaxy-induced contributions to $\Delta\tilde{\omega}$; these are discussed in TEGP 8.3.

The first term in Equation 33 is the classical relativistic perihelion shift, which depends upon the PPN parameters γ and β. The second term depends upon the ratio of the masses of the two bodies; it is zero in any fully conservative theory of gravity ($\alpha_1 \equiv \alpha_2 \equiv \alpha_3 \equiv \zeta_2 \equiv 0$); it is also negligible for Mercury, since $\mu/m \approx m_{\text{Merc}}/m_\odot \approx 2 \times 10^{-7}$. We shall drop this term henceforth. The third term depends upon the solar quadrupole moment J_2. For a Sun that rotates uniformly with its observed surface angular velocity, so that the quadrupole moment is produced by centrifugal flattening, one may estimate J_2 to be $\sim 1 \times 10^{-7}$. This actually agrees reasonably well with values inferred from rotating solar models that are in accord with observations of the normal modes of solar oscillations (helioseismology). Substituting standard orbital elements and physical constants for Mercury and the Sun we obtain the rate of perihelion shift $\dot{\tilde{\omega}}$, in seconds of arc per century,

$$\dot{\tilde{\omega}} = 42.''98 \left[\frac{1}{3}(2 + 2\gamma - \beta) + 3 \times 10^{-4}(J_2/10^{-7})\right]. \quad (34)$$

Now, the measured perihelion shift of Mercury is known accurately: after the perturbing effects of the other planets have been accounted for, the excess shift is known to about 0.1 percent from radar observations of Mercury since 1966. The solar oblateness effect is smaller than the observational error, so we obtain the PPN bound $|2\gamma - \beta - 1| < 3 \times 10^{-3}$.

3.6 Tests of the strong equivalence principle

The next class of solar-system experiments that test relativistic gravitational effects may be called tests of the strong equivalence principle (SEP). In Sec. 3.1.2, we pointed out that many metric theories of gravity (perhaps all except general relativity) can be expected to violate one or more aspects of SEP. Among the testable violations of SEP are a violation of the weak equivalence principle for gravitating bodies that leads to perturbations in the Earth-Moon orbit; preferred-location and preferred-frame effects in the locally measured gravitational constant that could produce observable geophysical effects; and possible variations in the gravitational constant over cosmological timescales.

The Nordtvedt effect and the lunar Eötvös Experiment

In a pioneering calculation using his early form of the PPN formalism, Nordtvedt showed that many metric theories of gravity predict that massive bodies violate the weak equivalence principle–that is, fall with different accelerations depending on their gravitational self-energy. For a spherically symmetric body, the acceleration from rest in an external gravitational potential U has the form

$$\mathbf{a} = (m_p/m)\nabla U$$
$$m_p/m = 1 - \eta(E_g/m)$$
$$\eta = 4\beta - \gamma - 3 - \frac{10}{3}\xi - \alpha_1 + \frac{2}{3}\alpha_2 - \frac{2}{3}\zeta_1 - \frac{1}{3}\zeta_2 \qquad (35)$$

where E_g is the negative of the gravitational self-energy of the body ($E_g > 0$). This violation of the massive-body equivalence principle is known as the 'Nordtvedt effect'. The effect is absent in general relativity ($\eta = 0$) but present in scalar-tensor theory ($\eta = 1/(2+\omega) + 4\Lambda$). The existence of the Nordtvedt effect does not violate the results of laboratory Eötvös experiments, since for laboratory-sized objects, $E_g/m \leq 10^{-27}$, far below the sensitivity of current or future experiments. However, for astronomical bodies, E_g/m may be significant (10^{-5} for the Sun, 10^{-8} for Jupiter, 4.6×10^{-10} for the Earth, 0.2×10^{-10} for the Moon). If the Nordtvedt effect is present ($\eta \neq 0$) then the Earth should fall toward the Sun with a slightly different acceleration than the Moon. This perturbation in the Earth-Moon orbit leads to a polarization of the orbit that is directed toward the Sun as it moves around the Earth-Moon system, as seen from Earth. This polarization represents a perturbation in the Earth-Moon distance of the form

$$\delta r = 13.1\eta \cos(\omega_0 - \omega_s)t \quad \text{m.} \qquad (36)$$

where ω_0 and ω_s are the angular frequencies of the orbits of the Moon and Sun around the Earth (see TEGP 8.1, for detailed derivations and references; for improved calculations of the numerical coefficient, see Nordtvedt 1995, Damour and Vokrouhlický 1995).

Since August 1969, when the first successful acquisition was made of a laser signal reflected from the Apollo 11 retroreflector on the Moon, the lunar laser-ranging experiment (LURE) has made regular measurements of the round-trip travel times of laser pulses between a network of observatories and the lunar retroreflectors, with accuracies that are approaching 50 ps (1 cm). These measurements are fit using the method of least-squares to a theoretical model for the lunar motion that takes into account perturbations due to the Sun and the other planets, tidal interactions, and post-Newtonian gravitational effects. The predicted round-trip travel times between retroreflector and telescope also take into account the librations of the Moon, the orientation of the Earth, the location of the observatory, and atmospheric effects on the signal propagation. The 'Nordtvedt' parameter η along with several other important parameters of the model are then estimated in the least-squares method.

Several independent analyses of the data found no evidence, within experimental uncertainty, for the Nordtvedt effect (for recent results see Dickey et al. 1994). Their results can be summarized by the bound $|\eta| < 0.001$. These results represent a limit on a possible violation of WEP for massive bodies of 7 parts in 10^{13} (compare Figure 1). For Brans-Dicke theory, these results force a lower limit on the coupling constant ω of 1000. Note that, at this level of precision, one cannot regard the results of lunar laser ranging as a 'clean' test of SEP because the precision exceeds that of laboratory tests of WEP. Because the chemical compositions of the Earth and Moon differ in the relative fractions of iron and silicates, an extrapolation from laboratory Eötvös-type experiments to the Earth-Moon systems using various non-metric couplings to matter yields bounds on violations of WEP only of the order of 2×10^{-12}. Thus if lunar laser ranging is to test SEP at higher accuracy, tests of WEP must keep pace; to this end, a proposed satellite test of the equivalence principle (Sec. 3.8.2) will be an important advance.

In general relativity, the Nordtvedt effect vanishes; at the level of several centimeters and below, a number of non-null general relativistic effects should be present (Nordtvedt 1995)

Preferred-frame and preferred-location Effects

Some theories of gravity violate SEP by predicting that the outcomes of local gravitational experiments may depend on the velocity of the laboratory relative to the mean rest frame of the universe (preferred-frame effects) or on the location of the laboratory relative to a nearby gravitating body (preferred-location effects). In the post-Newtonian limit, preferred-frame effects are governed by the values of the PPN parameters α_1, α_2, and α_3, and some preferred-location effects are governed by ξ (see Table 1).

The most important such effects are variations and anisotropies in the locally-measured value of the gravitational constant, which lead to anomalous Earth tides and variations in the Earth's rotation rate; anomalous contributions to the orbital dynamics of Mercury and Mars; self-accelerations of pulsars, and anomalous torques on the Sun that would cause its spin axis to be randomly oriented relative to the ecliptic (see TEGP 8.2, 8.3, 9.3 and 14.3(c)). An improved bound on α_3 of 4×10^{-16} from the period derivatives of 20 millisecond pulsars was reported by Bell (1995); Damour and Esposito-Farèse (1994) discussed the prospects for improving the bounds on α_1 us-

Method	$\dot{G}/G(10^{-12}\,\mathrm{yr}^{-1})$
Lunar Laser Ranging	0 ± 11
Viking Radar	2 ± 4
	-2 ± 10
Binary Pulsar	11 ± 11
Pulsar PSR 0655+64	< 55

Table 4. *Constancy of the gravitational constant. (The bounds for the pulsars are dependent upon the theory of gravity in strong-field regime and the neutron star equation of state).*

ing artificial Earth satellites. Negative searches for these effects have produced strong constraints on the PPN parameters (Table 3).

Constancy of the newtonian gravitational constant

Most theories of gravity that violate SEP predict that the locally measured Newtonian gravitational constant may vary with time as the universe evolves. For the scalar-tensor theories listed in Table 2, the predictions for $\dot{G}/G$ can be written in terms of time derivatives of the asymptotic scalar field. Where G does change with cosmic evolution, its rate of variation should be of the order of the expansion rate of the universe, i.e., $\dot{G}/G \sim H_0$, where H_0 is the Hubble expansion parameter and is given by $H_0 = 100h$ km s^{-1} Mpc$^{-1} = h \times 10^{-10}$ yr^{-1}, where current observations of the expansion of the universe give $\frac{1}{2} < h < 1$.

Several observational constraints can be placed on $\dot{G}/G$ using methods that include studies of the evolution of the Sun, observations of lunar occultations (including analyses of ancient eclipse data), lunar laser-ranging measurements, planetary radar-ranging measurements, and pulsar timing data. Laboratory experiments may one day lead to interesting limits (for review and references to past work see TEGP 8.4 and 14.3(c)). Recent results are shown in Table 4. The best limits on $\dot{G}/G$ still come from ranging measurements to the Viking landers and Lunar laser ranging measurements (Dickey et al. 1994). It has been suggested that radar observations of a Mercury orbiter over a two-year mission (30 cm accuracy in range) could yield $\Delta(\dot{G}/G) \sim 10^{-14}$ yr^{-1}.

Although bounds on $\dot{G}/G$ using solar-system measurements can be obtained in a phenomenological manner through the simple expedient of replacing G by $G_0 + \dot{G}_0(t-t_0)$ in Newton's equations of motion, the same does not hold true for pulsar and binary pulsar timing measurements. The reason is that, in theories of gravity that violate SEP, such as scalar-tensor theories, the 'mass' and moment of inertia of a gravitationally bound body may vary with variation in G. Because neutron stars are highly relativistic, the fractional variation in these quantities can be comparable to $\Delta G/G$, the precise variation depending both on the equation of state of neutron star matter and on the theory of gravity in the strong-field regime. The variation in the moment of inertia affects the spin rate of the pulsar, while the variation in the mass can affect the orbital period in a manner that can add to or subtract from the direct effect of a variation in G, given by $\dot{P}_b/P_b = -\frac{1}{2}\dot{G}/G$. Thus, the bounds quoted in Table 4 for the binary pulsar PSR 1913+16 and the pulsar PSR 0655+64 are theory-dependent and must be treated as merely suggestive.

3.7 Other tests of post-Newtonian gravity

Tests of post-Newtonian conservation laws

Of the five 'conservation law' PPN parameters ζ_1, ζ_2, ζ_3, ζ_4, and α_3, only three, ζ_2, ζ_3 and α_3, have been constrained directly with any precision; ζ_1 is constrained indirectly through its appearance in the Nordtvedt effect parameter η, Equation 35. There is strong theoretical evidence that ζ_4, which is related to the gravity generated by fluid pressure, is not really an independent parameter–in any reasonable theory of gravity there should be a connection between the gravity produced by kinetic energy (ρv^2), internal energy ($\rho\Pi$), and pressure (p). From such considerations, Will (1976) derived the additional theoretical constraint

$$6\zeta_4 = 3\alpha_3 + 2\zeta_1 - 3\zeta_3 . \tag{37}$$

A non-zero value for any of these parameters would result in a violation of conservation of momentum, or of Newton's third law in gravitating systems. An alternative statement of Newton's third law for gravitating systems is that the 'active gravitational mass', that is the mass that determines the gravitational potential exhibited by a body, should equal the 'passive gravitational mass', the mass that determines the force on a body in a gravitational field. Such an equality guarantees the equality of action and reaction and of conservation of momentum, at least in the Newtonian limit.

A classic test of Newton's third law for gravitating systems was carried out in 1968 by Kreuzer, in which the gravitational attraction of fluorine and bromine were compared to a precision of 5 parts in 10^5.

A remarkable planetary test of Newton's third law was reported by Bartlett and van Buren. They noted that current understanding of the structure of the Moon involves an iron-rich, aluminum-poor mantle whose center of mass is offset about 10 km from the center of mass of an aluminum-rich, iron-poor crust. The direction of offset is toward the Earth, about 14° to the east of the Earth-Moon line. Such a model accounts for the basaltic maria which face the Earth, and the aluminum-rich highlands on the Moon's far side, and for a 2 km offset between the observed center of mass and center of figure for the Moon. Because of this asymmetry, a violation of Newton's third law for aluminum and iron would result in a momentum non-conserving self-force on the Moon, whose component along the orbital direction would contribute to the secular acceleration of the lunar orbit. Improved knowledge of the lunar orbit through lunar laser ranging, and a better understanding of tidal effects in the Earth-Moon system (which also contribute to the secular acceleration) through satellite data, severely limit any anomalous secular acceleration, with the resulting limit

$$\left|\frac{(m_A/m_P)_{\text{Al}} - (m_A/m_P)_{\text{Fe}}}{(m_A/m_P)_{\text{Fe}}}\right| < 4\times10^{-12}. \quad (38)$$

According to the PPN formalism, in a theory of gravity that violates conservation of momentum, but that obeys the constraint of Equation 37, the electrostatic binding energy E_e of an atomic nucleus could make a contribution to the ratio of active to passive mass of the form

$$m_A = m_P + \frac{1}{2}\zeta_3 E_e/c^2. \quad (39)$$

The resulting limits on ζ_3 from the lunar experiment is $\zeta_3 < 1\times10^{-8}$ (TEGP 9.2, 14.3(d)).

Another consequence of a violation of conservation of momentum is a self-acceleration of the center of mass of a binary stellar system, given by

$$\mathbf{a}_{\text{CM}} = \frac{1}{2}(\zeta_2 + \alpha_3)\frac{m\,\mu}{a^2\,a}\frac{\delta m}{m}\frac{e}{(1-e^2)^{3/2}}\mathbf{n}_P \quad (40)$$

where $\delta m = m_1 - m_2$, a is the semi-major axis, and $\mathbf{n}_P$ is a unit vector directed from the center of mass to the point of periastron of m_1 (TEGP 9.3). A consequence of this acceleration would be non-vanishing values for d^2P/dt^2, where P denotes the period of any intrinsic process in the system (orbit, spectra, pulsar periods). The observed upper limit on d^2P_p/dt^2 of the binary pulsar PSR 1913+16 places a strong constraint on such an effect, resulting in the bound $|\alpha_3 + \zeta_2| < 4\times10^{-5}$. Since α_3 has already been constrained to be much less than this (Table 3), we obtain a strong bound on ζ_2 alone (Will 1992c).

Geodetic precession

A gyroscope moving through curved spacetime suffers a precession of its axis given by

$$d\mathbf{S}/d\tau = \mathbf{\Omega}_G \times \mathbf{S} \qquad \mathbf{\Omega}_G = (\gamma + \frac{1}{2})\mathbf{v} \times \mathbf{\nabla}U \quad (41)$$

where $\mathbf{v}$ is the velocity of the gyroscope, and U is the Newtonian gravitational potential of the source (TEGP 9.1). The Earth-Moon system can be considered as a 'gyroscope', with its axis perpendicular to the orbital plane. The predicted precession is about 2 arcseconds per century, an effect first calculated by de Sitter. This effect has been measured to about 2 percent using Lunar laser ranging data (TEGP 14.3(e)).

For a gyroscope orbiting the Earth, the precession is about 8 arcseconds per year. The Stanford Gyroscope Experiment has as one of its goals the measurement of this effect to 5×10^{-5} (see below).

3.8 The future of experimental gravitation

Although the golden era of experimental gravitation may be over, there remains considerable opportunity both for refining our knowledge of gravity, and for exploring new regimes of gravitational phenomena. Nowhere is the intellectual vigor and continuing excitement of this field more apparent than in the ideas that have been developed for experiments and observations to push us to the frontiers of knowledge. Unfortunately, these exciting ideas have run up against a severely restricted budgetary climate, especially in the US (and particularly at NASA), casting the future of many proposed and ongoing projects into doubt.

Search for gravitomagnetism

According to general relativity, moving or rotating matter should produce a contribution to the gravitational field that is the analogue of the magnetic field of a moving charge or a magnetic dipole. Although gravitomagnetism plays a role in a variety of measured relativistic effects, it has not been seen to date, isolated from other post-Newtonian effects. The Relativity Gyroscope Experiment (Gravity Probe B or GP-B) at Stanford University, in collaboration with NASA and Lockheed-Martin Corporation, is in the advanced stage of developing a space mission to detect this phenomenon directly. A set of four superconducting-niobium-coated, spherical quartz gyroscopes will be flown in a low polar Earth orbit, and the precession of the gyroscopes relative to the distant stars will be measured. In the PPN formalism, the predicted effect of gravitomagnetism is a precession (also known as the Lense-Thirring effect, or the dragging of inertial frames), given by

$$d\mathbf{S}/d\tau = \mathbf{\Omega}_{\rm LT} \times \mathbf{S} \qquad \mathbf{\Omega}_{\rm LT} = -\frac{1}{2}(1 + \gamma + \frac{1}{4}\alpha_1)[\mathbf{J} - 3\mathbf{n}(\mathbf{n}\cdot\mathbf{J})]/r^3 \qquad (42)$$

where $\mathbf{J}$ is the angular momentum of the Earth, $\mathbf{n}$ is a unit radial vector, and r is the distance from the center of the Earth (TEGP 9.1). For a polar orbit at about 650 km altitude, this leads to a secular angular precession at a rate $\frac{1}{2}(1 + \gamma + \frac{1}{4}\alpha_1)42 \times 10^{-3}$ arcsec/yr. The accuracy goal of the experiment is about 0.5 milliarcseconds per year. A full-size flight prototype of the instrument package has been tested as an integrated unit and a spacecraft is under construction. Unless budget pressures cause NASA to terminate the project, it is scheduled for launch around 1999.

Another proposal to look for an effect of gravitomagnetism is to measure the relative precession of the line of nodes of a pair of laser-ranged geodynamics satellites (LAGEOS), with supplementary inclination angles; the inclinations must be supplementary in order to cancel the dominant nodal precession caused by the Earth's Newtonian gravitational multipole moments. A third proposal envisages orbiting an array of three mutually orthogonal, superconducting gravity gradiometers around the Earth, to measure directly the contribution of the gravitomagnetic field to the tidal gravitational force. No commitment from any space agencies to launch either of these experiments exists at present (TEGP 14.4(a)).

Tests of the Einstein equivalence principle

The concept of an Eötvös experiment in space has been developed, with the potential to test WEP to 10^{-17}. Known generically as Satellite Test of the Equivalence Principle (STEP), various versions of the project are under consideration as joint efforts of NASA and the European Space Agency.

The gravitational redshift could be improved to the 10^{-9} level and second-order effects and the effects of J_2 of the Sun discerned by placing a hydrogen maser clock on board Solar Probe, a proposed spacecraft which would travel to within four solar radii of the Sun.

Improved PPN parameter values

A number of advanced space missions have been proposed in which spacecraft orbiters or landers and improved tracking capabilities could lead to significant improvements in values of the PPN parameters, of J_2 of the Sun, and of $\dot{G}/G$. For example, a Mercury orbiter, in a two-year experiment, with 3 cm range capability, could yield improvements in the perihelion shift to a part in 10^4, in γ to 4×10^{-5}, in $\dot{G}/G$ to 10^{-14} yr^{-1}, and in J_2 to a few parts in 10^8. Proposals are being developed, primarily in Europe, for advanced space missions which will have tests of PPN parameters as key components, including GAIA, a high-precision astrometric telescope, which could measure light-deflection and γ to the 10^{-6} level, and SORT, a solar orbit relativity test, which could measure γ to 10^{-7} from time delay and light deflection measurements.

Gravitational-wave astronomy

A significant part of the field of experimental gravitation is devoted to building and designing sensitive devices to detect gravitational radiation and to use gravity waves as a new astronomical tool. The centerpieces of this effort are the US Laser Interferometric Gravitational-wave Observatory (LIGO) and the European VIRGO projects, which are currently under construction. This important topic has been reviewed thoroughly elsewhere (Thorne 1987, 1995), and in this volume.

3.9 Rise and fall of the fifth force

A clear example of the role of 'opportunism' in experimental gravity since 1980 is the story of the 'fifth force'. In 1986, as a result of a detailed reanalysis of Eötvös' original data, Fischbach et al. suggested the existence of a fifth force of nature, with a strength of about a percent that of gravity, but with a range (as defined by the range λ of a Yukawa potential, $e^{-r/\lambda}/r$) of a few hundred meters. This proposal dovetailed with earlier hints of a deviation from the inverse-square law of Newtonian gravitation derived from measurements of the gravity profile down deep mines in Australia, and with ideas from particle physics suggesting the possible presence of very low-mass particles with gravitational-strength couplings. During the next four years numerous experiments looked for evidence of the fifth force by searching for composition-dependent differences in acceleration, with variants of the Eötvös experiment or with free-fall Galileo-type experiments. Although two early experiments reported positive evidence, the others all yielded null results. Over the range between one and 10^4 meters, the null experiments produced upper limits on the strength of a postulated fifth force between 10^{-3} and 10^{-6} of the strength of gravity. Interpreted as tests of WEP (corresponding to the limit of infinite-range forces), the results of the free-fall Galileo experiment, and of the

Eöt-Wash III experiment are shown in Figure 1. At the same time, tests of the inverse-square law of gravity were carried out by comparing variations in gravity measurements up tall towers or down mines or boreholes with gravity variations predicted using the inverse square law together with Earth models and surface gravity data mathematically 'continued' up the tower or down the hole. Despite early reports of anomalies, independent tower, borehole and seawater measurements now show no evidence of a deviation. The consensus at present is that there is no credible experimental evidence for a fifth force of nature. Some tests of composition-dependence of gravity continue with improving sensitivity in the hope of yielding tighter constraints on particle physics and on WEP. For reviews and bibliographies, see Fischbach and Talmadge (1992), Will (1990), Adelberger et al. (1991), and Fischbach et al. (1992).

4 Stellar system tests of gravitational theory

4.1 Binary pulsars and general relativity

The majority of tests of gravitational theory described so far have involved solar-system dynamics or laboratory experiments. Although the results confirm general relativity, they test only a limited portion of the 'space of predictions' of a theory. This portion corresponds to the weak-field, slow-motion, post-Newtonian limit. The 1974 discovery of the binary pulsar PSR 1913+16 by Taylor and Hulse opened up the possibility of probing new aspects of gravitational theory: the effects of strong relativistic internal gravitational fields on orbital dynamics, and the effects of gravitational radiation reaction. For reviews of the discovery and current status, see the published Nobel Prize lectures by Hulse (1994) and Taylor (1994).

The system consists of a pulsar of nominal period 59 ms in a close binary orbit with an as yet unseen companion. From detailed analyses of the arrival times of pulses (which amounts to an integrated version of the Doppler-shift methods used in spectroscopic binary systems), extremely accurate orbital and physical parameters for the system have been obtained (Table 5). Because the orbit is so close ($\approx 1 R_\odot$) and because there is no evidence of an eclipse of the pulsar signal or of mass transfer from the companion, it is generally believed that the companion is compact: evolutionary arguments suggest that it is most likely a dead pulsar. Thus the orbital motion is very clean, free from tidal or other complicating effects. Furthermore, the data acquisition is 'clean' in the sense that the observers can keep track of the pulsar phase with an accuracy of $15\mu s$, despite gaps of up to six months between observing sessions. The pulsar has shown no evidence of 'glitches' in its pulse period.

Three factors make this system an arena where relativistic celestial mechanics must be used: the relatively large size of relativistic effects [$v_{\rm orbit} \approx (m/r)^{1/2} \approx 10^{-3}$]; the short orbital period (8 hours), allowing secular effects to build up rapidly; and the cleanliness of the system, allowing accurate determinations of small effects. Just as Newtonian gravity is used as a tool for measuring astrophysical parameters of ordinary binary systems, so general relativity is used as a tool for measuring astrophysical parameters in the binary pulsar.

The observational parameters that are obtained from a least squares solution of the

Parameter	Symbol	Value
(i) 'Physical' Parameters		
Right Ascension	α	$19^h13^m12.^s46549(15)$
Declination	δ	$16°01'08.''189(3)$
Pulsar Period	P_p (ms)	$59.029997929883(7)$
Derivative of Period	$\dot{P}_p$	$8.62629(8)\times 10^{-18}$
(ii) 'Keplerian' Parameters		
Projected semimajor axis	$a_p \sin i$ (s)	$2.341759(3)$
Eccentricity	e	$0.6171309(6)$
Orbital Period	P_b (s)	$27906.9807807(9)$
Longitude of periastron	ω_0 (°)	$226.57531(9)$
Julian ephemeris date of periastron	T_0 (MJD)	$46443.99588321(5)$
(iii) 'Post-Keplerian' Parameters		
Mean rate of periastron advance	$\langle\dot\omega\rangle$ (° yr^{-1})	$4.226628(18)$
Gravitational redshift/time dilation	γ (ms)	$4.294(3)$
Orbital period derivative	$\dot{P}_b$ (10^{-12})	$-2.425(10)$

Table 5. *Parameters of the binary pulsar PSR 1913+16. (Numbers in parentheses denote errors in last digit).*

arrival time data fall into three groups: (i) non-orbital parameters, such as the pulsar period and its rate of change, and the position of the pulsar on the sky; (ii) five 'Keplerian' parameters, most closely related to those appropriate for standard Newtonian systems, such as the eccentricity e and the orbital period P_b; and (iii) five 'post-Keplerian' parameters. The five post-Keplerian parameters are $\langle\dot\omega\rangle$, the average rate of periastron advance; γ, the amplitude of delays in arrival of pulses caused by the varying effects of the gravitational redshift and time dilation as the pulsar moves in its elliptical orbit at varying distances from the companion and with varying speeds; $\dot{P}_b$, the rate of change of orbital period, caused predominantly by gravitational radiation damping; and r and $s = \sin i$, respectively the 'range' and 'shape' of the Shapiro time delay caused by the companion, where i is the angle of inclination of the orbit relative to the plane of the sky.

In general relativity, these post-Keplerian parameters can be related to the masses of the two bodies and to measured Keplerian parameters by the equations (TEGP 12.1, 14.6(a))

$$\langle\dot\omega\rangle = 3(2\pi/P_b)^{5/3}m^{2/3}(1-e^2)^{-1} \qquad (43)$$

$$\gamma = e(P_b/2\pi)^{1/3}m_2 m^{-1/3}(1+m_2/m) \qquad (44)$$

$$\dot{P}_b = -(192\pi/5)(2\pi m/P_b)^{5/3}(\mu/m)\left(1+\frac{73}{24}e^2+\frac{37}{96}e^4\right)(1-e^2)^{-7/2} \qquad (45)$$

$$s = \sin i \qquad (46)$$

$$r = m_2 \qquad (47)$$

where m_1 and m_2 denote the pulsar and companion masses, respectively, the total mass is $m = m_1 + m_2$ and $\mu = m_1 m_2/m$ is the reduced mass. The formula for $\langle \dot\omega \rangle$ ignores possible non-relativistic contributions to the periastron shift, such as tidally or rotationally induced effects caused by the companion (for discussion of these effects, see TEGP 12.1(c)). The formula for $\dot P_b$ represents the effect of energy loss through the emission of gravitational radiation, and makes use of the 'quadrupole formula' of general relativity (for a survey of the quadrupole and other approximations for gravitational radiation, see Damour 1987); it ignores other sources of energy loss, such as tidal dissipation (TEGP 12.1(f)).

The timing model that contains these parameters was developed by Damour, Deruelle and Taylor, superseding earlier treatments by Haugan, Blandford, Teukolsky and Epstein. The current values for Keplerian and post-Keplerian parameters are shown in Table 5.

The most convenient way to display these results is to plot the constraints they imply for the two masses m_1 and m_2, via Equations 43 and 44. These are shown in Figure 6. From $\langle \dot\omega \rangle$ and γ we obtain the values $m_1 = 1.4411 \pm 0.0007 M_\odot$ and $m_2 = 1.3873 \pm 0.0007 M_\odot$. Equation 45 then predicts the value $\dot P_b = -2.40243 \pm 0.00005 \times 10^{-12}$. In order to compare the predicted value for $\dot P_b$ with the observed value, it is necessary to take into account the effect of a relative acceleration between the binary pulsar system and the solar system caused by the differential rotation of the galaxy. This effect was previously considered unimportant when $\dot P_b$ was known only to 10 percent accuracy. Damour and Taylor carried out a careful estimate of this effect using data on the location and proper motion of the pulsar, combined with the best information available on galactic rotation, and found

$$\dot P_b^{\rm GAL} \simeq -(1.7 \pm 0.5) \times 10^{-14} \tag{48}$$

Subtracting this from the observed $\dot P_b$ (Table 5) gives the residual

$$\dot P_b^{\rm OBS} = -(2.408 \pm 0.010[{\rm OBS}] \pm 0.005[{\rm GAL}]) \times 10^{-12} \tag{49}$$

which agrees with the prediction, within the errors. In other words,

$$\frac{\dot P_b^{\rm GR}}{\dot P_b^{\rm OBS}} = 1.0023 \pm 0.0041[{\rm OBS}] \pm 0.0021[{\rm GAL}]. \tag{50}$$

The parameters r and s are not separately measurable with interesting accuracy for PSR 1913+16 because the orbit's 47° inclination does not lead to a substantial Shapiro delay.

The consistency among the measurements is also displayed in Figure 6, in which the regions allowed by the three most precise constraints have a single common overlap. This consistency provides a test of the assumption that the two bodies behave as 'point' masses, without complicated tidal effects, obeying the general relativistic equations of motion including gravitational radiation. It is also a test of the strong equivalence principle, in that the highly relativistic internal structure of the neutron star does not influence its orbital motion, as predicted by general relativity.

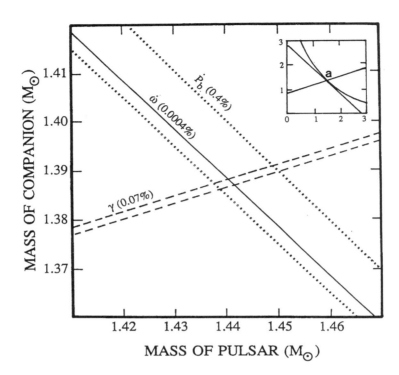

Figure 6. *Constraints on masses of pulsar and companion from data on PSR 1913+16, assuming general relativity to be valid. Width of each strip in the plane reflects observational accuracy, shown as a percentage. Inset shows the three constraints on the full mass plane; intersection region (a) has been magnified 400 times for the full figure.*

4.2 A population of binary pulsars?

In 1990, two new massive binary pulsars similar to PSR 1913+16 were discovered, leading to the possibility of new or improved tests of general relativity (see Taylor 1992 for a review of the prospects).

PSR 2127+11C.

This system appears to be a clone of the Hulse-Taylor binary pulsar: its parameters are listed in Table 6. The inferred total mass of the system is $2.706 \pm 0.011 M_\odot$. Because the system is in the globular cluster M15 (NGC 7078), it will suffer Doppler shifts resulting from local accelerations, either by the mean cluster gravitational field or by nearby stars, that are more difficult to estimate than was the case with the galactic system PSR 1913+16. This may limit the accuracy of measurement of the relativistic contribution to $\dot{P}_b$ to about 2 percent.

PSR 1534+12. This is a binary pulsar system in our galaxy. Its pulses are signifi-

Parameter	PSR 1534+12	PSR 2127+11C
(i) **'Physical' Parameters**		
Right Ascension	$15^h 34^m 47.^s 686(3)$	$21^h 27^m 36.^s 188(4)$
Declination	$12°05'45.''23(3)$	$11°57'26.''29(7)$
Pulsar Period	37.9044403665(4)	30.5292951285(9)
Derivative of Period	$2.43(8) \times 10^{-18}$	$4.99(5) \times 10^{-18}$
(ii) **'Keplerian' Parameters**		
Projected semimajor axis	3.729468(9)	2.520(3)
Eccentricity	0.2736779(6)	0.68141(2)
Orbital Period	36351.70270(3)	28968.3693(5)
Longitude of periastron	264.9721(16)	316.40(7)
Julian ephemeris date of periastron	48262.8434966(2)	47632.4672065(20)
(iii) **'Post-Keplerian' Parameters**[1]		
Mean rate of periastron advance	1.7558(1)	4.46(1)
Gravitational redshift/time dilation	2.08(3)	4.9(1.1)
Orbital period derivative	$-0.13(13)$	*
Range of Shapiro delay $r(\mu s)$	6.9(1.0)	*
Shape of Shapiro delay $s = \sin i$	0.983(5)	*

Table 6. *Parameters of new binary pulsars. ([1]From Wolszczan (1994). The asterisks indicate numbers not yet available from data.)*

cantly stronger and narrower than those of PSR 1913+16, so timing measurements are more precise, reaching $3\mu s$ accuracy. Its parameters are listed in Table 6. Because of the short data span, $\dot{P}_b$ is only on the verge of being measured, but it is expected that eventually the accuracy in its determination will exceed that of PSR 1913+16. The orbital plane appears to be almost edge-on relative to the line of sight ($i \simeq 80°$); as a result the Shapiro delay is substantial, and separate values of the parameters r and s have been obtained with interesting accuracy. This system may ultimately provide broader and more stringent tests of the consistency of general relativity than did the original binary pulsar (Damour and Taylor 1992).

4.3 Binary pulsars and alternative theories

Soon after the discovery of the binary pulsar it was widely hailed as a new testing ground for relativistic gravitational effects. As we have seen in the case of general relativity, in most respects, the system has lived up to, indeed exceeded, the early expectations.

In another respect, however, the system has only partially lived up to its promise, namely as a direct testing ground for alternative theories of gravity. The origin of this promise was the discovery that alternative theories of gravity generically predict the emission of dipole gravitational radiation from binary star systems. This additional form of gravitational radiation damping could, at least in principle, be significantly stronger than the usual quadrupole damping, because it depends on fewer powers of the parameter v/c, where v is the orbital velocity and c is the speed of light, and it depends on the gravitational binding energy per unit mass of the bodies, which, for neutron stars, could be as large as 40 per cent. As one fulfillment of this promise, Will and Eardley worked out in detail the effects of dipole gravitational radiation in the bimetric theory of Rosen, and, when the first observation of the decrease of the orbital period was announced in 1978, the Rosen theory suffered a terminal blow. It is expected that a wide class of alternative theories also fail the binary pulsar test because of dipole gravitational radiation.

On the other hand, the early observations already indicated that, in general relativity, the masses of the two bodies were nearly equal, so that, in theories of gravity that are in some sense 'close' to general relativity, dipole gravitational radiation would not be a strong effect, because of the apparent symmetry of the system (technically, the amount of dipole radiation depends on the difference between the gravitational binding energy per unit mass for the two bodies). The Rosen theory, and others like it, are not 'close' to general relativity, except in their predictions for the weak-field, slow-motion regime of the solar system. When relativistic neutron stars are present, theories like these can predict strong effects on the motion of the bodies resulting from their internal highly relativistic gravitational structure (violations of the Strong Equivalence Principle). As a consequence, the masses inferred from observations such as the periastron shift may be significantly different from those inferred using general relativity, and may be different from each other, leading to strong dipole gravitational radiation damping. By contrast, the Brans-Dicke theory, which was the basis for Eardley's discovery of the dipole radiation phenomenon, is 'close' to general relativity, roughly speaking within $1/\omega_{BD}$ of the predictions of the latter, for large values of the coupling constant ω_{BD} (henceforth this notation for the coupling constant is adopted to avoid confusion with the periastron angle). Thus, despite the presence of dipole gravitational radiation, the binary pulsar provides at present only a weak test of Brans-Dicke theory, not yet competitive with solar-system tests.

4.4 Binary pulsars and scalar-tensor theories

Making the usual assumption that both members of the system are neutron stars, and using the methods summarized in TEGP Chapters 10–12, one can obtain formulas for the periastron shift, the gravitational redshift/second-order Doppler shift parameter, and the rate of change of orbital period, analogous to Equations 43–45. These formulas depend on the masses of the two neutron stars, on their self-gravitational binding energy, represented by 'sensitivities' s and κ^* and on the Brans-Dicke coupling constant ω_{BD}. First, there is a modification of Kepler's third law, given by

$$P_b/2\pi = (a^3/\mathcal{G}m)^{1/2}. \tag{51}$$

Then, the predictions for $\langle\dot{\omega}\rangle$, γ and $\dot{P}_b$ are

$$\langle\dot{\omega}\rangle = 3(2\pi/P_b)^{5/3}m^{2/3}(1-e^2)^{-1}\mathcal{P}\mathcal{G}^{-4/3} \tag{52}$$

$$\gamma = e(P_b/2\pi)^{1/3}m_2 m^{-1/3}\mathcal{G}^{-1/3}(\alpha_2^* + \mathcal{G}m_2/m + \kappa_1^*\eta_2^*) \tag{53}$$

$$\dot{P}_b = -(192\pi/5)(2\pi m/P_b)^{5/3}(\mu/m)\mathcal{G}^{-4/3}F(e)$$
$$\quad -4\pi(2\pi m/P_b)(\mu/m)\xi\mathcal{S}^2 G(e) \tag{54}$$

where, to first order in $\xi \equiv (2+\omega_{BD})^{-1}$, we have

$$F(e) = \frac{1}{12}(1-e^2)^{-7/2}[\kappa_1(1+\frac{7}{2}e^2+\frac{1}{2}e^4) - \kappa_2(\frac{1}{2}e^2+\frac{1}{8}e^4)] \tag{55}$$

$$G(e) = (1-e^2)^{-5/2}(1+\frac{1}{2}e^2) \tag{56}$$

$$\mathcal{S} = s_1 - s_2 \tag{57}$$

$$\mathcal{G} = 1 - \xi(s_1 + s_2 - 2s_1 s_2) \tag{58}$$

$$\mathcal{P} = \mathcal{G}[1 - \frac{2}{3}\xi + \frac{1}{3}\xi(s_1+s_2-2s_1s_2)] \tag{59}$$

$$\alpha_2^* = 1 - \xi s_2 \tag{60}$$

$$\eta_2^* = (1-2s_2)\xi \tag{61}$$

$$\kappa_1 = \mathcal{G}^2[12(1-\frac{1}{2}\xi) + \xi\Gamma^2] \tag{62}$$

$$\kappa_2 = \mathcal{G}^2[11(1-\frac{1}{2}\xi) + \frac{1}{2}\xi(\Gamma^2 - 5\Gamma\Gamma' - \frac{15}{2}\Gamma'^2)] \tag{63}$$

$$\Gamma = 1 - 2(m_1 s_2 + m_2 s_1)/m \tag{64}$$

$$\Gamma' = 1 - s_1 - s_2. \tag{65}$$

The quantities s_a and κ_a^* are defined by

$$s_a = -\left(\frac{\partial(\ln m_a)}{\partial(\ln G)}\right)_N \qquad \kappa_a^* = -\left(\frac{\partial(\ln I_a)}{\partial(\ln G)}\right)_N \tag{66}$$

and measure the 'sensitivity' of the mass m_a and moment of inertia I_a of each body to changes in the scalar field (reflected in changes in G) for a fixed baryon number N (see TEGP 11, 12 and 14.6(c) for further details).

The first term in $\dot{P}_b$ is the effect of quadrupole and monopole gravitational radiation, while the second term is the effect of dipole radiation. Equations 51–57 are quite general, applying to a class of 'conservative' theories of gravity whose equations of motion for compact objects can be described by a Lagrangian (TEGP 11.3). The forms of the variables in the remaining equations are specific to Brans-Dicke theory. For a discussion of post-Keplerian parameters applicable to generalized scalar-tensor theories see Damour (1992).

In order to estimate the sensitivities s_1, s_2, $\mathcal{S}$ and κ^*, we must adopt an equation of state for the neutron stars. We restrict attention to relatively stiff neutron star equations of state in order to guarantee neutron stars of sufficient mass, approximately $1.4 M_\odot$; the M and O equations of state of Arnett and Bowers give similar results. The lower limit on ω_{BD} required to give consistency among the constraints on $\langle\dot{\omega}\rangle$, γ and $\dot{P}_b$ as in Figure 6 is 105. The combination of $\langle\dot{\omega}\rangle$ and γ give a constraint on the masses that is relatively weakly dependent on ξ, thus the constraint on ξ is dominated by $\dot{P}_b$ and is directly proportional to the measurement error in $\dot{P}_b$; in order to achieve a constraint comparable to the solar system value of 10^{-3}, the error in $\dot{P}_b^{OBS}$ would have to be reduced by a factor of ten.

Damour and Esposito-Farèse (1992) have devised a scalar-tensor theory in which two scalar fields are tuned so that their effects in the weak-field slow-motion regime of the solar system are suppressed, so that the theory is identical to general relativity in the post-Newtonian approximation. Yet in the regime appropriate to binary pulsars, it predicts strong-field SEP-violating effects and radiative effects that distinguish it from general relativity. It gives formulae for the post-Keplerian parameters of Equations 52–54 as well as for the parameters r and s that have corrections dependent upon the sensitivities of the relativistic neutron stars. The theory depends upon two arbitrary parameters β' and β''; general relativity corresponds to the values $\beta'=\beta''=0$. It turns out that the binary pulsar PSR 1913+16 alone constrains the two parameters to a narrow but long strip in the (β', β'')-plane that includes the origin (general relativity) but that could include some highly non-general relativistic theories. The sensitivity of PSR 1534+12 to r and s provides an orthogonal constraint that cuts the strip. In this class of theories, then, *both* binary pulsars are needed to provide a strong test. Interestingly, it is the strong-field aspects of PSR 1534+12, not gravitational radiation, that provide the non-trivial new constraint.

5 Conclusions

In 1995 we find that general relativity has held up under extensive experimental scrutiny. The question then arises, why bother to continue to test it? One reason is that gravity is a fundamental interaction of nature, and as such requires the most solid empirical underpinning we can provide. Another is that all attempts to quantize gravity and to unify it with the other forces suggest that gravity stands apart from the other interactions in many ways, thus the more deeply we understand gravity and its observational implications, the better we may be able to confront it with the other forces. Finally, and most importantly, the predictions of general relativity are fixed; the theory contains no adjustable constants so nothing can be changed. Thus every test of the theory is potentially a deadly test. A verified discrepancy between observation and prediction would kill the theory, and another would have to be substituted in its place. Although it is remarkable that this theory, born 80 years ago out of almost pure thought, has managed to survive every test, the possibility of suddenly finding a discrepancy will continue to drive experiments for years to come.

Acknowledgments

This work was supported in part by the National Science Foundation, Grant Number PHY 92-22902, and the National Aeronautics and Space Aministration, Grant Number NAGW 3874.

References

Adelberger E G, Heckel B R, Stubbs C W and Rogers W F, 1991, *Ann Rev Nucl Particle Sci* **41** 269.
Bell J F, 1995, *Astrophys. J.*, in press.
Damour T, 1987, in *300 Years of Gravitation*, eds Hawking S W and Israel W, 128–198 (Cambridge University Press, Cambridge).

Damour T, 1992, *Phil Trans R Soc London* **341** 135.
Damour T, 1995, in *Proc Seventh Marcel Grossman Meeting on General Relativity*, in press.
Damour T and Esposito-Farèse G, 1992, *Classical Quantum Grav* **9** 2093.
Damour T and Esposito-Farèse G, 1994, *Phys Rev D* **49** 1693.
Damour T and Nordtvedt K, 1993a, *Phys Rev Lett* **70** 2217.
Damour T and Nordtvedt K, 1993b, *Phys Rev D* **48** 3436.
Damour T and Taylor J H, 1992, *Phys Rev D* **45** 1840.
Damour T and Vokrouhlický D, 1995, *Phys Rev D*, in press.
Dickey G O, Bender P L, Faller J E, Newhall X X, Ricklefs R L, Ries J G, Shelus P J, Veillet C, Whipple A L, Wiant J R, Williams J G and Yoder C F, 1994, *Science* **265** 482.
Fischbach E, Gillies G T, Krause D E, Schwan J G, and Talmadge C, 1992, *Metrologia* **29** 213.
Fischbach E and Talmadge C, 1992, *Nature* **356** 207.
Godone A, Novero C and Tavella P, 1995, *Phys Rev D* **51** 319.
Hulse R A, 1994, *Rev Mod Phys* **66** 699.
Misner C W, Thorne K S and Wheeler J A, 1973 *Gravitation* (W H Freeman, San Francisco).
Nordtvedt K, 1995, *Icarus* **114** 51.
Prestage J D, Tjoelker R L and Maleki L, 1995, *Phys Rev Lett* **74** 3511.
Steinhardt P J and Will C M, 1995, *Phys Rev D*, in press.
Su Y, Heckel B R, Adelberger E G, Gundlach J H, Harris M, Smith G L and Swanson H E, 1995, *Phys Rev D* **50** 3614.
Taylor J H, 1992, *Phil Trans R Soc London* **341** 117.
Taylor J H, 1994, *Rev Mod Phys* **66** 711.
Thorne K S, 1987, in *300 Years of Gravitation*, eds Hawking S W and Israel W, 330–458 (Cambridge University Press, Cambridge).
Thorne K S, 1995, in *Proceedings of the Snowmass 94 Summer Study on Particle and Nuclear Astrophysics and Cosmology*, eds Kolb E W and Peccei R (World Scientific, Singapore), in press.
Weinberg S W, 1972, *Gravitation and Cosmology* (Wiley, New York).
Will C M, 1976, *Astrophys J* **204,** 224.
Will C M, 1987, in *300 Years of Gravitation*, eds Hawking S W and Israel W, 80–127 (Cambridge University Press, Cambridge).
Will C M 1990, *Sky and Telescope* **80** 472.
Will C M, 1992a, *Int J Mod Phys D* **1** 13.
Will C M, 1992b, *Phys Rev D* **45** 403.
Will C M, 1992c, *Astrophys J Lett* **393,** L59.
Will C M, 1993a, *Was Einstein Right?* (Basic Books, New York).
Will C M, 1993b, *Theory and Experiment in Gravitational Physics. Revised Edition* (Cambridge University Press, Cambridge), referred to as TEGP.
Wolszczan A, 1994, *Classical Quantum Grav* **6** A227.

Algebraic Computing in General Relativity

Ray d'Inverno

Faculty of Mathematical Studies
Southampton University

1 Ten Questions about Algebraic Computing

1.1 Why?

In 1916 Einstein proposed his full field equations for the gravitational field in the presence of matter, namely

$$G_{ab} = 8\pi T_{ab} \tag{1}$$

where G_{ab} is the Einstein tensor and T_{ab} the energy-momentum tensor for the matter present. (See, for example, d'Inverno (1992)). He was somewhat sceptical about the status of these equations but was less so about the equations in the absence of matter, that is the vacuum field equations, which are usually written in the form

$$R_{ab} = 0 \tag{2}$$

where R_{ab} is the Ricci tensor. These consist of ten non-linear partial differential equations for the space-time metric g_{ab}. The non-linearity means that they constitute a highly non-trivial set of equations and, indeed originally, Einstein thought that it would be impossible to find an exact solution of the equations. It came as something of a surprise that within a year of publication of the equations an exact solution was found. This was the famous Schwarzschild solution (1916)

$$ds^2 = \left(1 - \frac{2m}{r}\right) dt^2 - \left(1 - \frac{2m}{r}\right)^{-1} dr^2 - r^2 d\theta^2 - r^2 \sin^2\theta \, d\phi^2 \tag{3}$$

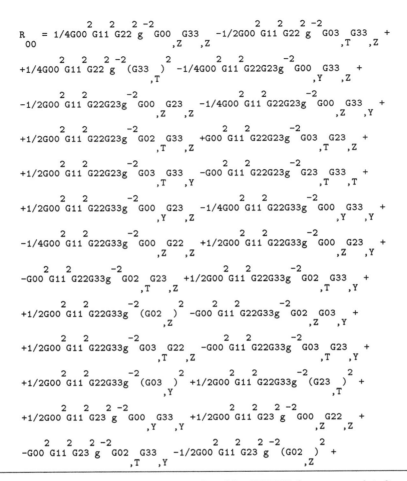

Figure 1. The first 26 terms in R_{00} produced by SHEEP for a general 4-dimensional metric

which was obtained by making the simplest assumptions possible of a static, spherically-symmetric solution. This was the start of the study of exact solutions in General Relativity which, even to-day, is probably the biggest field of enquiry in the classical theory.

The complexity of the field equations and the seemingly impossible task of solving them in general can best be appreciated by writing out the vacuum field equations (2) for a completely general metric

$$g_{ab} = g_{ab}(x^c) \tag{4}$$

which depends on all four coordinates $(x^c) = (x^0, x^1, x^2, x^3) = (t, x, y, z)$, say. The first 26 terms in the R_{00} equation are shown in Figure 1. Note that its general form is a sum of terms, where each term is a product of degree 8 (treating the inverse of the determinant of the metric g^{-1} as a factor of degree 1). R_{00} has some 8,575 terms in it and R_{ab} has over 100,000 terms in it. This calculation was carried out by the algebraic computing system SHEEP and took about 2 hours on a Sun 4 workstation. The thought of undertaking this calculation by hand is quite daunting: although the calculation is straightforward, that is to say it is quite clear what one needs to do at each stage. It would obviously be extremely tedious to undertake by hand as it would involve a massive calculational effort and, in all likelihood, errors would crop up which would lead to the wrong answer.

It was precisely for reasons like this that algebraic computing first came into existence. In the words of Jean Sammett in a general review article on the emerging field in 1966 (Sammet 1966) 'It has become obvious that there are a large number of problems requiring very tedious, time-consuming, error-prone and straightforward algebraic manipulation, and these characteristics make computer solution both necessary and desirable'.

1.2 What?

Algebraic computing is the field of using computers for carrying out algebraic calculations. It is also frequently referred to as computer algebra. The advent of high speed digital computers has revolutionized many fields of enquiry in the last 50 years. How does algebraic computing fit into computing generally? I have attempted a rough placement in Figure 2. Computers have principally been used for information processing, especially in the commercial arena, and number crunching, particularly in the scientific arena. There are a number of other applications which do not fit into these two general areas. Perhaps the most important application area within these is that of Artificial Intelligence (AI for short). AI in turn can be broken down into various specific sub-areas such as theorem proving, automatic programming, perception and pattern recognition, problem solving, game playing, natural language processing, robotics and expert systems. These divisions are somewhat artificial and are certainly not mutually exclusive. Algebraic computing currently plays an important rôle in at least the first four areas. It was certainly the province of AI historically. The high level programming language LISP is the most important language in AI and was developed in the late 50's by John McCarthy who had applications in algebra specifically in mind in its design. Again, early algebraic computing systems used standard AI techniques such as pattern recognition and heuristic search.

As these systems grew, better algorithms were developed. An algorithm is a step by step method for solving a particular problem. There is a purist school of thought in AI which argues that if a problem can be reduced to an algorithm then it is not part of the province of AI. This school of thought would argue that algebraic computing, now being algorithm based, is no longer part of AI. This would seem to be an extreme viewpoint since, in the end, computers can only implement algorithms and so nothing in that case would merit the term AI. However, even if algebraic computing is no longer a branch of AI, it originated from the field and has now become an important tool in several application areas.

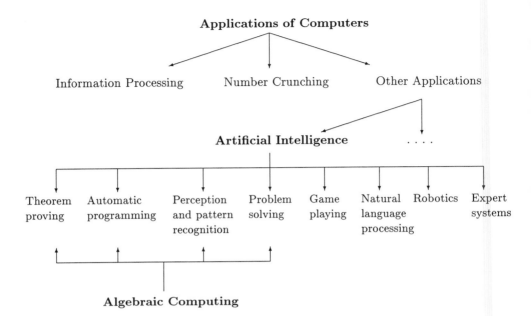

Figure 2. *Algebraic Computing in the arena of Applications of Computing*

1.3 When?

The earliest known program for carrying out algebra on a computer was one written in 1953, which was a simple program for carrying out symbolic differentiation. John McCarthy's LISP became generally available around 1960. However, it was not until the mid 60's that serious algebra systems were developed. They emerged from two quite distinct backgrounds. There were those which were perhaps more theoretically driven and came from a computer science background. The other camp consisted of theoretical physicists interested in getting answers to real problems. Included in this latter was Tony Hearn's REDUCE system, which was originally designed for applications concerned with computing Feynman diagrams in Quantum Electrodynamics. Another example was Barton and Fitch's CAMAL which was designed for calculations in Celestial Mechanics. My own system LAM (Lisp Algebraic Manipulator) was designed for standard calculations mostly connected with exact solutions in General Relativity. Perhaps not surprisingly, the theoretical systems did not cope well with real problems, whereas the theoretical physicists' systems coped extremely well. The two communities learned a lot from each other in subsequent years and the distinction between them blurred.

As the bigger systems grew in size and capability, newer, often special purpose, systems came along, designed for very specialized use and usually only employed by a handful of people. By around 1985 it was estimated that there were over 100 algebra systems in existence. To-day the field is dominated by the 'big four' general purpose systems REDUCE, MATHEMATICA, MAPLE and MACSYMA. In General Relativity, the special purpose system SHEEP, the successor to LAM, is one of the most frequently used systems. IBM invested a considerable effort over the years in the system SCRATCHPAD, but it was never commercially released. However, a modern variant of it called AXIOM is now available on a limited range of machines. AXIOM is rather different in character from other pre-existing systems and would appear to be the first of the next generation of algebra systems. It allows the user considerable flexibility and one can even have control at a very low level on the nature of the underlying algebraic structures. The systems divide into LISP-based ones and the rest (see Table I). LISP is the most frequently used base on which algebra systems are built and we next discuss briefly the reasons why.

	LISP-based	Other
General purpose Algebra Systems	REDUCE MACSYMA AXIOM DERIVE	MAPLE MATHEMATICA
Special Purpose General Relativity Systems & Packages	SHEEP CLASSI STENSOR ORTHOCARTAN EXCALC (REDUCE) REDTEN (REDUCE) TENSOR-UTILITY (DERIVE)	GRTENSOR (MAPLE) CARTAN (MAPLE) ITENSOR (MACSYMA) CARTAN (MATHEMATICA)

Table 1. *The major algebra systems in use in General Relativity*

1.4 LISP?

LISP stands for LISt Processing. A list is represented by an opening left bracket, a sequence of elements separated by spaces and a closing right bracket. Thus

$$(A \ B \ C)$$

represents the list of three elements: A, B and C in order. LISP is designed to create and modify such structures. We will not go into the detail of how it does so here (but see d'Inverno (1980) for a brief introduction). The first thing to note is that if we convert the standard mathematical infix notation for algebraic expressions into an equivalent prefix notation the conversion to a list structure is fairly automatic. For example, the expression

$$\frac{3}{4} + a + x^2 + \log(\sin(x)) - axy + a_{,y} \tag{5}$$

gets converted in LAM into

$$(+\ (3\ 4)\ A\ (**\ X\ (2\ 1))\ (\text{LOG}\ (\text{SIN}\ X))\ (-\ (*\ A\ X\ Y))\ (\&\ A\ 1)) \qquad (6)$$

Historically, lower case expressions were represented by their upper case equivalents. This is not necessary with current day systems, but we shall mostly stick to this convention. Note that in LAM the operator plus (+) takes an indefinite number of arguments, as does times (*); exponentiation (**), that is raising to a power, takes two, namely the base and the exponent; and minus (−), log (LOG) and sine (SIN) all take one argument. All numbers are rational pairs or algebraic concomitants of such pairs. The differentiation operator & is followed by the name of the function being differentiated and a list of numbers corresponding to the variables with respect to which the function is being differentiated.

A powerful feature of LISP is that it enables recursive functions to be defined easily. A recursive function is one defined in terms of itself. For example, to differentiate a sum of terms we differentiate each term in turn and add the result together. We represent this symbolically as

$$(+\ A1\ A2\ \ldots\ AN)' \longrightarrow (+\ A1'\ A2'\ \ldots\ AN')$$

where prime denotes differentiation and $A1, A2$, etc denote general algebraic expressions. (In fact, although it is natural to define many algebraic operations recursively, it usually turns out to be more efficient to convert the recursive definition into an equivalent iterative one—something which can always be accomplished).

The main reason for preferring LISP is that it possesses a garbage collector. The major difference in algebra calculations over other types of calculation is, as we shall see, that of 'intermediate term swell'. The intermediate terms that crop up in a calculation can grow to such an extent that they fill up the memory and cause the program to halt. However, many of these terms are only created in passing and are no longer required. They have become 'garbage'. A garbage collector reclaims the storage occupied by the garbage and, assuming that it has found some, allows the program to start up again. This process happens automatically in LISP.

Most LISP systems possess a compiler which translates LISP code into machine code and so allows the program to run significantly faster. The compiler works in two stages: first it converts LISP code to an equivalent assembly code and, in a second pass, converts the assembly code to machine code. In fact, the first serious version of LAM, called ALAM (Atlas LAM) (d'Inverno 1969), was written directly in assembly code for the Atlas 1 computer, which meant that it could be written very efficiently. Another advantage of using LISP is that it encourages the use of various programming paradigms. For example, LAM was one of the first systems to exploit the use of an object-centred paradigm in the way in which it carried out differentiation. Instead of writing a program which tests to see what sort of operator it is dealing with before dispatching expressions to the various rules which the different operators possess, the operators are themselves the names of the differentiation routines. Thus, in essence, LAM gets an expression to differentiate itself simply by 'evaluating' it.

1.5 Where?

Algebraic computing became established originally in physics. We have already mentioned REDUCE in connection with Quantum Electrodynamics, CAMAL with Celestial Mechanics and LAM with General Relativity. The use then spread to Quantum Mechanics, QCD and Fluid Mechanics. However, it was some time before its capabilities were appreciated in other fields such as engineering and mathematics. Given that much of the underlying theory is concerned with factorisation of polynomials and integration, it is rather surprising that it has only had an impact on mathematics in recent times.

1.6 On?

The major problem with algebra systems historically was that they were machine-specific. If you did not have access to the right kind of mainframe, then you could not use the systems. For example, ALAM only worked on the Atlas 1, CLAM (its immediate successor) only on the CDC 6600 range and ILAM only on the IBM 390 series. Moreover, because the systems make very big demands on memory they only ran on mainframes. The trend has been to make the systems more portable. Those written in C are highly portable, whereas those written in LISP generally are not. Unfortunately there are a large number of inequivalent LISP dialects in existence. Sometimes it is relatively easy to write a module which converts one dialect into another. These modules are typically composed of LISP macros. A macro is a device for converting one piece of LISP into another prior to evaluation. Thus macros can effectively replace a piece of code which works in one dialect by its equivalent in another dialect.

A move was made to make LISP more standard by defining a small subset called Standard LISP and it was hoped that it would become a target language for defining LISP systems. Standard LISP was superseded by PSL (Portable Standard Lisp), which it was hoped would live up to its name. The aspiration was only partially successful. For example, a typical SHEEP implementation requires a module to convert Standard LISP to SHEEP LISP (previously called EXLISP) which is the LISP dialect in which SHEEP itself is written. It also usually needs at least one module to convert Standard LISP into the underlying LISP system.

As the processing power of machines has increased and the cost of memory reduced, so algebra systems have been made to work on smaller machines. It is still probably the case that most serious systems require a minicomputer or workstation to reside on. However, there are some useful PC versions of most of the big systems in existence. It would only seem to be a question of time before algebra systems are available on pocket calculators (indeed Hewlett-Packard have had a prototype for some time).

1.7 How?

In this section we wish to describe in outline some of the issues which affect how algebra systems work. The central issue is the need for simplification. This is needed first of all for compactness, since humans can only make sense of relatively short expressions. We give a simple example of the way that intermediate expressions grow in a typical algebraic calculation, from which it should be clear that simplification is needed to reduce the expressions down again. Consider the recurrence relation for defining the nth Legendre polynomial

$$p_n = xp_{n-1} + \frac{x^2-1}{n}\frac{d}{dx}(p_{n-1}) \quad \text{where} \quad p_1 = x$$

A direct computation of p_3 might lead to expressions like the following:-

$$\begin{aligned}
p_3 &= xp_2 + \frac{x^2-1}{3}\frac{d}{dx}(p_2) \\
&= x\left(xp_1 + \frac{x^2-1}{2}\frac{d}{dx}(p_1)\right) + \frac{x^2-1}{3}\frac{d}{dx}\left(xp_1 + \frac{x^2-1}{2}\frac{d}{dx}(p_1)\right) \\
&= x\left(x.x + \frac{x^2-1}{2}.1\right) + \frac{x^2-1}{3}\frac{d}{dx}\left(x.x + \frac{x^2-1}{2}.1\right) \\
&= x\left(x^2 + \frac{x^2-1}{2}\right) + \frac{x^2-1}{3}\frac{d}{dx}\left(x^2 + \frac{x^2-1}{2}\right) \\
&= x\left(x^2 + \frac{x^2-1}{2}\right) + \frac{x^2-1}{3}\left(2x + \frac{2x}{2}\right) \\
&= x\left(\frac{3}{2}x^2 - \frac{1}{2}\right) + \frac{x^2-1}{3}(3x) \\
&= \frac{3}{2}x^3 - \frac{x}{2} + x^3 - x \\
&= \frac{5}{2}x^3 - \frac{3}{2}x
\end{aligned}$$

There are examples in which the number of intermediate terms have been in the tens of thousands and yet ultimately reduce to just one or two terms on simplification.

The other key reason for simplification is the zero-identity problem. From one point of view, much of science and mathematics is concerned with equations, and equations can always be rewritten so that one side of them is zero. In this sense, it is a central problem to decide whether or not an expression reduces to zero. This might be simple if there were a unique way of representing expressions, but there is not. For example, consider an expression involving powers of $\sin\theta$ and $\cos\theta$. There are clearly equivalent but different expressions which can be obtained from it by making different uses of the identity

$$\sin^2\theta + \cos^2\theta = 1 \tag{7}$$

Unfortunately the problem is even more profound because Richardson showed in 1966 that the problem is non-decidable. To be precise, he showed that if you construct an algebraic expression using the operators $+, *, /$, substitution, $\sin, \exp, \log$, and mod together with the integers, π and a variable x, then it is not possible to decide in a finite number of steps whether or not the expression is zero. This would seem to place an intrinsic limitation on the capability of an algebra system. There are two responses to this limitation. The first is theoretical and consists of attempting to define classes of algebraic expressions which possess canonical forms. Since a canonical form is unique it will determine unambiguously whether or not an expression is zero. An important example is the class of polynomials. Clearly a polynomial

$$a_0 + a_1x + a_2x^2 + \cdots + a_nx^n \tag{8}$$

will be zero if and only if all the coefficients $a_0, a_1, \ldots, a_n$ are zero. Many early algebra systems were polynomial manipulators, CAMAL being an example of one of the more successful of them. The other approach is rather more *ad hoc* and consists of trying to develop practical procedures which generally work. The basic design of LAM follows what a human does in a calculation, namely it first removes zeros and then expands everything out fully and collects up like terms. This procedure is supplemented by allowing the use of side relations or substitutions. This simple minded approach does not always produce the simplest answer, but it has certainly proved its worth in innumerable calculations.

There are two major areas of research in algebraic computing, namely factorisation of polynomials and integration. There have been considerable advances in devising efficient algorithms for factorising polynomials in recent years. Many of the algorithms involve factorisation over prime fields. All the big systems possess factorisation capabilities, but the algorithms used vary from system to system. It is fairly obvious that symbolic differentiation is algorithmic in character—it simply consists of a set of well defined rules. What is perhaps surprising is that the inverse operation of integration is, when suitably defined, algorithmic as well.

The earliest special purpose integration system, called SAINT, used AI techniques to search for integrals. Thus the system would attempt various approaches such as standard tables, integration by parts, substitution, etc and make heuristic estimates at each stage of which approach was most likely to lead to a successful conclusion. It could perform at a level comparable to a freshman university student. The situation was revolutionised by Risch who rediscovered some earlier work of Liouville. The resulting Risch algorithm is a decision algorithm which can determine whether or not an expression can be integrated within a given class of functions and, when it can, then the algorithm determines the integral. In essence, the algorithm guesses a general standard form for the integral and by differentiating back it is able to determine the unknown coefficients in the form. The algorithm has been developed further by Norman and Davenport in recent years using REDUCE as the platform (MacCallum and Wright 1991) and it is now extremely powerful (although it still has difficulties with nested radicals). It also utilizes polynomial factorisation.

The algebra system LAM was the first system to produce a two-dimensional output involving superscripts and subscripts. However, it required the user to convert input expressions from infix to prefix notation. This was rectified by CLAM (d'Inverno and Russell-Clark 1971a), where the user could input expressions in infix form, and was developed further in SHEEP which possesses a user-friendly, interactive environment for input. Once algebra systems became more widely available then other issues grew in importance such as the documentation available on the system, the system's user-friendliness and the status of its development. A modern system like MAPLE has a manual, a tutorial program and on-line help information. Basically you just need to type '?' to get into the help system, and a command like '? int' will produce help on integration. Moreover, it is continuously being developed and new versions of the system are released on a regular basis. The current status of SHEEP is that it was designed to be user-friendly, but it has virtually no help facilities and development of the system has virtually ceased. However, the status of documentation has improved radically with the work of MacCallum *et al.* (1994). This gives an excellent introduction to SHEEP, and the two extensions CLASSI and STENSOR, as well as including articles on the application of REDUCE and MAPLE to General Relativity.

1.8 Examples?

One of the most famous examples of the capabilities of algebra systems occurs in a calculation in Celestial Mechanics. Delaunay completed a hand calculation in 1867 to determine the moon's position as a function of time, which had taken 10 years to complete by hand and a further 10 years to check and publish. This was checked by an algebra system in 1970 and took about 20 hours to complete. It discovered one main error and two others which stemmed from this error. This main error was in a quantity called R. Delaunay had obtained

$$R = \frac{\mu}{2a} + \cdots + \frac{23}{16}\gamma^2 e'^2 + \cdots \qquad (9)$$

whereas the fraction in this term should have been $\frac{33}{16}$!

Tables of integrals are used in engineering on a frequent basis. These standard references were checked using MACSYMA and it was found that the typical error rates ran at about 10% and that there was one key reference which has a 23% error rate! Although many of these errors are in the detail, and involve such things as bounds on the constants of integration which occur, nonetheless it demonstrates the reliability of modern algebra systems.

When LAM was first developed it was used principally for calculations concerned with Bondi's work on radiation from an isolated source. Bondi's radiating metric is

$$ds^2 = \left(\frac{V}{r}e^{2\beta} - U^2 r^2 e^{2\gamma}\right) du^2 + 2e^{2\beta} du\, dr + 2Ur^2 e^{2\gamma} du\, d\theta - r^2 e^{2\gamma} d\theta^2 - r^2 e^{-2\gamma} \sin^2\theta d\phi^2 \qquad (10)$$

where $(x^a) = (x^0, x^1, x^2, x^3) = (u, r, \theta, \phi)$ and V, U, β and γ are arbitrary functions of u, r and θ. The original hand calculations had extended over a period of some 6 months. ALAM took about 4 minutes to process the metric and found 6 errors in the key reference Bondi, van der Burg and Metzner (1962). SHEEP takes about a second to carry out a similar calculation. In work with Tony Russell-Clark in 1970, 40 metrics originally obtained by B. K. Harrison were investigated using CLAM (d'Inverno and Russell-Clark 1971b). These were supposedly vacuum metrics, but since they had only been obtained by indirect methods there had not previously been a direct check of their properties. We were able to show that 4 of them were not in fact vacuum solutions. We estimated that this work represented a calculation which would take more than a life-time to complete by hand. With just one of the metrics having the complexity of that shown in Figure 3, it is easy to see why.

1.9 Significance?

Why should one bother to undertake huge algebraic calculations in the first place? First of all, it is often the case that these calculations lead to simple results, and simple

$$ds^2 = -\left(\frac{\sqrt{3}x_0}{3\ell} + \frac{x_1^2}{12\ell^2}\right)^{1+\sqrt{3}} \left(\exp\int\frac{vdx_3}{x_3}\right)^{(1+2\sqrt{2})/3} \left(\frac{x_3}{\ell}\right)^{-(1+\sqrt{3})/3} dx_0^2$$
$$+ \left(\frac{\sqrt{3}x_0}{3\ell} + \frac{x_1^2}{12\ell^2}\right)^{2+\sqrt{3}} \left(\exp\int\frac{vdx_3}{x_3}\right)^{(1+2\sqrt{3})/3} \left(\frac{x_3}{\ell}\right)^{-(1/\sqrt{3})} dx_1^2$$
$$+ \left(\frac{\sqrt{3}x_0}{3\ell} + \frac{x_1^2}{12\ell^2}\right)^{-\sqrt{3}} \left(\exp\int\frac{vdx_3}{x_3}\right)^{-1} \left(\frac{x_3}{\ell}\right)^{1/\sqrt{3}} dx_2^2$$
$$+ \left(\frac{\sqrt{3}x_0}{3\ell} + \frac{x_1^2}{12\ell^2}\right)^{3+\sqrt{3}} \left(\exp\int\frac{vdx_3}{x_3}\right)^{(1+2\sqrt{3})/3} \left(\frac{x_3}{\ell}\right)^{-(2+\sqrt{3})/3} \left(\frac{v^2-\ell}{x_3-\ell}\right) dx_3^2$$

where ℓ is a constant and $v = v(x_3)$ satisfies the ordinary differential equation

$$\frac{dv}{dx_3} = \left(\frac{v^2-1}{4x_3}\right)\left(\frac{2\ell v}{x_3-\ell} + \frac{4}{\sqrt{3}}\right)$$

Figure 3. *The Harrison metric II-A-7*

algebraic expressions can provide insight. One need only quote Newton's universal law of gravitation for the magnitude of the force between two particles of masses m_1 and m_2 separated by a distance r

$$F = \frac{Gm_1m_2}{r^2} \tag{11}$$

to understand the significance of this remark. Moreover, algebraic answers are always exact; there is no problem with the rounding errors which occur in numerical calculations. There are plenty of examples where huge calculations have resulted unexpectedly in simple answers. Further investigation has lead to the discovery that there is some underlying structure which had not been appreciated before, and the simple answer can be established by a different route. The point is that the complexity of calculations like these might well have not been tackled without the aid of algebra systems. So, in short, they expand the frontiers of research since they make previous intractable problems tractable and stimulate alternative methods. A typical scenario in which algebra systems can contribute is when problems can only be tackled by approximate methods such as asymptotic analyses. An algebra system can compute more terms in such an expansion and hence improve on the order of the approximation. There can also be situations where it is simply more economic to undertake a calculation (wholly or in part) algebraically rather than numerically.

What impact have these systems had to date? This is difficult to quantify, but let us just restrict attention to General Relativity. In the early days of using algebra systems it was quite normal to cite their use; but as the usage spread, there were less explicit references in the literature. This is not surprising since they have now become standard tools. If one were to attempt some quantification of this particular usage then I would guess that something like 5% of physicists work in General Relativity; of these perhaps 20% have worked in exact solutions at some time or another, and I would imagine that these days the majority would employ an algebra system. So in this area alone the rôle of the systems is quite central. I estimated in 1967 that it would take about 25 years before these systems became commonplace tools. I do not think I was that far out.

1.10 Future?

It would appear that AXIOM is the first of the next generation of algebra systems. It is based on abstract data types organized into algebraic hierarchies, and exploits object orientated programming technologies. It also provides a highly sophisticated working environment based around high quality graphics and on-line help systems. It has an extensive library whose topics gives a good idea of the current capabilities of algebra systems. The library includes:-

- Arbitrary precision integers, rationals and floating-point numbers
- Partial fractions, continued fractions
- Complex elementary and special functions
- Operators
- Orthogonal polynomials
- Polynomial and number factorisation
- Symbolic roots of polynomials
- Solution of systems of linear and polynomial equations
- Matrices and vectors
- Eigenvalues and eigenvectors
- Symbolic differentiation and integration
- Laplace transforms
- Power series
- Limits
- Symbolic solution of linear ODE's
- Power series solutions of nonlinear ODE's
- Finite fields (modular arithmetic, prime fields)
- Primary decomposition of polynomial ideals
- Computation of Galois groups
- Permutations and finite group theory
- Clifford algebras
- Padé approximants
- Number theory functions

Despite AXIOM's sophistication, it is still an algebra system which is designed for use in the traditional rôle of a calculational tool. The computer is simply being used in a passive rôle. In a scenario proposed by John Elvey, he looks forward to a time when computers are used for research in a much more active way. We shall only give an outline of Elvey's thinking. He calls the computing environment SIGMA (Simulation and Implementation of General Mathematical Activity). He sees it as a research tool with which one can experiment in the hopes of discovering new theorems or, if exact results are inappropriate to the domain of discourse, then the system allows one to resort to the appropriate approximation model. SIGMA is seen as a time-developing environment for investigation. Access to the system would be shared by a whole community of users. A menu would allow one to home in on the specific field of interest. It would have a highly sophisticated environment for communication, including the easy representation and manipulation of mathematical symbols. At its base the system would have an algebra system similar in character to AXIOM. It would include an AI processor possessing standard tools such as heuristic search and automatic theorem provers. In addition, it would possess an operational information bank containing, for example, definitions, theorems, procedures, comments and so on, as well as a special module for selecting and applying appropriate approximation modes. A typical usage might be in trying to establish a new theorem in some field of mathematics. On the basis of the system's current state of knowledge of the field, man-machine interaction would set up conjectures and try to prove them. Any resulting new theorem would then be added to the data base, illustrating its nature as a time developing environment for investigation.

2 Introduction to SHEEP

2.1 Why SHEEP?

SHEEP has probably been the most used algebra system for calculations in General Relativity. It is small and fast and is specifically designed for calculations in the exact solutions area, especially those concerned with classifying solutions. It was written by Inge Frick as a development of LAM and CLAM (Frick 1977). Although there are equivalent packages on other systems such as MAPLE and REDUCE, the sophistication of these systems often results in much bigger memory requirements and running times. For example, SHEEP does not possess any factorization capabilities, which means that there are certain applications where it is easier to use general purpose packages. However, even in these cases it is often possible to exploit the substitution facilities to achieve equivalent results. The usability of SHEEP has increased significantly with the work of MacCallum et al. (1994). Although SHEEP possesses a manual, it is not particularly easy to use and it is incomplete. The purpose of the rest of this chapter is to give you an idea of the capabilities of the system, but without going into all the details. My hope is that potential users will be persuaded to have a go at using the system. In the final analysis, most users get used to a particular system and are happy to stick with it. This is my position with SHEEP. Further details about its use can be found in MacCallum et al. (1994). For details about SHEEP distribution (including a freeware PC version) contact Malcolm MacCallum at Queen Mary and Westfield College, London University (M.A.H.MacCallum@maths.qmw.ac.uk).

2.2 The canonical example

SHEEP is built on an interactive dialect of LISP, which means that it accommodates a dialogue between the user and the system. When one is first learning to use the system it is usually best to exploit this conversational feature. However, as one gets better at using it, then most of the effort is transferred to developing files which, when loaded into SHEEP, accomplish the required calculation. We shall illustrate this by considering the two approaches for what has become the canonical example in the field: namely computing the Ricci tensor for Bondi's radiating metric (Equation 10). The system I have access to is version 2.57 of SHEEP, based on PSL 3.4, running on a Sun 4/75 in a unix environment of Sun OS 4.1.3_U1 (Solaris 1.1.1). Apart from a few local variations, interacting with any other SHEEP installation is virtually the same (see MacCallum et al. (1994) for more details).

At the unix prompt we invoke SHEEP by typing 'sheep'. The following is a typical response.

```
Loading file /home/sheep/057/ssheep.ini
File /home/sheep/057/ssheep.ini loaded
Loading file /home/sheep/057/shpsrc/sheep.ini
File /home/sheep/057/shpsrc/sheep.ini loaded
Baa... SHEEP 2 (version 57) welcomes you (18-Dec-91)
SHP>
```

The last line is the SHEEP prompt which is awaiting input from the user. SHEEP is modularly organized. The idea is that you can, in principle, only load those modules which are needed at any one time. Thus the system initially loaded is a very small version which can carry out basic algebraic manipulations, but has no knowledge of previously defined tensors. At the other extreme, if you want a system which knows about most tensors, then load the classification version of SHEEP called CLASSI. Since this is a much larger system, it occupies much more memory and will generally run slower as a consequence. We shall use the basic SHEEP system and so the first thing we do on entry is to load the module or package called CORD, which possesses the standard definitions of tensors in a coordinate base.

```
SHP> (LOAD CORD)
Loading file /home/sheep/057/shpbin/cord.b
File /home/sheep/057/shpbin/cord.b loaded
```

Let us dive straight in and ask the system to make the Ricci tensor which is called RIC.

```
SHP> (MAKE RIC)
```

g_{00}
=

In order for the system to start it needs first to know the components of the metric tensor. Algebraic expressions are read into SHEEP by a function called FOIN (FOrmula INput). The standard conventions for infix notation are used, except that the multiplication operator * needs to be included explicitly. We omit the details except to say that the generic method for getting out of problems which may occur in any sort of input activity is to invoke the function (RESET). This will always get you back to 'top level', that is to the SHP> prompt. The 00 component is the most complicated. We attempt to input it and then ask the system to echo back the attempt for checking. We

Algebraic Computing in General Relativity 297

do this by ending the input with a semicolon. We shall give the arbitrary functions β and γ the names B and G, respectively.

```
=E^(2*B)*V/R-E^(2*G)*R^2*U^2;
```

$$g_{00} = -R^2 U^2 e^{2G} + R^{-1} V e^{2B}$$

?

Note the changes SHEEP makes to the input expression. We get the system to accept an expression by typing a dollar. We shall next input the remaining components without checking.

```
?$
g_01
=E^(2*B)$
g_02
=E^(2*G)*R^2*U$
g_03
=0$
g_11
=0$
g_12
=0$
g_13
=0$
g_22
=-R^2*E^(2*G)$
g_23
=0$
g_33
=-R^2/E^(2*G)*SIN H^2$
GDF
Declare function R
now
```

The system does not know what R is and so assumes it is the name of a function. In fact it is one of the four coordinates. We give the four coordinates (x^0, x^1, x^2, x^3) the names (T,R,H,P) where we use H and P because they look very roughly like θ and ϕ. These are declared with the function VARS.

```
Declare function R
now(VARS T R H P)
Declare function U
now
```

Next we need to declare the four functions U, V, B and G to each be functions of the variables T, R and H. We use the function FUNS to do this.

```
Declare function U
now(FUNS (U T R H) (V T R H) (B T R H) (G T R H))
ended
GAM    ended
GDET   ended
GINV   ended
GDET   lost
GUU    ended
GDF    lost
GAMU   ended
RIE    ended
GAM    lost
GAMU   lost
RIC    ended
```

The Ricci tensor has now been constructed. Note that when SHEEP starts to construct a tensor it first prints its name out and, when the computation is completed, the word 'ended' is printed out. The system also loses some unwanted intermediate expressions in order to save on storage. The default settings can be overridden with explicit calls of the functions KEEP and LOSE. The definitions of the various tensors are given in MacCallum et al. (1994). We are now in a position to write out the components of the Ricci tensor. Since they are rather extensive let us first look at the shortest component.

```
SHP> (WRITE RIC 1 1)
```

$$R_{11} = -2(G_{,R})^2 + 4R^{-1} B_{,R}$$

SHEEP always assumes you are referring to the most recently cited tensor. So we can look at the next shortest component simply by typing

```
SHP> (WRITE 1 2)
```

$$R_{12} = -R^2 e^{-2B+2G} B_{,R} U_{,R} + R^2 e^{-2B+2G} G_{,R} U_{,R} + 1/2 R^2 e^{-2B+2G} U_{,RR} + 2Re^{-2B+2G} U_{,R} +$$
$$+ 2\cos(H)\sin^{-1}(H) G_{,R} - B_{,RH} - 2G_{,R} G_{,H} + G_{,RH} + 2R^{-1} B_{,H}$$

We could go on to inspect the rest of the components in a similar manner. The internal names of tensors are not always the same as the user names. This is known as tensor aliasing. It means that we can refer to the Ricci tensor RIC whether we are using a coordinate frame (where the internal name is RICC) or a user defined frame (internal name RICF). We can get the internal names of the tensors simply by typing TENSORS.

```
SHP> TENSORS
(DS2 GDET GINV RSCLC GDD GUU GDF GAMC GAMUC RIEC RICC RICMC EINC EINMC)
```

If we want to see the internal representation of the two components R_{11} and R_{12} then we can do this using the internal name RICC.

```
SHP> (RICC 1 1)
(PLUS (TIMES -2 (EXPT (DF1 G 1) 2)) (TIMES 4 (EXPT R -1) (DF1 B 1)))
SHP> (RICC 1 2)
(PLUS (TIMES (-1 . 2) (EXPT R 4) (EXP (PLUS (TIMES -4 B) (TIMES 4 G))) (
EXPT (DF1 U 1) 2)) (TIMES (EXPT R 2) U (EXP (PLUS (TIMES -2 B) (TIMES
2 G))) (COS H) (EXPT (SIN H) -1) (DF1 G 1)) (TIMES 2 (EXPT R 2) U (EXP (
```

Algebraic Computing in General Relativity 299

```
PLUS (TIMES -2 B) (TIMES 2 G))) (DF1 G 1 2)) (TIMES (EXPT R 2) (EXP (
PLUS (TIMES -2 B) (TIMES 2 G))) (DF1 G 1) (DF1 U 2)) (TIMES (EXPT R
2) (EXP (PLUS (TIMES -2 B) (TIMES 2 G))) (DF1 G 2) (DF1 U 1)) (TIMES
2 (EXPT R 2) (EXP (PLUS (TIMES -2 B) (TIMES 2 G))) (DF1 G 0 1)) (TIMES (
EXPT R 2) (EXP (PLUS (TIMES -2 B) (TIMES 2 G))) (DF1 U 1 2)) (TIMES R U (
EXP (PLUS (TIMES -2 B) (TIMES 2 G))) (COS H) (EXPT (SIN H) -1)) (TIMES
2 R U (EXP (PLUS (TIMES -2 B) (TIMES 2 G))) (DF1 G 2)) (TIMES -1 R V (
EXP (PLUS (TIMES -2 B) (TIMES 2 G))) (DF1 G 1 1)) (TIMES 2 R (EXP (PLUS (
TIMES -2 B) (TIMES 2 G))) (DF1 G 0)) (TIMES -1 R (EXP (PLUS (TIMES
-2 B) (TIMES 2 G))) (DF1 G 1) (DF1 V 1)) (TIMES 3 R (EXP (PLUS (TIMES
-2 B) (TIMES 2 G))) (DF1 U 2)) (TIMES -1 V (EXP (PLUS (TIMES -2 B) (
TIMES 2 G))) (DF1 G 1)) (TIMES 3 (COS H) (EXPT (SIN H) -1) (DF1 G 2)) (
TIMES -2 (EXPT (DF1 B 2) 2)) (TIMES 2 (DF1 B 2) (DF1 G 2)) (TIMES -2 (
DF1 B 2 2)) (TIMES -2 (EXPT (DF1 G 2) 2)) (DF1 G 2 2) 1 (TIMES -1 (EXP (
PLUS (TIMES -2 B) (TIMES 2 G))) (DF1 V 1)))
```

Note the similarities with the internal representation of LAM. However, rationals are represented by signed dotted pairs, e.g. $(-1 \, . \, 2)$ is the equivalent to $(-\,(1\,2))$ in LAM, and integral numbers are represented by signed integers. The operators are given their literal names and exponentiation is called EXPT. Finally the differentiation operator is called DF1 rather than & (the 1 refers to the first set of variables, since it is possible to have other sets). When we have computed some tensor of interest we usually want to send the output to a file in order to obtain a hard copy later. For example, we can save the two components of the Ricci tensor discussed above with

    SHP> (FILEWRITE BONDI RIC 1 1 RIC 2 2)
    RIC   ended
    RIC   ended

These are saved in the file bondi.LST. If you save long expressions then look out for occurrences of ^L (control L) in the file. These have been included by SHEEP in an attempt to produce page breaks in the right place. They usually need to be removed by an editor. The file bondi.LST looks like:-

$$R_{11} = -2(G_{,r})^2 + 4r^{-1} B_{,r}$$

$$R_{12} = -r^2 e^{-2B+2G} B_{,r} U_{,r} + r e^{2 -2B+2G} G_{,r} U_{,r} + 1/2 r^2 e^{-2B+2G} U_{,rr} + 2r e^{-2B+2G} U_{,r} +$$
$$+2\cos(H)\sin^{-1}(H)G_{,r} - B_{,rH} - 2G_{,r} G_{,H} + G_{,rH} + 2r^{-1} B_{,H}$$

Similarly, we may wish to save particular tensors or components of tensors in their internal representation for later use. For example, we may wish to reload the tensors later and pick up the computation from where we had left off. The function SAVETN does the saving and RESTORETN does the reloading. For example

    SHP> (SAVETN BONDI RIC 1 1 RIC 2 2)

saves the components in the file bondi.SHO which looks like:-

    %SHEEP save
    (DTINPUT (RICC 1 1)
    (PLUS (TIMES -2 (EXPT (DF1 G 1) 2)) (TIMES 4 (EXPT R -1) (DF1 B 1)))
)%

```
    (DTINPUT (RICC 1 2)
    (PLUS (TIMES -1 (EXPT R 2) (EXP (PLUS (TIMES -2 B) (TIMES 2 G))) (DF1 B
    1) (DF1 U 1)) (TIMES (EXPT R 2) (EXP (PLUS (TIMES -2 B) (TIMES 2 G))) (
    DF1 G 1) (DF1 U 1)) (TIMES (QUOT 1 2) (EXPT R 2) (EXP (PLUS (TIMES
    -2 B) (TIMES 2 G))) (DF1 U 1 1)) (TIMES 2 R (EXP (PLUS (TIMES -2 B) (
    TIMES 2 G))) (DF1 U 1)) (TIMES 2 (COS H) (EXPT (SIN H) -1) (DF1 G 1)) (
    TIMES -1 (DF1 B 1 2)) (TIMES -2 (DF1 G 1) (DF1 G 2)) (DF1 G 1 2) (TIMES
    2 (EXPT R -1) (DF1 B 2)))
    )%
```
Finally we quit from the system with
```
    SHP> (QUIT)
    Quitting
```
We have described the basic machinery for processing straightforward metrics which do not involve substitutions. This is a good way of familiarizing yourself with the system. It is not very efficient, since all of the information is lost when the system is quit. The standard paradigm for using SHEEP is to do so through prepared input files.

2.3 Metric Files

The Bondi metric, being a canonical example, possesses a metric file called bondi.crd which is distributed with the system. It looks like:-
```
    (TITLE "
    BONDI.CRD
    Bondi metric
    H. Bondi, M. van der Burg, A. Metzner,
    Proc. Roy. Soc. A, Vol. 269, p 21 (1962).")
    (OFF ALL) (ON NOZERO PTEVAR)
    (NAM T R)
    t $ r $
    (VARS T R H P)   % (t,r,theta,phi)
    (FUNS (B T R H) (G T R H) (U T R H) (V T R H))   % beta,gamma,U,V
    (RPL GDD)
    E^(2*B)*V/R-E^(2*G)*R^2*U^2 $   E^(2*B) $   E^(2*G)*R^2*U $  0           $
                                    0         $  0              $  0         $
                                                  -R^2*E^(2*G)  $  0         $
                                                                   -R^2/E^(2*G)*SIN H^2 $
```
It includes three functions we have not yet met. TITLE is followed by the name of the file, the name of the metric and the standard reference to the paper included between double quotes. This is used for indentification purposes, it is printed out when the file is loaded and becomes the header in any file produced by a call of FILEWRITE. The system possesses a large number of switches whose status is either ON or OFF. They control such things as output characteristics, substitutions and details about which tensors are computed and which are not. Most of these need not concern us in this introduction. (OFF ALL), as one might expect, switches off all switches, which is usually the safest thing to do. ON takes an indefinite number of arguments, so (ON NOZERO PTEVAR) switches on NOZERO (which inhibits the printing out of zero components) and

PTEVAR (which outputs indices with literal coordinate names rather than numeric ones: R_{tr} rather than R_{01}, for example). The switches available can be found by typing SWITCHES. There is also some limited help about switches which can be obtained with the function HELP. For example:-

```
SHP> SWITCHES
(ESUBS TPSSW CFILFO NOORDEXP NOLEXORD IMAGFIRST NOOPWEIGHT ORDPLMAIN
NOEXPAND NOSIMP POTSIM NOALLSIMP NOALLSUBS MSUBS SUBPOT NODGNV NONAM
NOPRT PSUBS PVARNUM RATPAR PTEVAR NOREDARY NOOFIT FITIN NOZERO NOCHKLONG
FILECHKLONG NOASKLONG TEXMODE NOREDSWITCHES REDSWIQUIET FACOUT FACIN
NOFACTORMSG EXPTEXP EXPNDLOG SEQSUBS NOAUTONAM AUTONAMF OUTLOWER
NONAMFOLD DONTRANS NOPRDIFIFVAL FUNSLET ALLMSG DOMAKE NOASCOP NOMSG
NOPREQ NODELK NOSEQUENCE NOSYMSG NOAUTOLOSDEP ASKLOSDEP DIAGONAL
PRWHATIS AUTOLOSDEP NOAUTOFRAME FORCEIT)
SHP> (HELP NOZERO)
NOZERO         Do not print zero components
SHP> (HELP PTEVAR)
PTEVAR         Use variable name in index output
```

The function NAM defines the output name of the variables T and R to be their lower case equivalent. The syntax of NAM is similar to that of FOIN. The replace function RPL is the main function for inputting user expressions. In this case the package CORD has the tensor GDD defined in it. The convention GDD stands for 'G', the name of the metric tensor, DD, 'Down Down', *i.e.* both indices are covariant ones. In 4 dimensions, the default, this tensor has 10 independent components, and hence (RPL GDD) is followed in the input file by the input version of these ten components for the Bondi metric. The normal practise when developing metric files is to load them in and check the metric by making and writing the line element which is called DS2. Thus:-

```
SHP> (LOAD "bondi.crd")
Loading file bondi.crd
 Loading file /home/sheep/057/shpbin/cord.b
 File /home/sheep/057/shpbin/cord.b loaded
BONDI.CRD
Bondi metric
H. Bondi, M. van der Burg, A. Metzner,
Proc. Roy. Soc. A, Vol. 269, p 21 (1962).
GDD   ended
File bondi.crd loaded
SHP> (WMAKE DS2)
   2      2 2 2G  -1   2B   2      2B          2  2G        2 2G  2
 ds  = (-r U e   +r  Ve  )dt  +2e   dtdr +2r Ue   dtdH -r e    dH  +
     2 -2G   2    2
   -r e    sin (H)dP
```

Note the usual syntax for loading a file. Normally user-defined metric files reside in the current directory. However, LOAD is an intelligent function and looks in various places for the file (including the metrics directory which is distributed with the system). File extensions play a very important rôle in SHEEP. For example, the extension 'crd' means that the metric is in a coordinate base. When SHEEP loads a file with the crd extension then the CORD package is loaded *automatically*. Other extensions lead to other modules being loaded automatically. Note also that WMAKE is a compound of write and make.

A useful switch for inputting diagonal metrics is **DIAGONAL** which, when on, only prompts for the diagonal components of the covariant metric. This can often be exploited to input almost diagonal metrics like the Bondi metric. In this case, with **DIAGONAL** switched on, we simply input the diagonal components and the two off-diagonal components g_{01} and g_{02} with explicit calls of (RPL GDD 0 1) and (RPL GDD 0 2).

2.4 Using substitutions

SHEEP possesses no factorization facilities (although there is a version of SHEEP called RSHEEP (MacCallum *et al.* 1994) which combines the SHEEP and REDUCE environments and hence allows the REDUCE factorization to be exploited). However, it is usually possible to overcome this restriction by a skillful use of substitution. We shall give an idea of the procedure involved by processing the Schwarzschild metric (1.3). The problem arises when SHEEP is asked to a simplify a term like

$$\left(1 - \frac{2m}{r}\right) / \left(1 - \frac{2m}{r}\right)$$

and produces

$$\left(1 - \frac{2m}{r}\right)^{-1} - \frac{2m}{r}\left(1 - \frac{2m}{r}\right)^{-1}$$

rather than unity. If we set

$$A = \left(1 - \frac{2m}{r}\right)$$

then SHEEP will correctly simplify A/A to 1. In fact, in this case, it proves better to take

$$A = r - 2m$$

We start with a pre-Schwarzschild file and do the rest interactively. The file is

```
(TITLE "
PRESCH.CRD
Pre-Schwarzschild file")
(OFF ALL) (ON DIAGONAL NOZERO)
(NAM T R M)
t $ r $ m $
(VARS T R H P)  % (t,r,theta,phi)
(FUNS (A R) (M))
(RPL GDD)
A/R$ -R/A$ -R^2$ -R^2*SIN H^2$
```

We have declared A to be a function of R and M to be a constant, *i.e.* a function of no variables. We have also used the function NAM to give the quantities T, R and M the lower case print names t, r and m, respectively. We then load the file and check the line element.

```
SHP> (LOAD "presch.crd")
Loading file presch.crd
 Loading file /home/sheep/057/shpbin/cord.b
 File /home/sheep/057/shpbin/cord.b loaded
```

```
PRESCH.CRD
Pre-Schwarzschild file
GDD   ended
File presch.crd loaded
SHP> (WMAKE DS2)
```
$$ds^2 = Ar^{-1}dt^2 - A^{-1}rdr^2 - r^2 dH^2 - r^2 \sin^2(H)dP^2$$

Another way to check the input is to use the function PRINPUT, (PRint INPUT).

```
SHP> (PRINPUT)
Variables from 0 to 3 : t r H P
A depends on r
m is a constant
```
$$g_{00} = Ar^{-1}$$
$$g_{11} = -A^{-1}r$$
$$g_{22} = -r^2$$
$$g_{33} = -r^2 \sin^2(H)$$

Now that we know the metric has been loaded successfully we make and write the Ricci tensor.

```
SHP> (WMAKE RIC)
GDF   ended
GAM   ended
GDET  ended
GINV  ended
GDET  lost
GUU   ended
GDF   lost
GAMU  ended
RIE   ended
GAM   lost
GAMU  lost
```
$$R_{00} = 1/2 Ar^{-2} A_{,rr}$$
$$R_{11} = -1/2 A^{-1} A_{,rr}$$
$$R_{22} = -A_{,r} + 1$$
$$R_{33} = -\sin^2(H) A_{,r} + \sin^2(H)$$

We now carry out the substitution for A interactively. Remember that in real applications, once the right substitutions have been found, then they should placed in the metric file.

```
SHP> (SETSUB ESUL (DIFF 2))
1 For
=A$
Substitute
=R-2M$
3
```

The function SETSUB takes as argument the name of the substitution list we are constructing or augmenting, ESUL in our case. The additional argument (DIFF 2) tells the function to compute the first and second derivatives of A as well, since they occur in the Ricci tensor. We input the function and its substituted value using FOIN. The value of SETSUB is 3 which is the number of the last substitution added to the list. PRSUB which prints out either nominated items, sections or all of the substitution list.

```
SHP> (PRSUB ALL)
ESUL
1 For A      substitute 0
      ,rr
2 For A      substitute 1
      ,r
3 For A substitute  -2m +r
```

Then (ON ESUBS) switches on the substitution device for the substitution list ESUL. Its value is the previous setting of the switch (which is NIL if the switch is off).

```
SHP> (ON ESUBS)
NIL
```

Substitutions are applied to particular tensors with the function LSIMP. So finally we substitute in the Ricci tensor and write out the results using WLSIMP.

```
SHP> (WLSIMP RIC)
    RIC all components zero
```

We have thus obtained the anticipated result. The actual metric file distributed with the system for the Schwarzschild metric is

```
(TITLE "
SCHWAR.CRD
Schwarzschild metric in Schwarzschild coordinates
empty space")
(OFF ALL) (ON NOZERO PTEVAR)
(NAM T R M)
t $ r $ m $
(VARS T R H P)       % (t,r,theta,phi)
(RPL A)
1 -2*M/R $
(FUNS (M) A)
(NEWSUL RICSUL)
A $  :A $
(USESUL RICSUL RICC RIEMC)
(RPL GDD)
A $    0   $   0   $    0          $
       -1/A $     0   $   0          $
              -R^2 $    0          $
                     -(R*SIN H)^2 $
```

It involves a couple of new functions. The first is NEWSUL which is used to make a user-defined substitution list called RICSUL. Then USESUL uses RICSUL to substitute in the Ricci tensor (RICC) and the Riemann tensor (RIEMC). In FOIN the expression :A means the value of A, *i.e.* the value $1 - 2m/r$ defined previously.

We have presented the same simple example of substitution as is given in MacCallum et al. (1994). For a much more complicated example involving the Hauser metric see d'Inverno and Frick (1982). We have only shown how to use SHEEP in a coordinate base. Most calculations are better carried out in standard frame. Again see Maccallum et al. (1994) for more details.

2.5 The Karlhede classification

One of the central problems in General Relativity is the equivalence problem: given two metrics, does there exist a local transformation which transforms one into the other? Cartan showed in some classic work that the problem is decidable but depends on computing the 10th covariant derivative of the Riemann tensor of each metric. Even with modern algebra systems this is out of the question. The work of Karlhede (Karlhede 1980), following on some previous ideas of Brans, significantly improved the situation. Karlhede's approach provides an invariant classification of a metric. Thus, if two metrics have different classifications then they are necessarily inequivalent, whereas if they have the same classification then they are candidates for equivalence. The problem then reduces to solving four algebraic equations. This is not an algorithmic process, although progress is often possible in particular applications. Karlhede's algorithm has made the computation viable since it reduces the derivative bounds significantly from the original 10 of Cartan. The actual order n of covariant derivative needed depends on the Petrov type, that is on the algebraic character of the Riemann tensor (see d'Inverno (1992) for further details). The Karlhede algorithm produces the bounds

Petrov type	Bound n
I, II, III	5
D, N	7
0	7

In fact, these bounds can be improved upon. In Collins et al. (1990) it is shown that vacuum Petrov type D can be reduced to $n=3$ in theory, and a direct computation of all such metrics only involves $n=2$. In Collins (1991) vacuum type N is reduced to $n=6$ and in Collins and d'Inverno (1993) non-vacuum type D is reduced to $n=6$. In some recent work of Ramos and Vickers (Machado Ramos & Vickers), it is shown that vacuum type N can be reduced further to $n=5$. Indeed, their methods look capable of reducing the bounds in most of the other cases. A specific case is known where $n=4$ is needed (Koutras 1992).

The Karlhede algorithm involves determining canonically defined frames for the covariant derivatives at each stage. Letting R^q denote the set of frames of the 0th, 1st, and so on, up to the qth covariant derivative of the Riemann tensor, then the classification proceeds as follows:-

1. Let $q = 0$

2. Compute R^q

3. Find the isotropy group H^q which leaves R^q invariant

4. Fix the frame up to H^q by choosing a canonical form for R^q

5. Find f_q the set of functionally independent functions in R^q and the number of elements t^q in this set

6. If $t_q \neq t_{q-1}$ or $H^q \neq H^{q-1}$ then set $q = q + 1$ and go to 2

7. Otherwise $\{H^q, t_q, f_q\}$ classifies the solution

The algorithm has been implemented in an extension of SHEEP called CLASSI by Jan Åman (Åman 1987, MacCallum et al. 1994). The code has been developed so that it is possible, at least in principle, to compute the 4th covariant derivatives of the Riemann tensor in CLASSI. We shall not look at the details of the implementation here, but simply illustrate the capabilities of the system by showing the output obtained when CLASSI is used to classify the Levi-Civita cylindrically symmetric static vacuum metric.

```
SHP> (LOAD "levi-ci.dia")
Loading file levi-ci.dia
 Loading file /home/sheep/057/clasrc/diainp.shp
 File /home/sheep/057/clasrc/diainp.shp loaded
LEVI-CI.DIA
        STATIC CYLINDRICALLY SYMMETRIC VACUUM FIELD.
        LEVI-CIVITA (1917-19) SOLUTION.
        Ref: D. Kramer, H. Stephani, M. MacCallum & E. Herlt.
        Exact Solutions of Einstein's Field Equations.
        Equation (20.8)
GD   ended
File levi-ci.dia loaded
SHP> (CLASSIFY)
PSL version 3.4  on  SUN 4
SHEEP 2 version 57 (18-Dec-91)
CLASSI made  18-Dec-91
Started   7-Jul-95  at  uk.ac.soton.mir
(TOTAL (TIME 85) (GCTIME 0) (NETTIME 85))
LEVI-CI.DIA
        STATIC CYLINDRICALLY SYMMETRIC VACUUM FIELD.
        LEVI-CIVITA (1917-19) SOLUTION.
        Ref: D. Kramer, H. Stephani, M. MacCallum & E. Herlt.
        Exact Solutions of Einstein's Field Equations.
        Equation (20.8)
The following switches are on:  POTSIM NOZERO SEQSUBS
Variables from 0 to 3 : T P F Z
M is a constant
FRAME = CONST
```

IFRAME = LORENTZ

$g_0 = P^M$

$g_1 = P^{M^2-M}$

$g_2 = P^{-M+1}$

$g_3 = P^{M^2-M}$

((TIME 34) (GCTIME 0) (NETTIME 34))
Changing to null frame
SHP> (WMAKE DS2)
LORTOIFR   ended
IZUD   ended
NULTOIFR   ended
NULTOIFRDET   ended
IFRTONULDET   ended
IFRTONUL   ended
NULTOFR   ended
TRZ   ended
ZUD   ended
HDD   ended
GDD   ended

$ds^2 = P^{2M} dT^2 - P^{2M^2-2M} dP^2 - P^{-2M+2} dF^2 - P^{2M^2-2M} dZ^2$

SHP> (WMAKE FORMSU)

$W^0 = 1/2(2)^{1/2} P^M dT + 1/2(2)^{1/2} P^{M^2-M} dP$

$W^1 = 1/2(2)^{1/2} P^M dT - 1/2(2)^{1/2} P^{M^2-M} dP$

$W^2 = 1/2(2)^{1/2} P^{-M+1} dF + 1/2(2)^{1/2} iP^{M^2-M} dZ$

$W^3 = 1/2(2)^{1/2} P^{-M+1} dF - 1/2(2)^{1/2} iP^{M^2-M} dZ$

SHP> (WMAKE DS2F)

$ds^2 = 2(1/2(2)^{1/2} P^M dT + 1/2(2)^{1/2} P^{M^2-M} dP)(1/2(2)^{1/2} P^M dT - 1/2(2)^{1/2} P^{M^2-M} dP) +$
$-2(1/2(2)^{1/2} P^{-M+1} dF + 1/2(2)^{1/2} iP^{M^2-M} dZ)(1/2(2)^{1/2} P^{-M+1} dF - 1/2(2)^{1/2} iP^{M^2-M} dZ)$

SHP> (WMAKE ZDET)

$ZDET = iP^{2M^2-2M+1}$

SHP> (WMAKE SIG)
ZINV   ended

ZDU ended
NULLTREQUIRE ended
SPINORREQUIRE ended

$s^0{}_{00'} = 1/2(2)^{1/2} P^{-M}$

$s^0{}_{11'} = 1/2(2)^{1/2} P^{-M}$

$s^1{}_{00'} = 1/2(2)^{1/2} P^{-M^2+M}$

$s^1{}_{11'} = -1/2(2)^{1/2} P^{-M^2+M}$

$s^2{}_{01'} = 1/2(2)^{1/2} P^{M-1}$

$s^2{}_{10'} = 1/2(2)^{1/2} P^{M-1}$

$s^3{}_{01'} = -1/2(2)^{1/2} iP^{-M^2+M}$

$s^3{}_{10'} = 1/2(2)^{1/2} iP^{-M^2+M}$

SHP> (WMAKE UNSGAM)
CONSTFRAME ended
ASD ended
APZ ended
LIE ended
ASD lost
APZ lost
GAM ended

$GAM^u{}_{0001'} = \text{sigma} = 1/4(2)^{1/2} M^2 P^{-M^2+M-1} - 1/4(2)^{1/2} P^{-M^2+M-1}$

$GAM^u{}_{0010'} = \text{rho} = -1/4(2)^{1/2} M^2 P^{-M^2+M-1} + 1/2(2)^{1/2} MP^{-M^2+M-1} +$

$-1/4(2)^{1/2} P^{-M^2+M-1}$

$GAM^u{}_{0100'} = \text{epsilon} = 1/4(2)^{1/2} MP^{-M^2+M-1}$

$GAM^u{}_{0111'} = \text{gamma} = 1/4(2)^{1/2} MP^{-M^2+M-1}$

$GAM^u{}_{1101'} = \text{mu} = -1/4(2)^{1/2} M^2 P^{-M^2+M-1} + 1/2(2)^{1/2} MP^{-M^2+M-1} +$

$$-1/4(2)^{1/2}P^{-M^2+M-1}$$

$$\text{GAM}^{u}{}_{1110} = \text{lambda} = 1/4(2)^{1/2}M^2 P^{-M^2+M-1} - 1/4(2)^{1/2}P^{-M^2+M-1}$$

((TIME 510) (GCTIME 0) (NETTIME 510))
SHP> (WMAKE PSI)
HDET   ended
HINV   ended
HUU   ended
RIE   ended
RIC   ended
RSCL   ended
HHD   ended
WEYL   ended
UNPSI ended

$$\text{PSI}_0 = -1/2 M^3 P^{-2M^2+2M-2} + 1/2 M P^{-2M^2+2M-2}$$

$$\text{PSI}_2 = -1/2 M^3 P^{-2M^2+2M-2} + M^2 P^{-2M^2+2M-2} - 1/2 M P^{-2M^2+2M-2}$$

$$\text{PSI}_4 = -1/2 M^3 P^{-2M^2+2M-2} + 1/2 M P^{-2M^2+2M-2}$$

SHP> (PETROV)
PSI4   ended
PSI0   ended
PSI3   ended
PSI1   ended
PSI2   ended
BPSI   ended
U1CRIT ended
U10CRIT ended
U11CRIT ended
Please check that PSI4 U10CRIT U11CRIT are really non-zero !
If so, Petrov type is I

$$\text{PSI}_4 = -1/2 M^3 P^{-2M^2+2M-2} + 1/2 M P^{-2M^2+2M-2}$$

$$U_{10} = -M^3 P^{-2M^2+2M-2} + 3 M^2 P^{-2M^2+2M-2} - 2 M P^{-2M^2+2M-2}$$

$$U_{11} = -2 M^3 P^{-2M^2+2M-2} + 3 M^2 P^{-2M^2+2M-2} - M P^{-2M^2+2M-2}$$

SHP> (ISOTST PSI)
Remaining isotropy group is:
none (0-dim), (shorthand notation: 0) and swap of null directions

This is so far a standard frame.
SHP> (FUNTST PSI)
New function, probably independent:
$f1 = \text{Re}(PSI_0)$
$= (M^3 P^{-2M^2+2M-2} - MP^{-2M^2+2M-2})$
Please check if JC1 really nonzero:
$JC1_1 = -2M^5 P^{-2M^2+2M-3} + 2M^4 P^{-2M^2+2M-3} - 2M^2 P^{-2M^2+2M-3} + 2MP^{-2M^2+2M-3}$

1 independent function found so far
SHP> (WMAKE IINV)
PSI2SQ   ended
B3CRIT   ended
B2CRIT   ended
$IINV = M^6 P^{-4M^2+4M-4} - 3M^5 P^{-4M^2+4M-4} + 4M^4 P^{-4M^2+4M-4} - 3M^3 P^{-4M^2+4M-4} +$
$+ M^2 P^{-4M^2+4M-4}$
SHP> (WMAKE WEYLSQ)
$C^{ABCD} C_{ABCD} = 16M^6 P^{-4M^2+4M-4} - 48M^5 P^{-4M^2+4M-4} + 64M^4 P^{-4M^2+4M-4} +$
$-48M^3 P^{-4M^2+4M-4} + 16M^2 P^{-4M^2+4M-4}$
SHP> (WMAKE PHI)
UNPHI   ended
 PHI all components zero
SHP> (WMAKE LAMBD)
$LAMBD = 0$
SHP> (WMAKE PHISTD)
 PHISTD all components zero
SHP> (SEGRE)
Vacuum solution
((TIME 935) (GCTIME 238) (NETTIME 697))
SHP> (WMAKE DPSI)
SGAM   ended
PSID   ended
$DPSI_{00'} = 1/2(2)^{1/2} M^5 P^{-3M^2+3M-3} - 1/2(2)^{1/2} MP^{-3M^2+3M-3}$
$DPSI_{11'} = -1/2(2)^{1/2} M^5 P^{-3M^2+3M-3} + (2)^{1/2} M^4 P^{-3M^2+3M-3} - (2)^{1/2} M^2 P^{-3M^2+3M-3} +$
$+1/2(2)^{1/2} MP^{-3M^2+3M-3}$

Algebraic Computing in General Relativity 311

$$DPSI_{20'} = 1/2(2)^{1/2} 5 M^{-3M^2+3M-3} P - 3/2(2)^{1/2} 4 M^{-3M^2+3M-3} P +$$

$$+2(2)^{1/2} 3 M^{-3M^2+3M-3} P - 3/2(2)^{1/2} 2 M^{-3M^2+3M-3} P + 1/2(2)^{1/2} M P^{-3M^2+3M-3}$$

$$DPSI_{31'} = -1/2(2)^{1/2} 5 M^{-3M^2+3M-3} P + 3/2(2)^{1/2} 4 M^{-3M^2+3M-3} P +$$

$$-2(2)^{1/2} 3 M^{-3M^2+3M-3} P + 3/2(2)^{1/2} 2 M^{-3M^2+3M-3} P - 1/2(2)^{1/2} M P^{-3M^2+3M-3}$$

$$DPSI_{40'} = 1/2(2)^{1/2} 5 M^{-3M^2+3M-3} P - (2)^{1/2} 4 M^{-3M^2+3M-3} P + (2)^{1/2} 2 M^{-3M^2+3M-3} P +$$

$$-1/2(2)^{1/2} MP^{-3M^2+3M-3}$$

$$DPSI_{51'} = -1/2(2)^{1/2} 5 M^{-3M^2+3M-3} P + 1/2(2)^{1/2} MP^{-3M^2+3M-3}$$

```
SHP> (FUNISOTST DPSI)
1 independent function found so far
Remaining isotropy group is:
none (0-dim), (shorthand notation: 0)
This is so far a standard frame.
SHP> (WMAKE DPSISQ)
```

$$PSI^{(abcd;e)f'} PSI_{(abcd;e)f'} = -16M^{10} P^{-6M^2+6M-6} + 80M^9 P^{-6M^2+6M-6} +$$

$$-190M^8 P^{-6M^2+6M-6} + 280M^7 P^{-6M^2+6M-6} - 308M^6 P^{-6M^2+6M-6} + 280M^5 P^{-6M^2+6M-6} +$$

$$-190M^4 P^{-6M^2+6M-6} + 80M^3 P^{-6M^2+6M-6} - 16M^2 P^{-6M^2+6M-6}$$

```
SHP> (WMAKE DWEYLSQ)
```

$$C^{ABCD;E} C_{ABCD;E} = -128M^{10} P^{-6M^2+6M-6} + 640M^9 P^{-6M^2+6M-6} - 1520M^8 P^{-6M^2+6M-6} +$$

$$+2240M^7 P^{-6M^2+6M-6} - 2464M^6 P^{-6M^2+6M-6} + 2240M^5 P^{-6M^2+6M-6} - 1520M^4 P^{-6M^2+6M-6} +$$

$$+640M^3 P^{-6M^2+6M-6} - 128M^2 P^{-6M^2+6M-6}$$

```
((TIME 476) (GCTIME 0) (NETTIME 476))
SHP> (ISOTROPY)
Isotropy group is:
none (0-dim), (shorthand notation: 0)
This is a standard frame.
Isometry group is of dimension 3.
LEVI-CI.DIA
        STATIC CYLINDRICALLY SYMMETRIC VACUUM FIELD.
        LEVI-CIVITA (1917-19) SOLUTION.
```

```
            Ref: D. Kramer, H. Stephani, M. MacCallum & E. Herlt.
            Exact Solutions of Einstein's Field Equations.
            Equation (20.8)
    (RUN (TIME 1972) (GCTIME 238) (NETTIME 1734))
    (TOTAL (TIME 2057) (GCTIME 238) (NETTIME 1819))
    SHP> (CLASSISUM)
       #     001  3 0 00--11--
```
The parent function is CLASSIFY, which can be broken down into a number of separate parts. Note that the user is prompted to check that certain important expressions are non-zero. It is possible, especially when there are side relations around, that expressions which are apparently non-zero can in fact be reduced to zero. If this happens then it will radically alter the ensuing classification. The classification closes with some timing statistics. The final function CLASSISUM produces a one line summary (MacCallum et al. 1994). The first three numbers, 001, indicates that the solution is vacuum, the Ricci scalar is zero and the Petrov type is I. This is followed by the dimensions of the isometry and isotropy groups, namely 3 and 0, respectively. Finally, 11--, indicates the dimensions of the isotropy groups H_0 and H_1 and 00-- indicates that t_0 and t_1, the number of independent functions which the groups possess, are both zero.

2.6 The computer data base of exact solutions

Although the field of exact solutions is one of the biggest fields in General Relativity, it has been, for much of its history, in something of a mess. One aspect of this has been the repeated discovery of 'new' solutions which then turn out to be known already. The worst example of this is the Schwarzschild metric, which has apparently been 'discovered' on at least 20 separate occasions. The literature was brought into a much greater state of coherence with the appearance of the Exact Solutions book in 1980 (Kramer et al. 1980). A more recent contribution has concentrated on the important case of inhomogeneous cosmologies (Krasinski 1993) (where many of the solutions considered were processed by the algebra system ORTHOCARTAN).

The Karlhede algorithm and its implementation in CLASSI has greatly improved the situation with regard to the equivalence problem (Åman and Karlhede 1980). Indeed, it has been used to show that three of the Harrison metrics (HIII9A, HIII9b and HIII9c) are in fact equivalent. This was not noticed in the original paper (Harrison 1959), nor in the paper (d'Inverno and Russell-Clark 1971b) which processed the metrics. The advent of the classification algorithm has lead to the setting up of a computer database of exact solutions through collaborative work between QMW (London), Stockholm and Southampton. At present some 200 metrics exist in the database. One of the goals is to 'put' the Exact Solutions book into the database. The ultimate hope is that it will contain all known solutions, fully documented and classified. Then any 'newly' discovered solution can be compared with the contents of the database and, if indeed it is new, then the database can be updated accordingly. Were this to be fully realised then it would provide a valuable resource for the international community of relativists. At present, the database is built on SHEEP and CLASSI, which is another motivating factor for individuals to become conversant with these systems.

3 Introduction to STENSOR

3.1 The status of STENSOR

STENSOR stands for Symbolic TENSOR manipulator and it is designed for handling indicial tensors rather than components of tensors which SHEEP operates on. To be precise, since neither STENSOR nor SHEEP knows what a tensor is (because the objects need not obey tensor transformation laws) STENSOR really only manipulates scalar and indexed quantities. STENSOR has been written by Lars Hörnfeldt of Stockholm University. The good news about STENSOR is that when it works, it works very well. Moreover, it has a track record of achievement (see MacCallum *et al.* (1994) and Hörnfeldt (1988) for more details). Unfortunately, STENSOR has a downside as well. The first problem is that there is no such thing as STENSOR—but rather there are a series of different versions spread across the globe. (Most of these were probably originally installed by Hörnfeldt himself on his various visits.) He has concentrated on developing the capabilities of the system rather than on producing a definitive version. As a consequence, many of the systems in existence, being developmental in character, not surprisingly possess bugs. Worse still, different systems possess different bugs and so something which works on one site may not work on another. As a consequence, a mythology has come into existence that if you really want to get STENSOR to process your problem successfully, then you probably need Hörnfeldt himself around. This is somewhat unfair. I use STENSOR unaided on a fairly regular basis and usually to good effect. You do, however, need a creative attitude to using it. Sometimes it will mysteriously not work, but persistence will often pay off.

I use the system most frequently for defining new tensors that neither SHEEP nor CLASSI know about. Once they are defined in STENSOR then the system can be used to create component versions which can be subsequently loaded into SHEEP or CLASSI for specific applications. I will give an extensive example below. It is best if you can use the system in a context where there are opportunities for checking that it is giving the right answers. Thus, for example, one might know what the answer should be in a simple case. Or again, it is possible to check that it is giving the right sort of expressions by making judicious use of the function SYMBOLIC. As its name suggests, SYMBOLIC makes the components of its arguments symbolic and so one can use it to inspect the general form of a new tensor. This is somewhat like the application in Section 1 where we were able to look at the general form of the Ricci tensor for a general metric (Figure 1).

Defining new tensors is only a small but important part of what STENSOR is capable of. It can deal with geometric objects other than tensors, such as spinors. It has a unique substitution feature which can optimize with respect to side relations like

$$\sin^2 \theta + \cos^2 \theta = 1 \qquad (12)$$

It can cope with non-commutative operators. It has special facilities for sub-space splitting with application areas like the 3+1 formalism, Kaluza-Klein theories and Kähler manifolds. It has a 'bucketing' feature which can divide very large expressions into parts stored on disk file. This means that it is possible to extend calculations substantially compared to the usual situation where all expressions need to be stored in memory. This facility was used in the first calculation of the Ricci tensor for a general

5 dimensional metric.

STENSOR possesses a 116 page manual (Hörnfeldt 1988). Unfortunately it is written in a very telegrammatic style (like much of the on-line documentation associated with the system) which I, for one, have great difficulty in understanding. In short, these features make it harder to become an STENSOR user than it does to become a SHEEP or CLASSI user. Nontheless, I would advocate that people should give it a try.

3.2 Geroch decomposition

The example is connected with work on Numerical Relativity. It involves the general problem of computing the field equations which arise when a Geroch decomposition is carried out relative to a non-null Killing direction ξ^a (Geroch 1970). The norm of the Killing vector is defined by

$$\lambda = \xi_a \xi^a \tag{13}$$

and its twist ω_a is defined by

$$\omega_a = \sqrt{-g}\, \epsilon_{abcd}\, \xi^b g^{ce} \nabla_e \xi^d \tag{14}$$

where ∇ denotes covariant derivative with respect to the metric connection (d'Inverno 1992). Let S denote the collection of all trajectories of ξ^a. We shall assume that S has been given the structure of a differentiable 3-manifold (which is always possible locally). In the case when ξ^a is hypersurface orthogonal then it is possible to represent S as one of the hypersurfaces in the 4-dimensional manifold M which is everywhere orthogonal to ξ^a. We can then define a tensor field h_{ab} on M which is a metric on S by

$$h_{ab} = g_{ab} - (\xi^c \xi_c)^{-1} \xi_a \xi_b \tag{15}$$

with inverse

$$h^{ab} = g^{ab} - (\xi^c \xi_c)^{-1} \xi^a \xi^b \tag{16}$$

We can also define a projection operator into S by

$$h_a{}^b = \delta_a{}^b - (\xi^c \xi_c)^{-1} \xi_a \xi^b \tag{17}$$

This projection operator can be used to define a covariant derivative D in S. If $T^{b...d}_{a...c}$ is any tensor field on S then D is defined by

$$D_e T^{b...d}_{a...c} = h_e{}^p h_a{}^m \cdots h_c{}^n h_r{}^b \cdots h_s{}^d \nabla_p T^{r...s}_{m...n} \tag{18}$$

This machinery can be used to carry out a '3+1' decomposition relative to ξ^a. In particular, in vacuum, we discover that ω_a is a gradient, i.e.

$$\omega_a = D_a \omega \tag{19}$$

and the vacuum field equations take on the equivalent form (see equation (A18) of Geroch (1970))

$$D^2 \lambda = \frac{1}{2\lambda}(D^m \lambda)(D_m \lambda) - \frac{1}{\lambda}(D^m \omega)(D_m \omega) \tag{20}$$

$$D^2 \omega = \frac{3}{2\lambda}(D^m \lambda)(D_m \omega) \tag{21}$$

$$\mathcal{R}_{ab} = \frac{1}{2\lambda^2}[(D_a \omega)(D_b \omega) - h_{ab}(D^m \omega)(D_m \omega)] + \frac{1}{2\lambda} D_a D_b \lambda - \frac{1}{4\lambda^2}(D_a \lambda)(D_b \lambda) \tag{22}$$

where $D^2 = h^{ab}D_a D_b$ and $\mathcal{R}_{ab}$ is the Ricci tensor of S built out of h_{ab}. We now discuss how these quantities are defined in STENSOR.

3.3 Using STENSOR to define new tensors

The file with all the definitions in is called defs.pdef and consists of
```
%Definition file for xid
(declt (xiu 1))
(pdef xid)
<gdd a b><xiu b>$
%Definition file for lambda
(pdef lambda)
<xiu a><xid a>$
%Definition file for norm la
(pdef la)
<lambda ;a>$
%Definition file for lla
(pdef lla s12)
<la a ;b>$
%Definition file for twist vector wa
(declt fac)
(pdef wa)
<fac><eps4 a b c d><xiu b><guu c p><xiu d ;p>$
%Definition file for wwa
(pdef wwa)
<wa a ;b>$
%Definition file for hdu
(declt (hdu 2) (delta 2))
(pdef hdu)
<delta a b> -1/<lambda><xid a><xiu b>$
%Definition file for D^2lambda
(pdef dsqlam)
<guu a b><hdu b c><hdu a d><la d ;c>$
%Definition file for D^2omega
(pdef dsqom)
<guu a b><hdu b c><hdu a d><wa d ;c>$
%Definition file for rhslam
(pdef rhslam)
1/(2<lambda>)<guu m n><hdu m a><la a><hdu n b><la b>
-1/(<lambda>)<guu m n><hdu m a><wa a><hdu n b><wa b>$
%Definition file for rhsom
(pdef rhsom)
3/(2<lambda>)<guu m n><hdu m a><la a><hdu n b><wa b>$
%Definition file for hdd
(declt  (hdd s12) )
(pdef hdd)
```

```
<gdd a b> -1/<lambda><xid a><xid b>$
%Definition file for rhsric
(pdef waa s12)
<wa m><wa n>$
(pdef laa s12)
<la m><la n>$
(pdef rhsric s12)
1/(2<lambda>^2)<hdu a c><hdu b d><waa c d>
-1/(2<lambda>^2)<hdd a b><guu m n><hdu m c><hdu n d><waa c d>
+1/(2<lambda>)<lla a b>
-1/(4<lambda>^2)<hdu a c><hdu b d><laa c d>$
```

Note that the file is written in the more modern lower case. In STENSOR scalar and tensor quantities begin and end with angular brackets. They enclose any tensor indices in the standard way, except that they are all assumed to be covariant. Contravariant indices are defined by the function DECLT (DECLare Tensor) which takes an indefinite number of arguments and declares, in particular, which indices are contravariant for each tensor. For example, the tensor XIU which represents ξ^a is contravariant and hence DECLT is followed by (XIU 1). Dummy indices are treated in the usual way. Ordinary derivatives and covariant derivatives are represented by comma amd semicolon, respectively. DECLT also declares any symmetries a tensor may possess. For example, LLA has the declared property s12 which means that it is symmetric on the first and second index. The function PDEF defines tensors and has an input syntax similar to that of FOIN. The tensors g_{ab} and h_{ab} are called GDD and HDD, respectively, and ϵ_{abcd} is EPS4. The tensors GDD and EPS4 are known to the STENSOR system. The system also knows about the Kronecker delta tensor (called DEL), but I shall define my own, called DELTA, because I have had difficulty with the system version (probably connected with symmetry inconsistencies). It should now be possible with this information to see how the definition file is able to represent the tensors defined in the Geroch decomposition.

We shall show the interaction with STENSOR in which this file is loaded and the tensors are defined for use by SHEEP. There are three key functions:

PDEF, which defines tensors
WDEF, which writes tensors out
EDEF, which evaluates tensors

When definitions are read in from a file they are automatically displayed by WDEF. Evaluating a tensor with EDEF is carried out interactively. We will not go through the options, although typing H (for Help) will list these. We shall use three responses only:

E, which evaluates the definitions
N (for No), to any request to plug one definition into another
Q (for Quit), at the end of the interaction with EDEF.

I have had problems with plugging definitions of tensors into others (again, probably to do with symmetry inconsistencies). Moreover, the system just crashes and gives no helpful information about what the problem is. The function TCOMP (Tensor COMpiler) produces the SHEEP code needed to calculate the components of a tensor in particular applications. Not only does it do this automatically, but the resulting code produced is extremely efficient (see Holmes and d'Inverno (1986) for an example). You may need to tell TCOMP about the rank of various quantities which occur in the definition. For example, scalars have rank 0 and GAMUC rank 3. The function TCOMPOUT also compiles a tensor but, in addition, it produces a file with the SHEEP code in. This is called tensor.blf where the extension blf stands for BuiLding Function. There may be a problem about the affine connection in these blf files. STENSOR uses the generic GAMU, whereas in an application GAMUC (or GAMUF if it is a frame application) is required. The function AFFINITY is supposed to define the specific name of the connection, but I have sometimes found that it does not appear to work (although I may well be doing something wrong and there can be a difficulty in detecting just exactly what the error may be). If, within SHEEP, you are trying to evaluate a tensor for a particular metric and you get the prompt

$$\text{GAM}^{0}{}_{00}$$
$$=$$

then SHEEP has not recognized the connection used in the definition file. Remember first to get back to top level with (RESET), then go to the appropriate .blf file and use an editor to replace GAMU with GAMUC (or GAMUF).

The system is loaded in Unix with the command stensor. Here is part of the resulting interaction.

```
Loading file /home/sheep/057/sstensor.ini
****************************************************************************
*** >>>>  NEW TOPLEVEL, with 'DWIM' and HISTORY !! <<<<***
*** For info on above:    TYPE ? (shows toplevel Menu,***
*** there among N for N-EWS)  TYPE N (shows News_topics,***
*** there among I for I-ntro) TYPE I (get later topics by <return>) ***
*** -ie type ? <return> N <return> I <return>NOW!!! ***
*** Try: (HELP),(SEE), (SEE identi) etc,   and (WTRIGLOOP)! ***
*** (EDSUBON tsul tsr) Editor of substitutions/SP-programs ***
***   Excise filegroups: Type (!%) for alternatives !%EXC=!%PDEFR+!%EXF ***
****************************************************************************
SLISPTOP
(********LOCALs= 331)REMOBed
File /home/sheep/057/sstensor.ini loaded
Loading file /home/sheep/057/stslib/stensor.ini
Put your PDEFs in a {\sc stensor}.INI on your local directory, for auto loading
File /home/sheep/057/stslib/stensor.ini loaded
Copyright "(C)" 1989 by Lars Hornfeldt, Phys. Dept., Stockholm Univ.
STENSOR 2.304 on Sheep 57  welcomes you (19-Dec-91)
STS> (AFFINITY GAMUC)
(GAMUC)
STS> (LOAD "defs.pdef")
Loading file defs.pdef
<XID A> defined
```

$$\text{XID}_a = g_{ab}\, \text{XIU}^b$$

`<LAMBDA> defined`

$$\text{LAMBDA} = \text{XIU}^a \text{XID}_a$$

`<LA A> defined`

$$\text{LA}_a = \text{LAMBDA}_{;a}$$

`<LLA A B> defined`

(LLA -Syms: S12)

$$\text{LLA}_{ab} = \text{LA}_{a;b}$$

`<WA A> defined`

$$\text{WA}_a = \text{XIU}^d \text{FACe}_{;p\ abcd} \text{XIU}^b g^{cp}$$

`<WWA A B> defined`

$$\text{WWA}_{ab} = \text{WA}_{a;b}$$

`<HDU A B> defined`

$$\text{HDU}_a^{\ b} = \text{DELTA}_a^{\ b} - \text{LAMBDA}^{-1} \text{XID}_a \text{XIU}^b$$

`<DSQLAM> defined`

$$\text{DSQLAM} = \text{LA}_{d;c} g^{ab} \text{HDU}_b^{\ c} \text{HDU}_a^{\ d}$$

`<DSQOM> defined`

$$\text{DSQOM} = \text{WA}_{d;c} g^{ab} \text{HDU}_b^{\ c} \text{HDU}_a^{\ d}$$

`<RHSLAM> defined`

$$\text{RHSLAM} = 1/2 \text{LAMBDA}^{-1} g^{mn} \text{HDU}_m^{\ a} \text{LA}_a \text{HDU}_n^{\ b} \text{LA}_b - \text{LAMBDA}^{-1} g^{mn} \text{HDU}_m^{\ a} \text{WA}_a \text{HDU}_n^{\ b} \text{WA}_b$$

`<RHSOM> defined`

$$\text{RHSOM} = 3/2 \text{LAMBDA}^{-1} g^{mn} \text{HDU}_m^{\ a} \text{LA}_a \text{HDU}_n^{\ b} \text{WA}_b$$

(HDD -Syms: S12)

`<HDD A B> defined`

$$\text{HDD}_{ab} = g_{ab} - \text{LAMBDA}^{-1} \text{XID}_a \text{XID}_b$$

(HDD -Syms: S12)

(LLA -Syms: S12)

`<WAA M N> defined`

(WAA -Syms: S12)

$$\text{WAA}_{mn} = \text{WA}_m \text{WA}_n$$

`<LAA M N> defined`

(LAA -Syms: S12)

$$\text{LAA}_{mn} = \text{LA}_m \text{LA}_n$$

`<RHSRIC A B> defined`

(RHSRIC -Syms: S12)

$$RHSRIC_{ab} = 1/2 LAMBDA^{-2} HDU^{c}_{a} HDU^{d}_{b} WAA_{cd} - 1/2 LAMBDA^{-2} HDD_{ab} g^{mn} HDU^{c}_{m} HDU^{d}_{n} WAA_{cd} +$$
$$+ 1/2 LAMBDA^{-1} LLA_{ab} - 1/4 LAMBDA^{-2} HDU^{c}_{a} HDU^{d}_{b} LAA_{cd}$$

```
File defs.pdef loaded
NIL
STS> (EDEF XID)
 XID 1/ 1TERMSXID Action?E
  1/ 1TERMSXID Action?Q
 (XID)
STS> (TCOMPOUT XID)
Loading file /home/sheep/057/stssrc/qtcomp.sts
*** Function 'TCOMPOUT' has been redefined
*** Function 'TCOMP' has been redefined
Value:
***** Can't find file PP
Done (PON NOGRIN) since no grind available
File /home/sheep/057/stssrc/qtcomp.sts loaded
 xid.BLF
 XID
 XID TCOMP-iled
 PREQ: (XIU GDD) DEP:  (LAMBDA HDU HDD)
 BULF:(PUTD (QUOTE CBXID) (QUOTE EXPR) (QUOTE (LAMBDA (INDX1) (PROG (ANS*) (
SETQ ANS* 0) (SUM*1 DEFAULTDIMENSION* (QUOTE (SSIMP (DOMULC (XIU *1) (GDD
INDX1 *1))))) (RETURN ANS*)))))
 (TYPE 1) CBXID= BULF PREQ DEP PRT PARSF ORGDEF DFUN
 XIU (TYPE 1) DEP PRT UPINDX PARSF DFUN
 GDD (TYPE (NOSYM (SYM 2) 0)) DEP PRT NAM PARSF DFUNNOGRIN was tmpry ON
NIL
STS> (EDEF LAMBDA)
 LAMBDA Plugin?N
 tsr:XID 1/ 1TERMS
 LAMBDA Action?E
  Plugin?N
 tsr:XID 1/ 1TERMS
LAMBDA Action?Q (LAMBDA)
STS> (TCOMPOUT LAMBDA)
 lambda.BLF
 LAMBDA
 LAMBDA TCOMP-iled
 PREQ: (XIU XID) DEP:  (LA HDU RHSLAM HDD)
 BULF:(PUTD (QUOTE CBLAMBDA) (QUOTE EXPR) (QUOTE (LAMBDA NIL (PROG (ANS*) (
SETQ ANS* 0) (SUM*1 DEFAULTDIMENSION* (QUOTE (SSIMP (DOMULC (XIU *1) (XID *1))))
) (RETURN ANS*)))))
 (TYPE 0) CBLAMBDA= BULF PREQ DEP PRT PARSF ORGDEF
 XIU (TYPE 1) DEP PRT UPINDX PARSF DFUN
 XID (TYPE 1) CBXID= BULF PREQ DEP PRT PARSF ORGDEF DFUN
```

```
   GDD (TYPE (NOSYM (SYM 2) 0)) DEP PRT NAM PARSF DFUNNOGRIN was tmpry ON
NIL
STS> (EDEF LA)
 LA 1/ 1TERMS
LA Action?E
  Plugin?N
 tsr:LAMBDA 1/ 1TERMS
LA Action?Q
 (LA)
STS> (TCOMPOUT LA)
 la.BLF
 LA
 LA TCOMP-iled
 PREQ:(LAMBDA) DEP:  (LLA DSQLAM RHSLAM RHSOM LAA)
 BULF:(PUTD (QUOTE CBLA) (QUOTE EXPR) (QUOTE (LAMBDA (INDX1) (SSIMP (SMDIFFC (
ZGETEN (QUOTE LAMBDA) NIL NIL NIL) INDX1 CRDIF*)))))
 (TYPE 1) CBLA= BULF PREQ DEP PRT PARSF ORGDEF DFUN
 LAMBDA (TYPE 0) CBLAMBDA= BULF PREQ DEP PRT PARSF ORGDEF
 XIU (TYPE 1) DEP PRT UPINDX PARSF DFUN
 XID (TYPE 1) CBXID= BULF PREQ DEP PRT PARSF ORGDEF DFUN
 GDD (TYPE (NOSYM (SYM 2) 0)) DEP PRT NAM PARSF DFUNNOGRIN was tmpry ON
NIL
STS> (EDEF LLA)
 LLA 2/ 2TERMS
LLA Action?E
  Plugin?N
 tsr:LA 2/ 2TERMS
LLA Action?Q
 (LLA)
STS> (TCOMPOUT LLA)
 lla.BLF
 LLA
 LLA (GAMUC no sym)
 Now: GAMUC-has what rank?3
 Type nbr:  TCOMP-iled
 PREQ: (GAMUC LA) DEP:  (RHSRIC)
 BULF:(PUTD (QUOTE CBLLA) (QUOTE EXPR) (QUOTE (LAMBDA (INDX1 INDX2) (PROG (
ANS*) (SETQ ANS* 0) (SUM*1 DEFAULTDIMENSION* (QUOTE (SMMINUS (SSIMP (DOMULC (
LA *1) (GAMUC *1 INDX1 INDX2)))))) (ADANZ (SSIMP (SMDIFFC (ZGETEN (QUOTE LA) (
LIST INDX1) NIL NIL) INDX2 CRDIF*))) (RETURN ANS*)))))
 (TYPE (NOSYM (SYM 2) 0)) CBLLA= BULF PREQ DEP PRT PARSF ORGDEF DFUN
 GAMUC (TYPE (NOSYM 3)) DEP PRT UPINDX PARSF DFUN
 LA (TYPE 1) CBLA= BULF PREQ DEP PRT PARSF ORGDEF DFUN
 LAMBDA (TYPE 0) CBLAMBDA= BULF PREQ DEP PRT PARSF ORGDEF
 XIU (TYPE 1) DEP PRT UPINDX PARSF DFUN
 XID (TYPE 1) CBXID= BULF PREQ DEP PRT PARSF ORGDEF DFUN
 GDD (TYPE (NOSYM (SYM 2) 0)) DEP PRT NAM PARSF DFUNNOGRIN was tmpry ON
```

```
            NIL
            STS> (EDEF WA)
            WA 2/ 2TERMS
            WA Action?E
              2/ 2TERMS
            WA Action?Q
             (WA)
            STS> (TCOMPOUT WA)
            wa.BLF
            WA
            WA
            Loading file /home/sheep/057/stssrc/qtcom2.sl
            File /home/sheep/057/stssrc/qtcom2.sl loaded
              1/ 1TERMS
            Loading file /home/sheep/057/stssrc/qtcomf.sl
              (EPS4 -Syms: (A 1 TO 4))
            *** Function 'EPS4' has been redefined
              (RDEL -Syms: S12)
              (EPS2 -Syms: A12)
              (EPS3 -Syms: (A 1 TO 3))
            File /home/sheep/057/stssrc/qtcomf.sl loaded
              6/ 6TERMS 1/ 1TERMS 6/ 6TERMS TCOMP-iled (won't handle TPS)
              PREQ: (GAMUC GUU FAC XIU) DEP:  (WWA DSQOM RHSLAM RHSOM WAA)
              BULF:(PUTD (QUOTE CBWA) (QUOTE EXPR) (QUOTE (LAMBDA (INDX1) (PROG (ANS*) (
              ....
```

We interrupt the interaction here because the `blf` files for `WA` and many of the succeeding tensors are rather long. Go back and have a look at the header display on entering this version of STENSOR to get an example of Hörnfeldt's telegrammatic style.

3.4 Using STENSOR's definitions in a SHEEP run

We shall now see an example of using the STENSOR tensor component definitions in a SHEEP run. We are interested in the case of a cylindrically symmetric line element with coordinates $(x^0, x^1, x^2, x^3) = (u, r, z, \phi)$, where u is a null coordinate. The Killing vector is

$$\xi^a = \delta_2{}^a = (0, 0, 1, 0)$$

corresponding to the translational invariance in the z-direction. The metric file is called `cyl7a.crd`:

```
            (TITLE "
            CYL7A.CRD
            Cylindrically symmetric solution (2 degrees of freedom psi and w)
            with null coordinate u")
            (OFF ALL)
            (ON NOZERO)
            (VARS U R Z P)  % (u,r,z,phi)
            (FUNS  (GA U R) (PS U R) (W U R))  % (gamma, psi, w)
```

```
(RPL GDD)
-E^(2GA-2PS)$   -E^(2GA-2PS)$   0$              0$
                0$              0$              0$
                                E^(2PS)$        W*E^(2PS)$
                                                R^2*E^(-2PS)+W^2*E^(2PS)$
```

Since we usually want repeated runs using different metrics, we construct a file which has all the commands in it to make the various quantities. The file is called **make.crd** and consists of:-

```
(load "cyl7a.crd")
(keep gdet)
(make gamu)
(load "xid.blf")
(rpl xiu)
0$ 0$ 1$ 0$
(wmake xid)
(load "lambda.blf")
(wmake lambda)
(load "lla.blf")
(wmake lla)
(load "wa.blf")
(rpl fac)
(-:gdet)^(1/2)$
(wmake wa)
(load "hhdu.blf")
(rpl delta)
1$0$0$0$
0$1$0$0$
0$0$1$0$
0$0$0$1$
(wmake hhdu)
(load "wwa.blf")
(wmake wwa)
(load "dsqlam.blf")
(wmake dsqlam)
(load "dsqom.blf")
(wmake dsqom)
(load "rhslam.blf")
(wmake rhslam)
(load "rhsom.blf")
(wmake rhsom)
(load "hhdd.blf")
(wmake hhdd)
(load "rhsric.blf")
(wmake rhsric)
```

It mostly consists of loading the various **blf** files and constructing the appropriate tensors. At the beginning **GDET** is kept because it is needed for the definition of the twist vector (14). The components of the Killing vector and the Kronecker delta are input explicitly in the file. Apart from a few messages and the input of all of the components of the Kronecker delta, here is the full resulting interaction.

```
SHP> (LOAD "cyl7a.crd")
Loading file cyl7a.crd
 Loading file /home/sheep/057/shpbin/cord.b
 File /home/sheep/057/shpbin/cord.b loaded
GDD   ended
File cyl7a.crd loaded
SHP> (WMAKE DS2)
```

$$ds^2 = -e^{2GA-2PS}dU^2 - 2e^{2GA-2PS}dUdR + e^{2PS}dZ^2 + 2We^{2PS}dZdP +$$
$$+(R^2 e^{-2PS} + W^2 e^{2PS})dP^2$$

```
SHP> (KEEP GDET)
(GDET)
SHP> (MAKE GAMU)
GDF   ended
GAM   ended
GDET  ended
GINV  ended
GUU   ended
GDF lost
GAMU  ended
SHP> (LOAD "xid.blf")
Loading file xid.blf
File xid.blf loaded
SHP> (RPL XIU)
```

XIU^0
=0$

XIU^1
=0$

XIU^2
=1$

XIU^3
=0$

```
SHP> (WMAKE XID)
```

$XID_2 = e^{2PS}$

$XID_3 = We^{2PS}$

```
SHP> (LOAD "la.blf")
Loading file la.blf
File la.blf loaded
SHP> (WMAKE LA)
```

$$LA_0 = 2e^{2PS} PS_{,U}$$

$$LA_1 = 2e^{2PS} PS_{,R}$$

SHP> (LOAD "lambda.blf")
Loading file lambda.blf
File lambda.blf loaded
SHP> (WMAKE LAMBDA)

$$LAMBDA = e^{2PS}$$

SHP> (LOAD "lla.blf")
Loading file lla.blf
File lla.blf loaded
SHP> (WMAKE LLA)

$$LLA_{00} = -4e^{2PS} GA\ PS_{,U,U} +2e^{2PS} GA\ PS_{,U,R} +2e^{2PS} GA\ PS_{,R,U} -2e^{2PS} GA\ PS_{,R,R} +$$
$$+8e^{2PS}(PS_{,U})^2 -4e^{2PS} PS_{,U} PS_{,R} +2e^{2PS}(PS_{,R})^2 +2e^{2PS} PS_{,UU}$$

$$LLA_{01} = -2e^{2PS} GA\ PS_{,R,R} +4e^{2PS} PS_{,U} PS_{,R} +2e^{2PS}(PS_{,R})^2 +2e^{2PS} PS_{,UR}$$

$$LLA_{11} = -4e^{2PS} GA\ PS_{,R,R} +8e^{2PS}(PS_{,R})^2 +2e^{2PS} PS_{,RR}$$

$$LLA_{22} = -4e^{-2GA+6PS} PS_{,U} PS_{,R} +2e^{-2GA+6PS}(PS_{,R})^2$$

$$LLA_{23} = -4We^{-2GA+6PS} PS_{,U} PS_{,R} +2We^{-2GA+6PS}(PS_{,R})^2 -e^{-2GA+6PS} PS_{,U} W_{,R} +$$
$$-e^{-2GA+6PS} PS_{,R} W_{,U} +e^{-2GA+6PS} PS_{,R} W_{,R}$$

$$LLA_{33} = 4R^2 e^{-2GA+2PS} PS_{,U} PS_{,R} -2R^2 e^{-2GA+2PS}(PS_{,R})^2 -2Re^{-2GA+2PS} PS_{,U} +$$
$$+2Re^{-2GA+2PS} PS_{,R} -4W^2 e^{-2GA+6PS} PS_{,U} PS_{,R} +2W^2 e^{-2GA+6PS}(PS_{,R})^2 +$$
$$-2We^{-2GA+6PS} PS_{,U} W_{,R} -2We^{-2GA+6PS} PS_{,R} W_{,U} +2We^{-2GA+6PS} PS_{,R} W_{,R}$$

SHP> (LOAD "wa.blf")
Loading file wa.blf
File wa.blf loaded
SHP> (RPL FAC)
FAC
=(-:GDET)^(1/2)$
SHP> (WMAKE WA)

$$WA_0 = R^{-1} e^{4PS} W_{,U} -R^{-1} e^{4PS} W_{,R}$$

$$WA_1 = -R^{-1} e^{4PS} W_{,R}$$

SHP> (load "hdu.blf")

```
Loading file hdu.blf
File hdu.blf loaded
SHP> (RPL DELTA)
```
DELTA^0_0
$=1\$$

DELTA^1_0
$=0\$$

. . .

DELTA^3_3
$=1\$$

```
SHP> (wmake hdu)
```
$\text{HDU}^0_0 = 1$

$\text{HDU}^1_1 = 1$

$\text{HDU}^2_3 = -W$

$\text{HDU}^3_3 = 1$

```
SHP> (LOAD "wwa.blf")
Loading file wwa.blf
File wwa.blf loaded
SHP> (WMAKE WWA)
```

$WWA_{00} = -2R\ e^{-1}\ GA\ W^{4PS}_{,U,U} + R\ e^{-1}\ GA\ W^{4PS}_{,U,R} + R\ e^{-1}\ GA\ W^{4PS}_{,R,U} + 6R\ e^{-1}\ PS\ W^{4PS}_{,U,U} +$
$-5R\ e^{-1}\ PS\ W^{4PS}_{,U,R} - R\ e^{-1}\ PS\ W^{4PS}_{,R,U} + R\ e^{-1}\ W^{4PS}_{,UU} - R\ e^{-1}\ W^{4PS}_{,UR}$

$WWA_{01} = R\ e^{-1}\ GA\ W^{4PS}_{,R,R} + 4R\ e^{-1}\ PS\ W^{4PS}_{,R,U} - 5R\ e^{-1}\ PS\ W^{4PS}_{,R,R} + R\ e^{-1}\ W^{4PS}_{,UR} +$
$-R\ e^{-1}\ W^{4PS}_{,RR} - R\ e^{-2}\ W^{4PS}_{,U} + R\ e^{-2}\ W^{4PS}_{,R}$

$WWA_{10} = R\ e^{-1}\ GA\ W^{4PS}_{,R,R} - 4R\ e^{-1}\ PS\ W^{4PS}_{,U,R} - R\ e^{-1}\ PS\ W^{4PS}_{,R,R} - R\ e^{-1}\ W^{4PS}_{,UR}$

$WWA_{11} = 2R\ e^{-1}\ GA\ W^{4PS}_{,R,R} - 6R\ e^{-1}\ PS\ W^{4PS}_{,R,R} - R\ e^{-1}\ W^{4PS}_{,RR} + R\ e^{-2}\ W^{4PS}_{,R}$

$WWA_{22} = R\ e^{-1\ -2GA+8PS}\ PS\ W_{,U,R} - R\ e^{-1\ -2GA+8PS}\ PS\ W_{,R,U}$

$WWA_{23} = R\ We^{-1\ -2GA+8PS}\ PS\ W_{,U,R} - R\ We^{-1\ -2GA+8PS}\ PS\ W_{,R,U}$

$WWA_{32} = R\ We^{-1\ -2GA+8PS}\ PS\ W_{,U,R} - R\ We^{-1\ -2GA+8PS}\ PS\ W_{,R,U}$

$$WWA_{33} = -Re^{-2GA+4PS}PS_{,U}W_{,R} + Re^{-2GA+4PS}PS_{,R}W_{,U} - e^{-2GA+4PS}W_{,U} +$$

$$+R^{-1}W^2e^{-2GA+8PS}PS_{,U}W_{,R} - R^{-1}W^2e^{-2GA+8PS}PS_{,R}W_{,U}$$

SHP> (LOAD "dsqlam.blf")
Loading file dsqlam.blf
File dsqlam.blf loaded
SHP> (WMAKE DSQLAM)

$$DSQLAM = -4e^{-2GA+4PS}PS_{,U}PS_{,R} + 2e^{-2GA+4PS}(PS_{,R})^2 - 4e^{-2GA+4PS}PS_{,UR} +$$

$$+2e^{-2GA+4PS}PS_{,RR} - 2R^{-1}e^{-2GA+4PS}PS_{,U} + 2R^{-1}e^{-2GA+4PS}PS_{,R}$$

SHP> (LOAD "dsqom.blf")
Loading file dsqom.blf
File dsqom.blf loaded
SHP> (WMAKE DSQOM)

$$DSQOM = 3R^{-1}e^{-2GA+6PS}PS_{,U}W_{,R} - 3R^{-1}e^{-2GA+6PS}PS_{,R}W_{,U}$$

SHP> (LOAD "rhslam.blf")
Loading file rhslam.blf
File rhslam.blf loaded
SHP> (WMAKE RHSLAM)

$$RHSLAM = -4e^{-2GA+4PS}PS_{,U}PS_{,R} + 2e^{-2GA+4PS}(PS_{,R})^2 - 2R^{-2}e^{-2GA+8PS}W_{,U}W_{,R} +$$

$$+R^{-2}e^{-2GA+8PS}(W_{,R})^2$$

SHP> (LOAD "rhsom.blf")
Loading file rhsom.blf
File rhsom.blf loaded
SHP> (WMAKE RHSOM)

$$RHSOM = 3R^{-1}e^{-2GA+6PS}PS_{,U}W_{,R} - 3R^{-1}e^{-2GA+6PS}PS_{,R}W_{,U}$$

SHP> (LOAD "hdd.blf")
Loading file hdd.blf
File hdd.blf loaded
SHP> (WMAKE HDD)

$$HDD_{00} = -e^{2GA-2PS}$$

$$HDD_{01} = -e^{2GA-2PS}$$

$$HDD_{33} = R^2 e^{-2PS}$$

SHP> (LOAD "rhsric.blf")
Loading file rhsric.blf
File rhsric.blf loaded
SHP> (WMAKE RHSRIC)

Algebraic Computing in General Relativity

```
LAA   ended
WAA   ended
```

$$\text{RHSRIC}_{00} = -2GA_{,U} PS_{,U} + GA_{,U} PS_{,R} + GA_{,R} PS_{,U} - GA_{,R} PS_{,R} + 3(PS_{,U})^2 - 2PS_{,U} PS_{,R}$$
$$+(PS_{,R})^2 + PS_{,UU} + 1/2 R^{-2} e^{4PS} (W_{,U})^2$$

$$\text{RHSRIC}_{01} = -GA_{,R} PS_{,R} + PS_{,U} PS_{,R} + (PS_{,R})^2 + PS_{,UR} + 1/2 R^{-2} e^{4PS} W_{,U} W_{,R}$$

$$\text{RHSRIC}_{11} = -2GA_{,R} PS_{,R} + 3(PS_{,R})^2 + PS_{,RR} + 1/2 R^{-2} e^{4PS} (W_{,R})^2$$

$$\text{RHSRIC}_{22} = -2e^{-2GA+4PS} PS_{,U} PS_{,R} + e^{-2GA+4PS} (PS_{,R})^2$$

$$\text{RHSRIC}_{23} = -2W e^{-2GA+4PS} PS_{,U} PS_{,R} + W e^{-2GA+4PS} (PS_{,R})^2 - 1/2 e^{-2GA+4PS} PS_{,U} W_{,R} +$$
$$-1/2 e^{-2GA+4PS} PS_{,R} W_{,U} + 1/2 e^{-2GA+4PS} PS_{,R} W_{,R}$$

$$\text{RHSRIC}_{33} = 2R^2 e^{-2GA} PS_{,U} PS_{,R} - R^2 e^{-2GA} (PS_{,R})^2 - Re^{-2GA} PS_{,U} + Re^{-2GA} PS_{,R} +$$
$$-2W^2 e^{-2GA+4PS} PS_{,U} PS_{,R} + W^2 e^{-2GA+4PS} (PS_{,R})^2 - We^{-2GA+4PS} PS_{,U} W_{,R} +$$
$$-We^{-2GA+4PS} PS_{,R} W_{,U} + We^{-2GA+4PS} PS_{,R} W_{,R} - e^{-2GA+4PS} W_{,U} W_{,R} + 1/2 e^{-2GA+4PS} (W_{,R})^2$$

We have now computed RHSLAM, RHSOM and RHSRIC which are all the terms on the right hand side of (20–22). We have also computed DSQLAM and DSQOM which are the terms on the left hand side of (20) and (21), respectively. The only outstanding term is the 3-dimensional Ricci tensor $\mathcal{R}_{ab}$. This is obtained from the 3-dimensional metric h_{ab}. Here is the corresponding metric file:

```
(TITLE "
CYL7B.CRD
Induced 3-dimensional metric form Geroch decomposition of
Cylindrically symmetric solution (2 degrees of freedom psi and w)
with null coordinate u")
(DIMENSION 3)
(OFF ALL)
(ON NOZERO)
(VARS U R P)   % (u,r,phi)
(FUNS (GA U R) (PS U R))  % (gamma, psi)
(RPL GDD)
-E^(2GA-2PS)$   -E^(2GA-2PS)$   0$
                0$              0$
                                R^2*E^(-2PS)$
```

The function DIMENSION defines the dimension required, where the default is 4. Here is the corresponding SHEEP output:-

```
SHP> (WMAKE DS2)
```

$$ds^2 = -e^{2GA-2PS} dU^2 - 2e^{2GA-2PS} dUdR + R^2 e^{-2PS} dP^2$$

```
SHP> (WMAKE RIC)
GDF   ended
GAM   ended
GDET  ended
GINV  ended
GDET  lost
GUU   ended
GDF   lost
GAMU  ended
RIE   ended
GAM   lost
GAMU  lost
```

$$R_{00} = -2GA_{,U} PS_{,U} + GA_{,U} PS_{,R} + GA_{,U} PS_{,R} - GA_{,R} PS_{,U} - 2GA_{,R} + GA_{,RR} + (PS_{,U})^2 +$$

$$-2PS_{,U} PS_{,R} + (PS_{,R})^2 + PS_{,UU} + 2PS_{,UR} - PS_{,RR} - R^{-1} GA_{,U} + R^{-1} GA_{,R} + R^{-1} PS_{,U} +$$

$$-R^{-1} PS_{,R}$$

$$R_{01} = -GA_{,R} PS_{,R} - 2GA_{,UR} + GA_{,RR} - PS_{,U} PS_{,R} + (PS_{,R})^2 + 3PS_{,UR} - PS_{,RR} + R^{-1} GA_{,R} +$$

$$+R^{-1} PS_{,U} - R^{-1} PS_{,R}$$

$$R_{11} = -2GA_{,R} PS_{,R} + (PS_{,R})^2 + PS_{,RR} + 2R^{-1} GA_{,R}$$

$$R_{22} = 2R^2 e^{-2GA} PS_{,U} PS_{,R} - R^2 e^{-2GA} (PS_{,R})^2 - 2R^2 e^{-2GA} PS_{,UR} + R^2 e^{-2GA} PS_{,RR} +$$

$$-2Re^{-2GA} PS_{,U} + 2Re^{-2GA} PS_{,R}$$

The important point to realise is that we have used STENSOR to define the tensors in complete generality. To apply the Geroch decomposition to any other metric we need only change the metric loaded in the file make.crd together with the components of the Killing vector XIU. The metric we have looked at is called cyl7a whose number is illustrative of the fact that various other cases have been looked at in the work on Numerical Relativity.

Acknowledgements

I would like to thank the Stockholm school of Ian Cohen, Inge Frick, Jan Åman and Lars Hörnfeldt under the direction of Bertel Laurent for the work they did in developing my original ideas. In the UK, I would also like to thank Tony Russell-Clark for his initial collaboration and subsequently Malcolm MacCallum and Jim Skea who took the work yet further.

References

Åman J E, 1987, "Manual for CLASSI: classification programs in general relativity (third provisional edition)". University of Stockholm, Institute of Theoretical Physics Report.

Åman J E and Karlhede A, 1980, *Phys. Lett. A*, **80**, 229.

Bondi H, van der Burg M and Metzner A, 1962, *Proc. Roy. Soc. London*, **A269**, 21.

Collins J M, d'Inverno R A and Vickers J A, 1990, *Class. Quant. Grav.*, **7**, 2005.

Collins J M, 1991, *Class. Quant. Grav.*, **8**, 1859.

Collins J M and d'Inverno R A, 1993, *Class. Quant. Grav.*, **10**, 343.

d'Inverno R A, 1969, *Comput. J*, **12**, 124.

d'Inverno R A, 1980, "A review of Algebraic Computing in General Relativity" in *General Relativity and Gravitation*, Volume 1, ed. A. Held, Plenum, 491.

d'Inverno R A, 1992, *Introducing Einstein's Relativity*, Oxford University Press

d'Inverno R A and Frick I, 1982, *Gen. Rel. Grav.*, **14**, 735.

d'Inverno R A and Russell-Clark R A, 1971a, *Comput. J*, **17**, 229.

d'Inverno R A and Russell-Clark R A, 1971b, *J Math. Phys.*, **12**, 1258.

Frick I, 1977, "The Computer Algebra Systen SHEEP. What it can and cannot do in General Relativity". University of Stockholm, Institute of Theoretical Physics Report 77–14.

Geroch R, 1970, *J. Math. Phys.*, **12**, 918.

Harrison B K, 1959, *Phys. Rev*, **116**, 1285.

Holmes G and d'Inverno R A, 1986, "Interacting with VFRAME", GR11 abstracts, eds. B Laurent and K Rosquist, University of Stockholm.

Hörnfeldt L, 1988, "STENSOR User Manual", University of Stockholm, Institute of Theoretical Physics.

Karlhede A, 1980, *Gen Rel. Grav.*, **12**, 693.

Koutras A, 1992, *Class. Quant. Grav.*, **9**, L143.

Kramer D, Stephani H, MacCallum M A H and Herlt E, 1980, *Exact solutions of Einstein's field equations*, Deutscher Verlag der Wissenscaften, Berlin, and Cambridge UP

Krasiński A, 1993, *Physics in an inhomogeneous universe*, (to appear).

MacCallum M A H and Wright F J, 1991, *Algebraic Computing with REDUCE*, Clarendon Press, Oxford.

MacCallum M A H, Skea J E F, McCrea J D and McLenaghan R G, 1994, *Algebraic Computing in General Relativity*, Clarendon Press, Oxford.

Machado Ramos M P and Vickers J A, "Invariant differential operators and the Karlhede classification of type N space-times", (in preparation).

Sammet J E, 1966, *Commun. ACM*, **9**, 555.

Numerical Computing in General Relativity

Ray d'Inverno

Faculty of Mathematical Studies
Southampton University

1 Overview of Numerical Relativity

1.1 The status of exact solutions

The study of exact solutions of Einstein's equations has been central to much of classical General Relativity and has lead to important advances in our understanding of the theory. There are now a huge number of exact solutions known. The exact solutions book (Kramer et al. 1980) has references running into four figures. However, although many solutions are known, significantly fewer of them are fully understood in the sense that we know their singularity, geodesic, causal and global structure. This is known for black holes, the simpler cosmological solutions and plane wave solutions, but not for many other cases.

As is well known, partial differential equations admit a wide class of solutions. Many of these are pathological in nature. To find physically realistic solutions, we need to augment the equations with appropriate initial and boundary conditions. It would appear that a large number of exact solutions in General Relativity are indeed pathological in character and do not represent or approximate to physically realistic scenarios. Many of then, for example, possess strange singularity structures. Worse still is the fact that we do not know any exact solutions which correspond to important physical situations such as 2-body or n-body systems, radiative sources, interiors of rotating sources undergoing gravitational collapse, and so on. Yet these are precisely the objects that are of interest to us, especially on the astrophysical scale. This is where Numerical Relativity comes in.

1.2 Why Numerical Relativity?

Numerical Relativity consists of solving Einstein's equations numerically on a computer. The standard approach, in outline, is to specify the 3-metric $\overset{3}{g}$—the intrinsic geometry—of some spacelike slice ($t = t_1$ = constant, say) together with its time derivative, and use the field equations to compute the 3-metric at some future time ($t=t_2>t_1$). This approach can then be used to model physically interesting scenarios. Indeed, given the freedom to vary the initial configuration, we can consider the resulting numerical simulations as being in the arena of experimental relativity. A well constructed numerical code can serve as this numerical laboratory in which one can investigate parameter dependence, extract generic behaviour, and so on. Of course, this requires that the code be general enough and can be run fast enough for such a use.

Again, when exact methods fail, recourse is often made to approximate methods. These rely on the existence of a small expansion parameter which is a physical quantity characteristic of the problem. However, for strong field, highly non-linear problems, all natural expansion parameters are large. Numerical Relativity can handle these problems since they rely on non-physical expansion parameters such as the grid spacing. Many situations of astronomical interest fall into this category.

Numerical Relativity is a young field and has only been in existence for some 20 years. As a consequence there are fewer people currently working in it, although it has become a major growth area in recent years. The significance of the field is likely to become more pronounced when the long awaited detection of gravitational waves is at last reported and we move into the era of gravitational astronomy. The need will then arise of finding theoretical justifications for actual observations, and this need is likely to push Numerical Relativity into the forefront of General Relativity.

1.3 The arena of Numerical Relativity

The majority of studies in Numerical Relativity have been reported in conferences devoted to the subject. The proceedings of five conferences which were held at:

- Batelle Seattle, US, 1978 (Smarr 1979)
- Les Houches, France, 1982 (Deruelle and Piran 1982)
- Drexel, US, 1985 (Centrella 1986)
- Illinois, US, 1988 (Evans et al. 1989)
- Southampton, UK, 1991 (d'Inverno 1992)

together survey most of the advances in the field of recent years. In Numerical Relativity, we are concerned with dynamical situations, and hence it is conventional to refer to the dimension of a problem in terms of its spatial dimension. Thus most investigations usually fall into one, two or three dimensional categories. Some pioneering work was carried out in the mid sixties by May and White (1966) on spherical collapse and Hahn and Lindquist (1964) on black hole collisions. However, it is generally acknowledged that the field only came of age in the mid seventies with the work of Smarr on two black hole collisions (Smarr 1979). There are now a large number of successful codes in existence including: spherical collapse, dust collapse, 2-dimensional axisymmetric neutrino star bounce, 2-dimensional black hole collision, primordial black holes, interaction of black holes with gravitational radiation, gravitational wave generation, colliding gravitational waves, Brill waves, Teukolsky waves, neutrino radiation, cylindrically symmetric solutions, planar symmetry solutions, stellar collapse, supernovae, accretion disks, shock waves, n-body calculations, collapse of massless scalar fields, evolution of 3-dimensional wave packets, 3-dimensional relativistic hydrodynamics, planar cosmologies, inhomogeneous cosmologies, inflationary cosmology and cosmic strings. Most of the 3-dimensional work undertaken to date has involved Newtonian models of one sort or another. However, there are several 3-dimensional fully relativistic codes which are in existence or are reaching completion. These codes make enormous demands, even by present day standards, on computer time and memory.

1.4 Techniques of Numerical Relativity

The standard formulation is that of a 3+1 ADM (Arnowitt, Deser and Misner; see Smarr (1979)) decomposition in which time is singled out as a privileged direction and spacetime is foliated by 3-dimensional spacelike hypersurfaces corresponding to constant time slices. The (2+1)+1 formulation (Maeda et al. 1980) uses Geroch decomposition to factor out a Killing direction. The ADM approach is then used to decompose the remaining 3-dimensional spacetime into a 2+1 form. The 2+2 formalism (d'Inverno 1984) decomposes spacetime into two families of 2-dimensional spacelike hypersurfaces. It is closely related to the characteristic approach (Gomez et al. 1986, Bishop 1992) which is specially suited to studying gravitational waves. These and the 3+1 approach will be discussed in ensuing chapters. Finally, there is the Regge calculus approach which is a discrete formulation of General Relativity where blocks of flat spacetime are 'glued' together and curvature exists only on the two-dimensional edges of the blocks (Regge 1961).

There are five categories of numerical techniques for solving partial differential equations which have been employed in Numerical Relativity. The most frequently employed method is that of finite differencing, which is essentially based on Taylor series expansions of functions (Ames 1969). We shall mostly concentrate on this method. The finite element method utilizes locally supported basis functions which may be chosen arbitrarily, but which are usually taken to be polynomials (Zienkiewicz 1977). Spectral methods are based on global expansions of functions in terms of orthogonal basis functions, often taken to be Chebyshev polynomials (Gottlieb and Orszag 1977). Characteristic methods are suitable for hyperbolic equations only, and reduce the PDEs to ODEs which are to be integrated along the characteristics (Fletcher 1991). Finally, there are the less commonly used particle methods e.g. PIC (Particle in Cell) or SPH (Smoothed Particle Hydrodynamics) in which the fluid is approximated by a collection of particles carrying mass, momentum and energy (Hockney and Eastward 1981).

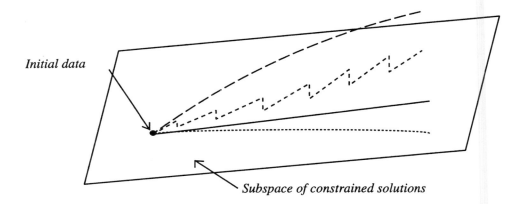

Figure 1. *Actual trajectory of solution (solid line); free evolution (dashes); chopped evolution (small dashes); fully constrained (dots)*

1.5 Problems in Numerical Relativity

As we shall see, the standard approach reduces the field equations into constraint equations and evolution equations. The finite difference version of Einstein's equations leads to an overdetermined system in which the constraints are most frequently ignored (free evolution) or artificially imposed. In the latter case, one method involves imposing the constraints after finite intervals of times (chopped evolution) and another is to impose them at every stage of integration (fully constrained evolution). Unfortunately, each method may have associated drawbacks. For example, computations with particular exact solutions suggest that a free evolution drifts further away from the true solution as it evolves in time. Similar problems may arise with chopped and fully constrained solutions. Piran has indicated this schematically in Figure 1, where the plane represents the subspace of solutions which satisfy the constraint equations. However, see the article of Choptuik (1991) for a more recent discussion of this issue. Other problems relate to the finite difference approximation. Unfortunately, there are an infinite number of possible finite difference schemes, each with its own solution, of which a large number will bear little resemblance to the exact solution of the original equations. Even if one is using a stable scheme, another major source of inaccuracy occurs in truncation errors. These latter errors stem from the fact that one is essentially approximating a function by a finite part of the Taylor series expansion. Other difficulties involve applying appropriate coordinate (or gauge) conditions and coordinate singularities. A major problem which plagues many codes is the necessity to impose artificial boundary conditions arising from the use of a finite numerical grid. For example, in codes modelling radiative isolated sources it is necessary to specify a boundary condition exterior to the source which it is hoped rules out incoming radiation. However, the new combined or mixed approaches, which we discuss later, offer the possibility of advancement here.

There is the problem of representing and interpreting the solution. For example, what are meaningful quantities for which one should produce a graphical display? How

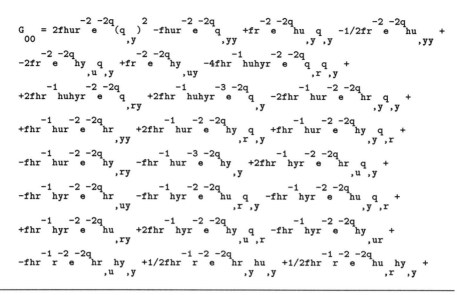

Figure 2. *SHEEP output for part of the zero zero component of the Einstein tensor*

does one display 3-dimensional quantities? Then there is the problem that Numerical Relativity usually makes enormous demands on computer time and memory, which produces limitations on what is attainable at any one time.

1.6 Computer aided relativity

Many of the formalisms employed involve long and complicated algebraic computations leading to lengthy expressions which often require conversion into a particular coding format—a process which could well introduce errors. In 1986, Nakamura (1986) used REDUCE to first generate such large algebraic expressions and then exploited REDUCE's ability to convert algebraic expressions into their FORTRAN equivalent, prior to numerical computation. Again, in some numerical work by the Southampton group on radiative systems, I wrote a program in LISP to automatically convert SHEEP output (Figure 2) into OCCAM code (Figure 3), which was then loaded onto a transputer system for processing. Here, computers are used both for algebraic and numeric work. Nakamura has proposed as a name for this combined area CAR—Computer Aided Relativity. (It is amusing to note that CAR is also the name of a basic function in LISP for picking up the first element in a list.)

1.7 The status of Numerical Relativity

In the article 'Algebraic Computing in General Relativity' in this volume, we discussed the origins of Algebraic Computing systems. Historically, relativists were originally distrustful of results obtained by a computer because they were not convinced that the results were reliable. It was only after very complicated calculations had been

```
chuc:=chuc+((-1.0000000E+00(REAL64))*(f*(hr*
(((1.0000000E+00(REAL64))/r)*(e2qa*(hy1*qa2))))))
chuc:=chuc+((-1.0000000E+00(REAL64))*(f*(hr*
(((1.0000000E+00(REAL64))/r)*(e2qa*(hy1*qa1))))))
chuc:=chuc+((5.0000000E-01(REAL64))*(f*(hr*
(((1.0000000E+00(REAL64))/r)*(e2qa*hy12))))) 
chuc:=chuc+(f*(hr*(((1.0000000E+00(REAL64))/(r*r))*
(e2qa*hy2))))
chuc:=chuc+((-2.0000000E+00(REAL64)0*(f*((hy*hy)*
(((1.0000000E+00(REAL64))/r)*(e2qa*(qa1*qa1))))))
chuc:=chuc+(f*((hy*hy)*(((1.0000000E+00(REAL64))/r)*
(e2qa*qa11))))
chuc:=chuc+((2.0000000E+00(REAL64))*(f*((hy*hy)*
(((1.0000000E+00(REAL64))/(r*r))*(e2qa*qa1)))))
chuc:=chuc+(f*((hy*hy)*(((1.0000000E+00(REAL64))/(r*(r*
r)))*e2qa)))
chuc:=chuc+((5.0000000E-01(REAL64))*(f*(hy*
(((1.0000000E+00(REAL64))/r)*(e2qa*hr12)))))
chuc:=chuc+((2.0000000E+00(REAL64))*(f*(hy*
(((1.0000000E+00(REAL64))/r)*(e2qa*(hy1*qa1))))))
chuc:=chuc+((-5.0000000E-01(REAL64))*(f*(hy*
(((1.0000000E+00(REAL64))/r)*(e2qa*hy11)))))
chuc:=chuc+((-1.0000000E+00(REAL64))*(f*(hy*
(((1.0000000E+00(REAL64))/(r*r))*(e2qa*hr2)))))
chuc:=chuc+((-1.0000000E+00(REAL64))*(f*(hy*
(((1.0000000E+00(REAL64))/(r*r))*(e2qa*hy1)))))
chuc:=chuc+((2.5000000E-01(REAL64))*(f*
(((1.0000000E+00(REAL64))/r)*(e2qa*(hr2*hr2)))))
chuc:=chuc+((-2.5000000E-01(REAL64))*(f*
(((1.0000000E+00(REAL64))/r)*(e2qa*(hy1*hy1)))))
chuc:=chuc+((-1.0000000E+00(REAL64))*(f*
(((1.0000000E+00(REAL64))/hr)*((hy*hy)*
(((1.0000000E+00(REAL64))/r)*(e2qa*(hr1*qa1))))))))
```

Figure 3. *Part of the LISP translation into OCCAM code*

checked successfully against each other using algebra systems based on different machines employing different software and design philosophies, that confidence was eventually established in the tool. A similar credibility problem seems to apply to Numerical Relativity. After all, the one thing you can pretty much guarantee about a numerical calculation is that it will produce a result; but is it the correct result? Because it is still a relatively small field less work has been done in calibrating codes and cross checking results. Centrella and others (1986) have suggested a series of test-bed calculations involving both 'analytic' and numerical solutions which may be used to demonstrate the reliability of numerical codes. Moreover, as Choptuik has repeatedly pointed out (d'Inverno 1992), codes should be checked for self-consistency by showing that they converge appropriately. There needs to be a greater move towards making numerical codes freely available to other workers so that they can be checked. Only in this way can we move to a situation where the mainstream community is convinced that what they are getting is science and not cookery.

2 The 3+1 approach to Numerical Relativity

2.1 The 3+1 decomposition of the metric

The 3+1 formalism is central to much of Numerical Relativity. You can find the formulae in many references (*e.g.* Smarr 1979). Since the decomposition of the metric is so important, and is also a prototype for other similar calculations, I shall present a detailed derivation of it here for completeness. We start with a foliation of spacetime into a family of spacelike hypersurfaces Σ given by

$$\phi(x^a) = \text{constant} \tag{1}$$

Then $w_a = \phi_{,a}$ is the covariant normal to the family and $w^a = g^{ab}w_b$ is the contravariant normal formed from it with the contravariant metric. We define n^a to be the unit normal to the foliation which is necessarily proportional to w^a and, being a unit normal, satisfies

$$n^a n_a = -1 \tag{2}$$

where we use the standard 3+1 signature of $(-1, +1, +1, +1)$. We use the unit normal to define the projection operator

$$B_a{}^b = \delta_a{}^b + n_a n^b \tag{3}$$

which can be used to project tensors into the foliation. Clearly $B_a{}^b n^a = 0$ by (2) and (3). Moreover, we can project the 4-dimensional metric into the foliation Σ to obtain the induced 3-dimensional metric ${}^3\!g_{ab}$, namely

$${}^3\!g_{ab} = B_a{}^c B_b{}^d g_{cd} \tag{4}$$

Using (2) and (3) we find that

$${}^3\!g_{ab} = g_{ab} + n_a n_b \tag{5}$$

or, equivalently,

$${}^3\!g^{ab} = g^{ab} + n^a n^b \tag{6}$$

Next, we consider any rigging vector field t^a which transvects the foliation (*i.e.* lies nowhere in the foliation). Then we can decompose it into a component parallel and a component orthogonal to n^a, specifically

$$t^a = \alpha n^a + \beta^a \tag{7}$$

where α is a proportionality factor called the lapse and β^a is called the shift vector (see Figure 4) and satisfies

$$\beta^a n_a = 0 \tag{8}$$

Thus, from (6) and (7), we find

$$g^{ab} = -\frac{1}{\alpha}(t^a - \beta^a)\frac{1}{\alpha}(t^b - \beta^b) + {}^3\!g^{ab} \tag{9}$$

Finally, we introduce coordinates

$$(x^a) = (x^0, x^\mu) = (t, x^1, x^2, x^3) \tag{10}$$

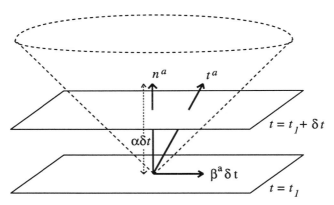

Figure 4. *Two neighbouring slices of the foliation Σ*

(where Greek indices run from 1 to 3), adapted to the foliation and the rigging, in which the hypersurfaces $\Sigma(t)$ are the constant time slices

$$t = \text{constant} \tag{11}$$

and the rigging vector field is $\frac{\partial}{\partial t}$. In these coordinates, since the induced 3-metric and the shift vector are purely spatial, we can write

$${}^3g^{ab} = \delta_\mu{}^a \delta_\nu{}^b \gamma^{\mu\nu} \tag{12}$$

say, and

$$\beta^a = \delta_\mu{}^a \beta^\mu \tag{13}$$

We therefore find from (9), (12) and (13) that the contravariant metric can be written in the form

$$g^{ab} = \begin{pmatrix} -\dfrac{1}{\alpha^2} & \dfrac{\beta^\mu}{\alpha^2} \\ \dfrac{\beta^\mu}{\alpha^2} & \gamma^{\mu\nu} - \dfrac{\beta^\mu \beta^\nu}{\alpha^2} \end{pmatrix} \tag{14}$$

Defining the induced covariant metric $\gamma_{\mu\nu}$ by

$$\gamma_{\mu\nu} \gamma^{\nu\sigma} = \delta_\mu{}^\sigma \tag{15}$$

then we can raise and lower Greek indices with $\gamma^{\mu\nu}$ and $\gamma_{\mu\nu}$, respectively, and so

$$\beta_\mu = \gamma_{\mu\nu} \beta^\nu \tag{16}$$

The covariant metric is

$$g_{ab} = \begin{pmatrix} -\alpha^2 + \beta^\sigma \beta_\sigma & \beta_\mu \\ \beta_\mu & \gamma_{\mu\nu} \end{pmatrix} \tag{17}$$

since (14) and (17) satisfy

$$g_{ab} g^{bc} = \delta_a^c \tag{18}$$

The line element in canonical 3+1 form is therefore

$$ds^2 = -(\alpha^2 - \beta^\sigma \beta_\sigma) dt^2 + 2\beta_\mu dt dx^\mu + \gamma_{\mu\nu} dx^\mu dx^\nu \tag{19}$$

2.2 The extrinsic curvature

The extrinsic curvature is given by

$$K_{ab} = -2B_a{}^c B_b{}^d \nabla_{(c} n_{d)} \tag{20}$$

Since $K_{ab} = K_{ba}$ and $K_{ab}n^a = K_{ab}n^b = 0$ the extrinsic curvature is purely a spatial quantity $K_{\mu\nu}$. It follows from the previous equations that

$$K_{\mu\nu} = \frac{1}{2\alpha}(-\frac{\partial \gamma_{\mu\nu}}{\partial t} + D_\nu \beta_\mu + D_\mu \beta_\nu) \tag{21}$$

with

$$D_\nu \beta_\mu = \beta_{\mu,\nu} - {}^3\Gamma^\sigma_{\mu\nu}\beta_\sigma \tag{22}$$

where D is the 3-covariant derivative operator defined in terms of the induced 3-metric $\gamma_{\mu\nu}$ (or alternatively can be obtained by projection of the 4-covariant derivative ∇). Then

$$K^\mu{}_\nu = \gamma^{\mu\sigma} K_{\sigma\nu} = \frac{1}{2\alpha}(-\gamma^{\mu\sigma}\frac{\partial \gamma_{\sigma\nu}}{\partial t} + D_\nu \beta^\mu + D^\mu \beta_\nu) \tag{23}$$

and

$$K = \gamma^{\mu\nu} K_{\mu\nu} = \frac{1}{2\alpha}(-\gamma^{\mu\nu}\frac{\partial \gamma_{\mu\nu}}{\partial t} - 2D_\mu \beta^\mu) \tag{24}$$

2.3 The 3+1 form of Einstein's equations

The full Einstein equations

$$G_{ab} = 8\pi T_{ab} \tag{25}$$

give rise to the constraint equations, consisting of the Hamiltonian constraint (obtained from $G_{ab}n^a n^b$))

$$\tfrac{1}{2}({}^3R + K^2 - K_{\mu\nu}K^{\mu\nu}) = 8\pi \rho \tag{26}$$

and the momentum constraints (obtained from $B_a{}^\mu n^b G_b{}^a$)

$$D_\nu(K^{\mu\nu} - \gamma^{\mu\nu} K) = 8\pi j^\mu \tag{27}$$

together with the dynamical equations (obtained from $B_\mu{}^a B^{\nu b}(T_{ab} - \tfrac{1}{2}g_{ab}g^{cd}T_{cd})$)

$$\begin{aligned}\frac{\partial K_\mu{}^\nu}{\partial t} &= -D^\nu D_\mu \alpha + \beta^\sigma D_\sigma K_\mu{}^\nu + K_\sigma{}^\nu D_\mu \beta^\sigma - K_\mu{}^\sigma D_\sigma \beta^\nu \\ &\quad + \alpha[{}^3R_\mu{}^\nu + KK_\mu{}^\nu - 3\pi S_\mu{}^\nu - 4\pi\delta_\mu{}^\nu(\rho - S)]\end{aligned} \tag{28}$$

where

$$T_{ab}n^a n^b = \rho \tag{29}$$
$$B_a{}^\mu T_b{}^a n^b = -j^\mu \tag{30}$$
$$B_\mu{}^a B_\nu{}^b T_{ab} = S_{\mu\nu} \tag{31}$$

It is possible to make these equations look less coordinate-dependent and more manifestly covariant by, for example, replacing partial derivatives with respect to t by Lie derivatives with respect to t^a, but we shall not pursue this further here.

2.4 Interpretation of the 3+1 approach

From a purely geometrical point of view the 3+1 formalism is simply a decomposition relative to a foliation of spacetime by spacelike hypersurfaces transvected by a timelike fibration *i.e.* the vector t^a is tangent to a congruence of curves threading the foliation. The extrinsic curvature provides information on the rate of change of the 3-metric with respect to local orthogonal proper time. The extrinsic curvature of a given $\Sigma(t)$ also describes the manner in which the hypersurface is embedded in the enveloping 4-geometry.

An alternative 3+1 viewpoint regards General Relativity as a dynamical theory (geometrodynamics) in which spacetime is comprised of the 'time history' of a spacelike hypersurface $\Sigma(t)$ regarded as an 'instant of time'. The geometry of the spacetime is regarded as the 4-metric g_{ab} and the geometry of Σ is described by the 3-metric $\gamma_{\mu\nu}$. There are then two types of variables: the four functions comprising the lapse α and shift β^μ are kinematical and are essentially freely specifiable, since they embody the 4-fold coordinate freedom of General Relativity. The 6 functions comprising the 3-metric $\gamma_{\mu\nu}$ are the dynamical variables. From a Hamiltonian viewpoint the metric components $\gamma_{\mu\nu}$ are the generalized coordinates and the generalized momenta are the conjugate variables $K_{\mu\nu}$. This formulation casts Einstein's equations into a set of equations which are first order in time.

2.5 The Cauchy initial value problem for general relativity

It is clear from the form of the field equations that in vacuum the set of 12 variables $\{\gamma_{\mu\nu}, K_{\mu\nu}\}$ constitute a complete set of variables for the initial value problem. In the case of the full field equations the set $\{\rho, j^\mu, S_{\mu\nu}\}$ must be obtained by the dynamical field equations from the conservation equations

$$\nabla_a T^{ab} = 0 \qquad (32)$$

together with any other required relations like an equation of state and conservation of baryon number in the case of a fluid. We shall, for simplicity, restrict attention to the case of the vacuum field from now on, although the story for the full field equations is not very different. Then the constraint equations reveal that $\{\gamma_{\mu\nu}, K_{\mu\nu}\}$ cannot be chosen arbitrarily but must satisfy these equations on all hypersurfaces $\Sigma(t)$. However, the field equations are not independent because they must satisfy the contracted Bianchi identities

$$\nabla_a G^{ab} \equiv 0 \qquad (33)$$

These can be used to establish the Lichnerowicz lemma: if the evolution equations are satisfied generally and the constraints are also satisfied initially (that is on $\Sigma(0)$) then the constraints will automatically be satisfied at all later times.

Thus the initial value problem for General Relativity involves a specification of the initial topology of $\Sigma(0)$ together with a specification of the dynamical variables $\{\gamma_{\mu\nu}, K_{\mu\nu}\}$ such that the constraints are initially satisfied on $\Sigma(0)$. As a consequence, we may only choose 8 of the 12 dynamical variables freely, since the other 4 are fixed by the 4 constraints. Thus the question arises as to which variables can be specified a priori and which can be determined by the constraints.

2.6 The conformal approach

The key idea of the conformal approach (York and ŌMurchadha (1974)) is to introduce conformal scalings so that the constraint equations are cast into a set of 4 quasi-linear elliptic partial differential equations for 4 gravitational 'potentials'. This idea facilitates both theoretical analysis as well as providing a numerical technique. We introduce a conformal factor ψ and write the 3-metric $\gamma_{\mu\nu}$ in the form

$$\gamma_{\mu\nu} = \psi^4 \, \widehat{\gamma}_{\mu\nu} \tag{34}$$

The conformal factor is one of the 'potentials' which will be fixed by the Hamiltonian constraint. Then, among other things, the scalar curvature transforms as

$$R = \psi^{-4}\widehat{R} - 8\psi^{-5}(\widehat{D}^\mu \widehat{D}_\mu)\psi \tag{35}$$

We then perform a transverse traceless decomposition of the extrinsic curvature tensor which introduces three additional 'potentials' X^μ which will be fixed by the momentum constraints. Defining the trace-free part of the extrinsic curvature by

$$A^{\mu\nu} = K^{\mu\nu} - \tfrac{1}{3}\gamma^{\mu\nu} K \tag{36}$$

then the choice

$$A^{\mu\nu} = \psi^{-10}\widehat{A}^{\mu\nu} \tag{37}$$

results in the property

$$D_\nu A^{\mu\nu} = \psi^{-10}\widehat{D}_\nu \widehat{A}^{\mu\nu} \tag{38}$$

As with any traceless symmetric tensor, $\widehat{A}^{\mu\nu}$ can be decomposed into a part $\widehat{A}^{\mu\nu}_{\text{TT}}$ with vanishing divergence and trace, and another trace-free part which can be obtained from differentiating a vector potential W^μ, namely

$$\widehat{A}^{\mu\nu} = \widehat{A}^{\mu\nu}_{\text{TT}} + (\widehat{\ell}W)^{\mu\nu} \tag{39}$$

where

$$(\widehat{\ell}W)^{\mu\nu} = \widehat{D}^\mu W^\nu + \widehat{D}^\nu W^\mu - \tfrac{2}{3}\widehat{\gamma}^{\mu\nu}\widehat{D}_\sigma W^\sigma \tag{40}$$

and the TT (Transverse-Traceless) part $\widehat{A}^{\mu\nu}_{\text{TT}}$ satisfies

$$\widehat{D}_\nu \widehat{A}^{\mu\nu}_{\text{TT}} = 0 \tag{41}$$

In practice, it will generally be inconvenient to give the freely specifiable part of the conformally scaled extrinsic curvature in terms of a transverse-traceless tensor. So we 'reverse decompose' $\widehat{A}^{\mu\nu}_{\text{TT}}$ as

$$\widehat{A}^{\mu\nu}_{\text{TT}} = \widehat{T}^{\mu\nu} - (\widehat{\ell}V)^{\mu\nu} \tag{42}$$

where the traceless, symmetric tensor $\widehat{T}^{\mu\nu}$ is freely specifiable and V^μ is another vector field. Then

$$\widehat{A}^{\mu\nu} = \widehat{T}^{\mu\nu} + (\widehat{\ell}W)^{\mu\nu} - (\widehat{\ell}V)^{\mu\nu} \equiv \widehat{T}^{\mu\nu} + (\widehat{\ell}X)^{\mu\nu}$$

The Hamiltonian and momentum constraints become, in this approach,

$$\widehat{\triangle}\psi \equiv (\widehat{D}^\mu \widehat{D}_\mu)\psi = \tfrac{1}{8}\widehat{R}\psi + \tfrac{1}{12}K^2\psi^5 - \tfrac{1}{8}\left(\widehat{T}^{\mu\nu} + (\widehat{\ell}X)^{\mu\nu}\right)^2 \psi^{-7} \tag{43}$$

$$(\widehat{\triangle}_\ell X)^\mu \equiv \widehat{D}_\nu(\widehat{\ell}X)^{\mu\nu} = -\widehat{D}_\nu \widehat{T}^{\mu\nu} + \tfrac{2}{3}\psi^6 \widehat{D}^\mu K \tag{44}$$

These equations are a set of 4 quasi-linear, coupled elliptic PDE's for the 4 gravitational potentials $\{\psi, X^\mu\}$. So, to summarize, the procedure is:

- Freely specify $\{\widehat{\gamma}^{\mu\nu}, K, \widehat{T}^{\mu\nu}\}$
- Solve the constraints for the potentials $\{\psi, X^\mu\}$
- Construct physical initial data using

$$\gamma_{\mu\nu} = \psi^4\, \widehat{\gamma}_{\mu\nu} \qquad (45)$$
$$K^{\mu\nu} = \left(\widehat{T}^{\mu\nu} + (\widehat{\ell}X)^{\mu\nu}\right)\psi^{-10} + \tfrac{1}{3}K\psi^{-4}\widehat{\gamma}^{\mu\nu} \qquad (46)$$

A particularly simple choice for solving the constraints is:

- Introduce Cartesian coordinates (x, y, z) and take the conformal metric to be flat i.e. $\widehat{\gamma}_{\mu\nu} = \text{diag}(1,1,1)$
- Choose the initial slice to be maximal i.e. $K = 0$
- Choose a minimal radiation condition i.e. $\widehat{T}^{\mu\nu} = 0$

With the above choices the constraints become

$$\widehat{\Delta}\psi = -\tfrac{1}{8}(\widehat{\ell}X)^{\mu\nu}(\widehat{\ell}X)_{\mu\nu}\psi^{-7} \qquad (47)$$
$$(\widehat{\Delta}_\ell X)^\mu = 0 \qquad (48)$$

This greatly simplifies the problem since the momentum constraint is decoupled from the Hamiltonian constraint and is linear. It has been possible this way to even find analytic solutions of the momentum constraints, for example, corresponding to one or more black holes with freely specified linear and angular momentum.

2.7 Gauge conditions

There are two conditions which motivate the choice of a particular gauge:-

- The avoidance of both coordinate and physical singularities. The latter can be avoided by slowing down the evolution of the spatial region near the singularity. This is controlled by α.

- We would like to make the Einstein evolution equations as simple as possible, so that a numerical solution is not unduly complicated. This is often controlled by β^μ. For example a good choice can lead to several of the components of $\gamma_{\mu\nu}$ vanishing which reduces the size of expressions like ^3R.

Choices for the lapse

Geodesic slicing. If $\alpha=1$ is combined with $\beta^\mu=0$ then we are using Eulerian observers who are freely falling. The spatial hypersurfaces are geodesically parallel. This slicing is singularity seeking.

Lagrangian slicing. In spherical symmetry we can use $\alpha U^t = 1$ combined with $\beta^\mu = 0$, where U^t is the time-component of the fluid 4-velocity. Then $U^\mu = 0$ and fluid world lines are orthogonal to the spatial hypersurfaces (which is possible because there is no vorticity). Since $\beta^\mu = 0$ the coordinates follow the matter. The fluid world lines will focus towards any singularity and so this slicing is again singularity seeking.

Maximal slicing. We can avoid the focusing of world lines towards singularities by choosing α in such a way that K remains zero on each hypersurface if it is zero on the initial one. Physically K measures the expansion of a congruence of world lines normal to the foliation. Substituting both $K = 0$ and $\partial K/\partial t = 0$ into the evolution equations gives

$$\gamma^{\mu\nu} D_\mu D_\nu \alpha - \alpha^3 R = 4\pi\alpha(S - 3\rho) \qquad (49)$$

This is an elliptic equation for α to be solved on each slice $\Sigma(t)$, which means that it can be computationally expensive.

Constant mean curvature slicing. This slicing condition is $K = C(t)$ where C is a function which is constant on each slice. It is a generalization of the natural time coordinate in homogeneous cosmologies and has been extensively used for cosmological problems.

Algebraic and hypergeometric slicing. The slicing condition

$$\frac{1}{r^2} \frac{d}{dr}\left(r^2 \frac{d}{dr}\right)\alpha = V_0 \operatorname{sech}^2(mr)\alpha \qquad (50)$$

can be made to simulate the singularity avoiding features of maximal slicing without the need to solve an elliptic equation on each slice. With the boundary condition $\alpha \to 1$ as $r \to \infty$ and $d\alpha/dr = 0$ at $r = 0$ it is possible to express α directly in terms of hypergeometric functions. The free parameters V_0 and m provide the latitude for adjustment and experimentation. Nakamura (1981) used this slicing for rotating axisymmetric stellar collapse. For disc-like configurations it was found to be superior to maximal slicing.

Polar slicing. In spherical polar coordinates the condition can be expressed as $K = K_r{}^r$, where r is the radial coordinate. Substituting into the evolution equations yields a parabolic equation for α which can be integrated inwards on a single sweep of the grid. The single boundary condition is $\alpha \to 1$ as $r \to \infty$. Although a significant time saving can be had over maximal slicing, it is irregular at $r = 0$ and so must be smoothly joined to some other slicing at finite R. When combined with the radial gauge the singularity avoidance is very strong, the hypersurfaces never cross the horizon. If it is combined with the radial gauge and axial symmetry then the metric can be written explicitly in terms of six arbitrary functions (d'Inverno 1995).

Choices for the shift

Eulerian gauge. We simply set $\beta^\mu = 0$ so that the coordinate congruence is normal to the foliation. Early stellar collapse codes used this gauge.

Lagrangian gauge. For spacetimes containing matter we can set $\beta^\mu = U^\mu/U^t$ where U^a is the fluid 4-velocity. Thus the coordinate congruence coincides with the congruence of fluid world lines. For one dimensional flows this is a convenient choice, but in two or three dimensions where vorticity may be present the coordinate grid can become severely distorted leading to a loss of accuracy. Taub has given the general formalism for this choice (Taub 1978).

Isothermal and radial gauge. A particular choice of β^μ can simplify Einstein's equations by making certain components of the 3-metric zero. Three conditions on β^μ enable three components to be eliminated giving, for example, a diagonal 3-metric with line element

$$^3ds^2 = \gamma_{xx}dx^2 + \gamma_{yy}dy^2 + \gamma_{zz}dz^2 \tag{51}$$

In an isothermal gauge $\gamma_{r\theta} = 0$ and $\gamma^{rr} = \gamma^{\theta\theta}$. In a radial gauge $\gamma_{r\theta} = \gamma_{r\phi} = 0$ and $\gamma_{\theta\theta}\gamma_{\phi\phi} - \gamma_{\theta\phi}^2 = r^4 \sin^2\theta$, and the 3-metric line element has the form

$$^3ds^2 = A^2 dr^2 + r^2 B^{-2} d\theta^2 + r^2 B^2 (\sin\theta d\phi + \xi d\theta)^2 \tag{52}$$

where A, B and ξ are metric functions to be determined. This is a particularly useful gauge for studying radiation in asymptotically flat spacetimes.

Minimal distortion gauge. The shift vector is found by imposing the condition

$$D_\mu \left(\frac{\partial \tilde{\gamma}^{\mu\nu}}{\partial t}\right) = 0 \tag{53}$$

where $\tilde{\gamma}_{\mu\nu} = \gamma_{\mu\nu}/\gamma^{1/3}$. In linearized theory this reduces to a transverse-traceless gauge and also eliminates non-physical gravitational waves due to coordinate effects. It is not often employed because a complicated vector elliptic equation must be solved on each slice to find the shift, and it offers no simplification of the Einstein equations.

3 The 2+2 and characteristic approaches

3.1 Limitations of the 3+1 approach

There are two major limitations of the 3+1 approach. The first is that the initial data is not freely specifiable, but must satisfy the constraints. The conformal approach is a powerful technique for achieving this, but it does not reveal what the freely specifiable initial data—the true gravitational degrees of General Relativity—are in clear geometrical terms.

Let us do some counting in the 3+1 regime. Restricting attention to the vacuum case, then we start off with 10 unknowns, namely the components of the 4-dimensional metric, and 10 vacuum field equations. However, 4 of these unknowns may be prescribed arbitrarily because of the 4-fold coordinate freedom, leaving 6 components of the 4-metric freely specifiable. Moreover, the field equations are not independent but satisfy 4 differential constraints, namely the contracted Bianchi identities. The Lichnerowicz lemma reveals that if the constraints are satisfied initially and the evolution equations hold generally, then the constraints are satisfied for all time by virtue of the contracted Bianchi identities. Then, as we have seen in the 3+1 first-order formulation of the initial value problem, we end up needing to specify on an initial slice the 6 components of the 3-metric $\gamma_{\mu\nu}$ together with the corresponding 6 variables $K_{\mu\nu}$ subject to the 4 constraints. So this leaves 8 variables freely specifiable. However, there exists a 3-fold coordinate freedom within the initial slice. This can be used, for example, to specify 3 of the $\gamma_{\mu\nu}$. This leaves 5 variables free. Finally there is a condition which describes the embedding of the initial slice into the 4-geometry. This is a little harder to see, but we

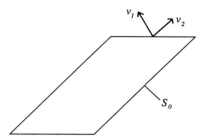

Figure 5. *2-dimensional submanifold and two transvecting submanifolds*

have already met examples of conditions like this such as maximal slicing $K = 0$ and constant mean curvature slicing $K = C(t)$. The point is that relationships like these are constraints between the $\gamma_{\mu\nu}$ and the $K_{\mu\nu}$ which encode the nature of the embedding. Such a constraint finally reduces the number of freely specifiable data to 4. These can be thought of, from a Lagrangian point of view, as being 2 q's and 2 $\dot{q}$'s, that is 2 pieces of information encoded in the metric and 2 pieces in its time derivative (or equivalently 2 pieces of information in the extrinsic curvature). It is in this sense that we say the gravitational field has two dynamical degrees of freedom. But what are they explicitly in the 3+1 case? Moreover, why are there 6 evolution equations rather than the 2 you would expect for a system with 2 degrees of freedom? The 2+2 approach answers these questions in a transparent way.

The other problem is that the 3+1 approach fails if the foliation goes null (although this may be difficult to detect precisely in a numerical regime where one cannot easily characterize vanishing quantities because of rounding errors). Yet null foliations are important in their own right as we shall see.

3.2 The 2+2 approach

The basis of this approach is to decompose spacetime into two families of spacelike 2-surfaces. We can view this as a constructive procedure in which an initial 2-dimensional submanifold S_0 is chosen in a bare manifold, together with two vector fields v_1 and v_2 which transvect the submanifold everywhere (Figure 5). The two vector fields can then be used to drag the initial 2-surface out into two foliations of 3-surfaces. The character of these 3-surfaces will depend in turn on the character of the two vector fields. The most important cases are when at least one of the vector fields is taken to be null. For example, if the two vector fields are null we obtain a double-null foliation (indicated schematically in Figure 6), or if one is null and the other is timelike we obtain a null-timelike foliation (Figure 7). An example of a non-null foliation is when one vector field is timelike and the other is spacelike in which case we obtain a timelike-spacelike foliation (Figure 8).

The most elegant way of proceeding is to introduce a formalism which is manifestly covariant and which uses projection operators and Lie derivatives associated with the two vector fields. The resulting formalism is called the 2+2 formalism (Smallwood 1983,

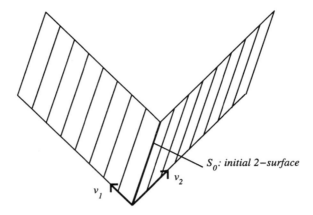

Figure 6. *Double null foliation*

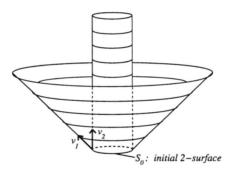

Figure 7. *Null-timelike foliation*

d'Inverno and Smallwood 1980). When the vector fields are of a particular geometric character, then this can be refined further into a 2+(1+1) formalism. Finally, in analogy to the conformal approach of the last chapter, one extracts a conformal factor from the spacelike 2-geometries to isolate the gravitational degrees of freedom.

The resulting formalism leads to a number of advantages. First of all, it identifies the two gravitational degrees of freedom in an explicit geometrical way as residing in the conformal 2-geometry (d'Inverno and Stachel 1978). Secondly, the data is unconstrained and satisfies two dynamical equations which are simply ODEs along the vector fields. Most importantly, the formalism applies to situations where the foliation either is or becomes null. Such initial value problems are called null or characteristic initial value problems. They are the natural vehicle for studying gravitational radiation problems (since gravitational radiation propagates along null geodesics), asymptotics of isolated systems (since future and past null infinity are null hypersurfaces) and problems in

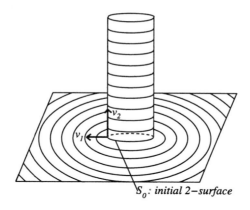

Figure 8. *Timelike-spacelike foliation*

cosmology (since we gain information about the universe along our past null cone). From a calculational viewpoint, this formalism allows null infinity to be incorporated into the calculational domain and so allows one to define gravitational radiation in an unambiguous manner.

The characteristic approach, however, suffers from one main drawback resulting from the fact that, in general, null hypersurfaces develop caustics. There are two quite distinct ways of proceeding. One approach is to develop techniques for generating solutions through caustics (Corkhill and Stewart 1983). The other is to restrict attention to caustic-free regimes such as the far zone in asymptotically flat regions or systems which are sufficiently close to spherical symmetry. These latter cases still include a number of astronomically interesting scenarios.

We shall not develop the 2+2 formalism in the same detail as we did with the 3+1 formalism in the last chapter. This is partly because it is rather complicated-looking at first sight. However, much of the procedure is analogous to that of the 3+1 decomposition, and largely rests on the use of projection operators and Lie derivatives. The only new entities are tensors which essentially encode two dimensional Lie derivatives (Smallwood 1983). However, we shall look in detail at the 2+2 decomposition of the metric so that we can compare and contrast it with the 3+1 case.

3.3 The 2+2 metric decomposition

In this section only, Greek indices run from 0 to 3, early Latin indices $(a, b, \ldots)$ run from 0 to 1, middle Latin indices $(i, j, \ldots)$ run from 2 to 3, uppercase Latin indices $(A, B, \ldots)$ run from 1 to 3 (see the notation used in d'Inverno and Vickers (1995)). Let M be a four-dimensional orientable manifold with metric g of signature $(+1, -1, -1, -1)$. A foliation of codimension two can be described by two closed 1-forms n^0 and n^1. Thus

locally
$$dn^a = 0 \iff n^a = d\phi^a \tag{54}$$
The two 1-forms generate hypersurfaces defined by
$$\begin{aligned}\{\Sigma_0\} &: \phi^0(x^\alpha) = \text{constant} \\ \{\Sigma_1\} &: \phi^1(x^\alpha) = \text{constant}\end{aligned}$$
respectively. These hypersurfaces define a family of 2-surfaces $\{S\}$ by
$$\{S\} = \{\Sigma_0\} \cap \{\Sigma_1\}$$
We restrict attention to the case when $\{S\}$ is spacelike and denote the family of two dimensional timelike spaces orthogonal to $\{S\}$ at each point by $\{T\}$ (Figure 9). Let n_a be the dyad basis of vectors dual to n^a in $\{T\}$, so that
$$n_a{}^\alpha n^b{}_\alpha = \delta_a^b \tag{55}$$
Since n^a is a 1-form basis for $\{T\}$, the vectors n_a form a basis of vectors for the span of $\{T\}$. Note that, in general, $[n_0, n_1] \neq 0$ so that $\{T\}$ does not form an integrable distribution. If, however, the Lie bracket vanishes then $\{T\}$ forms a 2-dimensional subspace of M and is said to be holonomic. We use the n_a to define a 2×2 matrix of scalars N_{ab} by
$$N_{ab} = g_{\alpha\beta} n_a{}^\alpha n_b{}^\beta \tag{56}$$
with inverse N^{ab}. We may use N_{ab} to relate n^a and n_a since
$$n_a{}^\alpha = g^{\alpha\beta} N_{ab} n^b{}_\beta \tag{57}$$
and
$$n^a{}_\alpha = g_{\alpha\beta} N^{ab} n_b{}^\beta \tag{58}$$
We define projection operators into $\{S\}$ and $\{T\}$ by
$$\begin{aligned} B^\alpha_\beta &= \delta^\alpha_\beta - n_a{}^\alpha n^a{}_\beta \tag{59}\\ T^\alpha_\beta &= n_a{}^\alpha n^a{}_\beta \tag{60}\end{aligned}$$
The 2-metric induced on $\{S\}$ is given by the projection
$$^2g_{\alpha\beta} = B^\gamma_\alpha B^\delta_\beta g_{\gamma\delta} = B_{\alpha\delta} B^\delta{}_\beta = B_{\alpha\beta}$$
Similarly, the 2-metric induced on $\{T\}$ is given by the projection
$$h_{\alpha\beta} = T^\gamma_\alpha T^\delta_\beta g_{\gamma\delta} = T_{\alpha\gamma} T^\gamma_\beta = T_{\alpha\beta}$$
Note that the tetrad components of $h_{\alpha\beta}$ are just N_{ab} since
$$h_{ab} = h_{\alpha\beta} n_a{}^\alpha n_b{}^\beta = g_{\alpha\beta} n_a{}^\alpha n_b{}^\beta = N_{ab}$$
In particular, the elements N_{00} and N_{11} define the lapses of $\{S\}$ in $\{\Sigma_0\}$ and $\{\Sigma_1\}$, respectively. We now choose a pair of vectors E_a which connect neighbouring 2-surfaces

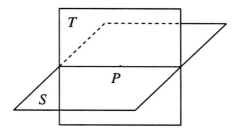

Figure 9. *The timelike 2-space $\{T\}$ orthogonal to $\{S\}$ at P*

in $\{S\}$. We choose them such that

$$n^a{}_\alpha E_b{}^\alpha = \delta^a_b \tag{61}$$

which defines E_a up to an arbitrary shift vector b_a, *i.e.*

$$E_a = n_a + b_a \tag{62}$$

with

$$n^a{}_\alpha b_c{}^\alpha = 0 \tag{63}$$

Although, in general, the n_a do not commute, it is always possible to choose b_a so that $[E_0, E_1] = 0$. Thus, each E_a is tangent to a congruence of curves in $\{\Sigma_a\}$ parametrized by $\phi^a(x^\alpha)$. We may, therefore, choose coordinates such that $\phi^0(x^\alpha) = x^0$, $\phi^1(x^\alpha) = x^1$ with x^2 and x^3 being constant along the congruence of curves.

In these coordinates

$$n^0 = dx^0, \quad n^1 = dx^1$$

and

$$E_0 = \frac{\partial}{\partial x^0}, \quad E_1 = \frac{\partial}{\partial x^1}$$

so that

$$n_0 = E_0 - b_0 = (1, 0, b^i{}_0)$$
$$n_1 = E_1 - b_1 = (0, 1, b^i{}_1)$$

This results in the 2+2 decomposition of the contravariant metric

$$g^{\alpha\beta} = \begin{pmatrix} N^{ab} & -N^{ab}b^i{}_b \\ -N^{ab}b^i{}_b & {}^2g^{ij} + N^{ab}b^i{}_a b^j{}_b \end{pmatrix} \tag{64}$$

with inverse

$$g_{\alpha\beta} = \begin{pmatrix} N_{ab} + {}^2g_{ij}b^i{}_a b^j{}_b & {}^2g_{ij}b^j{}_a \\ {}^2g_{ij}b^j{}_a & {}^2g_{ij} \end{pmatrix} \tag{65}$$

where

$$N_{ab}N^{bc} = \delta^c_a \tag{66}$$

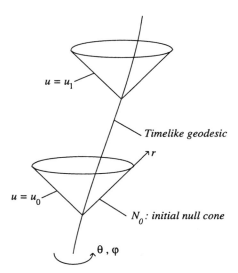

Figure 10. *The Bondi-type coordinates*

Compare and contrast (64), (65) and (66) with (14), (17) and (15), respectively. Note that in the 2+2 case the lapse function becomes a 2×2 lapse matrix and there are two shift vectors.

In the 2+2 formalism, the next procedure is to extract the conformal factor γ given by

$$\overset{2}{g}_{ij} = \gamma^2 \bar{g}_{ij} \tag{67}$$

where

$$\gamma = |\overset{2}{g}_{ij}| \tag{68}$$

An analysis of the field equations goes on to show that the two gravitational degrees of freedom may be chosen to lie in the conformal 2-structure $\bar{g}_{ij}$. We will not pursue the matter further here. We mention that in a recent application the formalism has been used to provide a 2+2 decomposition of the Ashtekar variables (d'Inverno and Vickers 1995).

3.4 The characteristic approach

This approach is based on the pioneering work of Winicour and collaborators (Isaacson *et al.* 1983, Gómez and Winicour 1992) together with work based on their approach by Bishop, Clarke and collaborators (Bishop *et al.* 1990a, 1990b). We restrict attention to a class of problems which are sufficiently close to spherical symmetry that no caustics develop in the setup we now describe. We introduce Bondi-type coordinates based on a family of outgoing null cones emanating from a central geodesic (Figure 10). The proper time, u, along the geodesic labels the null cones. The radial parameter, r, is the luminosity distance measured along null rays emanating from the geodesic. Finally, the angular coordinates θ, ϕ are introduced in the usual way. Again, for simplicity, we restrict attention to an axially symmetric non-rotating system with perfect fluid source. Writing $y = \cos\theta$, for later convenience, then in the coordinates (u, r, y, ϕ) the 4-dimensional line element becomes

$$ds^2 = hu\, du^2 + 2hr\, du dr + hy\, du dy + r^2 e^{2q} f^{-1} dy^2 + r^2 e^{-2q} f d\phi^2 \qquad (69)$$

where $f = (1 - y^2)$ and hu, hr, hy and q are all functions of u, r and y. The perfect fluid source has energy-momentum tensor

$$T_{ab} = (\rho + p) w_a w_b - p g_{ab} \qquad (70)$$

with equation of state $p = p(\rho)$ and unit 4-velocity

$$w_a = (wu, wr, wy, 0) \qquad (71)$$

where wu, wr and wy are all functions of u, r and y, and $w_a w^a = 1$.

We define the quantity E_{ab} by

$$E_{ab} \equiv G_{ab} - 8\pi T_{ab} \qquad (72)$$

Then the field equations break up into the following integration hierarchy

$$E_{11} = 0 \qquad (73)$$
$$E_{12} = 0 \qquad (74)$$
$$E_{01} = 0 \qquad (75)$$
$$E_{22} = \tfrac{1}{2} g_{22}(g^{22} E_{22} + g^{33} E_{33}) \qquad (76)$$
$$T_a{}^b{}_{;b} = 0 \qquad (77)$$

We now consider the characteristic initial value problem for this setup. We choose a null cone, N_0, as an initial null cone, and prescribe on it the initial data set $\{q, \rho, wr, wy\}$. Then the first three equations in the hierarchy reduce to linear ODEs for hr, hy and hu along the null rays in N_0. Equation (76) then determines $q_{,u}$ on N_0 and the three independent equations in (77) serve to determine the u-derivatives of the remaining matter variables, namely $\rho_{,u}, wr_{,u}$ and $wy_{,u}$. Thus, the u-derivatives of the initial data are known on N_0, and hence the initial data are again known on the next neighbouring null hypersurface, N_1, say. We can now repeat the process on N_1 and continue in this way. Thus the hierarchy of equations constitutes an integration schema for determining the solution to the future of N_0.

3.5 The combined approach

The 3+1 method has been successful in developing powerful codes, especially for investigating central matter distributions. The main drawback of this approach is that it relies on somewhat *ad hoc* boundary conditions imposed at the edge of the numerical grid. In many cases it is hoped that these conditions encode the absence of incoming gravitational radiation. The fact that the definitions which are used in practise do

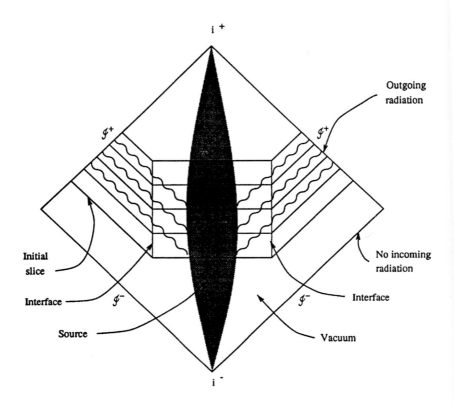

Figure 11. *Combined Cauchy-characteristic surfaces for an isolated radiative system*

not achieve this limits the usefulness of many existing codes. (This is not always so apparent in the literature since results are sometimes presented in a regime where problems associated with the use of incorrect boundary conditions have not had a chance to reveal themselves). A case where this problem is going to be of central importance is in the US Grand Challenge problem of numerically simulating coalescing black hole binaries. Now definitions of gravitational radiation are only really currently fully understood asymptotically or, more precisely, at null infinity. Thus it would appear that an essential requirement for investigating gravitational radiation is that it is first necessary to carry out a Penrose conformal compactification of the solution.

In recent years, the idea of adopting characteristic or 2+2 codes for investigating the asymptotics of isolated systems has been proposed (d'Inverno 1992). A typical scenario consists of an isolated source (perhaps initially close to spherical symmetry) which has been quiescent for a semi-infinite period, and which undergoes a disturbance (such as may result from a starquake, for example) and which causes it to radiate. The central idea underlying the combined approach for investigating scenarios like this is to combine a 3+1 code for determining the dynamics of the central source out to some finite distance surrounding the source, together with an exterior characteristic code for determining the behaviour of radiation emitted from the source. Moreover, the characteristic code needs to encompass future null infinity where gravitational radiation is unambiguously defined. A compactified picture illustrating schematically this scenario is given in Figure 11. The initial data is prescribed on the initial combined Cauchy-characteristic surface and is then evolved to some future combined surface.

The problem falls into three parts: (i) the central Cauchy region, (ii) the asymptotic characteristic region, (iii) the interface. The approach is directed at handling the exchange of information between the two regions at the interface in such a way as to effect an evolution. In this way we have dispensed with the need for a boundary condition at the interface, which would arise were we to consider separately either a Cauchy or a characteristic code.

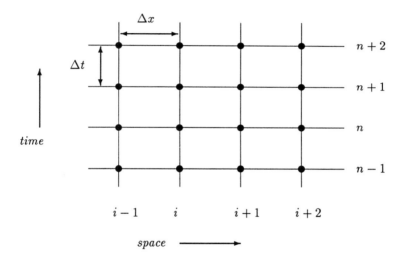

Figure 12. *Two dimensional grid in t and x*

4 Numerical issues

4.1 Order of truncation error

We consider the question of truncation error in finite difference methods where, for simplicity, we restrict attention to a function F which depends only on the time t and one spatial dimension x. We assume that the grid is uniform with steps Δt and Δx in the timelike and spatial directions (see Figure 12). We can use the Taylor expansions to write

$$F_{i+1}^n = \sum_{m=0}^{\infty} \frac{\Delta x^m}{m!} \left(\frac{\partial^m F}{\partial x^m}\right)_i^n \tag{78}$$

and
$$F_i^{n+1} = \sum_{m=0}^{\infty} \frac{\Delta t^m}{m!} \left(\frac{\partial^m F}{\partial t^m}\right)_i^n \tag{79}$$

The series can be truncated at any point with the error being dominated by the next term in the expansion if $\Delta x \to 0$ and $\Delta t \to 0$. We can write

$$F_{i+1}^n = F_i^n + \Delta x \left(\frac{\partial F}{\partial x}\right)_i^n + \frac{\Delta x^2}{2}\left(\frac{\partial^2 F}{\partial x^2}\right)_i^n + O(\Delta x^3) \tag{80}$$

where $O(\Delta x^3)$ means that a positive constant $K(F)$ exists such that the difference between F_{i+1}^n and the first three terms on the right hand side is no larger in absolute value than $K(F_i^n)\,\Delta x^3$ for sufficiently small Δx. It follows that

$$\left(\frac{\partial F}{\partial x}\right)_i^n = \frac{F_{i+1}^n - F_i^n}{\Delta x} - \tfrac{1}{2}\Delta x \left(\frac{\partial^2 F}{\partial x^2}\right)_i^n + \cdots \tag{81}$$

and so

$$\left(\frac{\partial F}{\partial x}\right)_i \approx \frac{F_{i+1}^n - F_i^n}{\Delta x} \tag{82}$$

with a truncation error of $O(\Delta x)$. In other words the finite difference expression is first-order accurate.

A more general technique is to start off from an expression of the type

$$\left(\frac{\partial F}{\partial x}\right)_i^n = a F_{i-1}^n + b F_i^n + c F_{i+1}^n + O(\Delta x^m) \tag{83}$$

where a, b and c are to be determined. From Taylor expansions we find

$$aF_{i-1}^n + bF_i^n + cF_{i+1}^n = (a+b+c)F_i^n + (-a+c)\Delta x \left(\frac{\partial F}{\partial x}\right)_i^n$$
$$+ (a+c)\frac{\Delta x^2}{2}\left(\frac{\partial^2 F}{\partial x^2}\right)_i^n + (-a+c)\frac{\Delta x^3}{6}\left(\frac{\partial^3 F}{\partial x^3}\right)_i^n + \cdots \tag{84}$$

Then from (83) and (84) it follows that

$$a + b + c = 0 \tag{85}$$
$$(-a + c)\Delta x = 1 \tag{86}$$
$$a + c = 0 \tag{87}$$

The solution of these equations is $a = -c = -\tfrac{1}{2}\Delta x$ and $b = 0$. We then have

$$\left(\frac{\partial F}{\partial x}\right)_i^n = \frac{1}{2\Delta x}(-F_{i-1}^n + F_{i+1}^n) - \frac{\Delta x^2}{6}\left(\frac{\partial^3 F}{\partial x^3}\right)_i^n + \cdots \tag{88}$$

so that

$$\left(\frac{\partial F}{\partial x}\right)_i \approx \frac{F_{i+1}^n - F_{i-1}^n}{2\Delta x} \tag{89}$$

which has a truncation error of $O(\Delta x^2)$, that is it is second-order accurate. This is the most accurate approximation possible with three arbitrary parameters.

This is a general technique which can be used to:-

- provide approximations of derivatives to arbitrary order of truncation error (provided enough arbitrary constants are used),
- produce one-sided finite difference approximations,
- produce finite difference formulae on non-uniform grids,
- produce multi-dimensional finite difference formulae.

4.2 Consistency

The system of algebraic equations generated by the discretization procedure is said to be *consistent* if, in the limit $\triangle x \to 0$, $\triangle t \to 0$, the system is equivalent to the partial differential equation at each grid point. Testing for consistency requires the substitution of the exact equation into the algebraic equations and expanding all grid point values as a Taylor expansion about a single grid point. For consistency the result must be the original differential equation plus a remainder (the truncation error) which must reduce to zero as $\triangle x \to 0$, $\triangle t \to 0$.

4.3 Stability

The concept of stability is concerned with the growth or decay of errors (which are always present) at any stage of the computation. The two most common methods of stability analysis are:-

Matrix method. The set of equations governing the error propagation is put into a matrix equation of the form

$$e^{n+1} = Ae^n \tag{90}$$

It can be shown (Fletcher 1991) that the errors e^n are bounded as n increases if the eigenvalues λ_i of A are all distinct and satisfy $|\lambda_i| \leq 1$.

von Neumann method. This is a commonly used method and is both the easiest to apply and the most dependable. Unfortunately it can only be used for analyzing the stability of linear initial value problems with constant coefficients. Linearized versions of the field equations can be analyzed. Stability is determined by checking whether separate Fourier components of the error distribution decay or amplify in progressing to the next time level. Initially for each grid point we write

$$F_j^n = \tilde{F}^n(m) \exp(i\theta_m j) \tag{91}$$

where $i = \sqrt{-1}$ and $\theta_m = m\pi \triangle x$. We need only choose one mode m and substitute it into the finite difference formula to obtain an equation of the form

$$\tilde{F}^{n+1}(m) = G^n(m)\tilde{F}^n(m) \tag{92}$$

The quantity $G^n(m)$ is the amplification factor for the mth Fourier mode, at time level n, of the error distribution. The errors will be bounded if, for all Fourier modes m, we have

$$|G^n(m)| \leq 1 \tag{93}$$

The stability requirements $|\lambda_i| \leq 1$ or $|G^n(m)| \leq 1$ usually result in limitations on the timestep Δt for explicit finite difference schemes, or identifies those schemes which are always stable or unstable regardless of Δt.

4.4 Convergence

The finite difference equations which approximate the differential equations are locally convergent if the solution error

$$e_i^n = \overline{F}(t^n, x_i) - F_i^n \qquad (94)$$

satisfies $e_i^n \to 0$ as $\Delta x \to 0$ and $\Delta t \to 0$. Here $\overline{F}$ is the 'true' (exact) value of the variable F at position x and time t. In general, the proof that the finite difference system of equations converges to the differential one is very difficult. Unless an exact solution is known the solution error e_i^n cannot be found. In practise convergence can be shown by solving the same problem at differing grid resolutions of, say, $(\Delta t, \Delta x), (\Delta t/2, \Delta x/2), (\Delta t/4, \Delta x/4), \ldots$ and using the finest grid as an 'exact' solution. Local convergence is shown if e_i^n decreases as grid resolution increases.

Given a properly posed linear initial value problem and a finite difference approximation to it that satisfies the consistency condition, then the Lax equivalence theorem (Richtmeyer and Morton 1967) establishes that stability is a necessary and sufficient condition for convergence. This is important because stability and convergence are easier to show. For most real (non-linear) problems the Lax theorem cannot be applied rigorously, but it is useful for excluding inconsistent discretizations and unstable algorithms.

4.5 Some rules of thumb

In the case of a coupled PDE, a second-order finite difference method is normally most useful. Higher order methods can be very accurate, but only when Δx is sufficiently small. In fact, for a given Δx the truncation error for a second-order scheme may be smaller than the truncation error for a fourth-order scheme. As $\Delta x \to 0$ the error in the fourth-order scheme drops more rapidly. In addition, higher order schemes involve more grid points. This makes them computationally quite expensive and they can cause problems at boundaries. They are also more prone to instabilities. First-order schemes are the most stable. Second-order schemes can be stabilized with small amounts of artificial viscosity.

If care is not taken the order of a scheme can be reduced by:

- improper treatment of boundary conditions, for example differencing them at a lower order,

- using a non-uniform grid (in time as well as space),

- the addition of artificial viscosity.

In general, it is best to aim for second-order accuracy in space and time. Artificial viscosity should only be used where necessary, for example in the treatment of shocks.

When differencing products the averaging should have precedence over other operations. For example, if $p(t,x)$ and $q(t,x)$ are defined at grid points (t^n, x_i), and pq is required at $(t^n, x_{i+1/2})$ then

$$(pq)^n_{i+1/2} = \tfrac{1}{4}(p^n_{i+1} + p^n_i)(q^n_{i+1} + q^n_i) \tag{95}$$

should be used rather than

$$(pq)^n_{i+1/2} = \tfrac{1}{2}(p^n_{i+1}q^n_{i+1} + p^n_i q^n_i) \tag{96}$$

If a particular behaviour of pq is known then this can modify the above statement.

4.6 Code Structure

Most numerical codes are written in FORTRAN, although C is becoming more popular. In commercial applications the programmer is concerned with making the code robust against misuse. Numerical codes are only likely to be used by a very small number of people, who usually know what they are doing, and so the emphasis is on getting stable and accurate results. Although efficiency is important it is not so at the expense of writing excessively complicated codes. Here are some general considerations in writing codes (we restrict attention to FORTRAN):-

- Use **real*8** to employ 64 bit floating point arithmetic to reduce round-off errors. In particular, unity is represented in double precision by **1.d0**.

- It is best to avoid **common** blocks and pass all variables required through a subroutine call. This makes the code more modular and easier to modify.

- Use **implicit none** so that all variables must be declared. This makes for code which is easier to understand and less prone to bugs caused by undefined variables and typing errors.

- Use meaningful variable names.

- The main evolution loop should be a series of subroutine calls.

- Vectors are to be preferred over arrays, since arrays are more prone to error when being passed through subroutines and are more prone to memory page faulting which can slow code down.

- Most benefit can be gained from optimizing inner loops which are called most often.

- Division is expensive and should only be used where nothing else will do. For example, use **0.5d*a** rather than **a/2.d0**.

- Exponentiation is also expensive. For square roots use **sqrt(a)** rather than the exponentiation **a**0.5**. For taking a square **a*a** is faster than **a**2**.

- Short inner do loops can be rolled out to make them faster. For example

  ```
  do i = 1, 10
  a(i) = b(i) * c(i)
  end do
  ```

  can be written out as ten lines of the type `a(1) = b(1) * c(1)`. Some compilers will do this automatically.

- The appearance of a printed code should be like that of a document, *i.e.* it should be easy to read. Use white space and indentation.

4.7 Code verification

A standard method for checking codes is to try and reproduce known analytic solutions. This can be useful, but the analytic solution may only be valid under special circumstances and for a particular example it may be that some portions of the code never get used. More powerful code checks concern internal consistency. For example:

- The ADM 3+1 formulation provides constraint and evolution equations. Since the constraints must be satisfied on each spatial hypersurface these provide useful internal consistency checks.

- Many spacetimes have a conserved energy or mass. A check on such conserved quantities can be very useful.

Another important check is on convergence of solutions, which can often provide information on code errors, etc. Convergence testing is best achieved by outputting data at specific times, then decreasing the grid spacing by 2 and outputting the data at the same times. This is repeated a few times. The highest resolution can be taken as a reference solution to which all the coarser solutions can be compared. For example, if $\{F(h)_i^n\}$ is the set of Fs at time level n for the highest resolution (grid spacing h), then the L_2 error norm for F at the same time level for grid spacing $2h$ is

$$L_2 = \frac{1}{N} \sum_{i=1}^{M} (F(2h)_i^n - F(h)_{2i}^n)^2 \tag{97}$$

where N is the number of grid points at resolution $2h$. The error norm should decrease steadily as the grid spacing gets smaller. The degree to which constraints and energies are conserved should improve as the grid becomes finer.

4.8 Visualization

It is quite important to have good graphics early on in the debugging and testing stages. It is far easier to spot problems with a graph than from looking at a list of numbers. The main problem with graphics is that they tend to be hardware specific and there is often a substantial learning curve for the use of any graphics package. One dimensional plots are easy to deal with. The problem gets harder in 2D where more data is involved, but contour imaging and surface plots provide a good representation of the data. Things become considerably more difficult in 3D. Here techniques such as translucent volume contouring, voxel imaging and rendering are useful, but this requires advanced graphics workstations. The most useful visualization tools are those that allow real time management of the data, allowing one to change a point of view, zoom in on a specific area of the data and animate the data in time. Such packages are generally not useful for hardcopy output.

We mention several visualization packages explicitly. The simplest to use is GNU-PLOT which is freely available by anonymous ftp and can run on anything from MS Windows to X Windows and Tektronix terminals. The 1D routines and 2D contouring is good, but surface plots are of low quality. There are no real 3D utilities. Hard copy output from GNUPLOT is not of publication quality. Data can be read in by MAPLE and MATHEMATICA and can be manipulated, but these packages are not really designed for specialist manipulation. Animation of 2D data is very slow and memory intensive.

At Southampton we have access to a specialist graphics package developed by Matt Choptuik over a number of years. It is not commercially available. It only runs on Silicon Graphics machines and makes use of their special graphics library. It is geared towards 1D data and has an extensive number of facilities for data management, but there is no documentation. AVS (Advanced Visualization Systems) is a very expensive commercial package which runs on any workstation with X Windows. It takes some time to learn, but can deal with large amounts of data and has good 3D data manipulation routines. Once your data is in the correct format, you can connect a network of modules to animate, scale and so forth. The package will also perform volume rendering. IDL and PV-Wave are also commercial interactive visualization packages. They are easier to use than AVS but do not have the same 3D facilities. These packages do not require the user to write any programs. On the other hand, for hardcopy output, packages like NAG Graphics and UNIRAS are available, but you need to write a program for this.

4.9 Closing comments

There are many other numerical issues which we will not discuss in this introduction to Numerical Relativity. These include important issues to do with grids (including multi-grids and adaptive grids), the difference between explicit and implicit codes, artificial viscosity, imaging techniques, outgoing radiation conditions and many other issues. Of particular importance is the fact that the methods which have been developed for elliptic, parabolic and hyperbolic PDEs are very different in character. We will meet some new ideas in the example of the next chapter including the leapfrog method and the CFL (Courant-Friedrichs-Lewry) condition for stability in a hyperbolic code, namely that for a fixed Δt the finite difference cone of influence must contain the physical propagation cone. There are also important issues concerning computer processing power. It is clear that fully three dimensional codes require supercomputers of either the vector or parallel processor type, and even then these codes will push the present generation of computers to the limit. In some work at Southampton (d'Inverno 1992), we have shown that transputer arrays may provide a cheap alternative.

5 An example of a combined code

5.1 Introduction

As we discussed in Section 1, there are a large number of successful codes in existence. For details of these codes consult the conference proceedings listed in Section 1 together with references therein. In this section we shall look in some detail at a code which the group at Southampton (consisting of Mark Dubal, Chris Clarke, James Vickers and myself) has been developing (Clarke *et al.* 1995, Dubal *et al.* 1995). It serves as a useful example because it brings together a number of ideas which we have discussed in our survey of computer methods in relativity, including: the use of the algebra systems MAPLE, SHEEP and STENSOR to derive the field equations; as well as the use of 3+1, characteristic and combined approaches; the problems of code development, comparison with exact solutions and visualization.

We shall describe in some detail a numerical approach for passing gravitational wave information back and forth between an inner Cauchy (or 3+1) region and an outer, compactified, characteristic region via an interface. The point of this technique is that it allows for the inclusion of future null infinity into the numerical domain. In this developmental work we look first at vacuum cylindrical symmetry. Although physically unrealistic, this spacetime offers a simple, yet non-trivial, regime in which to develop and test the techniques necessary to implement an interface with the required properties. In future work we hope to investigate more realistic two and three-dimensional spacetimes with matter. We first outline the overall numerical approach to combining the 3+1 and characteristic methods. For the most part we use standard finite-difference schemes and aim for second-order accuracy in the space and time discretizations. We then discuss the implementation of the interface where connections are established between the fundamental variables in the two regions. A 'spacetime' interpolative technique is described by which function values and their derivatives at the interface can be obtained from local information in both regions. Such values and derivatives serve as 'boundary conditions' for the outer edge of the Cauchy region and the inner edge of the characteristic region. Particular emphasis is given to the numerical implementation with regard to maintaining the order of truncation error in the finite difference method. A problem regarding the 'start-up' of the evolution is discussed and a resolution presented. We show a comparison of the code results with an analytic solution and perform convergence testing to gauge its overall accuracy. A number of runs of the code have been made with a variety of initial data involving Gaussian wave packets.

5.2 The Cauchy 3+1 region

The Jordan-Ehlers-Kompaneets (Jordan *et al.* 1960, Kompaneets 1958) form of the general vacuum cylindrical line element is

$$ds^2 = -e^{2(\gamma-\psi)}(dt^2 - dr^2) + e^{2\psi}(dz + \omega d\phi)^2 + r^2 e^{-2\psi} d\phi^2 \tag{98}$$

where (r, ϕ, z) are the usual cylindrical coordinates, t is the coordinate time and γ, ψ and ω are functions of r and t only.

We use the ADM 3+1 formulation. A package written in MAPLE for this formalism was used to carry out the calculations. We start by imposing the slicing condition,

$$K^\phi_\phi + K^z_z = 0 \qquad (99)$$

Then the Hamiltonian constraint equation for (98) is

$$e^{-2\psi}[\tfrac{1}{4}(\omega_{,r})^2 e^{4\psi} - r(\chi_{,r}+\psi_{,r}) + r^2(\psi_{,r})^2] + r^2(K^\phi_\phi - \omega K^\phi_z)^2 + r^4 e^{-4\psi}(K^\phi_z)^2 = 0 \qquad (100)$$

and the single momentum constraint is

$$(K^\phi_\phi - \omega K^\phi_z)(1 - 2r\psi_{,r}) - K^r_r + r\omega_{,r} K^\phi_z = 0 \qquad (101)$$

where $\chi = \gamma - \psi$ so that $e^\chi = \alpha$ is the lapse function. The extrinsic curvature is defined using (21) and for the line element (98) the shift vector has all components set to zero. This definition gives rise to the following metric evolution equations

$$\chi_{,t} = -e^\chi K^r_r \qquad (102)$$
$$\psi_{,t} = e^\chi (K^\phi_\phi - \omega K^\phi_z) \qquad (103)$$
$$\omega_{,t} = -2r^2 e^{-4\psi} e^\chi K^\phi_z \qquad (104)$$

The dynamical 3+1 Einstein equations are

$$K^r_{r,t} = -e^{-\chi}[\chi_{,rr} + \tfrac{1}{2r^2} e^{4\psi}\omega_{,r}^2 - \tfrac{1}{r}(\chi_{,r} + 2\psi_{,r}) + 2\psi_{,r}^2 - e^{2\chi}(K^r_r)^2] \qquad (105)$$

$$K^\phi_{z,t} = e^{-\chi}[e^{4\psi}(\tfrac{1}{2}\omega_{,r} - \tfrac{1}{2}r\omega_{,rr} - 2r\psi_{,r}\omega_{,r})\tfrac{1}{r^3} + e^{2\chi} K^\phi_z K^r_r] \qquad (106)$$

$$K^\phi_{\phi,t} = e^{-\chi}[\omega e^{4\psi}(\tfrac{1}{2}\omega_{,r} - \tfrac{1}{2}r\omega_{,rr} - 2r\psi_{,r}\omega_{,r})\tfrac{1}{r^3} + \psi_{,rr} + \tfrac{1}{r}\psi_{,r}$$
$$- \tfrac{1}{2r^2} e^{4\psi}\omega_{,r}^2 + e^{2\chi} K^\phi_\phi K^r_r] \qquad (107)$$

The constraint and evolution equations are a set of eight equations for the six unknown quantities χ, ψ, ω, K^r_r, K^ϕ_z and K^ϕ_ϕ.

If we define the quantities,

$$L^r_r = re^\chi K^r_r, \qquad L^\phi_z = r^2 e^\chi K^\phi_z, \qquad \tilde{L} = re^\chi (K^\phi_\phi - \omega K^\phi_z) \qquad (108)$$

then the above 3+1 equations may be written as

$$\psi_{,t} = \tfrac{1}{r}\tilde{L} \qquad (109)$$
$$\omega_{,t} = -2e^{-4\psi} L^\phi_z \qquad (110)$$
$$L^\phi_{z,t} = \tfrac{1}{r} e^{4\psi}(\tfrac{1}{2}\omega_{,r} - \tfrac{1}{2}r\omega_{,rr} - 2r\psi_{,r}\omega_{,r}) \qquad (111)$$
$$\tilde{L}_{,t} = \tfrac{1}{r}[r^2\psi_{,rr} + r\psi_{,r} - \tfrac{1}{2} e^{4\psi}\omega_{,r}^2 + 2e^{-4\psi}(L^\phi_z)^2] \qquad (112)$$

which is an independent set of dynamical equations for the variables ψ, ω, L^ϕ_z and $\tilde{L}$. The constraint equations can be written as

$$\chi_{,r} = \tfrac{1}{4r} e^{4\psi}\omega_{,r}^2 - \psi_{,r} + r\psi_{,r}^2 + \tfrac{1}{r}[\tilde{L}^2 + e^{-4\psi}(L^\phi_z)^2] \qquad (113)$$
$$L^r_r = \tilde{L}(1 - 2r\psi_{,r}) + \omega_{,r} L^\phi_z \qquad (114)$$

from which the variables χ and L_r^r may be obtained. Alternatively these variables may be obtained via the extra dynamical equations,

$$\chi_{,t} = -\frac{1}{r}L_r^r \tag{115}$$

$$L_{r,t}^r = -\frac{1}{r}[r^2\chi_{,rr} + \frac{1}{2}e^{4\psi}\omega_{,r}^2 - r(\chi_{,r} + 2\psi_{,r}) + 2r^2\psi_{,r}^2] \tag{116}$$

The boundary conditions at $r = 0$ for the transformed extrinsic curvature components are simply

$$L_r^r = L_z^\phi = L_\phi^\phi = \tilde{L} = 0 \tag{117}$$

while conditions of regularity and flatness at the origin require that

$$\omega = \omega_{,r} = \psi_{,r} = \chi_{,r} = 0 \tag{118}$$

and

$$\psi = -\chi \tag{119}$$

A polynomial expansion of the form

$$\psi = a + br^2 \tag{120}$$

at $r = 0$ provides the correct behaviour for ψ and its radial derivative.

In the actual numerical code we solve equations (109–112) and periodically solve (113) as a check on how well the constraints are satisfied by the numerical evolution solution. Thus the Cauchy portion of the code is in 'free evolution'.

5.3 The characteristic region

In the characteristic region we introduce the null coordinate

$$u = t - r \tag{121}$$

so that the line element (98) becomes

$$ds^2 = -e^{2(\gamma-\psi)}(du^2 + 2dudr) + e^{2\psi}(dz + \omega d\phi)^2 + r^2 e^{-2\psi}d\phi^2 \tag{122}$$

and thus the curves

$$u = \text{const} \qquad \phi = \text{const} \qquad z = \text{const} \tag{123}$$

are outgoing null geodesics. The Einstein equations are to be solved asymptotically, *i.e.* on constant u hypersurfaces as $r \to \infty$, however the cylindrical symmetry complicates compactification. The complication is due to the infinite extent in the z-direction, and therefore, to proceed we use the decomposition technique due to Geroch, described in Section 3.2 of the article 'Algebraic Computing in General Relativity' in this volume, to factor out the z-direction, which is a Killing direction. All the calculations of this section were performed with the aid of SHEEP and STENSOR, as described in that article. The Geroch twist for this metric (which we now call o to distinguish it from the metric function ω) is

$$o_a = \frac{1}{r}e^{4\psi}(\omega_{,r} - \omega_{,u}, \ \omega_{,r}, \ 0, \ 0), \tag{124}$$

The metric induced on the resulting three-dimensional spacetime (cyl7b of the previous article) has line element

$$d\sigma^2 = -e^{2(\gamma-\psi)}du^2 - 2e^{2(\gamma-\psi)}dudr + r^2 e^{-2\psi}d\phi^2 \qquad (125)$$

where the coordinates $(x^a) = (x^0, x^1, x^2) = (u, r, \phi)$.

The characteristic region of interest to us consists of $1 \leq r \leq \infty$ integrated between an initial u_0 and final u_f, i.e. we will set the interface at $r = 1$ and include future null infinity in the characteristic region. To compactify the region we use a new radial coordinate $x = 1/r$ resulting in the conformal metric

$$d\sigma^2 = \frac{1}{\lambda}(-x^2 e^{2\gamma}du^2 + 2dudx + d\phi^2) \qquad (126)$$

the conformal factor being $\Omega = 1/x$. If we now set

$$\lambda = 1 + ym, \qquad w = o/y \qquad (127)$$

where

$$y = x^{1/2} \quad \text{and} \quad m = (e^{2\psi} - 1)/y \qquad (128)$$

then the Einstein equations can be written in the first-order form

$$m_{,u} = \lambda M \qquad (129)$$
$$w_{,u} = \lambda W \qquad (130)$$
$$M_{,y} = -\frac{1}{\lambda}(yw)_{,y} W + \frac{1}{4\lambda}[-y(m + y^2 m_{,yy} + 3ym_{,y}) +$$
$$\frac{1}{\lambda}y^2(m^2 + 2ymm_{,y} - w^2 - 2yww_{,y} + y^2 m^2_{,y} - y^2 w^2_{,y})] \qquad (131)$$
$$W_{,y} = \frac{1}{\lambda}(yw)_{,y} M + \frac{1}{4\lambda}[-y(w + y^2 w_{,yy} + 3yw_{,y}) +$$
$$\frac{1}{\lambda}2y^2(mw + ymw_{,y} + ywm_{,y} + y^2 m_{,y} w_{,y})] \qquad (132)$$

Equations (129–132) constitute an independent set of four equations for the variables m, w, M and W, which can be related to the 3+1 variables ψ, ω, L_z^ϕ and $\tilde{L}$. As in the Cauchy region the metric function γ can be obtained via the constraint equation,

$$\gamma_{,y} = -\frac{1}{8\lambda^2} y[m^2 + w^2 + 2y(mm_{,y} + ww_{,y}) + y^2(m^2_{,y} + w^2_{,y})] \qquad (133)$$

or via the dynamical equation

$$\gamma_{,u} = -\frac{1}{2}\{M^2 + W^2 + \frac{1}{2\lambda} y^2 [(m + ym_{,y})M + (w + yw_{,y})W]\} \qquad (134)$$

The equations (129–132) are those used in the numerical code. We also periodically solve (133). The region over which these equations are solved is now

$$u_0 \leq u \leq u_f \qquad 0 \leq y \leq 1 \qquad (135)$$

(see Figure 13) and clearly the equations are regular as $y \to 0$ i.e. at future null infinity. Equations (131) and (132) are integrated from $y = 1$, where boundary data is provided via the interface, out to $y = 0$.

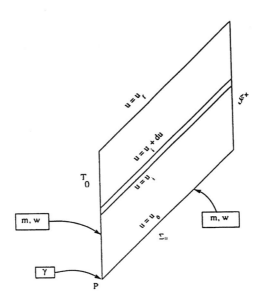

Figure 13. *The initial data set for a combined Cauchy-characteristic evolution*

5.4 The general numerical approach

In differencing the hyperbolic equations, *i.e.* (109–112), (129) and (130) we use an explicit, second-order (in space and time) leapfrog technique (Press *et al.* 1986). The grids have a regular spacing of Δr in the Cauchy region and Δy on the characteristic side. For convenience the interface is set at a radial position of $r_I = y_I = 1$ (see Figure 14). The coordinate timesteps in the two regions can be related using (121). Thus

$$u^{n+1} = u^n + \Delta u = t^n - r_I + \Delta u \tag{136}$$

but

$$u^{n+1} = t^{n+1} - r_I \quad \text{and} \quad t^{n+1} = t^n + \Delta t \tag{137}$$

so that

$$\Delta u = \Delta t \tag{138}$$

for the two slices to stay in step. For stability of the leapfrog scheme we require the CFL (Courant-Friedrichs-Lewry) condition to hold in both regions, *i.e.*

$$\left|\frac{\Delta t}{\Delta r}\right| \leq 1 \quad \text{and} \quad \left|\frac{\Delta t}{\Delta y}\right| \leq 1 \tag{139}$$

For this reason we take the timestep to be

$$\Delta t = C \min\{|\Delta r|, |\Delta y|\} \tag{140}$$

Numerical Computing in General Relativity 365

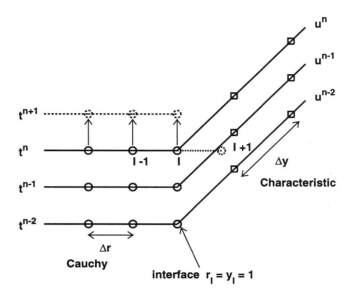

Figure 14. *Grid structure near the interface. The interface is fixed at grid point I where $r = y = 1$. Information is updated on the Cauchy side by extending the grid a distance Δr into the characteristic region and interpolating using information from the three characteristic surfaces.*

where $C \leq 1$ is a Courant factor.

The parabolic equations (131) and (132) are coupled and need to be integrated together on the characteristic surface. These equations are finite-differenced to second-order using mid-point averaging and are solved using a Newton-Raphson iteration technique. Convergence appears to be rapid and robust.

5.5 The interface

The first task in constructing the interface is to establish relations among the variables in the two coordinate systems. The right hand side of the Cauchy dynamical equations (109–112) involve the seven quantities

$$\{\psi, \psi_{,r}, \psi_{,rr}, \omega_{,r}, \omega_{,rr}, L_z^\phi, \tilde{L}\}$$

We need expressions for these quantities in terms of the characteristic quantities at the interface. To obtain radial derivatives in terms of characteristic variables, we make use of the chain rule in the form

$$\frac{\partial}{\partial r} f(u, y) = -\frac{\partial f}{\partial u} - \frac{y^3}{2} \frac{\partial f}{\partial y} \qquad (141)$$

using (121) and (128). We then find, using (127–132), that the required relationships are

$$\psi = \frac{1}{2}\log(1+my) \tag{142}$$

$$\psi_{,r} = -\frac{1}{2}yM - \frac{1}{4}\frac{y^3}{\lambda}(ym)_{,y} \tag{143}$$

$$\psi_{,rr} = \frac{1}{2}yM_{,u} + \frac{1}{2}y^3(yM)_{,y} - \frac{1}{8}\frac{y^6}{\lambda^2}[(ym)_{,y}]^2 + \frac{1}{8}\frac{y^5}{\lambda}\{2(ym)_{,y} + [y(ym)_{,y}]_{,y}\} \tag{144}$$

$$\tilde{L} = \frac{1}{2}\frac{M}{y} \tag{145}$$

$$\omega_{,r} = \frac{W}{y\lambda} \tag{146}$$

$$\omega_{,rr} = -\frac{W_{,u}}{y\lambda} + \frac{MW}{\lambda} + \frac{yW}{2\lambda} + \frac{1}{2}\frac{y^2 W_{,y}}{2\lambda^2}[(ym)_{,y} - \lambda] \tag{147}$$

$$L_z^\phi = \frac{1}{2}\frac{\lambda W}{y} + \frac{1}{4}y(yw)_{,y} \tag{148}$$

Similarly, the right hand side of the characteristic dynamical equations (129–132) involve the nine quantities

$$\{\lambda, m, m_{,y}, m_{,yy}, \omega, \omega_{,y}, \omega_{,yy}, W, M\}$$

The Geroch twist potential can be found from the twist vector at any point P by using

$$o(u,r) = \int_F^P (o_{,u}du + o_{,r}dr) \tag{149}$$

where F is some fixed point and the integration is along any path in the (u,r) plane connecting F to P. Using (124), this can be written in terms of ω as

$$o = \int r^{-1}e^{4\psi(u,r)}(\omega(u,r)_{,r} - \omega(u,r)_{,u})du - \int r^{-1}e^{4\psi(u,r)}\omega(u,r)_{,r}\,dr \tag{150}$$

Transforming this to the (t,r) coordinates used in the Cauchy region and, using (127), we find

$$w = \int r^{-1/2}e^{4\psi(t,r)}\omega(t,r)_{,r}\,dt + r^{1/2}\int r^{-1}e^{4\psi(t,r)}\omega(t,r)_{,t}\,dr \tag{151}$$

However we can, without loss of generality, integrate along an $r = $ constant line, in which case this reduces to

$$w = \int r^{-1/2}e^{4\psi(t,r)}\omega(t,r)_{,r}\,dt \tag{152}$$

We also make use of the chain rule, this time in the form

$$\frac{\partial}{\partial y}f(t,r) = -2r^{-3/2}\left(\frac{\partial f}{\partial t} + \frac{\partial f}{\partial r}\right) \tag{153}$$

Then the complete set of relationships can now be written

$$\lambda = e^{2\psi} \tag{154}$$

$$m = r^{1/2}(e^{2\psi} - 1) \tag{155}$$

$$m_{,y} = -r^2 e^{2\psi}(4\psi_{,t} + 4\psi_{,r} + r^{-1}) + r \tag{156}$$

$$m_{,yy} = 2r^{7/2}e^{2\psi}(4\psi_{,rr} + 8\psi_{,tr} + 8\psi_{,t}^2 + 10r^{-1}\psi_{,t} + 16\psi_{,t}\psi_{,r} + 4\psi_{,rr}$$
$$+ 8\psi_{,r}^2 + 10r^{-1}\psi_{,r} + r^{-2}) - 2r^{3/2} \tag{157}$$

$$M = 2r^{-1}\tilde{L} \tag{158}$$

$$w = \int r^{-1/2} e^{4\psi(t,r)} \omega(t,r)_{,r}\, dt \tag{159}$$

$$w_{,y} = 4rL_z^\phi - 2re^{4\psi}\omega_{,r} - \int e^{4\psi(t,r)}\omega(t,r)_{,r}\, dt \tag{160}$$

$$w_{,yy} = 2r^{5/2}e^{4\psi}(8\psi_{,t}\omega_{,r} + 2\omega_{,tr} + 3r^{-1}\omega_{,r} + 8\psi_{,r}\omega_{,r} + 2\omega_{,rr} + 2r^{-1}\omega_{,t}$$
$$- 8r^{5/2}(L_{z,t}^\phi + L_{z,r}^\phi + r^{-1}L_t^\phi) + 2\int r^{1/2}e^{4\psi}\omega_{,r} dt \tag{161}$$

$$W = r^{-1/2}e^{2\psi}\omega_{,r} \tag{162}$$

For simplicity we describe the implementation of the interface when $\omega = 0$ for all time, i.e. the system is reduced so that only one polarization mode of the radiation exists. In this case we need consider only the variables ψ, $\tilde{L}$, m and M.

Assuming evolutions on the Cauchy side of the spacetime to a coordinate time t^n and the corresponding evolution of the characteristic portion of spacetime to u^n have been performed, then ψ is evolved first to time $t^{n+1} = t^n + \Delta t$ in the Cauchy region using (109). Then $\tilde{L}$ can be evolved to t^{n+1}, again in the Cauchy region, using (112), except for the grid point I on the interface where $(\psi_{,r})_I^n$ and $(\psi_{,rr})_I^n$ are required. A second-order, centred finite difference approximation to these derivatives can be found if a value for ψ on the t^n Cauchy slice can be obtained one grid point beyond the interface, i.e. at a point within the characteristic region (see Figure 14). Using three characteristic slices at u^n, u^{n-1} and u^{n-2} and fourth-order interpolation (necessary to maintain a second-order truncation error in the derivatives) a value for m can be estimated at the appropriate position. The relation (142) then gives a value for ψ_{I+1}^n so that $\tilde{L}$ can be updated to time t^{n+1} on the interface.

For the characteristic portion we first evolve m, using (129), to $u^{n+1} = u^n + \Delta u$. It is then necessary to radially integrate M from $y = 1$ to $y = 0$. A starting value for M can be found using the relation (158) and the value $\tilde{L}_I^{n+1}$ which has previously been computed on the Cauchy side. However, in order to maintain a centred, second-order scheme it is necessary to approximate the derivative $(m_{,yy})_I^n$ in (131) using a value of m at one grid point beyond the interface, i.e. at a point within the Cauchy region as shown in Figure 15. This requirement can be seen as follows; consider a second-order finite difference approximation to (131), taking $w = W = 0$. When centred at half grid points we have,

$$M_{i+1} = M_i + \frac{1}{8}\frac{\Delta y_{i+1}}{\mu \lambda_{i+1}}\left\{-\mu y_{i+1}\left[\mu m_{i+1} + \frac{1}{4}(\Delta y_{i+1})^2 \mu\left(\frac{\Delta^2 m_{i+1}}{\Delta y_{i+2}\Delta y_{i+1}}\right) + 3\mu y_{i+1}\frac{\Delta m_{i+1}}{\Delta y_{i+1}}\right]\right.$$
$$\left. + \frac{(\mu y_{i+1})^2}{\mu \lambda_{i+1}}\left[\frac{1}{2}(\mu m_{i+1})^2 + \mu y_{i+1}\mu m_{i+1}\frac{\Delta m_{i+1}}{\Delta y_{i+1}} + \frac{1}{2}\left(\mu y_{i+1}\frac{\Delta m_{i+1}}{\Delta y_{i+1}}\right)^2\right]\right\} \tag{163}$$

where the difference and averaging operators are,

$$\Delta f_{i+1} = f_{i+1} - f_i \qquad \Delta^2 f_i = f_{i+1} - 2f_i + f_{i-1} \qquad \mu f_{i+1} = f_{i+1} + f_i \tag{164}$$

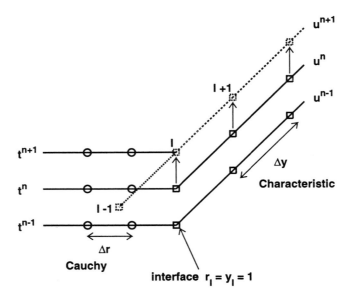

Figure 15. *Information on the characteristic side can be updated by extending the grid a distance Δy into the Cauchy region and using the most recent time-levels to interpolate data to the ghost grid point $I - 1$.*

Therefore at the interface ($i = I$) we require m_{I-1} due to the $\Delta^2 m_{i+1}$ term, *i.e.* the characteristic slice should be extended past the interface and into the Cauchy region. As for the estimate of ψ, obtained from data in the characteristic region, we can obtain m_{I-1} by interpolating ψ in space and time (using Cauchy slices t^{n+1}, t^n and t^{n-1}) at an appropriate position of Δy beyond the interface on the characteristic slice. Then relations (155) provide the required value for m_{I-1} so that M_{I+1} can be found using (163). When ω is non-zero similar procedures involving extensions of the Cauchy and characteristic grids combined with interpolation are used.

5.6 The start-up problem

Since three hypersurfaces in both the Cauchy and characteristic regions are required to evolve the system there is an obvious start-up problem. Using the 3+1 approach initial data can be set on a single spatial hypersurface. Likewise, for the characteristic region initial data is set up on a single null surface. In order to evolve forward we require information on hypersurfaces below the initial data surfaces. Moreover, such information is needed in any case to start the leapfrog evolution scheme. The solution to the problem is a 'thick sandwich' set-up which we now describe.

Assuming once again for simplicity that $\omega = 0$, we need only deal with the ψ, $\tilde{L}$, m and M variables. The initial data on the Cauchy side consists of specifying $\psi(r)$ and $\psi_{,t}(r)$, or $\tilde{L}(r)$, and on the characteristic side of specifying $m(y)$ and $M(y = 1)$, *i.e.* the

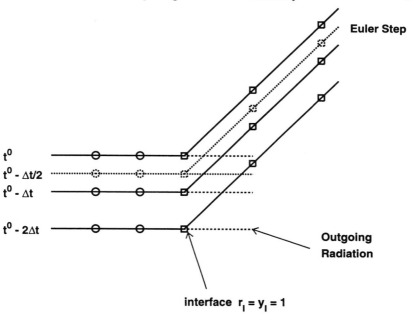

Figure 16. *The 'thick-sandwich' start-up method. Second-order accuracy is maintained using a combination of first-order Euler steps and second-order leapfrog steps. An artificial outgoing wave boundary condition is applied to an extended Cauchy grid to facilitate the start-up.*

value of M at the interface, which must satisfy (158) to be compatible with the Cauchy data, and from which $M(y)$ can be found by integrating equation (131). To start the evolution of the Cauchy portion of spacetime we ignore the presence of the characteristic hypersurface and extend the Cauchy surface beyond the interface point using around 10-15 extra grid points. An out-going radiation boundary condition is applied at the last grid point (see Figure 16). Now, to begin the leapfrog scheme and maintain second-order accuracy in time, we evolve the initial Cauchy slice backwards by $\frac{1}{2}\Delta t$ using an explicit first-order Euler method. This half time step slice is then used to evolve backwards by Δt via the second order leapfrog method. If the Cauchy region were present in isolation this would be sufficient to begin the evolution. However, for interpolation purposes, we require three initial hypersurfaces, and so using the initial data slice and the slice evolved backward by Δt we construct a further slice evolved backward by $2\Delta t$, once again using the leapfrog algorithm (see Figure 16). The characteristic region can be dealt with in a similar manner to that just described, but care is required at the interface since the Cauchy surfaces cannot be ignored. As for the Cauchy region we first evolve m backwards by $\frac{1}{2}\Delta u$, while M at the interface can be obtained from the value of $\tilde{L}$ at the interface on the corresponding time level. However, in order to integrate M along the characteristic surface we require the second derivative $m_{,yy}$ at the interface, as discussed previously, which involves information in the Cauchy region. For the start-up only we use the derivative relations (157) to obtain $m_{,yy}$ in terms of $\psi_{,r}$, $\psi_{,rr}$, $\psi_{,t}$ and $\psi_{,tt}$. All of these derivatives are known at the interface except for $\psi_{,tt}$ which we can estimate using information on the three Cauchy slices already constructed. Thus we can now use the leapfrog scheme to evolve backwards by Δu and $2\Delta u$ producing the required three hypersurfaces. This now enables the full evolution to proceed as described in the previous subsection.

Note that once this 'thick sandwich' has been built no further use is made of the extra grid points which extend the Cauchy region. It is clear that, provided a sufficiently large number of grid points are used in the extension, no information could have reached the interface from the outgoing radiation boundary condition. Thus there is no risk of this artificial boundary contaminating the initial data. The main defect in this procedure is in initially ignoring the characteristic surface and any incoming data that may be present on it. However, as the next section will show this initial error is very small and does not unduly affect the solution.

5.7 Comparison with an exact solution

When $\omega = 0$ a number of exact solutions for the vacuum cylindrical spacetime can be obtained. A particularly interesting one with which to test our combined code is the Weber-Wheeler wave (Weber and Wheeler 1957). This consists of an ingoing gravitational wave which hits the axis of symmetry and rebounds back out to infinity. Note that Piran *et al.* have found an exact solution for non-zero ω (Piran *et al.* 1986), however the Weber-Wheeler wave is sufficient for our use as a test.

In order to make a comparison with the code results we must first express the Weber-Wheeler solution in terms of variables used in the two coordinate systems of the combined code. For the Cauchy part we can write

$$\psi(r,t) = \frac{\sqrt{2}C\{(a^2+r^2-t^2)+[(a^2+r^2-t^2)^2+4a^2t^2]^{1/2}\}^{1/2}}{[(a^2+r^2-t^2)^2+4a^2t^2]^{1/2}} \quad (165)$$

and

$$\gamma(r,t) = \frac{1}{2}C^2\left[\frac{1}{a^2} - \frac{2r^2[(a^2+r^2-t^2)^2-4a^2t^2]}{[(a^2+r^2-t^2)^2+4a^2t^2]^2} - \frac{t^2+a^2-r^2}{[(a^2+r^2-t^2)^2+4a^2t^2]^{1/2}}\right] \quad (166)$$

where C and a are the amplitude and width of the wave respectively. In the characteristic region the solution becomes

$$\psi(y,u) = \frac{\sqrt{2}Cy\{(a^2y^2-u^2y^2-2u)+[(a^2y^2+u^2y^2+2u)^2+4a^2]^{1/2}\}^{1/2}}{[(a^2y^2+u^2y^2+2u)^2+4a^2]^{1/2}} \quad (167)$$

and

$$\gamma(y,u) = \frac{1}{2}C^2\left[\frac{1}{a^2} - \frac{2[(a^2y^2-u^2y^2-2u)^2-4a^2(uy^2+1)^2]}{[(a^2y^2+u^2y^2+2u)^2+4a^2]^2} \right.$$
$$\left. - \frac{a^2y^2+u^2y^2+2u}{[(a^2y^2+u^2y^2+2u)^2+4a^2]^{1/2}}\right] \quad (168)$$

The numerical solutions for ψ and γ when $C = 1$ and $a = 1$ are shown in Figure 17. The Cauchy coordinate time ranges from $t = -2.5$ to $t = 3.5$. The solution (in the

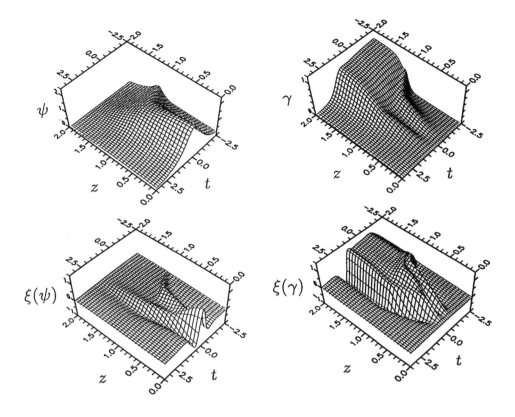

Figure 17. *Comparison of code results with the analytic Weber-Wheeler solution. A wave moves in from infinity, hits the $r = 0$ axis at $t = 0$ and bounces out again. Time evolution surface plots are shown for the metric functions ψ and γ and for the absolute errors $\xi(\psi)$ and $\xi(\gamma)$ (multiplied by 10^3) in these quantities. The interface is at $z = 1$.*

Cauchy region) is time symmetric about $t = 0$. Here we have used 300 points in each of the Cauchy and characteristic grids. Note that for convenience of plotting we have made use of a spatial coordinate parameter z defined by

$$z = \begin{cases} r & \text{for } 0 \leq z \leq 1 \\ 2 - y & \text{for } 1 \leq z \leq 2 \end{cases} \tag{169}$$

In this case $z = 2$ is future null infinity, while the interface is at $z = 1$. Information originally present on the characteristic hypersurface passes smoothly into the Cauchy region with little error despite the rather artificial start-up procedure of initially ignoring the characteristic surface. Some care is required in interpreting these plots because of the change of coordinate system at $z = 1$. For the characteristic region the null-

Resolution	Norm l_2 for ψ	Factor	Norm l_2 for γ	Factor
75	$6.586 \times 10(-5)$	–	$4.800 \times 10(-5)$	–
150	$4.261 \times 10(-6)$	15.4	$3.103 \times 10(-6)$	15.4
300	$2.667 \times 10(-7)$	15.9	$1.947 \times 10(-7)$	15.9
600	$1.662 \times 10(-8)$	16.0	$1.216 \times 10(-8)$	16.0

Table 1. *Convergence tests for the combined code results of the Weber-Wheeler solution at time t=0. Resolution denotes the number of grid points in each of the Cauchy and characteristic regions. Factor indicates the reduction in the norm from the previous resolution*

coordinate is $u = t - 1$ rather than the time t shown in the figures.

Below the numerical solutions we show the absolute pointwise errors defined, at the i-th grid point, by

$$\xi_i = (f(z_i)_{\text{exact}} - f(z_i)_{\text{computed}}) \qquad (170)$$

The numerical solutions for ψ and γ are both accurate to better than 1 part in 10^4 at this resolution. Note the slightly larger than expected error in γ at $t = 0$ on the characteristic side. This is due to ignoring the data on the characteristic region for the first three time steps when setting up the Cauchy initial hypersurfaces. However, this approximation does not cause a serious degradation of the solution at later times. The errors appear to be oscillatory with low growth rate. In addition little numerical noise is introduced by the interface.

As mentioned in the previous section, we have attempted to maintain second-order truncation errors in the space and time differencing. In order to verify this and as a further test of the code we have performed convergence testing. The Weber-Wheeler solution was obtained at four regular grid resolutions of 75, 150, 300 and 600 points in each of the coordinate regions. Beginning the evolutions at $t = -2$, at time $t = 0$ we construct a norm of the average absolute error for each grid resolution using

$$l_2 = \frac{1}{N} \sum_{i=1}^{N} \xi_i^2 \qquad (171)$$

where N is the total number of grid points and ξ_i is the pointwise error defined by (170). For true second-order convergence the norm should decrease by a factor of four each time the grid resolution is doubled and another factor of four due to the halfing of the time-step. Therefore the overall decrease in the error should be a factor of sixteen. In Table II we show the results obtained from the code. It is clear that second-order accuracy in space and time has been obtained.

It is interesting to compare the performance of the interface with the usual method of applying outgoing radiation boundary conditions (ORBCs) at the edge of the Cauchy grid (Piran 1980). To this end we have conducted the following numerical experiment: We evolve the Weber-Wheeler initial data described above (with $C = a = 1$) using the combined code (with the interface set at $r = 1$), and using the Cauchy portion of the code with ORBCs set at radii $r = 1$, $r = 5$ and $r = 25$. The quality of the solution is

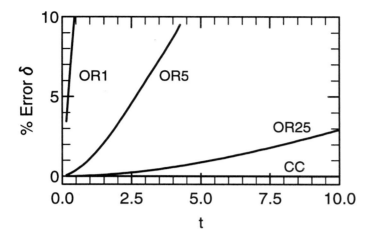

Figure 18. *Comparison of combined Cauchy-characteristic interface condition (label CC) with outgoing radiation boundary conditions placed at the edge of the Cauchy grid at $r = 1$ (OR1), $r = 5$ (OR5) and $r = 25$ (OR25). Plot shows the evolution of δ, the pointwise percentage error in ψ at the edge of the Cauchy grid.*

estimated by comparing the computed value of ψ at the edge of the Cauchy grid (in fact one grid point in from the edge) with the analytic value. The pointwise percentage error is given by

$$\delta = 100(\psi_{\text{exact}} - \psi_{\text{computed}})/\psi_{\text{exact}} \tag{172}$$

For this comparison we maintain a fixed resolution in radius, that is we use 300 constant spaced grid points between $r = 0$ and $r = 1$ so that the $r = 5$ Cauchy run has 1500 grid points while the $r = 25$ run has 7500. For $t < 0$ the Weber-Wheeler wave is ingoing and therefore we must start these evolutions from $t = 0$. We will attempt to run the evolutions up to $t = 10$. Since the wave is infinite in extent there is continuous outgoing radiation and this severely tests the boundary conditions. In the case of more realistic outgoing waves one would expect the error resulting from the ORBCs to be less marked.

The results from this experiment are shown in Figure 18. This shows the pointwise error δ as a function of time. The Cauchy run with ORBCs at $r = 1$ (labelled OR1 in the figure) develops a large error very quickly. This is not surprising since we are attempting to pass a strong wave out of the grid at a small radius. The ORBCs are really only effective for weak waves at large radii. The effect of moving the outer boundary to $r = 5$ (label OR5) and then $r = 25$ (label OR25) is to reduce the error growth significantly. It is clear that further reductions in error growth can be achieved by ever increasing the radius at which the ORBCs are applied. In the case of the combined code (label CC) the error in ψ never goes above 0.04% even though the strong wave is passing out of the Cauchy grid at a small radius. This experiment indicates that the characteristic matching technique can provide very effective outer boundary conditions for the Cauchy evolution even in a region with strong gravitational waves.

Since the conformal compactification leads to a stretching of the distance in the physical spacetime, we are able to achieve a small error without having to use very large numbers of grid points. This is something of crucial importance in Numerical Relativity codes, particularly in two and three dimensions.

5.8 Setting general initial data

For the metric (98) we can set initial data in terms of the following quantities

$$A(t,r) = 2(\psi_{,t} + \psi_{,r}) = 2(\tilde{L}/r + \psi_{,r}) \qquad (173)$$

$$B(t,r) = 2(\psi_{,t} - \psi_{,r}) = 2(\tilde{L}/r - \psi_{,r}) \qquad (174)$$

$$C(t,r) = \frac{e^{2\psi}}{r}(\omega_{,t} + \omega_{,r}) = -\frac{e^{2\psi}}{r}\left(2e^{-4\psi}L_z^\phi - \omega_{,r}\right) \qquad (175)$$

$$D(t,r) = \frac{e^{2\psi}}{r}(\omega_{,t} - \omega_{,r}) = -\frac{e^{2\psi}}{r}\left(2e^{-4\psi}L_z^\phi + \omega_{,r}\right) \qquad (176)$$

on the Cauchy portion. Following Piran et al. (1985) we take the quantities A and B represent, respectively, the ingoing and outgoing radiation in the + polarization mode, whereas C and D represent the ingoing and outgoing radiation in the × mode. Using (173–176) the Hamiltonian constraint equation (113) can be written as

$$\gamma_{,r} = \frac{r}{8}\left(A^2 + B^2 + C^2 + D^2\right) \qquad (177)$$

Since $\gamma(t,r)$ represents the total gravitational energy per unit length within a radius r at coordinate time t (Piran et al. 1985) we can determine the energy within a radius, say $r_{\max}$, for each ingoing/outgoing wave of either polarization mode as

$$\gamma_I = \int_0^{r_{\max}} \frac{r}{8} I^2 dr \qquad \text{where} \quad I = \{A, B, C, D\} \qquad (178)$$

Alternatively we can look at the radial distribution of the energy in the separate modes as

$$(\gamma_I)_{,r} = U_I = \frac{r}{8} I^2 \qquad (179)$$

Thus we are able to set up, on the initial Cauchy hypersurface, any particular combination and energy distribution of ingoing/outgoing waves with + or × polarization since we are free to set $\tilde{L}$, L_z^ϕ, $\psi_{,r}$ and $\omega_{,r}$ arbitrarily.

We have used this approach to investigate the combined Cauchy-characteristic code for various initial data sets. In general we took $A=C=0$ on both the Cauchy and characteristic hypersurfaces since there then exists no incoming radiation. Such a situation would be realistic for a star which suddenly begins to emit gravitational radiation due to collapse and supernova explosion. For calculational convenience the quantities B and D were specified initially as Gaussian pulses, i.e.

$$I = I_0 \exp[-(r - r_0)^2/\sigma^2] \qquad \text{where} \quad I = \{B, D\} \qquad (180)$$

where the amplitude, I_0, width, σ, and position, r_0 are user specified parameters.

We have undertaken a detailed examination of the parameter space for this problem (Dubal et al. 1995). We shall not discuss the details except to say that the combined approach has produced encouraging results when applied to vacuum cylindrical spacetimes. The numerical experiments have shown that gravitational wave information can pass smoothly between the two regions with little error introduced by the interface. In addition the presence of the interface appears to have no adverse affect on the long term evolution of the system.

A next step in the development of this work is an application to axi-symmetric spacetimes. Since the asymptotic behaviour of such spacetimes near future null infinity is better understood than for cylindrical symmetry some aspects of the method should be simpler. There would no longer be a need for the Geroch type decomposition which caused problems with very small, but finite, spurious reflections from the interface. Moreover, axi-symmetry allows the use of Bondi coordinates in the characteristic region with all the advantages of constructing invariant energy measurements at null infinity.

Acknowledgements

I would like to thank my colleague Mark Dubal who taught me most of what I know about Numerical Relativity during the tenure of a SERC grant, Number GR/J33883.

References

Ames W F, 1969, *Numerical Methods for Partial Differential Equations*, Barnes and Noble.
Bishop N T, 1992, in *Approaches to Numerical Relativity*, ed. R A d'Inverno, Cambridge University Press, 20.
Bishop N T, Clarke C J S and d'Inverno R A, 1990a. *Class Quant Grav Lett*, **7**, L23.
Bishop N T, Clarke C J S and d'Inverno R A, 1990b, *S A Jnl Sci*, **86**, 64.
Centrella J M (ed.), 1986, *Dynamical Spacetimes and Numerical Relativity*, Cambridge University Press.
Choptuik M, 1991, *Phys. Rev. D*, **44**, 3124.
Clarke C J S, d'Inverno R A and Vickers J A, 1995, *Phys. Rev. D*, **52**, 6863.
Corkhill R N and Stewart J M, 1983, *Proc Roy Soc*, **A386**, 373.
Deruelle N and Piran T (eds.), 1982, *Gravitational Radiation*, North Holland, Amsterdam.
d'Inverno R A, 1995, *Class Quant Grav*, **12**, L75.
d'Inverno R A (ed.), 1992, *Approaches to Numerical Relativity*, Cambridge University Press.
d'Inverno R A and Smallwood J, 1980, *Phys. Rev. D*, **22**, 1233.
d'Inverno R A and Stachel J, 1978, *J Math Phs*, **19**, 2447.
d'Inverno R A and Vickers J A, 1995, *Class Quant Grav*
d'Inverno R A, 1984, in *Problems of collapse and numerical relativity*, ed. D. Bancel and M. Signore, Reidel, Dordrecht, 221.
Dubal M R, d'Inverno R A and Clarke C J S, 1995, *Phys. Rev. D*, **52**, 6868.
Evans C R, Finn L S and Hobill D W (eds.), 1989, *Frontiers in Numerical Relativity*, Cambridge University Press.
Fletcher C A J, 1991, *Computational Techniques for Fluid Dynamics*, **Vol 1**, Springer Verlag.
Gottlieb D and Orszag S, 1977, *Numerical Analysis of Spectral Methods: Theory and Applications*, SIAM.
Gómez R and Winicour J, 1992, in *Approaches to Numerical Relativity*, ed. R A d'Inverno, Cambridge University Press, 143.

Gómez R, Isaacson R A, Welling J S and Winicour J, 1986, in *Dynamical Spacetimes and Numerical Relativity*, (ed.) J M Centrella, Cambridge University Press, 236.
Hahn S G and Lindquist R W, 1964, *Ann Phys*, **29**, 304.
Hockney R and Eastward J, 1981, *Computer Simulations using Particles*, McGraw-Hill.
Isaacson R A, Welling J S and Winicour J, 1983, *J Math Phys*, **24**, 1824.
Jordan P, Ehlers J and Kundt W, 1960, *Abh Wiss Mainz Math Nat*, **K1**, 2.
Kompaneets A S, 1958, *Sov Phys JEPT*, **7**, 659.
Kramer D, Stephani H, MacCallum M A H and Herlt E, 1980, *Exact solutions of Einstein's field equations*, Deutscher Verlag der Wissenscaften, Berlin, and Cambridge University Press.
Maeda K, Sasaki M, Nakamura T and Miyama S, 1980, *Prog Theo Phys*, **63**, 919.
May M M and White R H, 1966, *Phys Rev*, **141**, 1232.
Nakamura T, 1981, *Prog Theo Phys*, **65**, 76.
Nakamura T, 1986, in *Proceedings of the 14th Yamada Conference on Gravitational Collapse and Relativity*, eds. H, Sata and T. Nakamura, World Scientific, Singapore, 295.
Ō Murchadha N and York J W, 1974, *Phys Rev D*, **10**, 428.
Piran T, 1980, *J Comp Phys*, **35**, 254.
Piran T, Safier P N and Katz J, 1986, *Phys Rev D*, **34**, 331.
Piran T, Safier P N and Stark R F, 1985, *Phys Rev D*, **32**, 3101.
Press W H, Flannery B P, Teukolsky S A and Vetterling W T, 1986, *Numerical Recipes: The Art of Scientific Computing*, Cambridge University Press.
Regge T, 1961, *Nuov Cim*, **19**, 558.
Richtmeyer R D and Morton K W, 1967, *Difference Methods for Initial Value Problems*, Wiley.
Smallwood J, 1983, *J. Math. Phys.*, **24**, 599.
Smarr L (ed.), 1979, *Sources of Gravitational Radiation*, Cambridge University Press.
Taub A, 1978, *Ann Rev Fluid Mech*, **10**, 301.
Weber J and Wheeler J A, 1957, *Rev Mod Phys*, **29**, 509.
Zienkiewicz O C, 1977, *The Finite Element Method*, McGraw-Hill.

Quantum Gravity

Malcolm Perry

DAMTP
University of Cambridge

1 Introduction

There is no complete and consistent quantum theory of gravitation. What follows in these lectures is an historically motivated account of current thinking on the subject. The starting point is a review of canonical gravity and a brief discussion of minisuperspace methods. Then, we will look at how to derive the covariant path integral from the canonical formulation, and look at the issue of the failure of gravitation theories to be renormalizable even though semi-classically they can be used to make some physical predictions. Next, we tackle some of the interesting problems posed by the fact that black holes are quantum mechanically unstable by virtue of their Hawking temperature. Finally, we move on to a discussion of string theory which presently seems to be the best, or perhaps only, candidate for a theory of gravity that is consistent with the ideas of quantum mechanics. There is, of course, much more to quantum gravity than can be contained in these lectures.

The starting point for any discussion of gravitation is the classical theory of gravitation, namely Einstein's theory of general relativity. Here, spacetime is taken to be a differentiable manifold of dimension $d \geq 4$, with a Lorentz metric g_{ab} of signature $(-++\ldots)$, which obeys the Einstein equation

$$R_{ab} - \frac{1}{2}Rg_{ab} + \Lambda g_{ab} = 8\pi G T_{ab} \tag{1}$$

where Λ is the cosmological constant which is observed to be very small, or quite possibly zero, G is Newton's constant and T_{ab} is the energy-momentum tensor of any matter fields.

We can try to treat this system directly by identifying the true physical degrees of freedom in the theory. In the canonical approach, this means finding the reduced phase space and then replacing the canonical coordinates and momenta by the appropriate operators, and Poisson brackets by commutators. Next, one needs to identify the states of the theory; that is the Hilbert space in which these operators act. Finally one needs to be able to calculate probability amplitudes for events one has interest in. However, the route to the end point of this construction is littered with various hazards. We will discuss some of the difficulties as they arise in our treatment.

2 The action

The very first step in the construction of a quantum theory based on a given classical theory is to identify the correct action. What is required is a functional of the metric and various fields which reproduces Einstein's equations when this functional is extremized with respect to variations of the metric subject to the appropriate boundary conditions. Suppose that we are considering some domain D in spacetime, with boundary ∂D. The analogue of the Dirichlet problem is to specify the induced metric on ∂D, together with the projection of any other fields in to this surface. Components of the metric and fields normal to the surface will not be specified as part of the boundary conditions.

Suppose that the metric on D is g_{ab}. Then one can construct a unit normal vector n^a to the surface ∂D that is timelike (or spacelike) depending on whether ∂D is spacelike (or timelike) and so is normalised with $n^a n_a = \pm 1$. We will not treat the case of ∂D being null here. The induced metric on ∂D is then

$$h_{ab} = g_{ab} \mp n_a n_b. \tag{2}$$

The induced metric is also a projection into the surface as can be seen by the fact that h_{ab} annihilates n_a, $h_a{}^b n_b = 0$, and it is a projection operator, $h_a{}^b h_b{}^c = h_a{}^c$.

Information on the shape of the surface is contained entirely in h_{ab} which can be used to compute the curvature of the surface ∂D in the usual way. However, specification of the shape of the surface does not tell us how the surface is embedded in the spacetime (Spivak 1979). The information about the embedding is contained in the second fundamental form K_{ab}, which is a tensor defined on the surface ∂D. It tells us how the normal vector varies from point to point on the surface and is defined by

$$K_{ab} = h_a^c h_b^d \nabla_{(c} n_{d)}. \tag{3}$$

The action is now

$$I[g, \text{matter}] = \frac{1}{16\pi G} \int_D (R - 2\Lambda) g^{\frac{1}{2}} d^d x + \frac{1}{8\pi G} \int_{\partial D} K h^{\frac{1}{2}} d^{d-1} x + F[h] + I[\text{matter}]. \tag{4}$$

The first term here is the usual Einstein-Hilbert term where R is the Ricci scalar of metric g on the spacetime manifold with coordinates x^a, and Λ is the cosmological constant. In order to have the action extremised subject to the boundary conditions we have specified, we must make sure that boundary terms in the variation are of the correct form. If we just had the first term, then we would discover that it is necessary to fix not only h_{ab} on ∂D but also the normal components of the metric and their derivatives. This is a consequence of the well-known surface term in the variation of the Einstein-Hilbert action. These surfaces terms are cancelled by the addition of the second term, the Gibbons-Hawking-York boundary term (Gibbons and Hawking 1977, York 1972). To this we can, of course, add any functional of the metric on ∂D because this does not affect the variation. Finally, variation of the matter action, $I[\text{matter}]$, with respect to the metric simply gives the energy-momentum tensor of the matter T_{ab} since by definition

$$T_{ab} = 2\frac{\delta I[\text{matter}]}{\delta g_{ab}}. \tag{5}$$

Thus, the vanishing of the variation of the complete action given by (4),

$$\frac{\delta I[g, \text{matter}]}{\delta g_{ab}} = 0, \tag{6}$$

reproduces the Einstein equation (1).

3 The space-time split

The initial supposition in the canonical treatment is that it is possible to foliate spacetime so that spacetime can be regarded as a sequence of spacelike surfaces labelled by a time coordinate t. If this foliation cannot be carried out, the canonical quantization procedure must somehow break down. For the time being we will suppose that this does not happen, but will return to this issue when we discuss the question of black hole evaporation.

If we follow these assumptions, then we can write the spacetime line element in the form

$$-\alpha^2 dt^2 + \gamma_{ij}(dx^i + \beta^i dt)(dx^j + \beta^j dt). \tag{7}$$

Here γ_{ij} is a positive definite three-metric which is the induced metric on the spacelike surfaces labelled by t with coordinates x^i. With respect to diffeomorphisms on these surfaces, α and β^i are a scalar and a vector field respectively. The metric tensor is then

$$g_{ab} = \begin{pmatrix} -\alpha^2 + \beta^i \beta_i & \beta_j \\ \beta_k & \gamma_{jk} \end{pmatrix}. \tag{8}$$

In the above expression for the metric, the tensor manipulations are with respect to the spatial metric tensor γ_{ij}. The inverse metric is then easily found to be

$$g^{ab} = \begin{pmatrix} -1/\alpha^2 & \beta^j/\alpha^2 \\ \beta^k/\alpha^2 & \gamma^{jk} - \beta^j \beta^k/\alpha^2 \end{pmatrix}. \tag{9}$$

The determinant of the metric is given simply by $\det g_{ab} = -\alpha^2 \det \gamma_{ij}$.

The surfaces of constant t have, as well as the induced metric γ_{ij} on them, the second fundamental form K_{ij}. In terms of the metric variables given, this takes the form

$$K_{ij} = \frac{1}{2\alpha}(D_i\beta_j + D_j\beta_i - \dot{\gamma}_{ij}) \tag{10}$$

where the operator D is the covariant derivative formed from the spacelike metric γ_{ij} using the usual symmetric metric connection, and the dot means the time derivative $\frac{\partial}{\partial t}$. We can use this decomposition of the metric to rewrite the action as

$$\frac{1}{16\pi G}\int \mathcal{L} dt \tag{11}$$

where $\mathcal{L}$ is the lagrangian for the system. It takes the explicit form

$$\mathcal{L} = \alpha\gamma^{\frac{1}{2}}({}^{(3)}R - K_{ij}K^{ij} + K^2 - 2\Lambda) + \text{matter terms} + \text{boundary terms}. \tag{12}$$

Here ${}^{(3)}R$ is the Ricci scalar of the spatial surfaces, and K is the trace of the second fundamental form $K = K_{ij}\gamma^{ij}$. The matter field contributions are those of whatever matter fields one wishes to include. The boundary terms do not concern us as far as the canonical treatment is concerned.

4 The Hamiltonian approach

The idea is now to investigate the Hamiltonian treatment of this dynamical system as a prelude to attempting to quantize this system. We will follow the approach of Dirac for dealing with constrained Hamiltonian systems. An extremely comprehensive reference on such matters is the book of Henneaux and Teitelboim (1992). For simplicity we will only deal explicitly with the gravitational fields. Matter fields may be included and are treated in exactly the same way.

From the form of the action (12) we first calculate the canonical momenta conjugate to the canonical coordinates which are α, β^i and γ_{ij} at each point in space (de Witt 1967a).

The momenta are defined as being the functional derivatives of the lagrangian with respect to the time derivatives of the canonical coordinates. Thus

$$\pi = \frac{\delta\mathcal{L}}{\delta\dot{\alpha}} = 0, \qquad \pi^i = \frac{\delta\mathcal{L}}{\delta\dot{\beta}_i} = 0, \tag{13}$$

$$\pi^{ij} = \frac{d\mathcal{L}}{\delta\dot{\gamma}_{ij}} = \gamma^{\frac{1}{2}}(K^{ij} - \gamma^{ij}K). \tag{14}$$

It should be noted that the momenta are all densities of weight one. The vanishing of both π and π^i indicates that we are dealing with a constrained system, and that these are the primary constraints. We can now attempt to construct the hamiltonian in the usual canonical fashion. However, we will need to make certain that these primary constraints are preserved under the time evolution of the system.

Our first guess for the hamiltonian is therefore

$$H = \int d^3x \, (\pi\dot{\alpha} + \pi^i\dot{\beta}_i + \pi^{ij}\dot{\gamma}_{ij} - \mathcal{L}). \tag{15}$$

The hamiltonian now needs to be expressed in terms of the canonical variables so we rewrite it in the usual way.

$$H = \int d^3x \, (\pi\dot{\alpha} + \pi^i\dot{\beta}_i + \alpha\mathcal{H} + \beta_i\chi^i). \tag{16}$$

The two new quantities that we have defined, $\mathcal{H}$ and χ_i, play a central role in the theory. Writing them in terms of the geometrical variables of the space and time decomposition yields

$$\mathcal{H} = \gamma^{\frac{1}{2}}({}^{(3)}R - K^2 + K_{ij}K^{ij} + 16\pi G T_{ab}t^a t^b), \tag{17}$$

where t^a is the unit timelike vector normal to the surface $t = $ constant, or in terms of the canonical variables,

$$\mathcal{H} = \gamma^{-\frac{1}{2}}(\gamma\,^{(3)}R + (\gamma_{ik}\gamma_{jl} + \gamma_{il}\gamma_{jk} - \gamma_{ij}\gamma_{kl})\pi^{ij}\pi^{kl} + 16\pi G\gamma T_{ab}t^a t^b), \qquad (18)$$

and $\chi^i = -2D_j\pi^{ij}$.

Requiring that the primary constraints apply for all time is the same as asking for both $\mathcal{H}$ and χ^i to vanish. Thus the classical theory just boils down to a set of constraints, together with the whatever physics is contained in the energy-momentum tensor. The consequent vanishing of the hamiltonian is a generic phenomenon in theories that are generally covariant.

The Poisson bracket algebra is, for the canonical variables,

$$\{\pi(x), \alpha(x')\} = -\delta(x, x') \qquad (19)$$
$$\{\pi^i, \beta_j\} = -\delta^i_j(x, x') \qquad (20)$$
$$\{\pi^{ij}, \gamma_{kl}\} = -(\delta^i_k\delta^j_l + \delta^i_l\delta^j_k)\delta(x, x'). \qquad (21)$$

This induces the following algebra of constraints

$$\{\mathcal{H}(x), \mathcal{H}(x')\} = 2\chi^i\delta_{,i}(x, x') + D_i\chi^i\delta(x, x') \qquad (22)$$
$$\{\mathcal{H}(x), \chi_i(x')\} = \mathcal{H}(x)D_i\delta(x, x') \qquad (23)$$
$$\{\chi_i(x), \chi_j(x')\} = \chi_j(x)D_i\delta(x, x') + \chi_i(x')D_j\delta(x, x'). \qquad (24)$$

To go with each constraint $\mathcal{C}$, one must introduce a gauge fixing condition $\mathcal{G}$ by picking a set of conditions such that $\det\{\mathcal{C}, \mathcal{G}\} \neq 0$.

5 Canonical quantization

In principle, it is entirely straightforward to quantize this system canonically. However, technical complications prevent this from being a straightforward exercise. In all probability, a quantum theory of gravity based on general relativity only makes sense semi-classically. The failure of the usual quantization techniques can either be blamed on our inability to make sense of theory that is quantum mechanically sound or on a fundamental deficiency of the theory. We will adopt the second point of view in these lectures. Nevertheless, it is worth pursuing traditional methods to see what lessons can be learnt. Canonical quantization requires us to replace the canonical variables by operators that act on the space of states of the theory. For each of the Poisson brackets, one replaces them by commutators in the usual way. However, there is one subtlety to be faced up to and that is the nature of the space of states. The states of the gravitational field are described by the geometry of the spatial slices. This is represented by a positive definite three metric. Classically, there is no difficulty with this, but on quantization, the operator version of the Poisson brackets is given by

$$[\gamma_{ij}, \pi_k{}^l] = i(\delta^l_i\gamma_{jk} + \delta^l_j\gamma_{ik})\delta(x, x') \qquad (25)$$
$$[\pi_i{}^j, \pi_k{}^l] = i(\delta^l_i\pi_k{}^j - \delta^j_k\pi_i{}^l)\delta(x, x') \qquad (26)$$
$$[\gamma_{ij}, \gamma_{kl}] = 0. \qquad (27)$$

The Hilbert space of states based on this algebra, a generalization of the affine algebra, is restricted to those metrics that are positive definite (Isham 1984). The states of the theory are those that are annihilated by constraints of the theory. Suppose that we have a state $|\Psi\rangle$. Then, for it to be a state of the theory, it must be annihilated by the constraints. Thus

$$\mathcal{H}|\Psi\rangle = \chi^i|\Psi\rangle = \pi|\Psi\rangle = \pi^i|\Psi\rangle = 0, \tag{28}$$

in which one represents the momenta by the operator versions so that

$$\pi = -i\frac{\delta}{\delta\alpha}, \qquad \pi^i = -i\frac{\delta}{\delta\beta_i}, \qquad \pi_i^j = -i\gamma_{ik}\frac{\delta}{\delta\gamma_{kj}}. \tag{29}$$

There are of course other ways of realizing the operator algebra in question, but the one given here, the metric representation, is in some sense the most intuitively appealing one. The set of constraints equations defining the physical states are sometimes known as the Wheeler-de Witt equations.

One can of course go ahead and carry out various formal manipulations with these expressions. However, such manipulations are beset with operator-ordering ambiguities, and presently unresolved fundamental difficulties with the idea of defining products of operators. We will not pursue in detail these questions.

6 Minisuperspace

Rather than attempting to quantize all the degrees of freedom of the gravitational field, the idea of minisuperspace is to quantize only a few of them. One selects those that are expected by some semi-classical reasoning to be the most important.

A very simple example is motivated by cosmology. Consider a Friedmann-Robertson-Walker spacetime with $k=0$ (Kakas and Isham 1984). Its line element is given by

$$ds^2 = -\alpha(t)^2 dt^2 + R(t)^2 d\Omega^2. \tag{30}$$

Thus space is three-dimensional Euclidean space, with positive scale factor $R(t)$. The function α controls the definition of the cosmic time coordinate. The simplest matter that could live in this universe is a massless scalar field ϕ. A simple calculation reveals that the lagrangian for this problem is

$$L = -\frac{R\dot{R}}{2\alpha} + 8\pi G \frac{R^3 \dot{\phi}^2}{\alpha}. \tag{31}$$

The classical solutions can be found easily, provided one fixes the gauge. A suitable choice of gauge is $\phi = t$ and $\alpha = 8\pi G R^3$. Then the form of $R(t)$ is

$$R(t) = R_0 \exp(\pm\sqrt{16\pi G}\, t) \tag{32}$$

where R_0 is an arbitrary constant of integration. Quantization of this system is straightforward. One needs to identify the canonical variables and constraints, and quantize in the usual way. The momenta conjugate to the canonical coordinates are

$$\pi_\alpha = 0, \qquad \pi_R = -\frac{R\dot{R}}{\alpha}, \qquad \pi_\phi = 16\pi G \frac{R^3}{\alpha}. \tag{33}$$

Thus the first guess at the hamiltonian is

$$H = \mu_1 \pi_\alpha + \alpha \left(\frac{\pi_\phi^2}{32\pi G R^3} - \frac{\pi_R^2}{2R} \right). \tag{34}$$

Here, μ_1 is a Lagrange multiplier for the primary constraint $\pi_\alpha = 0$. Corresponding to the requirement that $\dot{\pi}_\alpha = \{\pi_\alpha, H\} = 0$ there is a secondary constraint

$$C = \frac{\pi_\phi^2}{32\pi G R^3} - \frac{\pi_R^2}{2R}. \tag{35}$$

The hamiltonian then becomes

$$H = \mu_1 \pi_\alpha + \mu_2\, C. \tag{36}$$

The Hamiltonian equations of motion that emerge are precisely equivalent to those from the lagrangian of the problem. One sees that the structure of the problem is basically the same as that for general quantum gravity. The problem has collapsed down to a collection of constraints that must be imposed on the Hilbert space of the theory after replacing the classical variables by their quantum mechanical counterparts. Suppose that the state vector is $|\Psi\rangle$ and we use the same gauge as we choose classically. Then $|\Psi\rangle$ is a function only of R and ϕ. The only non-trivial constraint equation is then

$$C|\Psi\rangle = 0. \tag{37}$$

As before, this is usually referred to as the Wheeler-de Witt equation. Explicitly, it becomes

$$\left(\frac{\partial^2}{\partial \phi^2} - 16\pi G R \frac{\partial}{\partial R} R \frac{\partial}{\partial R} \right) |\Psi\rangle = 0. \tag{38}$$

It should be noted that in this equation we made an assumption about how to define π_R^2/R in terms of the ordering of the corresponding operators. Solutions of this equation are easy to find by separation of variables. However, the physical meaning of $|\Psi\rangle$ is not entirely obvious. The solutions of the Wheeler-de Witt equation result in the amplitude $\langle R, \phi | \psi \rangle$ which is the probability amplitude for finding the universe in the state of radius R with the scalar field taking the value ϕ, or equivalently when the cosmic time coordinate takes the value ϕ. Within the context of cosmology it is not clear what this means. However, transition amplitudes have a very clear meaning. Thus for example $\langle R_2, \phi_2 | R_1, \phi_1 \rangle$ is the probability amplitude for the state of the universe to have radius R_2 at time ϕ_2 given that had radius R_1 at time ϕ_1. Such an amplitude can be calculated by the usual Green function methods of quantum mechanics for this particular version of the Wheeler-de Witt equation. After a certain effort, we discover that (Linden and Perry 1991)

$$\langle R_2, \phi_2 | R_1, \phi_1 \rangle = \frac{i}{\pi} \frac{\Delta \phi}{\ln^2(R_2/R_1) - 16\pi G(\Delta\phi - i\epsilon)^2} \tag{39}$$

with $\delta\phi = \phi_2 - \phi_1$. This is the propagator for the entire universe for this minisuperspace model. To use it, we must supplement it by the usual Feynman $i\epsilon$ prescription. Also, it is worth noting that it is singular on the classical solution to the problem that can found from directly from the Wheeler-de Witt equation.

7 The path integral

Starting from the canonical formulation, it is possible to derive a path integral formulation of quantum gravity. The idea is to start from the transition amplitude $\langle \gamma_f, t_f | \gamma_i, t_i \rangle$. This is the probability amplitude to go from some three metric, representing the state of the gravitational field initially γ_i at time t_i, to a three metric γ_f at time t_f. Then, the interval between t_i and t_f is divided up into $N+1$ equal intervals, with the intermediate times being $t_1, t_2, \ldots t_N$. The amplitude is then rewritten as

$$\langle \gamma_f, t_f | \gamma_i, t_i \rangle = \sum_{\{\gamma, \pi\}} \langle \gamma_f, t_f | \pi_f, t_f \rangle \langle \pi_f, t_f | e^{iH(t_f - t_N)} \Lambda | \gamma_N, t_N \rangle$$
$$\ldots \langle \pi_1, t_1 | e^{iH(t_1 - t_i)} | \gamma_i, t_i \rangle. \qquad (40)$$

Here Λ projects onto the physical states, that is those that obey the constraint equations. The states of the gravitational field are represented either by the metric or the momentum conjugate to it at each intermediate time. Explicitly, the projector onto the physical states can be written as the integral

$$\Lambda = \int D\beta_i D\alpha \exp\left(i \int dx \beta_i \chi^i + \alpha \mathcal{H}\right). \qquad (41)$$

The integrals here are taken over all β_i and α at each point of space. Using the representations of the matrix element

$$\langle \gamma | \pi \rangle = \exp\left(i \int \pi^{ij} \gamma_{ij} dx\right) \qquad (42)$$

we can substitute in and derive the path integral in the usual fashion. Thus

$$\langle \gamma_f, t_f | \gamma_i, t_i \rangle = \int \prod_{\text{each time step}} D\gamma D\pi \, \Lambda \, \exp\left(i \int \pi^{ij}(\gamma_{ij}(t) - \gamma_{ij}(t - \epsilon))\right) \qquad (43)$$

where ϵ is the width of the time interval. One can now take the limit as $N \to \infty$ and write the path integral as

$$\text{Amplitude} \sim \int D[\gamma, \pi, \alpha, \beta] \exp\left(i \int dx dt \, (\alpha \mathcal{H} + \beta_i \chi^i + \pi^{ij} \dot{\gamma}_{ij})\right). \qquad (44)$$

Finally, this path integral can be brought into the form of the conventional covariant path integral by integrating out the momenta π_{ij}, resulting in

$$\text{Amplitude} \sim \int D[g] e^{iI[g]}. \qquad (45)$$

The last step is carried out by performing the quadratic integrals over the momenta, and realizing that the variables $\alpha, \beta_i, \gamma_{ij}$ can be reassembled into the four-dimensional metric tensor g_{ab}. The treatment here parallels the usual arguments for gauge theories in flat space, and therefore presumes, unlike that case, that all the operator ordering problems can be somehow resolved, and that field products can be sensibly defined. In view of the well-known difficulties with defining quantum gravity, particularly with respect to the its renormalizability, it seems to be a considerable challenge to produce a derivation that is more mathematically satisfying than the one sketched here. The path integral we have just constructed of course has its own set of usual problems. The first is what metrics are to be included in the path integral. One answer to this question is those that are gauge inequivalent. Thus one is led to the usual Faddeev-Popov gauge-fixing procedure together with the corresponding ghost fields. Such a description of the path integral is guaranteed to lead to a set of Feynman rules that produce unitary amplitudes in the perturbative expansion of the path integral. This is however only part of the answer. We also need to ask whether metrics on spacetimes of different topologies need to be included. Here, no unambiguous answer can be given, but the overall consensus is that they should be. Finally, it is still tricky to define the path integral for metrics of Lorentz signature, to do so requires an extension of the Feynman $i\epsilon$ prescription beyond its usual flat space setting. The extension is fraught with ambiguity, and one way of resolving the issue is a suggestion of Hawking that the path integral should be defined over all Riemannian metrics, and then the results should be analytically continued back to Lorentzian signatured spacetime at the end of the calculation. The status of this suggestion is presently unclear, but there is no doubt that it produces useful results in the discussion of black hole equilibrium processes.

8 Perturbation theory and unrenormalizability

The path integral for gravitation is thus

$$Z \sim \int D[g] e^{iI[g]} \quad (46)$$

where $I[g]$ is the action for the gravitational field. One can add various matter fields to the action in a straightforward way. The way in which we treat path integrals of this type is to decompose the metric into two components, one which will be treated as a classical field $g_{ab}^{(0)}$, and one which we will treat as a quantum field h_{ab} propagating on the classical background. The complete metric g_{ab} is thus given by

$$g_{ab} = g_{ab}^{(0)} + h_{ab}. \quad (47)$$

The action is then expanded in powers of h_{ab}. So

$$I[g^{(0)} + h] = I[g^{(0)}] + \frac{\delta I}{\delta g_{ab}} h_{ab} + \frac{1}{2} \frac{\delta^2 I}{\delta g_{ab} \delta g_{cd}} h^{ab} h^{cd} + \cdots \quad (48)$$

The first term is the classical action evaluated for the metric $g_{ab}^{(0)}$. The second term vanishes if the classical metric $g_{ab}^{(0)}$ satisfies the Einstein equations. This is usually taken to be the case. The third term defines what is usually thought of as the propagator for the gravitational field. Finally, the dots denote the higher order terms which control the interactions of the gravitational field. One can see on purely dimensional grounds that each of these terms must contain two derivatives. This expansion never terminates. The procedure that decomposes the metric into its classical and quantum pieces is sometimes called the background field expansion (de Witt 1967b, 1967c, 't Hooft and Veltman 1970). To define the diagrammatic expansion, one must first fix the gauge. In the Faddeev-Popov method, this can be done straightforwardly. Suppose that the gauge condition applied to the perturbation is

$$G_a[h_{bc}] = 0. \quad (49)$$

A typical covariant gauge is the transverse gauge where $G_a = \nabla^b h_{ab}$ with ∇ being defined with respect to the classical metric $g_{ab}^{(0)}$. Then one simply adds two terms to the lagrangian. The first is just $\frac{1}{2} G_a G^a$. The second term is that for the Faddeev-Popov ghosts and antighosts, η_a and $\bar{\eta}^a$, which are anticommuting vector fields,

$$\bar{\eta}^a G_a [\nabla_c \eta_d + \nabla_d \eta_c]. \tag{50}$$

Assuming that we have chosen a sensible gauge condition, the part of the lagrangian that is quadratic in the perturbations can be written now as

$$\frac{1}{2} h \mathcal{O} h \tag{51}$$

where $\mathcal{O}$ is an invertible second order differential operator. Similarly the quadratic part of the ghost action will take the form

$$\bar{\eta} \mathcal{P} \eta \tag{52}$$

with $\mathcal{P}$ also being an invertible second order differential operator. This results in the propagators for both ghosts and metric perturbations, or gravitons, taking the form

$$\frac{T}{k^2} \tag{53}$$

where T is the polarization tensor which is dimensionless, and k is the momentum of either the ghost or the graviton.

One can now simply construct, at least in principle, all the Feynman diagrams of the theory, and evaluate them. However, it is not necessary to do any detailed calculations to see that there is potential for serious difficulties if this route is followed. Suppose that a Feynman diagram has E external lines, I internal lines, L loops and V_n n-point vertices. Then the primitive degree of divergence, D, for that diagram is

$$D = -2I + 4L + 2 \sum_{n \geq 3} V_n. \tag{54}$$

If one imposed an ultraviolet cut-off Λ in the theory, then a diagram with primitive degree of divergence D would diverge like Λ^D as Λ is taken to infinity. If D is unbounded above then, in the absence of any unaccounted for symmetries, the theory is unrenormalizable. There are two topological identities that apply to the Feynman graphs and which will allow us to decide whether this is the case here. The first results from the fact that each internal line in the diagram has two ends, and each external line has only a single end. Counting up the ends of the lines results in

$$2I + E = \sum_n n V_n. \tag{55}$$

The second comes from counting the number of loops in a diagram. The number of loops in a Feynman diagram is the number of momenta that are undetermined by the external momenta, and hence must be integrated over. Thus

$$L = I - E - \sum_n V_n. \tag{56}$$

Consequently,
$$D = 2L - 2E. \tag{57}$$

We see therefore that the degree of divergence increases unboundedly with the number of loops. This conclusion is not modified by the inclusion of any matter fields. It indicates that the theory is unrenormalizable.

An explicit calculation within the context of dimensional regularization yields a one-loop counterterm of
$$\frac{1}{\epsilon} \frac{1}{2880\pi^2} \int (R_{abcd} R^{abcd} + \cdots) \tag{58}$$

where $\epsilon = n - 4$ with n being the dimension of spacetime. The curvature tensor is evaluated with respect to the background metric $g^{(0)}$ and the dots denote terms that vanish by virtue of choosing $g^{(0)}$ to satisfy the classical equations motion. Whilst this counterterm is not of the original form, it does not immediately lead to a catastrophe. The Gauss-Bonnet theorem in four dimensions is a formula that relates the curvature to the Euler character χ, a topological invariant. Explicitly,
$$\chi = \frac{1}{32\pi^2} \int (R_{abcd} R^{abcd} - 4R_{ab} R^{ab} + R^2) + \text{boundary terms}. \tag{59}$$

Since χ is topological and the counterterm is, up to the classical equations of motion, just χ such a counterterm does not change the dynamics of the theory. However, at two loops, a counterterm is found (Goroff and Sagnotti 1986):
$$\frac{1}{\epsilon} \frac{209}{737280\pi^4} \int d^4x \, g^{\frac{1}{2}} R^{ab}{}_{cd} R^{cd}{}_{ef} R^{ef}{}_{ab} \tag{60}$$

up to terms that vanish by the equations of motion. There is no way to rescue the theory from such a counterterm. Thus, as a semi-classical theory (one-loop) Einstein gravitation is well-defined. However, because of its higher-order unrenormalizability, it is believed to be ultimately quantum-mechanically inconsistent.

Two ideas that have their roots in conventional quantum field theory of point particles have been suggested as ways of fixing these difficulties. The first is a class of modifications to the Einstein action in which one adds terms that are quadratic in the Riemann tensor to the action. Such theories contain fourth derivatives of the fields in their actions and so to have a sensible initial value formulation one must specify not only the fields and their first derivatives, but also the second and third derivatives. Ignoring this difficulty, however, leads to a theory that is renormalizable. This is because the graviton propagators now have a large momentum dependence of the form k^{-4} instead of k^{-2}, and so the primitive degree of divergence is now
$$D = -4I + 4L + 4\sum_n V_n. \tag{61}$$

Using the same identity as before for the number of loops results in
$$D = -4E. \tag{62}$$

The worst diagrams, the vacuum diagrams for which $E = 0$, are at most logarithmically divergent and so are essentially harmless. Such divergences can be absorbed by the usual type of renormalization procedure. However, we have traded one problem for another. Consider the typical graviton propagator, which is found by looking at the quadratic part of the action. In theories such as this the quadratic terms for h have the following form (schematically, ignoring the tensor structure)

$$h\Box\Box h + \lambda h \Box h \tag{63}$$

where λ is a dimensionful coupling that is of either sign. It cannot be zero since it comes from the Einstein term which must be important at low energies. The propagator for such a field will be of the form

$$\frac{1}{k^4 - \lambda k^2}. \tag{64}$$

This describes two particles, the usual massless graviton together with a massive partner. To see this, factorize the propagator into the sum of its pole terms, giving

$$\frac{1/\lambda}{k^2 - \lambda} - \frac{1/\lambda}{k^2}. \tag{65}$$

Depending on the sign of λ either the graviton or its massive partner will have a residue that is negative. This means that such particles are 'ghosts' because they have negative norm in the space of states in the theory. One generally supposes that such behavior is forbidden by conservation of probability, or unitarity. Furthermore, if $\lambda > 0$ then the massive partner is a tachyon which is presumably forbidden by causality. It is not known if such difficulties can be resolved within the context of field theory, and one must be realistic and suppose that the outlook for such modifications is rather bleak.

A second type of modification is to add matter to the theory in the hope of cancelling the divergences due to the gravitational field. If one adds ordinary matter in a general way suggested by classical general relativity, one finds that the situation deteriorates since the one-loop terms cease to be harmless because those terms that could be ignored by virtue of the classical equations of motion now turn out to ruin renormalizability at even the one loop order. On the other hand, if one adds matter to the theory in a supersymmetric way, there is some significant improvement in the situation. At one loop, the counterterms found remain harmless. Suppose that the theory consists of supergravity with N supersymmetries, and a bare cosmological constant Λ. Then the one loop counterterms are all of the form of the Euler character or the renormalization of the cosmological constant. Explicitly, they can be written as

$$\frac{1}{15\epsilon}\left(A\chi + \frac{4B}{3\pi^2}\Lambda^2 V\right) \tag{66}$$

where the term in B is the counterterm that renormalizes the cosmological constant. In its integrated form this is multiplied by the volume of spacetime. The results for the various possibilities are tabulated below (Christensen et al. 1980). Only for $N \leq 4$ is B shown, because supergravity with $N > 4$ cannot be consistently coupled to a cosmological constant.

N	0	1	2	3	4	5	6	7	8
A	$\frac{106}{45}$	$\frac{41}{24}$	$\frac{11}{12}$	0	-1	-2	-3	-4	-5
B	$-\frac{87}{8}$	$-\frac{77}{12}$	$-\frac{13}{3}$	$-\frac{5}{2}$	-1				

It is an intriguing unexplained fact that the last six entries in the first row are integers. Such theories are consistent at the one loop level. In fact, they are also consistent at the two loop level since there are no possible counterterms that are consistent with the requisite supersymmetry. However, for $N=1$ supergravity, unrenormalizable counterterms can appear at the three loop level. The coefficient of such a term could vanish as the consequence of some unsuspected symmetry, but it is generally expected that this will not happen. Calculations of the coefficient of such a term are being carried out. For higher values of N, counterterms are expected to appear at higher loops. Supergravity therefore appears to improve the situation, but certainly does not cure the divergence problem.

9 Black holes

Suppose one is interested in a statistical system. Then physical quantities can be derived from the partition function defined by

$$Z = \text{Tr}\, e^{-\beta H} \tag{67}$$

where H is the hamiltonian of the system, and β is the inverse temperature. Using an explicit representation of the trace, we see that

$$Z = \sum_{\text{states}} \langle \psi(t) | e^{-\beta H} | \psi(t) \rangle. \tag{68}$$

Since the hamiltonian is the time translation operator,

$$Z = \sum \langle \psi(t) | \psi(t + i\beta) \rangle. \tag{69}$$

A state will not contribute to this sum unless it is periodic in imaginary time with period β. We infer from this that periodic states are those that are characterized by a temperature $T = \beta^{-1}$.

Now consider the Schwarzschild black hole. In the usual Schwarzschild coordinates the metric takes the form

$$ds^2 = -\left(1 - \frac{2M}{r}\right) dt^2 + \left(1 - \frac{2M}{r}\right)^{-1} dr^2 + r^2 d\theta^2 + r^2 \sin^2\theta d\phi^2. \tag{70}$$

To investigate what happens in imaginary time, we continue to the 'Euclidean' domain by $t \to i\tau$. The metric then becomes

$$ds^2 = \left(1 - \frac{2M}{r}\right) d\tau^2 + \left(1 - \frac{2M}{r}\right)^{-1} dr^2 + r^2 d\theta^2 + r^2 \sin^2\theta d\phi^2. \tag{71}$$

There is a coordinate singularity at $r = 2M$ which can resolved by making a coordinate transformation

$$r = 2M + \frac{x^2}{8M}. \tag{72}$$

Near $x = 0$, the metric is

$$ds^2 = dx^2 + x^2 d\left(\frac{\tau}{4M}\right)^2 + 16M^2 d\Omega^2 \tag{73}$$

where $d\Omega$ is the line element on the unit two sphere. There is a conical singularity at $x = 0$ unless τ is identified with the appropriate period, $\tau \to \tau + 8\pi M$.

The Schwarzschild black hole has a temperature T given by

$$T = \frac{1}{8\pi M}. \tag{74}$$

A consequence of this temperature is that black holes, objects that are classically stable, are unstable quantum mechanically (Hawking 1975). A black hole will radiate particles like a black body. The radiation will have a Planckian spectrum and will appear to come from close to the event horizon. Since the area of the event horizon is of the order of M^2, it follows that the rate at which the black hole loses energy is given by

$$\dot{M} \sim -cM^2T^4 \tag{75}$$

where c is some purely numerical constant. Thus

$$\dot{M} = -kM^2. \tag{76}$$

Supposing that the black hole has an initial mass of M_0, the mass as a function of time, as measured at infinity, is given by

$$M(t) = (M_0^3 - 3kt)^{1/3}. \tag{77}$$

After a time $M_0^3/3k$ the black hole has completely disappeared. In classical general relativity, if a black hole disappears, the event horizon has a future end-point, which is inherently a naked singularity. The black hole ceases to exist at this point. It is not clear how this translates into quantum mechanical language. To put these results into perspective, we tabulate some typical masses, lifetimes and temperatures of various black holes.

Mass	Temperature	Lifetime
$1 M_\odot$	10^{-7} K	10^{64} yrs
10^{15} gms	10^{12} K	10^{10} yrs
10^{-5} gms	10^{32} K	10^{-43} secs

The overall picture presented here leads to one of the most important unresolved paradoxes in gravitational physics. The black hole can be formed out of some pure quantum mechanical state. Such a state is described by a density matrix ρ_i. The entropy of this state is given in terms of the density matrix by

$$S = -\operatorname{Tr}(\rho \ln \rho). \tag{78}$$

If it is a pure state, then the entropy will be initially zero. The black hole evaporates into thermal radiation, which has a large entropy. In itself, this is not paradoxical as the quantum mechanical information could be hidden behind the event horizon. However, the moment the black hole disappears there is a paradox since quantum mechanical evolution is thought to be incompatible with a change in the entropy. If the time evolution of the density matrix is given by the von-Neumann formula

$$\dot{\rho} = i[H, \rho] \tag{79}$$

where H is the hamiltonian, then it has a formal solution

$$\rho(t) = e^{-iHt}\rho(0)e^{iHt}. \tag{80}$$

The entropy is preserved under this evolution law. Thus, quantum mechanics appears to be incompatible with the semi-classical picture we have of black hole evaporation (Hawking 1977).

There seem to be broadly three possible types of resolution, none of which is currently accepted universally as the true answer. Firstly, it might be that the laws of quantum mechanics are in need of modification to allow this type of evolution. It is, however, difficult to develop a coherent formalism in which this happens. A second possibility is that the black hole does not completely evaporate, but leaves behind some kind of remnant. The difficulty here is that the remnant, however small it may be, must be able to contain all of the information that the black hole contained initially. One would expect that our semi-classical picture was reliable down to scales of the Planck length, and so any possible remnant would be quite light, presumably Planckian. It must therefore have an enormous density of states which would lead to such things being pair produced copiously in the early universe simply because their phase space factors are so large. This does not appear to have been the case. Finally, it has been speculated that the information slowly leaks out of the black hole with the thermal radiation. This means that the radiation is not precisely thermal. It is hard to reconcile this last possibility with causality. My personal feeling is that the first possibility is the correct one, but an understanding of how it happens is a long way off.

Part of the resolution of this problem is to understand if black holes themselves have any entropy. It is easy to evaluate this by semi-classical methods from the path integral. The thermodynamic partition function for gravity is given by

$$Z = \int D[g] e^{-I[g]} \tag{81}$$

where the integral is over all metrics that are periodic with imaginary time with the correct period. This is the path integral version of the general expression for the partition function given earlier. We can start to evaluate this using the background field method. The classical solution that is relevant is the Euclidean Schwarzschild metric. Expanding about it yields

$$Z = e^{-I[g^{(0)}]} \times \text{(quantum corrections)}. \tag{82}$$

The quantum corrections represent the effect of thermal gravitons propagating on the background of the black hole spacetime. One would expect that these represent thermal radiation in equilibrium with the black hole. The classical action for the Schwarzschild solution (Gibbons and Hawking 1977) is given entirely by its surface term

$$I_{\text{classical}} = \frac{M}{2T}. \tag{83}$$

Ignoring for the moment the quantum corrections, since the partition function is the exponential of the Helmholtz free energy F, $Z = e^{-\beta F}$, we see that

$$F = \frac{M}{2}. \tag{84}$$

The second law of thermodynamics can then be used to find the entropy since

$$S = -\frac{\partial F}{\partial T} = 4\pi M^2. \tag{85}$$

A black hole therefore has a very large entropy, which is consistent with the picture that emerges from the black hole uniqueness theorems.

The quantum corrections to Z are, however, infinite in a new way. Roughly speaking, the radiation must be in local thermodynamic equilibrium. Thus, the radiation close to the hole must in equilibrium with radiation at large distances from the hole. At infinity, let us suppose that the temperature is T_0 given by the traditional Hawking expression. To be in equilibrium close to the hole requires it to have a local temperature that is blue-shifted by the Tolman factor. This can be thought of as coming about because of the time dilation between infinity and regions close to the hole. The local temperature T_l is given by

$$T_l = \frac{T_0}{\sqrt{g_{00}}} \tag{86}$$

where g_{00} is the time-time component of the metric tensor. Close to the horizon, since g_{00} vanishes, the local temperature diverges. Such a divergence contributes an infinite amount to the physical energy. Clearly, something must be wrong for this to occur. One simple resolution is to suppose that there is a minimum length, as appears to be the case in string theory, and that it is impossible to probe arbitrarily close to the hole. It should be pointed out that this is a geometrical type of divergence that is entirely distinct from the typical ultraviolet divergences encountered in field theory. Again, no clear resolution of this problem is known.

There is another class of black holes that have some interest in quantum gravity. These can be thought of as non-perturbative states of some quantum theory. Suppose one thinks about charged black holes belonging to the Reissner-Nordstrom sequence of solutions rather than the Schwarzschild family. These are described by isolated objects of mass M and either electric or magnetic charge Q. The metric is given by

$$ds^2 = -\left(1 - \frac{2M}{r} + \frac{Q^2}{r^2}\right) dt^2 + \left(1 - \frac{2M}{r} + \frac{Q^2}{r^2}\right)^{-1} dr^2 + r^2 d\theta^2 + r^2 \sin^2\theta d\phi^2. \tag{87}$$

The same type of calculation as before gives the temperature as

$$T = \frac{\sqrt{M^2 - Q^2}}{4\pi M(M + \sqrt{M^2 - Q^2})} \tag{88}$$

and the entropy

$$S = 4\pi \left(M + \sqrt{M^2 - Q^2}\right)^2. \tag{89}$$

These formulae are valid for $M > |Q|$. If $M = |Q|$, then the temperature is zero, and the expression for the entropy is no longer valid. When proper account is taken of the analytic continuation of the zero temperature horizon, one finds that the entropy of this black hole is zero (Hawking *et al.* 1995). Such a black hole is a stable isolated lump, and so behaves very much like a soliton. Even though such an object does not suffer Hawking evaporation, one could still imagine that it could somehow decay. Under certain circumstances even this is impossible. Suppose that the object has magnetic charge (as opposed to electric charge), then one could imagine that it is the lightest state in the theory with that charge. Such an object would then be absolutely stable.

These zero temperature objects also have another extremely interesting property. Suppose that instead of thinking about general relativity we thought about $N=2$ supergravity theory. The theory contains a graviton, two gravitinos, and a photon in its $N=2$ supermultiplet. The action for the bosonic fields is that of the Einstein-Maxwell theory. A solution to the Einstein-Maxwell theory is said to be supersymmetric if a supersymmetry transformation maps the solution into itself. The condition for this is the existence of a Killing spinor (Gibbons and Hull 1982). Killing spinors in this theory satisfy the equation

$$\nabla_a \epsilon - \frac{1}{4}\sigma^{cd} F_{cd} \gamma_a \epsilon = 0. \tag{90}$$

The extreme Reissner-Nordstrom black hole is the unique asymptotically flat spacetime that does not contain a naked singularity and that is consistent with this equation having a solution. Physically, this means that an extreme Reissner-Nordstrom black hole belongs to a supermultiplet of the theory. The supersymmetry has been broken down to $N=1$ supersymmetry, and this means that the multiplet contains four distinct states. Thus, if one evaluates the one-loop quantum corrections to the classically vanishing entropy, one finds that there is a contribution from the Killing spinors, whilst all other quantum corrections vanish by virtue of the supersymmetry of the system. The result is that the entropy is given by

$$S = \ln 4. \tag{91}$$

Similar objects exist in more complicated supersymmetric theories. The existence of such states in both supergravity and string theories indicates that there is much still to be learnt about the physics of black holes.

10 String theory

String theory can be thought of as a generalization of particle theory to a situation in which the worldline of a particle, a one-dimensional timelike manifold, is generalized into a two-dimensional worldsheet of Lorentz signature. What emerges from this picture, and is quite remarkable, is a theory that contains general relativity as a classical limit, and is a finite quantum theory. We start our exploration of string theory by looking in a fairly detailed way at the bosonic string. It has many of the features of realistic superstring theories and has the advantage of being relatively simple to understand (Green *et al.* 1987).

The starting point of string theory is to consider a two-dimensional surface propagating in a d-dimensional flat Minkowski spacetime. This surface, the string worldsheet

Σ, will have induced on it by virtue of its embedding in spacetime a metric $\gamma_{\mu\nu}$. A simple action for such a system is the Nambu-Goto action

$$I = \frac{1}{2\pi\alpha'} \int_\Sigma \gamma^{1/2} d^2\xi \qquad (92)$$

where α' is the only dimensionful parameter in the theory, the inverse string tension. The worldsheet has been coordinatized by ξ^μ. String theories are usually characterised by discrete information and for this case we have implicitly assumed that the worldsheet is orientable so that we can define integration in a straightforward way. However, there is an additional topological choice to be made. There are two types of bosonic string, one in which the spatial sections of Σ are line segments, the open string, and one in which the line segments are just circles, the closed string. We will consider here just the case of closed strings.

The Nambu-Goto action is rather difficult to work with, but there is an equivalent alternative to it due to Polyakov (1981a, 1981b). Suppose that the coordinates in spacetime of points on the string worldsheet are given by $X^a(\xi)$. Then the induced metric on the worldsheet is given by

$$\gamma_{\mu\nu} \sim \partial_\mu X^a \partial_\nu X^b \eta_{ab}. \qquad (93)$$

One can then rewrite the action as

$$I = \frac{1}{4\pi\alpha'} \int \gamma^{1/2} d^2\xi \, \gamma^{\mu\nu} \partial_\mu X^a \partial_\nu X^b \eta_{ab}. \qquad (94)$$

This action has three distinct types of symmetries. The first is spacetime Poincaré invariance. If one makes a Poincaré transformation of the spacetime coordinates by

$$X^a \to \Lambda^a{}_b X^b + C^a \qquad (95)$$

with $\Lambda^a{}_b$ being a Lorentz rotation and C^a being a translation, then the action remains invariant. If one makes a worldsheet coordinate transformation, then since one is integrating a world sheet density, the action also remains invariant. Finally, the action is invariant under worldsheet conformal transformations in which the worldsheet metric undergoes a Weyl transformation

$$\gamma_{\mu\nu} \to \Omega^2(\xi) \gamma_{\mu\nu}. \qquad (96)$$

These last two symmetries indicate that the action, when treated as a functional of both the X's and the metric, is in fact independent of the worldsheet metric, apart from possible global considerations. Consequently, we are dealing with a system with a set of constraints, rather like the one we were dealing with before for general relativity. Quantization can then attempt to follow broadly similar lines.

The field equations that follow from this action are found by varying with respect to $\gamma_{\mu\nu}$ and X^a. Variation with respect to the worldsheet metric leads to the vanishing of the worldsheet energy-momentum tensor $\Theta_{\mu\nu}$,

$$\Theta_{\mu\nu} = \partial_\mu X^a \partial_\nu X^b \eta_{ab} - \frac{1}{2}\gamma_{\mu\nu} \partial_\rho X^a \partial_\sigma X^b \gamma^{\rho\sigma} \eta_{ab}. \qquad (97)$$

One can regard this as determining the metric, up to a conformal factor, in terms of the X's. The field equation for X is just the wave equation for the X's on the string worldsheet,

$$\gamma^{\mu\nu}\nabla_\mu\nabla_\nu X^a = 0. \tag{98}$$

In the canonical approach, we need first of all to consider dynamics on the string worldsheet. To analyse this, it is convenient to pick a coordinate system and metric on the string worldsheet. We will assume that the worldsheet has the topology of a cylinder so that we can pick the worldsheet metric to be flat. Then we choose a time coordinate τ, which runs from $-\infty$ to ∞, together with a spatial coordinate σ which runs from 0 to 2π. Sometimes it is useful to define lightcone coordinates on the string worldsheet, and these we will call $\xi_\pm = \tau \pm \sigma$. The thing to notice is that the worldsheet metric does not involve any derivatives, and therefore the momentum conjugate to $\gamma_{\mu\nu}$ vanishes. This is a primary constraint and, in order for this to be true for all time, one finds a secondary constraint that the hamiltonian must be independent of the metric. This is most easily expressed as being equivalent to the vanishing of the string energy-momentum tensor $\Theta_{\mu\nu}$. It should be noticed that although a symmetric tensor in two dimensions is expected to have three independent components, $\Theta_{\mu\nu}$ is tracefree identically, $\Theta_{\mu\nu}\gamma^{\mu\nu} \equiv 0$, and thus there are only two independent components in this expression. Whilst $\Theta_{\mu\nu}$ is simply expressed in terms of the fields, it does not lend itself easily to quantization. A simpler collection of objects to examine are the Fourier components of $\Theta_{\mu\nu}$. Denote the components of $\Theta_{\mu\nu}$ by $\Theta_{\pm\pm}$ meaning the components in the $\xi^\pm$ directions. Then the Fourier components, usually called the Virasoro coefficients, are given by

$$L_n = \frac{1}{8\pi\alpha'}\int d\sigma \Theta_{--}e^{-in\sigma}, \tag{99}$$

$$\tilde{L}_n = \frac{1}{8\pi\alpha'}\int d\sigma \Theta_{++}e^{in\sigma}. \tag{100}$$

Classically, each of the L_n and $\tilde{L}_n$ must vanish. They obey the Poisson bracket algebra

$$\{L_m, L_n\} = i(n-m)L_{n+m}. \tag{101}$$

It is simple to solve the wave equation in this coordinate system. It has solutions

$$X^a = f^a(\sigma - \tau) + g^a(\sigma + \tau) \tag{102}$$

where f and g are arbitrary functions governing the right and left moving waves of the string worldsheet. Again, the easiest thing to do in order to quantize this system is to decompose each of the X's into its Fourier modes. thus

$$X^a = X_0^a + \alpha' p^a \tau + i\sqrt{\alpha'} \sum_{\substack{n=-\infty \\ n\neq 0}}^{\infty} \left(\frac{1}{n}\alpha_n^a e^{-in(\tau-\sigma)} + \frac{1}{n}\tilde{\alpha}_n^a e^{-in(\tau+\sigma)}\right). \tag{103}$$

Since we require X to be real, it follows that

$$\alpha_n^a = (\alpha_{-n}^a)^\dagger \tag{104}$$

and similarly for the tilded quantities. X_0^a is the centre of mass of the string at time zero, and p_0^a is the centre of mass momentum governing the motion of the string in spacetime. After a canonical treatment of the X's, we find that the non-trivial Poisson brackets of the α's are

$$\{\alpha_n^a, \alpha_m^b\} = im\, \delta_{n+m,0} \eta^{ab} \tag{105}$$

together with a similar expression for the tilded quantities, and

$$\{x_0^a, p_b\} = \delta_b^a. \tag{106}$$

There is still some residual gauge freedom left in this system. This can be seen intuitively by observing that any change of X that lies in the worldsheet amounts to a worldsheet coordinate transformation, whereas a change in X normal to the worldsheet is a physical deformation of the worldsheet. Accordingly, we can choose the so-called light cone gauge to carry out quantization. Pick two null directions in Minkowski space and coordinatize them by $X^\pm$. The remaining directions orthogonal to $X^\pm$ are denoted by X^I, with $I = 1, \ldots, d-2$. The lightcone gauge consists of the choice

$$X^+ = x_0^+ + p^+ \tau \tag{107}$$

together with allowing X^- to be determined solely by asking that the Virasoro constraint equations be satisfied. In this gauge, it is clear that the only physical degrees of freedom are the transverse X's, all of which just behave like free fields.

One now straightforwardly quantizes this system by replacing the Poisson brackets by commutators, and the coefficients $\alpha \ldots$ by operators acting on the Hilbert space of states of the string. Thus, the coefficients for the transverse coordinates are operators obeying

$$[\alpha_m^I, \alpha_n^J] = m \delta_{n+m,0} \delta^{IJ} \tag{108}$$

together with a similar relation for the tilded quantities. Similarly, one must replace the coefficients for the centre of mass motion by operators obeying

$$[x_0^I, p^J] = \delta^{IJ}. \tag{109}$$

It will simplify many expressions in the future if we define

$$\alpha_0^I = \tilde{\alpha}_0^I = \sqrt{2\alpha'}\, p^I. \tag{110}$$

One can interpret the operators as being just a collection of simple harmonic oscillators, one for each Fourier mode of the string. In fact, if one rescales these operators by factors of $\sqrt{m}$, then this is precisely the algebra obeyed by the conventionally normalized harmonic oscillator creation and annihilation operators. However, we still have to worry about the constraints. These were classically that $L_n = \tilde{L}_n = 0$. Quantum mechanically, we therefore expect that we need to impose the condition that that these relations are true for expectation values of state vectors. A minimal necessary condition for this is that

$$L_n |\psi\rangle = \tilde{L}_n |\psi\rangle \tag{111}$$

for $n \geq 0$. Since the Virasoro operators obey the hermiticity conditions

$$L_n = (L_{-n})^\dagger, \qquad \tilde{L}_n = (\tilde{L}_{-n})^\dagger \tag{112}$$

this is sufficient to guarantee the expected constraints on the physical Hilbert space, that is the set of all possible states vectors $|\psi\rangle$. The L_n have no operator ordering ambiguities except for L_0 which can be defined so that

$$L_0 = \sum_m : \alpha_m^I \alpha_{-m}^J : \delta_{IJ} \tag{113}$$

where the semicolons indicate that the product should be normal ordered. Again similar considerations apply for the tilded quantities. The normal ordering ambiguity means that we should replace the L_0 and $\tilde{L}_0$ constraints by

$$(L_0 - a)|\psi\rangle = (\tilde{L}_0 - a)|\psi\rangle = 0 \tag{114}$$

where a is an as yet undetermined constant. Whilst most of the constraints are trivial in the light cone gauge, two of them have some physical content. We can rewrite the two constraints involving L_0 and $\tilde{L}_0$ as

$$(L_0 + \tilde{L}_0 - 2a)|\psi\rangle = 0 \tag{115}$$

and

$$(L_0 - \tilde{L}_0)|\psi\rangle = 0. \tag{116}$$

One way of interpreting L_0 is to observe that, apart from the centre of mass motion, it is the sum over all of the Fourier modes, summed over all transverse directions, of the number operator for each mode weighted by the Fourier weight of each mode. Given that, the first equation can be interpreted as saying that the centre of mass momentum squared, hence the rest mass of a given state, is given precisely by that number operator summed for the left- and right-moving modes. Thus,

$$\alpha' m^2 = -4a + 2\sum \left(:\alpha_m \alpha_{-m}: + :\tilde{\alpha}_m \tilde{\alpha}_{-m}: \right) \tag{117}$$

We will see later that a must take the value 1, and we will assume that this is so in what follows. The second equation says that the number operator for the left-moving excitations is the same as the number operator for the right-moving excitations. We can now simply construct the states of the theory, in the centre of mass frame.

Firstly, there is a vacuum state that is annihilated by all $\alpha_n^I, \tilde{\alpha}_n^I$ for $n > 0$. This state we will call $|0\rangle$. It has $m^2 = -4/\alpha'$, and thus represents a tachyon. The next $(d-2)^2$ states are of the form $\alpha_1^I \tilde{\alpha}_1^J |0\rangle$. These are all massless. The trace part of this object is a scalar, the symmetric tracefree part has $\frac{1}{2}d(d-3)$ independent components, and thus has the correct quantum numbers to describe the graviton. The antisymmetric part has $\frac{1}{2}(d-2)(d-3)$ independent components and so describes a 2-form potential that gives rise to a 3-form field strength. Next, there are a collection of states with mass given by $m^2 = 4/\alpha'$. There are quite a large number of them of the form

$$\alpha_{-2}^I \tilde{\alpha}_{-2}^J |0\rangle, \quad \alpha_{-2}^I \tilde{\alpha}_{-1}^{(J} \tilde{\alpha}_{-1}^{K)} |0\rangle, \quad \alpha_{-1}^{(I} \alpha_{-1}^{J)} \tilde{\alpha}_{-2}^K |0\rangle, \quad \alpha_{-1}^{(I} \alpha_{-1}^{J)} \tilde{\alpha}_{-1}^{(K} \tilde{\alpha}_{-1}^{L)} |0\rangle. \tag{118}$$

One can continue analyzing the states this way, but the key observations are that the theory contains a small number of massless states, followed by an infinite collection of massive states. It is the massive states that make string theory dramatically different to particle theories. Some of the massless states look as if they are trying to describe things that look like the graviton.

One might worry that the account given so far is profoundly noncovariant. One can get around this difficulty by using the path integral and the Faddeev-Popov scheme for quantizing systems with gauge degrees of freedom. Starting from the Polyakov form of the action,

$$I = \frac{1}{4\pi\alpha'}\int \gamma^{\frac{1}{2}} d^2\xi \; \partial_\mu X^a \partial_\nu X^b \gamma^{\mu\nu} \eta_{ab}. \tag{119}$$

we notice that this action is invariant under infinitesimal gauge transformations of the metric of the form

$$\gamma_{\mu\nu} \to \gamma_{\mu\nu} + \nabla_\mu \lambda_\nu + \nabla_\nu \lambda_\mu - \gamma_{\mu\nu} \nabla_\sigma \lambda^\sigma. \tag{120}$$

These are just infinitesimal coordinate transformations up to a conformal transformation. The standard Faddeev-Popov procedure then requires us to add to the lagrangian a term

$$b^{\mu\nu} \nabla_\mu c_\nu \tag{121}$$

where c^μ is a vector ghost, and $b_{\mu\nu}$ is a tracefree symmetric tensor antighost, both of which have odd Grassmann parity. We can now treat the system completely covariantly. The ghost and antighost fields can be treated as worldsheet fields in exactly the same way as the X's, so that, for cylindrical worldsheets, they can be decomposed into their Fourier modes. Thus, since the fields obey the field equations

$$\nabla_\mu b^{\mu\nu} = 0, \qquad \nabla_\mu c_\nu + \nabla_\nu c_\mu - \gamma_{\mu\nu} \nabla_\sigma c^\sigma = 0 \tag{122}$$

we can write the ghost and antighost fields as

$$c_- = \sum c_n e^{-in(\tau-\sigma)}, \qquad c_+ = \sum \tilde{c}_n e^{-in(\tau+\sigma)}, \tag{123}$$

and

$$b_{--} = \sum b_n e^{-in(\tau-\sigma)}, \qquad b_{++} = \sum \tilde{b}_n e^{-in(\tau+\sigma)}. \tag{124}$$

The anticommutation rules for these operators are

$$\{b_n, c_m\} = \delta_{n+m,0} \qquad \{\tilde{b}_n, \tilde{c}_m\} = \delta_{n+m,0} \tag{125}$$

with all other anticommutators vanishing.

The ghost and antighost contribute now to the energy-momentum tensor of the string worldsheet. This can be summarized by observing that Virasoro operators pick up an extra term,

$$L_n^{\text{total}} = L_n + \sum_m (m-n) b_{n+m} c_{-m} \tag{126}$$

with a similar term for the left movers. To make things slightly simpler, we will now absorb the coefficient a into L_0, so that $L_0 \to L_0 - a$. We then discover after a careful calculation that the algebra of the Virasoro operators is anomalous. To be precise, if we compute the commutator, we find that

$$[L_m, L_n] = (n-m) L_{n+m} + \left[\frac{d}{12}(m^3 - m) + \frac{1}{6}(m - 13m^3) + 2am\right] \delta_{m+n,0}. \tag{127}$$

The first term in square brackets comes from the X's and the second term comes from the ghost-antighost pair. Taking the expectation value of the commutator in an arbitrary physical state, the left hand side of this expression will vanish, and therefore so must the right hand side. Thus, the term proportional to the Kronecker delta must vanish. This will only happen if $d = 26$ and $a = 1$. Thus, string theory predicts the dimensionality of Minkowski spacetime together with the fact that there are massless particles that look like gravitons in the theory.

A calculation that has considerable interest, but whose meaning is not yet clear, is to ask about the density of states in the theory as a function of the mass. This was first carried out by Huang and Weinberg (1970), who showed, based on light cone gauge calculations similar to those outlined earlier, that the number of states in the mass range m to $m + dm$ is given by

$$\rho(m)dm = m^{-25/2} e^{m/T_0} dm \tag{128}$$

where T_0 is given by

$$T_0 = \frac{1}{2\pi\sqrt{\alpha'}}. \tag{129}$$

T_0 is sometimes called the Hagedorn temperature. It is a temperature beyond which the canonical ensemble cannot exist for strings since, if $T > T_0$, the Boltzmann factor fails to bound the density of states. It has been speculated that this signifies some sort of phase transition (Atick and Witten 1988) to a new phase of string theory that is not yet understood.

The idea of a phase transition is an analog of what happens if one attempts to use string theory to describe the physics of hadrons. There, the Hagedorn temperature corresponds to the deconfinement transition. At energy scales below T_0, string theory can be used quite successfully perturbatively to describe the physics of mesons, and non-perturbatively to describe the physics of baryons. The Hagedorn temperature marks the transition to where QCD is a more apt description and the physics of hadrons is better described by quarks and gluons. In the gravitational version of string theory we have no idea what the analog of QCD is.

If one wants to describe the whole of nature, three obvious defects of the bosonic string are the dimensionality of spacetime, the presence of a tachyon, and the absence of spacetime fermions. The last two of these can be cured, modifying the worldsheet physics and introducing worldsheet supersymmetry. Thus, for each of the X's, there is a corresponding worldsheet spinor, ψ^a, which is a superpartner of the X^a. One must also introduce a worldsheet gravitino to act as the superpartner of the worldsheet metric, although it can be set to zero by an appropriate choice of gauge for cylindrical worldsheets. In such a gauge, and where the worldsheet metric is taken to be flat, the action is

$$I = \frac{1}{4\pi\alpha'} \int \gamma^{\frac{1}{2}} d^2\xi \left(\partial_\mu X^a \partial_\nu X^b \gamma^{\mu\nu} - i\bar\psi^a \gamma^\mu \partial_\mu \psi^b \right) \eta_{ab}. \tag{130}$$

One can then check that under a supersymmetry transformation with a constant worldsheet spinor ϵ, the action is invariant under

$$\delta X^a = \bar\epsilon \psi^a \qquad \delta \psi^a = -i\gamma^\mu \partial_\mu X^a \epsilon. \tag{131}$$

In these formulae, the gamma matrices are the usual gamma matrices defined on the worldsheet.

Quantization of the superstring is rather complicated, and we will not go into detail here. Rather, we focus on the principal differences between the bosonic string and the superstring. Again, we consider here only closed superstrings. The first difference is that the dimensionality of spacetime turns out to be ten rather than 26. Secondly, the spectrum of particles is rather different. The theory does not contain a tachyon, and the massless states now contain both fermions and bosons. There are in fact two distinct types of string theory in ten dimensions usually called the type IIA string and type IIB string. The massless states of the type IIA string are, for the bosons: a graviton, a scalar usually called the dilaton, and 1-form, 2-form and 3-form potentials for 2-form, 3-form and 4-form field strengths. In addition, there are two Majorana-Weyl gravitinos and two spin-1/2 fermions of opposite chirality. The massless states of the type IIB string are: a graviton, a scalar, a pseudoscalar, two 2-form potentials giving rise to 3-form field strengths, and a 4-form potential that gives rise to a self-dual 5-form field strength. The fermions of the theory are the same as before except that they all have the same chirality. Again, both types of string theory have an infinite number of massive states and an associated Hagedorn temperature.

11 Background fields

The existence of fields that have the quantum numbers of the graviton is far from being sufficient evidence that the theory is a theory of gravitation with a curved spacetime. In fact, one knows that Lorentz invariant theories of gravitation based on flat spacetime are inconsistent with the observation of gravitational redshifting. For this reason, we will now investigate how strings propagate in backgrounds where some of the massless string fields have classical expectation values. We will focus for the moment on those fields which are common to all string theories, namely the gravitational field described a symmetric spacetime tensor g_{ab}, a 2-form potential B_{ab} and a scalar dilaton Φ. A string propagating in a background of these fields is described by a modest extension of the Polyakov form of the action,

$$I = \frac{1}{4\pi\alpha'} \int \sqrt{\gamma} d^2\xi \left(g_{ab}(X) \partial_\mu X^a \partial_\nu X^b \gamma^{\mu\nu} + B_{ab}(X) \partial_\mu X^a \partial_\nu X^b \epsilon^{\mu\nu} - \frac{1}{2} \alpha'\,^{(2)}R\Phi(X) \right) \tag{132}$$

where $\epsilon_{\mu\nu}$ is the alternating tensor on the string worldsheet, and $^{(2)}R$ is the Ricci scalar formed from the metric $\gamma_{\mu\nu}$.

This formula shows that the physical meaning of the dilaton is more than simply that of a scalar field in spacetime. Suppose that one is in a region of spacetime in which the dilaton takes a constant value Φ_0. Then, the dilaton part of the action is just

$$-\frac{1}{8\pi}\Phi_0 \int \sqrt{\gamma}\,^{(2)}R. \tag{133}$$

By virtue of the two-dimensional Gauss-Bonnet formula, the action becomes

$$-\frac{1}{2}\Phi_0 \chi \tag{134}$$

where χ is the Euler character of the string worldsheet. For a surface with genus g and e ends, χ is given by

$$\chi = 2 - 2g - e. \tag{135}$$

Thus, for a fixed number of ends, each time the genus increased by one unit the action picks up an additive factor of Φ_0. Thus, for each addition of a string loop, the path integral picks up a factor of $e^{-\Phi_0}$. Hence, $\mu = e^{-\Phi_0}$ can be viewed as the string loop coupling constant.

This form of the action has to lowest order in α' the property of conformal invariance, but the term involving the dilaton, and which is of higher order in α', does not. On the face of it, this means that this new form of the action violates one of the symmetries that the string is supposed to possess. In this picture however, the spacetime fields g_{ab}, B_{ab}, and Φ are all X-dependent, and that means that X is no longer a free quantum field. In fact these variables all act as coupling constants in the theory. It is well-known that if one attempts to quantize interacting conformally invariant fields, the conformal symmetry will in general be violated by the quantum corrections to the classical theory. Lastly, if we try to quantize the action given here, it should be noted that α' is the loop counting parameter for the loop expansion of the fields. Thus, we can ask for the conditions that the string maintain conformal invariance order-by-order in powers of α'. This is equivalent to asking that the beta-functions of the renormalization group for the coupling constants of the theory, namely g_{ab}, B_{ab} and Φ all vanish (Callan et al. 1985). These have been calculated and the results, to lowest order in α' are

$$\beta_g: \quad R_{ab} - \nabla_a \nabla_b \Phi + \frac{1}{4} H_{acd} H_b{}^{cd} = O(\alpha') \tag{136}$$

$$\beta_B: \quad \nabla_a H^{abc} + \nabla_a \Phi H^{abc} = O(\alpha') \tag{137}$$

and

$$\beta_\Phi: \quad \frac{2}{3}\frac{(d-26)}{\alpha'} + (-R + \nabla_a \Phi \nabla^a \Phi + 2\Box\Phi + \frac{1}{12} H_{abc} H^{abc}) = O(\alpha') \tag{138}$$

where H_{abc} is the field strength coming from the potential B_{ab} and is given by

$$H_{abc} = \partial_a B_{bc} + \partial_b B_{ca} + \partial_c B_{ab}. \tag{139}$$

These equations are the stringy version of Einstein's equations, and the terms higher order in α' can be regarded as being the stringy corrections to Einstein's theory. Although on dimensional grounds these extra terms will have higher derivatives in them, they should not give any cause for concern about the Cauchy problem. The reason for this is that in this picture all of the infinite number of massive modes of the string have been effectively integrated out. This will always cause apparently higher derivative type terms to appear in the action; a familiar example from the theory of quantum electrodynamics being the Euler-Heisenberg action. Such additional terms do not cause fundamental problems for the theory.

One key point to realize here is that Einstein's equations, together with the field equations for the dilaton and B_{ab}, have been derived in string theory from a symmetry principle. The Einstein equations together with the other field equations have come about by demanding that the string be conformally invariant. It should also be noticed that the restriction on dimension here has been somewhat loosened.

One might worry about whether these equations are consistent with the Bianchi identities since it is the Ricci tensor and not the Einstein tensor that appears in the analog of the Einstein equation. It turns out that the Bianchi identities, the Einstein equation and the field equation for B imply that the gradient of the dilaton equation of motion must vanish. Thus, these equations form a consistent set.

Another way of coming to the same conclusion is to ask if it is possible to derive these equations from an effective spacetime action. The action

$$S_S = \int d^d x g^{\frac{1}{2}} e^{\Phi} \left[\frac{2(d-26)}{3\alpha'} - R - \nabla_a \Phi \nabla^a \Phi + \frac{1}{12} H_{abc} H^{abc} \right] \qquad (140)$$

when varied with respect to g_{ab}, B_{ab} and Φ reproduces the three beta-function equations. (We are ignoring possible boundary terms here.) This does not look much like the action for Einsteinian gravitation coupled to a 2-form and a scalar because of the overall presence of the exponential of the dilaton. Such terms are however typical of gravitational theories with massless scalars. The exponential can be eliminated by making a spacetime conformal transformation taking us from what is usually called the string conformal frame to the Einstein conformal frame. Thus, if we transform the metric by

$$g_{ab} \to e^{-\frac{2\Phi}{d-2}} g_{ab} \qquad (141)$$

we can rewrite the action in the more familiar form

$$S_E = \int d^d x g^{\frac{1}{2}} \left[\frac{2(d-26)}{3\alpha'} e^{-\frac{2\Phi}{d-2}} - R + \frac{1}{d-2} \nabla_a \Phi \nabla^a \Phi + \frac{1}{12} e^{\frac{4\Phi}{d-2}} H_{abc} H^{abc} \right]. \qquad (142)$$

This kind of treatment can be extended to the superstring although the calculations become somewhat more involved. We will not investigate the details here.

12 T-duality

T-duality is an exact symmetry in string theory. The basic result is that certain pairs of spacetimes result in the same string theory. This can be thought of as a hidden symmetry, and constitutes a derivation of what in the relativity literature is usually called Ehlers' symmetry. Suppose that one has a string theory described in some spacetime that satisfies the beta function equations, and also has a Killing vector k^a. Suppose that the Killing vector is $\frac{\partial}{\partial x^0}$, so that the metric, dilaton and B_{ab} can all be chosen to be independent of x^0. We can now decompose the coordinates so that $x^a = (x^0, x^i)$. Using this decomposition, the string action is

$$I = \frac{1}{4\pi\alpha'} \int d^2 \xi \sqrt{\gamma} \{ \gamma^{\mu\nu} \left(g_{00} V_\mu V_\nu + 2 g_{0i} V_\mu \partial_\nu X^i + g_{ij} \partial_\mu X^i \partial_\nu X^j \right) \qquad (143)$$
$$+ \epsilon^{\mu\nu} \left(\hat{x}^0 \partial_\mu V_\nu + 2 B_{0i} V_\mu \partial_\nu X^i + B_{ij} \partial_\mu X^i \partial_\nu X^j \right) - \frac{1}{2} \alpha'^{(2)} R \Phi \}.$$

This is really the string action written in first order form. One can recover the original form of the action in two different ways. The first is to look at the equation of motion

for $\hat{x}^0$. It is
$$\epsilon^{\mu\nu}\partial_\mu V_\nu = 0. \tag{144}$$
This can be solved by writing $V_\nu = \partial_\nu X^0$. If this is substituted back into the action, then the original action is recovered.

A second possibility is to carry out the integration over the field V_μ as there are no derivatives of V_μ in the action (Buscher 1987, 1988). Solving for V_μ leads to
$$2g_{00}V^\mu b = -2g_{0i}\gamma^{\mu\nu}\partial_\nu X^i + \epsilon^{\mu\nu}\partial_\nu \hat{x}^0 - B_{0i}\epsilon^{\mu\nu}\partial_\nu X^i. \tag{145}$$
Substituting this back into the action for V_μ leads to the same functional form as the original action, but with the g_{ab} and B_{ab} replaced by $\hat{g}_{ab}$ and $\hat{B}_{ab}$, where
$$\hat{g}_{00} = \frac{1}{g_{00}}, \qquad \hat{g}_{0i} = \frac{B_{0i}}{g_{00}}, \tag{146}$$
$$\hat{g}_{ij} = g_{ij} - \frac{1}{g_{00}}\Big(g_{0i}g_{0j} - B_{0i}B_{0j}\Big) \tag{147}$$
and
$$\hat{B}_{0i} = \frac{g_{0i}}{g_{00}}, \qquad \hat{B}_{ij} = B_{ij} + \frac{1}{g_{00}}\Big(g_{0i}B_{0j} - B_{0i}g_{0j}\Big). \tag{148}$$
Coming from the measure in the path integral of the string, there is an anomalous shift in the dilaton which is entirely of quantum mechanical origin. This amounts to shifting Φ to $\hat{\Phi}$ where
$$\hat{\Phi} = \Phi + \frac{1}{2}\ln|g_{00}|. \tag{149}$$
From the point of view of string physics, the unhatted background is therefore precisely equivalent to the hatted background. The dual two backgrounds are said to be T-dual to each other. If one carries out the T-duality operation twice, one recovers the original background. A tedious calculation will reveal that if the unhatted background satisfies the beta function equations, then the hatted background will also. However, from the stringy point of view, this transformation is simply a trivial field redefinition.

13 Two-dimensional black holes in string theory

In two spacetime dimensions, it is easy to find solutions of the beta function equations. The first is the linear dilaton background which is just Minkowski spacetime
$$ds^2 = -dt^2 + dr^2 \tag{150}$$
with vanishing B_{ab} and a linear dilaton
$$\Phi = \Phi_0 + \lambda r \tag{151}$$
where Φ_0 is an arbitrary constant, and λ is fixed by the dilaton equation of motion.

A second solution (Witten 1991) is a black hole. The metric here is
$$ds^2 = -\tanh^2\lambda r\, dt^2 + dr^2 \tag{152}$$

with the dilaton given by $\Phi = \Phi_0 + \ln \cosh \lambda r$ and vanishing B_{ab}. Again, λ is determined by the dilaton equation of motion and takes the same numerical value as in the linear dilaton background. In this solution $\frac{\partial}{\partial t}$ is a Killing vector. The surface $r = 0$ is a Killing horizon. As $r \to \infty$, the metric and dilaton approach the linear dilaton background.

One can define analogs of Kruskal-Szekeres coordinates in this spacetime and then one finds a maximal analytic extension precisely analogous to the maximal extension of the Schwarzschild solution. The metric in such a coordinate system is

$$ds^2 = \frac{dU\,dV}{1 - UV}. \tag{153}$$

The horizons are the surfaces $U=0$ or $V=0$. The surfaces $UV=1$ are the past and future singularities.

If one applies the T-duality operation to the original metric exterior to the horizon, one discovers a new spacetime with metric

$$ds^2 = -\coth^2 \lambda r\, dt^2 + dr^2 \tag{154}$$

and with dilaton $\Phi = \Phi_0 + \ln \sinh \lambda r$. This spacetime is also asymptotic to the linear dilaton background but, instead of being a black hole, the surface $r = 0$ is a naked singularity. We have discovered that as far as string theory is concerned, there is no difference between the horizon and this naked singularity.

One can also find the metric in the region interior to the black hole in static form. It is

$$ds^2 = -\tan^2 \lambda r\, dt^2 + dr^2 \tag{155}$$

with $\Phi = \Phi_0 + \ln \cos \lambda r$. The surface $r = 0$ is the horizon, and $r = (\pi/2\lambda)$ is the singularity. Applying T-duality to this results in the metric and dilaton becoming

$$ds^2 = -\cot^2 \lambda r\, dt^2 + dr^2 \tag{156}$$

and $\Phi = \Phi_0 + \ln \sin \lambda r$. The effect of T-duality has been to effect a coordinate transformation

$$r \to r + \frac{\pi}{2\lambda} \tag{157}$$

which swaps the horizon and the singularity. The physical meaning of this is presently unclear, but holds promise for a future understanding of some of the problems associated with black hole physics.

It is believed that this metric is exact for the case of the superstring. However, for the bosonic string, the exact metric (as opposed to that which is a solution of the lowest order form of the beta function equations) is slightly different in an interesting way (Perry and Teo 1993). Here the metric is given by

$$ds^2 = \frac{-dt^2}{\coth^2 \lambda r - (8/9)} + dr^2. \tag{158}$$

This spacetime has quite different global structure to the previous one. Whilst there are still horizons at $r=0$, it turns out that this spacetime does not contain any singularities. Replacing the singularity in both the past and future is a surface of time symmetry on which the dilaton blows up. The resultant maximal extension is then an infinite series of asymptotically flat regions linked by a series of throats containing horizons. The overall picture is somewhat similar to the Reissner-Nordstrom black hole metrics, but without any timelike singularities. One might worry about the surface on which Φ is blowing up. However, within the context of string theory, since Φ determines the string loop coupling constant, it presumably indicates that on these surfaces of time symmetry, strong coupling string physics takes over. We do not understand the physics of such a regime. However it would appear to be more benign than a singularity where spacetime has a boundary.

14 S-duality

S-duality is a conjectured symmetry of string theory which has a field theoretic remnant that can be seen in the effective action for string theory. We will look at the simplest example first. In ten dimensions, the type IIB superstring displays S-duality (Julia 1985). Consider only that part of the action that describes the gravitational field, the scalar and the pseudoscalar. The spacetime action that describes their physics is, in the Einstein conformal frame,

$$S = \int g^{\frac{1}{2}} \, d^{10}x \left[-R + \frac{1}{2} \nabla_a \Phi \nabla^a \Phi + \frac{1}{2} e^{2\Phi} \partial_a \psi \partial^a \psi \right]. \tag{159}$$

One can straightforwardly unify the pseudoscalar ψ and the dilaton into a complex scalar field

$$\lambda = \psi + i e^{\Phi}. \tag{160}$$

The action becomes

$$S = \int g^{\frac{1}{2}} \, d^{10}x \left[-R + 2 \frac{\partial_a \lambda \partial^a \bar{\lambda}}{(\lambda - \bar{\lambda})^2} \right]. \tag{161}$$

In the field theoretic version of S-duality, the action is invariant under an $SL(2,\mathbf{R})$ transformation,

$$\lambda \to \frac{a\lambda + b}{c\lambda + d} \qquad ad - bc = 1 \quad a,b,c,d \in \mathbf{R} \tag{162}$$

In string theory, it is believed that this symmetry still holds but is broken down to an $SL(2,\mathbf{Z})$ symmetry, so that a,b,c and d are now all integers. This is a remarkable symmetry since, if it holds, it would relate strong string coupling to weak coupling. Recall that the dilaton is the string coupling constant. Amongst the transformations is that with $a = d = 0$ and $b = -c = 1$ that sends Φ to $-\Phi$. The string coupling constant μ is thus transformed to $1/\mu$. There is growing evidence for such a symmetry which would necessarily be connected to some interesting non-perturbative phenomena in string theory. This is an area of very active current research.

Acknowledgement

This work has been supported in part by the US Department of Energy under grant DE-FG03-91ER40546.

References

Atick J J and Witten E, 1988, *Nucl Phys*, **B310**, 291.
Buscher T H, 1987, *Phys Lett*, **B194**, 59.
Buscher T H, 1988, *Phys Lett*, **B201**, 466.
Callan C G, Friedan D, Martinec E and Perry M J, 1985, *Nucl Phys*, **B262**, 593.
Christensen S M, Duff M J, Gibbons G W and Rocek M, 1980, *Phys Rev Lett*, **45**, 161.
Gibbons G W and Hawking S W, 1977, *Phys Rev* **D15**, 2738.
Gibbons G W and Hull C M, 1982, *Phys Lett*, **109B**, 190.
Goroff M H and Sagnotti A, 1986, *Nucl Phys.*, **B266**, 709.
Green M B, Schwarz J H and Witten E, 1987, *Superstring Theory*, Vols I and II, Cambridge University Press, Cambridge.
Hawking S W, 1975, *Comm Math Phys*, **43**, 199.
Hawking S W, 1977, *Phys Rev*, **D14**, 2460.
Hawking S W, Horowitz G T and Ross S F, 1995, *Phys Rev*, **D51**, 4302.
Henneaux M and Teitelboim C, 1992, *The Quantization of Gauge Fields*, Princeton University Press, Princeton, New Jersey.
't Hooft G and Veltman M, 1970, *Ann Inst H Poincaré*, **A20**, 69.
Huang K and Weinberg S, 1970, *Phys Rev Lett*, **25**, 895.
Isham C J, 1984, in *Relativity, Groups and Topology II*, eds de Witt B S and Stora R, North-Holland, Amsterdam.
Julia B, 1985, 'Kac-Moody symmetry of Gravitation and Supergravity Theories,' in Lectures in Applied Mathematics, **21**, American Mathematical Society.
Kakas A and Isham C J, 1984, *Class and Quant Grav*, **1**, 621, 633.
Linden N and Perry M J, 1991, *Nucl Phys*, **B357**, 289.
Perry M J and Teo E H K, 1993, *Phys Rev Lett*, **70**, 2669.
Polyakov A M, 1981a, *Phys Lett*, **103B**, 207.
Polyakov A M, 1981b, *Phys Lett*, **103B**, 211.
Spivak M, 1979, *A Comprehensive Introduction to Differential Geometry*, Publish or Perish Inc, Berkeley.
de Witt B S, 1967a, *Phys Rev*, **160**, 1113.
de Witt B S, 1967b, *Phys Rev*, **162**, 1195.
de Witt B S, 1967c, *Phys Rev*, **162**, 1239.
Witten E, 1991, *Phys Rev*, **D44**, 314.
York J W, 1972, *Phys Rev Lett*, **28**, 1082.

Poster Abstracts

A poster session was held in the second week of the SUSSP46 Summer School in which the students at the School were given an opportunity to report on their own research. This section provides abstracts of the material displayed in the session. Addresses of the authors can be found at the end of this volume. Where the papers or work reported on involve collaboration with people not present at the Summer School these additional authors are given in brackets.

- **Black holes, boxes, and the Boulware state**
Warren Anderson
Far from a black hole of mass M, fill a box of dimension $\ell \ll M$ with matter having energy E_0 and entropy S. If the box is slowly lowered as close to the black hole as possible, and the matter released, it will increase the black hole entropy by ΔS_{BH} $8\pi M E_0 \sqrt{1 - 2M/r(\ell)}$. In 1981 Bekenstein, who first postulated this gedanken experiment in classical black hole physics, pointed out that, *a priori*, ΔS_{BH} can be less than S, violating the second law of thermodynamics. The second law can, however, be saved by considering quantum effects, as demonstrated by Unruh and Wald in 1982. I review these arguments as well as presenting a new perspective on the quantum state inside the box. Also, I use the equivalence principle to show an exact correspondence between quantum effects in Minkowski and Schwarzschild spacetimes.

- **Asymptotics of monopole solutions of Bogomolny equations with solvable gauge group.**
Arthur M. Aslanyan
We investigated the asymptotic behaviour of monopole solutions of Bogomolny equations with solvable gauge group. We say the solution is of monopole type if it is regular in all of 3-space and has a finite energy. For a compact gauge group any monopole solution is asymptotically flat. If we want to choose a non-compact gauge group we should have a non-degenerate invariant bilinear form on the corresponding Lie algebra, which is a restriction on the structure of the algebra. We deal with the so-called Generalized Harmonic Oscillator Lie algebra GHO(N,R), which is quasi-nilpotent and admits such a form, so we can define a notion of energy. But in this case there is no restriction on the existence of asymptotically non-flat monopoles. Our main result is that any asymptotically flat solution with GHO(N,R) as gauge group is trivial. This can be deduced by the investigation of the asymptotic behaviour of Bogomolny equations. So the main interest is concentrated on the investigation of non-trivial asymptotic behaviour of solutions, which necessarily contain divergencies at infinity. The divergencies are of exponential type, because the radial part of the gauge field is proportional to the Infeld function. At present we have one result concerning non-trivial asymptotics: *In the case of complete integrability of the Bogomolny equation there is just one monopole solution—the trivial one.* Note that in the general case the Bogomolny equation is not completely integrable.

- **QED blue-sheet effects inside black holes**
  Lior Burko
  The cosmic background radiation photons which are captured by a Reissner-Nordström black hole are infinitely blue-shifted at the inner horizon. We evaluate the flux of incident photons hitting an infalling object attempting to traverse the inner horizon. The probability for interactions is dominated by pair-production in the nuclei's field. For a microscopically small object, this probability is extremely small. Larger objects are heated by momentum transfer during the pair-production processes. This effect is dominated by the interactions with the atomic electrons. Extrapolating QED cross section up to a cutoff introduced at Planck energy, we evaluate the increase in temperature. For typical parameters this increase is bounded and small. However, it is fatal for a human being. Other objects, of typically smaller size, might survive it and arrive at the inner horizon. Our results do not provide support to the idea that the inner horizon is a non-traversable wall.

- **Stability Of the Cauchy horizon in Kerr-de Sitter spacetimes**
  Chris Chambers
  In this poster we present a program of work aimed at studying the stability of the Cauchy horizon (CH) in Kerr–de Sitter spacetimes. As in the Reissner–Nordström–de Sitter black hole, the CH is stable provided that the surface gravity at the CH is less than that at the cosmological event horizon, *i.e.* $\kappa_{\rm CH} < \kappa_{\rm CEH}$. The space of parameters (a, M, Λ) for which this condition holds is obtained and found to be of a small but non-zero measure. Though a non-zero cosmological constant is unphysical it does not violate any known law of physics, and is demanded in some models of the early universe. As such, the problem, at the least, provides a valid gedanken experiment in which the predictive power of General Relativity breaks down, Cosmic Censorship is violated and closed timelike curves are accessible to an observer.

- **Covariant double null dynamics of Einstein gravity**
  Serge Droz and Sharon Morsink (with P R Brady and W Israel)
  In this poster a new $2+2$ null splitting formalism of four dimensional spacetime is presented. The present formalism explicitly maintains four- and two-dimensional covariance while operating on objects with a direct geometrical meaning. As a consequence Einstein's equations can be written in a concise and transparent form: the ten components of the Ricci tensor can be written as a set of only three two-dimensionally covariant equations.

- **Global topological defects & anisotropies of the cosmic microwave background**
  Alejandro Gangui
  We review recent work aimed at showing how global topological defects influence the shape of the angular power spectrum of the cosmic microwave background radiation on small scales. While Sachs–Wolfe fluctuations give the dominant contribution on angular scales larger than about a few degrees, on intermediate scales, $0.1° \lesssim \theta \lesssim 2°$, the main role is played by coherent oscillations in the baryon radiation plasma before recombination. In standard cosmological models these oscillations lead to the 'Doppler peaks' in the angular power spectrum. Inflation-based cold dark matter models predict the location of the first peak to be at approximately $220/\sqrt{\Omega_0}$, with a height which is a few times the amplitude at large scales. Here we focus on perturbations induced by global textures. We find that the first Doppler peak is reduced to an amplitude comparable to that of the Sachs–Wolfe contribution, and that it is shifted to $\ell \sim 350$. We briefly comment on the relation between our results and what open models predict. This report is based on joint work done in collaboration with R. Durrer and M. Sakellariadou.

- **Constraining analytic topological defect models with the CMB collapsed three–point function**
Alejandro Gangui (with Silvia Mollerach)
   The analytic modelling of topological defects, like cosmic strings and textures, captures the basic features of their cosmological predictions. We review recent work done towards finding to what extent cosmic microwave background (CMB) non–Gaussian signatures can further constrain these models. We derive both analytic expressions and numerical estimates for the collapsed three–point correlation function in the anisotropies of CMB temperature for cosmic strings and textures, as well as for their RMS collapsed three–point function (cosmic variance). Applying our results to the COBE-DMR two–year data we show that the non-Gaussian features of the CMB predicted in each case are consistent with the theoretical uncertainties present in the maps, and therefore large scale anisotropies alone cannot place further constraints on these models.

- **Vacuum energy in ultralocal metrics for TT tensors with Gaussian wave functionals**
Remo Garattini
   Using variational methods we calculate, in a class of gauge invariant functionals, the difference of vacuum energy between two different backgrounds: Schwarzschild and Flat Space. We perform this evaluation in a Hamiltonian formulation of Quantum Gravity by a standard '3 + 1' decomposition. After the decomposition the scalar curvature is expanded up to second order with respect to the Schwarzschild metric. We evaluate this energy difference in a momentum space, in the lowest possible state regardless (in this paper) of any negative mode. We find a singular behaviour due to the horizon presence in the case of UV limit only if $r = 2m$, otherwise for $r > 2m$ this singular behaviour disappears in agreement with various models presented in the literature.

- **Initial data for general relativity with toroidal conformall Symmetry**
Sascha Husa
   A new class of time-symmetric solutions to the initial value constraints of vacuum General Relativity is introduced. The data is globally regular, asymptotically flat (with possibly several asymptotic regions) and in general has *no isometries*, but a $U(1) \times U(1)$ group of *conformal isometries*. The manifold might also have a boundary that is left invariant by the conformal isometries. In this case an especially interesting boundary condition is that of vanishing mean curvature. After decomposing the Lichnerowicz conformal factor in a double Fourier series on the group orbits, the solutions are given in terms of a countable family of uncoupled ODE's on the orbit space. Existence of positive solutions can be checked by computing the sign of the first eigenvalue of the conformal Laplacian, which is a pure ODE problem in our case.
   This work was supported by FWF, Project Number P09376-PHY.

- **A strange solution of vacuum Einstein equations**
  Radu Ionicioiu (with Dan Şelaru)

  An exact solution of Einstein vacuum equations $R_{\mu\nu} = 0$ is proposed. The metric in cylindrical coordinates (t, r, z, φ), has the form:

  $$ds^2 = -\frac{a z + b}{r^2} dt^2 + \frac{\delta}{r^4} e^{-4z/a} dr^2 + \frac{\delta}{r^2 (a z + b)} e^{-4z/a} dz^2 + r^2 d\varphi^2$$

  with a, b and δ constants, $a \neq 0$, $\delta > 0$. The metric is Petrov type I, static, axisymmetric and has some unusual properties. The plane $z = -b/a$ is a null surface and acts as a one-way membrane; only future directed null and timelike geodesics can cross it from $z > -b/a$ to $z < -b/a$. Crossing this surface, a signature change occurs $(-+++) \to (++-+)$, the time coordinate becomes spacelike and the z-coordinate timelike. Unlike the Schwarzschild case where the corresponding horizon $r = 2M$ is closed (a 2-sphere S_2), here the surface $z = -b/a$ is a plane and therefore open. The speed of light increases indefinitely for $z \to +\infty$, and therefore the point $z = +\infty$ can be reached in a finite time by a test particle. For the special cases z=const and r=const we derive the geodesics and we analyze the corresponding orbits.

- **'Microcanonical' actions, Noether charges and black hole entropy**
  Vivek Iyer (with Robert M Wald)

  Consider an arbitrary diffeomorphism-invariant Lagrangian gravity theory. We introduce a microcanonical action, I_M and a quantity E, interpreted as a quasilocal energy for the theory. The action I_M, is 'microcanonical' in the sense that (given the theory satisfies certain technical conditions) I_M is extremised when varied around a solution to the equations of motion, provided E (along with a small number of other quantities) is fixed by the variations on the (timelike) boundary of the spacetime. When evaluated on stationary 'Euclidean' black hole solutions, I_M yields the Noether-charge entropy of the black hole; this generalises the results of Brown and York to a much wider class of diffeomorphism-invariant theories. Work remains to be done: we have yet to find boundary conditions on fields and their variations (in higher order gravity theories) which are 'natural' and still allow the definition of E. Morover the meaning and severity of the technical conditions guaranteeing that I_M is indeed 'microcanonical' are poorly understood. Work on these issues is in progress.

- **On the existence of C^∞ solutions of the asymptotic characteristic initial value problem in general relativity**
  János Kánnár

  In the case of the asymptotic characteristic initial value problem in general relativity the initial values (in our case they are C^∞) are given on a null hypersurface and on a part of null infinity, which intersect in a two-dimensional spacelike surface diffeomorphic to S^2 (the result is true for more complicated topologies). We prove the existence and uniqueness of C^∞ solutions in a neighbourhood of the intersection surface. Our method is based on the following results. The conformal Einstein vacuum field equations are (at least formally) singular over null infinity. To avoid this problem Friedrich derived an equivalent regular system, the 'reduced conformal vacuum field equations'. This system is first order quasilinear and symmetric hyperbolic. For this kind of equation Rendall has given a general method, with the help of which one can trace the existence and uniqueness of the C^∞ solutions of the characteristic initial value problem back to that of the ordinary Cauchy problem, which is proved for general first order quasilinear symmetric hyperbolic systems. The proof is essentially based on this method, but requires careful analysis of the equations and consideration of the topology of the problem.

- **A relation between quantization of a relativistic particle and scalar field theory**
  Pavel Krtous
  A relation between path integral quantization of a relativistic particle and scalar field theory is investigated. It is shown that both theories give the same predictions and so it is possible to reconstruct one of the theories from the other. During a derivation of this equivalence different propagators for relativistic particle restricted to a spacetime domain are defined, including 'the first crossing propagator' and a new derivation of the formula for this propagator is presented.

- **Point splitting calculations using *Mathematica* and *MathTensor***
  Pavel Krtous
  Most bitensors (e.g. Green functions, HaMiDeW coefficients or the heat kernel of a wave operator) in a general curved background are not computable exactly. But often it is enough to know a coincidence limit or a covariant expansion of these objects. Derivation of such quantities involves straightforward, but difficult and long computations. It would be helpful to have a table of the quantities involved in these calculations. A partial copy of my World Wide Web pages containing such tables is presented here. Most derivations were done on a computer using *Mathematica* and *MathTensor*. The WWW pages also contain *Mathematica* packages which can be used for further point-splitting calculations on a computer. WWW pages can be found on URL:
  http://fermi.phys.ualberta.ca/krtous/Mathematica/pointsplitting/

- **Entropy and topology for manifolds with boundaries**
  Stefano Liberati and Giuseppe Pollifrone
  In this work a deep relation between the topological and thermodynamical features of manifolds with boundaries is shown. The expression for the Euler characteristic, through the Gauss-Bonnet integral, and the one for the entropy of gravitational instantons are proposed in a form which makes the relation between them self-evident. We obtain a generalization of the Bekenstein-Hawking formula, in which the entropy and Euler characteristic are related in the form $S = \chi A/8$. This formula also gives the correct result for extreme black holes, where the Bekenstein-Hawking one fails ($S=0$ but $A \neq 0$). In such a way it recovers a unified picture for the black hole entropy law. Moreover, it is proved that such a relation can be generalized to a wide class of manifolds with boundaries which are described by spherically symmetric metrics (e.g. Schwarzschild, Reissner-Nordström, static de Sitter).

- **'Why two makes it more exciting than one' or quantum Bianchi models in N=2 supergravity**
  Paulo Moniz (with A D Y Cheng)
  The theory of $N=2$ supergravity is applied to Bianchi models. Their canonical formulation is addressed for the case which contains a *global* $O(2)$ internal symmetry. No cosmological constant or mass-like term for the gravitinos is required here. The presence of the Maxwell field in the supersymmetry equations leads to a non-conservation of fermionic number. The possible quantum physical states are discussed regarding other results using Ashtekar variables. This work was supported in part by the Croucher Foundation of Hong Kong (ADYC) and a Human Capital and Mobility Fellowship from the European Union.

- **One-loop quantum amplitudes in Euclidean quantum gravity**
  Guiseppe Pollifrone (with G Esposito, A Yu. Kamenshchik and I V Mishakov)

  This paper studies the linearized gravitational field in the presence of boundaries. For this purpose, zeta-function regularization is used to perform the mode-by-mode evaluation of BRST-invariant Faddeev-Popov amplitudes in the limiting case of flat Euclidean four-space bounded by only one three-sphere. On choosing the de Donder gauge-averaging term, the resulting $\zeta(0)$ value is found to agree with the space-time covariant calculation of the same amplitudes, which relies on the recently corrected formulae for the asymptotic heat kernel in the case of mixed boundary conditions. Two sets of mixed boundary conditions for Euclidean quantum gravity are then compared in detail. The whole analysis proves that one cannot restrict the path-integral measure to transverse-traceless perturbations. By contrast, gauge-invariant amplitudes are only obtained by considering from the beginning all perturbative modes of the gravitational field, jointly with ghost modes.

- **Perfect fluid space-times admitting a 3-dimensional conformal group**
  Alicia Sintes (with J Carot)

  Perfect fluid space-times admitting a 3-parameter Lie group of conformal motions containing a 2-dimensional Abelian subgroup of isometries are studied. We assume that the Killing orbits are spacelike, diffeomorphic to $\mathbf{R}^2$ and admit orthogonal 2-surfaces. Demanding that the conformal Killing vector be proper, all such space-times are classified according to the structure of their corresponding 3-dimensional conformal Lie algebra and the nature of their corresponding orbits (timelike, spacelike or null). Each metric may then be explicitly displayed in coordinates adapted to the symmetry vectors. In the case of non-null orbits, attention is restricted to the diagonal case, and exact perfect fluid solutions are obtained in the cases in which the 4-velocity is tangential or orthogonal to the conformal orbits, as well as in the more general 'tilting' case. In the null case, not all the algebraic structures are possible and perfect fluid solutions exist in only a few cases.

- **Changing the spatial topology in time**
  Lukas Ziewer

  The path-integral approach turns out to provide both motivation and tools for topology change. In canonical gravity, though, the mere kinematics of such events is problematical; the configuration space comprises the Riemannian metrics on a fixed 3-manifold. Therefore, topologically different manifolds lead, *prima facie*, to separate systems. We propose the following method to join the configuration spaces of the Bianchi I model and thereby allow topology-changing trajectories: Certain degenerate metrics in a configuration space can be viewed as metrics on submanifolds of the fixed 3-manifold. When degenerate metrics on different 3-manifolds yield isometric Riemannian submanifolds, we postulate that the respective configuration spaces are 'linked' at these elements. Unfortunately, however, such a postulate is highly speculative, and the dynamics appears to be flawed by ambiguities. Moreover, our construction of unique Riemannian submanifolds associated with degenerate metrics depends on the classification of flat manifolds.

Participants

- Mr Warren G Anderson
  Department of Physics
  University of Alberta
  Edmonton
  Alberta T6G-2J1
  Canada

- Mr Fredrik Andersson
  Department of Mathematics
  Linköping University
  58183 Linköping
  Sweden

- Mr Marcus Ansorg
  School of Mathematical Sciences
  Queen Mary and Westfield College
  University of London
  Mile End Rd
  London E1 4NS
  UK

- Mr Arthur M Aslanyan
  Dept of General Relativity and Gravity
  Kazan State University
  Lenina Str. 18
  Kazan 420008
  Russia

- Mr John G Baker
  104 Davey Lab
  Penn State University
  University Park
  PA 16802
  USA

- Dr Marcelo S Berman
  INPE – Divisão de Astrofísica
  Caixa Postal 515
  São José Dos Campos – SP
  12201-970
  Brazil

- Dr Udayan Bhaduri
  University of Miami
  Coral Gables
  FL 33124
  USA

- Mr Paul A Blaga
  University of Cluj
  Faculty of Mathematics
  1, Kogalniceanu Street
  3400 Cluj-Napoca
  Romania

- Dr Antonella Borrelli
  Università degli Studi di Pavia
  Via Cocco Ortu 14
  00139 Roma
  Italy

- Mr Georges Bressange
  Laboratoire de Modèles de Physique Mathematique
  Faculté des Sciences et Techniques de Tours
  Parc de Grandmont
  37200 Tours
  France

- Dr Peter Bueken
  Department of Mathematical Sciences
  University of Aberdeen
  Edward Wright Building
  Dunbar Street
  Aberdeen AB24 3QY
  Scotland

- Mr Lior M Burko
  Department of Physics
  Technion-Israel Institute of Technology
  32000 Haifa
  Israel

- Mr Paweł Caban
  Department of Theoretical Physics
  University of Łódź
  Pomorska 149/153
  90-236 Łódź
  Poland

- Mr Michael Capocci
  Department of Mathematical Sciences
  University of Aberdeen
  Edward Wright Building
  Dunbar Street
  Aberdeen AB24 3QY
  Scotland

- Prof Bernard Carr
  School of Mathematical Sciences
  Queen Mary and Westfield College
  University of London
  Mile End Road
  London E1 4NS
  UK

- Ms Carmen Casares Antón
  Dept Física Teórica II
  Facultad de Ciencias Físicas
  Universidad Complutense de Madrid
  28040 Madrid
  Spain

- Mr Chris M Chambers
  Department of Physics
  The University
  Newcastle Upon Tyne NE1 7RU
  UK

- Dr Roger C Clark
  Department of Mathematical Sciences
  University of Aberdeen
  Edward Wright Building
  Dunbar Street
  Aberdeen AB24 3QY
  Scotland

- Dr José da Costa
  Departmento de Física
  Universidade da Madeira
  Colégio dos Jesuítas
  Praça do Município
  9000 Funchal
  Portugal

- Mr Konstantinos Dimopoulos
  Darwin College
  University of Cambridge
  Silver Street
  Cambridge CB3 9EU
  UK

- Mr Serge Droz
  Department of Physics
  University of Alberta
  Edmonton AB T6G 2J1
  Canada

- Mr Gavin Duffy
  Department of Mathematical Physics
  University College Dublin
  Belfield
  Dublin 4
  Ireland

- Mr Chris Fama
  Centre for Mathematics and
  Applications
  School of Mathematical Sciences
  Australian National University
  Canberra 0200
  Australia

- Mr Ralph M Gailis
  School of Physics
  University of Melbourne
  Parkville 3052
  Australia

- Mr Alejandro Gangui
  SISSA
  Via Beirut 2
  34013 Trieste
  Italy

- Dr Remo Garattini
  Università degli Studi di Bergamo
  Facoltà di Ingegneria
  Via Marconi 5
  24044 Dalmine (BG)
  Italy

- Prof Robert Geroch
  Enrico Fermi Institute
  University of Chicago
  5640 S. Ellis Ave.
  Chicago IL 60637
  USA

- Mr Martin Goliath
  Stockholm University
  Fysikum
  Box 6370
  S-113 85 Stockholm
  Sweden

- Dr Barry Haddow
  Department of Mathematics
  Trinity College
  Dublin 2
  Ireland

- Prof Graham S Hall
  Department of Mathematical Sciences
  Aberdeen University
  Edward Wright Building
  Dunbar Street
  Aberdeen AB24 3QY
  Scotland

- Mr Jose Luis Hernandez-Pastora
  Area de Fisica Teorica
  Facultad de Fisicas
  Edificio Trilingüe
  Universidad de Salamanca
  37008 Salamanca
  Spain

- Mag. Sascha Husa
  Institute for Theoretical Physics
  University of Vienna
  Boltzmanngasse 5
  A-1090 Wien
  Austria

- Dr Ray d'Inverno
  Faculty of Mathematical Studies
  University of Southampton
  Southampton SO17 1BJ
  UK

- Mr Radu Ionicioiu
  Institute of Gravitation and Space Sciences
  21 Ana Ipatescu Blvd.
  71111 Bucharest
  Romania

- Mr Vivek Iyer
  Enrico Fermi Institute
  University of Chicago
  5640 S. Ellis Ave.
  Chicago IL 60637
  USA

- Mr Bjørn Jensen
  Institute of Physics
  University of Oslo
  N-0314 Blindern
  Oslo 3
  Norway

- Mr János Kánnár
  Max-Planck-Institut für Gravitationsphysik
  Albert–Einstein–Institut
  Schlaatzweg 1
  D-14473
  Potsdam
  Germany

- Miss Panagiota Kanti
  The University of Ioannina
  Physics Department
  Division of Theoretical Physics
  PO Box 1186 GR-45110
  Ioannina
  Greece

- Mr Andreas Kneip
  SQN 210 Bloco F Apto 601
  70.862–020
  Brasília DF
  Brazil

- Dr Andreas Koutras
  Alopekis 6, 106-75
  Athens
  Greece

- Mr Gunar Krenzer
  Maurerstrasse 38
  Jena, 07749
  Germany

- Mr Pavel Krtous
  412 Avadh Bhatia Physics Lab
  University of Alberta
  Edmonton
  Alberta T6G 2J1
  Canada

- Dr Stefano Liberati
  SISSA/ISAS
  Via Beirut 2-4
  34014
  Trieste
  Italy

- Prof Alberto Lobo
  Departament de Física Fonamental
  Facultat de Física
  Universitat de Barcelona
  Avda. Diagonal 647
  E-08028 Barcelona
  Spain

- Dr Doug MacIntire
  Physics Department CB390
  University of Colorado
  Boulder
  CO 80309
  USA

- Mr Seth A Major
  Center for Gravitational Physics and Geometry
  104 Davey Lab.
  Penn State University
  University Park
  PA 16802-6300
  USA

- Mr Mattias Marklund
  Dept of Plasma Physics
  Umeå University
  S-90187 Umeå
  Sweden

- Miss Fotini G Markopoulou Kalamara
  Theoretical Physics
  Blackett Lab
  Imperial College
  Prince Consort Rd
  London SW7 2BZ
  UK

- Mr José M Martín-García
  LAEFF
  Apartado 50727
  E-28080 Madrid
  Spain

- Mr Luís M Estanqueiro Mendes
  Departamento de Física
  Instituto Superior Técnico
  Av. Rovisco Pais
  1096 LISBOA Codex
  Portugal

- Dr Paulo R L Vargas Moniz
  DAMTP
  University of Cambridge
  Silver St
  Cambridge CB3 9EW
  UK

- Ms Sharon Morsink
  Department of Physics
  University of Alberta
  Edmonton
  Alberta
  TG6 2JI
  Canada

- Mr John Myritzis
  Department of Mathematics
  University of the Aegean
  83200 Karlovassi
  Samos
  Greece

- Prof Gernot Neugebauer
  Theoretisch-Physikalisches Institut
  Friedrich-Schiller-Universitaet
  Max-Wien-Platz 1
  Sektion Physik
  D-07743 Jena
  Germany

- Mr Ulf Nilsson
  Dept of Physics
  Stockholm University
  Box 6730
  S-11385 Stockholm
  Sweden

- Mr José A Ortega
  Departament de Física Fonamental
  Facultat de Física
  Universitat de Barcelona
  Avda. Diagonal 647
  E-08028 Barcelona
  Spain

- Dr Philippos Papadopoulos
  508 Davey Lab
  Penn State University
  University Park
  PA 16802
  USA

- Prof Zoltán Perjés
  MTA KFKI Research Inst.
  H-1525 Budapest 114
  P.O.B. 49
  Hungary

- Dr Malcolm Perry
  DAMTP
  University of Cambridge
  Silver Street
  Cambridge CB3 9EW
  UK
  and
  Department of Physics 0319
  University of California at San Diego
  9500 Gilman Drive
  La Jolla
  California 92093
  USA

- Mr Suresh Pillai
  Department of Physics
  University of Alberta
  Edmonton
  Alberta
  T6G 2J1
  Canada

- Dr António Manuel Ramalho Pires
  Dpto Física
  Universidade da Madeira
  Colégio dos Jesuítas
  Praça do Município
  9000 Funchal
  Portugal

- Mr Giuseppe Pollifrone
  Dipartimento di Fisica
  Università di Roma 'La Sapienza' and
  INFN, Sezione di Roma
  P.le Aldo Moro 2
  00185 Roma
  Italy

- Dr John R Pulham
  Department of Mathematical Sciences
  University of Aberdeen
  Edward Wright Building
  Dunbar Street
  Aberdeen AB24 3QY
  Scotand

- Mr Anthony Rizzi
  Department of Physics
  Princeton University
  Princeton
  NJ 08540
  USA

- Dr Norna Robertson
  Department of Physics and Astronomy
  University of Glasgow
  Glasgow G12 8QQ
  Scotland

- Mr Ian Roy
  Department of Mathematical Sciences
  University of Aberdeen
  Edward Wright Building
  Dunbar Street
  Aberdeen AB24 3QY
  Scotland

- Prof Bernd Schmidt
  Max-Planck-Institut für
  Gravitationsphysik
  Albert–Einstein–Institut
  Schlaatzweg 1
  D-14473
  Potsdam
  Germany

- Mr Emre Sermutlu
  Department of Mathematics
  Bilkent University 13/12
  06533 Ankara
  Turkey

- Miss Alicia M Sintes Olives
  Departament de Física
  Universitat de Les Illes Balears
  Cra. Valldemossa km 7.5
  E-07071 Palma de Mallorca
  Spain

- Dr Jim E F Skea
  Symbolic Computation Group
  Departamento de Física Teórica
  Instituto de Física
  Universidade do Estado do Rio de Janeiro
  Rua São Francisco Xavier 524
  Maracanã
  20559-900 Rio de Janeiro – RJ
  Brazil

- Mr Carlos F Sopuerta
  Departament de Física Fonamental
  Facultat de Física
  Universitat de Barcelona
  Avda. Diagonal 647
  E-08028 Barcelona
  Spain

- Mr Fredrik Ståhl
  Dept of Mathematics
  University of Umeå
  S-90187 Umeå
  Sweden

- Ms Maria Süveges
  Central Research Institute for Physics
  Research Institute for Particle and
  Nuclear Physics
  H–1525 Budapest 114
  P.O.B. 49
  Hungary

- Prof John Wainwright
  Department of Applied Mathematics
  University of Waterloo
  Waterloo Ontario N2L 3G1
  Canada

- Prof Clifford Will
  McDonnell Center for the Space
  Sciences
  Department of Physics
  Washington University
  St Louis MO 63130
  USA

- Mr Adrian Wisdom
  Department of Mathematics
  King's College London
  Strand
  London WC2R 2LS
  UK

- Mr Elias Zafiris
  Blackett Laboratory
  Imperial College
  London SW7 2BZ
  UK

- Mr Jose A Zapata
  104 Davey Lab
  Penn State University
  University Park
  PA 16802
  USA

- Mr Lukas Ziewer
  Institut für Theoretische Physik
  Sidlerstrasse 5
  3012 Bern
  Switzerland

Index

action integral, 378
active galactic nuclei (AGN), 174
active galaxies, 174
age inequality, 115
age of the universe, 120
age parameter, 117
AGN, active galactic nuclei, 174
algebraic computing, 283
algebraic slicing, 343
analytic solutions, 1
anisotropies in Hubble flow, 136
astrophysical sources of GW, 215
axial symmetry, 91
AXIOM, 294

Bäcklund transformations, 71
Bianchi models, 110
big-bang nucleosynthesis (BBN), 108, 120
binary pulsar, 213, 229, 273
binary systems, 212
Birkhoff theorem, 84
black holes, 389
    accretion, 171
    area theorem, 161
    charged, 160
    dark matter, 181
    density of distribution, 150
    dynamical constraints, 185
    effects of, 151
    formation, 145
    general features, 144
    in binary systems, 173
    in cosmology, 174
    in galactic nuclei, 178
    instability, 390
    location, 147
    quantum effects, 187
    rotating, 73
    string theory, 403
    temperature, 390
    when they form, 148
Blandford-Znajek mechanism, 176

brown dwarfs, 162

c^2 formalism, 250
C^∞ solutions of vacuum field equations, 5
canonical quantization, 381
Cauchy $3+1$ region, 360
Cauchy initial value problem in GR, 340
Cauchy problem, 1
Cauchy-Kowalevkaya theorem, 2
characteristic approach, numerical GR, 350
characteristic region, 362
charged field equations, 49
Christodoulou-Klainerman theorem, 9
CLASSI, 306
CMB, cosmic microwave background, 134
coalescing binaries, 217
collapse theorems, 160
combined approach in numerical relativity, 351
    an example, 360
    combined numerical code, 351
    numerical approach, 364
    start-up problem, 368
    the interface, 365
combined systems, 32
compactified density parameter, 124
competing theories of gravity, 257
computer algebra systems, 287
conformal scaling, 341
conformastat spaces, 98
connection, on bundle, 23
conservation equation, 112
consistency of numerical methods, 355
constant mean curvature slicing, 343
constraints, 28
constraints, complete, 29
constraints, Hamiltonian, 3
constraints, momentum, 3
convergence of numerical methods, 356
cosmic censorship, 160
cosmic microwave background (CMB), 134
cosmological principle, 108

cross-section, of bundle, 22
Curzon solution, 93
cylindrical symmetry, 91

dark matter, 121
data analysis, 236
data filtering, 220
de Sitter universe, 114
deceleration parameter, 115
density parameter, 116
disk, rigidly rotating dust, 76
distance scale, 119
distribution of galaxies, 137
doppler tracking, 228
dust equations, 49
dynamical systems, 110, 121

EEP, Einstein equivalence principle, 241
Eötvös experiment, 241
eigendirections, 99
Einstein equivalence principle (EEP), 241
Einstein static universe, 115
Einstein's equations, 3 + 1 form, 339
Einstein-Maxwell fields , 103
elastic solid equations, 47
electromagnetism, 43
electrovacuum fields, 103
equivalence principle, 240
equivalence problem for metrics, 305
ergosphere, in Kerr solution, 159
Ernst equation, 66, 95
Eulerian gauge, 343
exact solutions, 61
exact solutions, data base, 312
existence and uniqueness theorem for
    symmetric, hyperbolic systems, 55
extrinsic curvature, 339

Feynman diagrams, 386
fibre bundle, 21
    connection, 23
    cross-section, 22
    fibre, 22
    horizontal vector, 23
    quotient, 37
    reduction, 39
    vertical vector, 22
fifth force, 272
FL, Friedmann-Lemaitre, 108
frame theory, 11
Friedmann equation, 112

Friedmann-Lemaitre cosmologies, 108
    flat universe, 114
    redshift, 113
    temperature, 113

gauge conditions, 342
gauge freedom, 24
Gauss-Bonnet theorem, 387
GEO-600, 234
geodesic deviation, 208
geodesic slicing, 342
geodetic precession, 270
Geroch decomposition, 314
global hyperbolicity, 8
global results, 9
gravitational constant, constancy of, 268
gravitational deflection of light, 261
gravitational field, 45
gravitational lensing effects, 182
gravitational redshift experiment, 245
gravitational waves (GW), 204
    detection, 223
    sources of, 203
    astrophysical sources, 215
    signals, short, 216
    signals, long, 218
    signals, stochastic, 219
gravitomagnetism, 271
gravitostatics, 62

hamiltonian for quantization, 380
horizontal vector, on bundle, 23
Hubble constant, 115
Hubble distance, 118
Hubble time, 118
Hubble variable, 115
hyperbolicity, 6
hyperbolization, 25

inflation, 109
inhomogeneities, 137
interactions, 37
inverse scattering method, 70
isothermal and radial gauge, 344
isotropic coordinates, 98

Karlhede classification, 305
Kastor and Traschen solution, 88
Kerr solution, 157
Kerr-NUT solution, 73
Killing trajectories, space of, 95

Index

kinematic properties of models, 131
kinetic theory, 51
Klein-Gordon equations, 43

Lagrangian gauge, 343
Lagrangian slicing, 342
Lagrangian systems, 51
lapse, 337
laser interferometric detector, 225
lensing effects, gravitational, 182
Levi-Civita metrics, 99
lifting of a diffeomorphism, 33
light, gravitational deflection of, 261
LIGO, 234
limits on PPN parameters, 265
linear theory of gravitational waves, 205
LISA, 227
LISP, 287
LLI, local Lorentz invariance, 241
local Lorentz invariance, 241
local position invariance, 241
low temperature detectors, 225
LPI, local position invariance, 241
luminosity distance, 119
lunar Eötvös experiment, 266

Maclaurin disk, 76
mass anisotropy experiments, 242
massive dark object (MDO), 178
massive object (MO), 145
matched filter, 220
maximal slicing, 343
MDO, massive dark object, 178
metric 2 + 2 decomposition, 333, 345
metric 3 + 1 decomposition, 333, 337
metric files (SHEEP), 300
metric theories of gravity, 252
Michelson interferometer, 225
Milne universe, 114
minimal distortion gauge, 344
minimal surfaces, 64
minisuperspace, 382
MO, massive object, 145

n-fluid models, 116
Nambu-Goto action, 394
neutron stars, 161
Newtonian limit, 14
no hair theorem, 161
Nordtvedt effect, 266
nucleosynthesis, big-bang (BBN), 108, 120

numerical code, verification, 358
numerical codes, issues, 357
numerical methods, convergence of, 356
numerical methods, stability of, 355
numerical relativity, 331
  numerical issues, 353
  status of, 335
  techniques, 333

observational parameters, 115
one-fluid models, 121
orbital period decay, 214

Papapetrou-Majumdar solutions, 103
parametrized post-Newtonian (PPN)
  formalism, 254
partial differential equations, 19
particle horizon, 117
PBH, primordial black hole, 146
peculiar velocities, 136
perfect fluid equations, 44
perihelion shift of Mercury, 264
perturbation theory, 385
photon noise, 229
plane waves, 207
polar slicing, 343
polarisation states, 209
Polyakov action, 394
post-Newtonian approximations, 16
post-Newtonian gravity, tests, 252
power recycling, 230
PPN formalism, 254
PPN parameters, current limits, 265
primordial black holes (PBH), 146, 187
  constraints on, 196
  cosmic rays from, 194
  evaporation of, 192
  formation of, 190
prior geometric theory, 252
pulsar, 212
pulsar timing, 228
pulse time of arrival, 213
purely dynamic metric theory, 252

quadrupole moments, 204
quantum gravity, 377
  background fields, 400
  black holes, 389
  path integral, 384
  perturbation theory, 385
  thermodynamic partition function, 391

quasars, 174
quotient bundle, 37

Raychaudhuri equation, 112
redshift parameter, 113
reduction of bundle, 39
regular conformal field equations, 9
Reissner-Nordstrøm solution, 85
resonant bar detector, 224
Ricci identity, 83
Riemann-Hilbert problems, 74
rigidly rotating dust disk, 76
Robertson-Walker metric, 111
Robinson-Bertotti metrics, 86
rotating black hole, 73
rotating bodies, 62

S-duality in string theory, 405
Sachs-Wolfe effect, 134
scalar field, 113
scalar-tensor theories, 259
Schiff's conjecture, 247
Schwarzschild solution, 84, 153
seismic noise, 231
SEP, strong equivalence principle, 252
SHEEP, 295
shift, 337
signal recycling, 231
signal template, 221
singularity theorems, 11, 160
size of the universe, 118
SMO, supermassive object, 146
space based GW detectors, 227
spherical symmetry, 84
spherically symmetric models, 110
spin-s systems, 46
stability of numerical methods, 355
standard model, 108
star clusters, gravitational collapse, 169
static space-times, 94
stationary fields, 93
stationary space-times, 93
stellar system tests of gravitational theory, 273
STENSOR, 313
string theory, 393
strong equivalence principle (SEP), 252
supermassive objects (SMO), 146, 164
supernovae, 216
supersymmetries, 388
Szekeres models, 110

T-duality in string theory, 402
temperature anisotropy, 135
tests of post-Newtonian conservation laws, 269
tests of strong equivalence principle (SEP), 266
TH$\epsilon\mu$ formalism, 248
thermal noise, 233
thick sandwich start-up method, 368
time-delay of light, 262
Tolman-Oppenheimer-Volkov equations, 66
Tomimatsu-Sato solutions, 74
transverse-traceless (TT) gauge, 208
triads, 96
truncation errors, 353
two-fluid models, 126

vertical vector, on bundle, 22
very massive objects (VMO), 146, 164
visualization of numerical data, 358
VIRGO, 234
VMO, very massive object, 146

weak equivalence principle (WEP), 241
Weyl curvature, 132
Weyl solutions, 91
WEP, weak equivalence principle, 241
Wheeler-de Witt equations, 382
white dwarfs, 161
worldsheet, 393